ISDMM 2007

Defect and Material Mechanics

Proceedings of the International Symposium on Defect and Material Mechanics (ISDMM), held in Aussois, France, March 25–29, 2007

Edited by

CRISTIAN DASCALU

University J. Fourier, Grenoble, France

GÉRARD A. MAUGIN

University Pierre et Marie Curie, Paris, France

and

CLAUDE STOLZ

Ecole Polytechnique, Palaiseau, France

Reprinted from *International Journal of Fracture*
Volume 147, Nos. 1–4 (2007)

Published by Springer,
P.O. Box 17, 3300 AA Dordrecht, The Netherlands

Sold and distributed in North, Central and South America
By Springer,
101 Philip Drive, Norwell, MA 02061, USA

In all other countries, sold and distributed
By Springer
P.O. Box 322, 3300 AH Dordrecht, The Netherlands

Library of Congress Control Number: 2008924449

ISBN-13: 978-1-4020-6928-4 e-ISBN-13: 978-1-4020-6929-1

Printed on acid-free paper

Springer.com

Table of Contents

Preface

C. Dascalu · G. A. Maugin

The volume presents recent developments in the theory of defects and the mechanics of material forces. Most of the contributions were presented at the International Symposium on Defect and Material Forces (ISDMM2007), held in Aussois, France, March 25–29, 2007.

Originated in the works of Eshelby, the Material or Configurational Mechanics experienced a remarkable revival over the last two decades. When the mechanics of continua is fully expressed on the material manifold, it captures the material inhomogeneities. The driving (material) forces on inhomogeneities appear naturally in this framework and are requesting for constitutive modeling of the evolution of inhomogeneities through kinetic laws.

In this way, a general scheme for describing structural changes in continua is obtained. The Eshelbian mechanics formulation comes up with a unifying treatment of different phenomena like fracture and damage evolution, phase transitions, plasticity and dislocation motion, etc.

This special issue aims at bringing together recent developments in Material Mechanics and the more classical Defect Mechanics approaches. The contributions are highlighting recent research on topics like: fracture and damage, electromagnetoelasticity, plasticity, distributed dislocations, thermodynamics, poroelasticity, generalized continua, structural optimization, conservation laws and symmetries, multiscale approaches and numerical solution strategies.

We expect the present volume to be a valuable resource for researchers in the field of Mechanics of Defects in Solids.

We dedicate this special issue to the memory of the late Professor George Herrmann (1921–2007). G. Herrmann was a prestigious scientist, well-known in the international mechanics community. In the last years, he was an active researcher in the field of Material Mechanics. George Herrmann supported the organization and registered for attending the ISDMM2007 Symposium in Aussois, before his sudden dead on January 7, 2007. None of us will forget his passion for mechanical sciences, his enthusiasm and generosity.

C. Dascalu (✉)
Laboratoire Sols Solides Structures, Université Joseph Fourier, Grenoble, Domaine Universitaire, B.P. 53, 38041 Grenoble cedex 9, France
e-mail: cristian.dascalu@hmg.inpg.fr

G. A. Maugin
Université Pierre et Marie Curie, Institut Jean Le Rond d'Alembert, Case 152, 4 place Jussieu, 75252 Paris cedex 05, France
e-mail: gam@ccr.jussieu.fr

Defect and Material Mechanics. C. Dascalu, G.A. Maugin & C. Stolz (eds.),
doi: 10.1007/978-1-4020-6929-1_1, © Springer Science+Business Media B.V. 2008

Reciprocity in fracture and defect mechanics

R. Kienzler

Abstract For defects in solids, when displaced within the material, reciprocity relations have been established recently similar to the theorems attributed to Betti and Maxwell. These theorems are applied to crack- and defect-interaction problems.

Keywords Reciprocity · Fracture · Defect interaction · Material forces

1 Introduction

When treating problems of linear elastic systems, such as beams, frames or two- and three-dimensional continuous elastic solids, the reciprocity theorems associated with the names of Betti and Maxwell have proven to be quite valuable. In its simplest form, Betti's theorem states that if a linear elastic body is supported properly such that rigid body displacements are precluded and if an external force $\boldsymbol{F}_1$ at point 1 which produces a displacement $\boldsymbol{u}_{21}$ at some other point 2, then a force $\boldsymbol{F}_2$ at 2 would produce a displacement $\boldsymbol{u}_{12}$ at 1 where (cf. e.g., Marguerre 1962)

$$\boldsymbol{F}_1 \cdot \boldsymbol{u}_{12} = \boldsymbol{F}_2 \cdot \boldsymbol{u}_{21}. \tag{1}$$

The contents of the present paper has been developed together with Prof. Dr. Dr. h. c. George Herrmann, Stanford University, California, who passed away on January 7, 2007.

R. Kienzler (✉)
Department of Production Engineering,
University of Bremen, Bremen, Germany
e-mail: rkienzler@uni-bremen.de

The dot marks the scalar product between the two vectors. In double-indexed terms, the first index indicates the position at which the quantity is measured (effect), and the second index indicates the cause due to which this quantity occurs.

To reach a scalar version of Betti's theorem the displacement component of $\boldsymbol{u}_{12}$ in the direction of $\boldsymbol{F}_1$ is introduced as u^P_{12} and the component of $\boldsymbol{u}_{21}$ in the direction of $\boldsymbol{F}_2$ as u^P_{21} (cf. Fig. 1). With the magnitude F_1 and F_2 of $\boldsymbol{F}_1$ and $\boldsymbol{F}_2$, respectively, it is

$$F_1 u^P_{12} = F_2 u^P_{21}. \tag{2}$$

Since u^P_{12} is proportional to F_2 and u^P_{21} is proportional to F_1 influence coefficient may be defined as

$$u^P_{12} = \delta_{12} F_2, \tag{3a}$$
$$u^P_{21} = \delta_{21} F_1, \tag{3b}$$

and according to Marguerre (1962), Maxwell's theorem states

$$\delta_{12} = \delta_{21}. \tag{4}$$

The reciprocity relations are based on the result that the energy stored in an elastic body after application of two forces is independent of their sequence of application, and equals the external work done on the body. Various applications of these theorems are to be found in , e.g., Timoshenko and Goodier (1970), Barber (2002).

During the recent decades a new topic has emerged in mechanics of elastically deformable media which is variously described as *Defect Mechanics*, *Fracture Mechanics*, *Configurational Mechanics*, *Mechanics in*

doi: 10.1007/978-1-4020-6929-1_2,

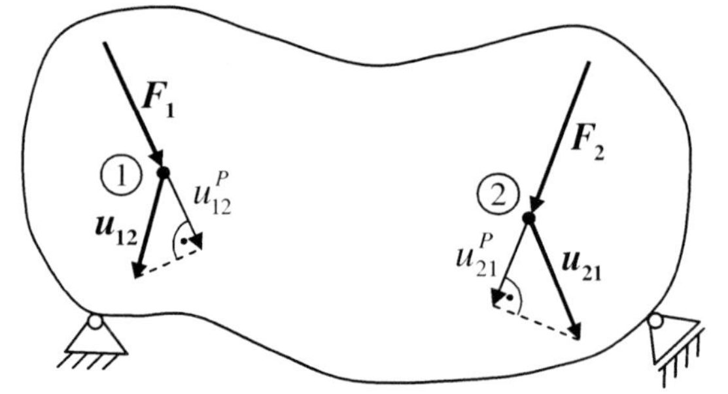

Fig. 1 Elastic body subjected to two forces

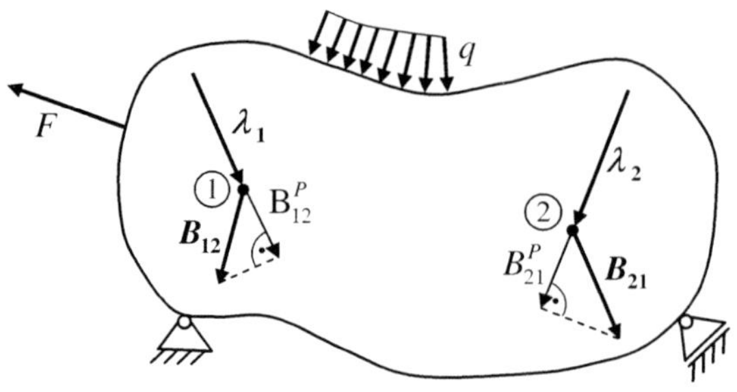

Fig. 2 Stressed elastic body containing two point defects

Material Space or *Eshelbian Mechanics* (cf. Maugin 1993; Gurtin 2000; Kienzler and Herrmann 2000). The importance of this topic is based on the necessity of improved material modeling of defects of various types and scales in deformable solids. And this necessity in turn is the result of developing new technologies in a variety of applied fields such as devices in IT, aerospace and energy sectors.

In approaching configurational mechanics one can say that a material force is associated with a defect in a stressed elastic body, because if this defect were displaced within the body, the total energy of the system would be changed and the negative ratio of this change and the displacement, in the limit as the displacement is made to approach zero, is the material force on the defect.

Let us consider a linearly elastic body of arbitrary geometry properly supported and subjected to surface tractions and/or body forces. And let us assume that the body contains an arbitrary number and type of distributed or localized (point) defects, such as dislocations, cracks, inclusions, cavities, etc. Let us focus attention on two localized defects placed at positions 1 and 2 and let the associated material forces be the vectors $\boldsymbol{B}_{10}$ and $\boldsymbol{B}_{20}$ respectively. Now, defect 1 will be displaced by an amount $\boldsymbol{\lambda}_1$ relatively to the material in which the defect is positioned (material displacement). In turn the material force at 1 will be changed by an amount $\boldsymbol{B}_{11}$ and the material force 2 will be changed by $\boldsymbol{B}_{21}$. The material displacement is assumed to be small in the sense that linearity is implied between material displacements and material forces, e.g., $\boldsymbol{B}_{21} \propto \boldsymbol{\lambda}_1$. If defect 2 would be materially displaced by $\boldsymbol{\lambda}_2$ the material force at 1 and 2 would be changed by $\boldsymbol{B}_{12}$ and $\boldsymbol{B}_{22}$, respectively.

Based on the similar argument (as applied in *Physical Space*) that the change of energy stored in the body after application of two material displacements is independent of their sequence of application, and equals the work done of the material forces in the material displacements, a material, Betti-like reciprocity theorem is established (Herrmann and Kienzler 2007a) as

$$\boldsymbol{\lambda}_2 \cdot \boldsymbol{B}_{21} = \boldsymbol{\lambda}_1 \cdot \boldsymbol{B}_{12}, \tag{5}$$

stating that the work in the material translation at 2 of the change of the material force due to a material translation at 1 equals the work in the material translation at 1 of the change of the material force due to a material translation at 2.

Note the difference between Physical and Material Space: in Physical Space, the applied physical forces are the causes of (the change of) physical displacements (effects) whereas in Material Space the material displacement cause (changes of) material forces (effects). In analogy, a scalar version of Eq. 5 is reached by introducing the component of $\boldsymbol{B}_{21}$ in the direction of $\boldsymbol{\lambda}_2$ as B^P_{21}, the component of $\boldsymbol{B}_{12}$ in the direction of $\boldsymbol{\lambda}_1$ as B^P_{12} (cf. Fig. 2) and the magnitudes of $\boldsymbol{\lambda}_1$ and $\boldsymbol{\lambda}_2$ as λ_1 and λ_2, respectively, and if follows

$$\lambda_2 B^P_{21} = \lambda_1 B^P_{12}. \tag{6}$$

Since linearity is implied, material influence coefficients may be defined as

$$B^P_{21} = \beta_{21}\lambda_1, \tag{7a}$$

$$B^P_{12} = \beta_{12}\lambda_2, \tag{7b}$$

and a material, Maxwell-like reciprocity theorem is obtained (Herrmann and Kienzler 2007a)

$$\beta_{21} = \beta_{12}. \tag{8}$$

It states that the change of the material force at 2 in the direction of the material displacement $\boldsymbol{\lambda}_2$ due to a

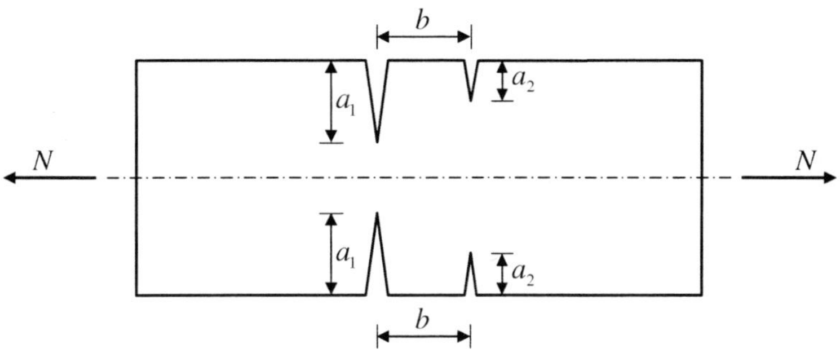

Fig. 3 Crack configuration within a bar in tension

unit-material translation at 1 equals the change of the material force at 1 in the direction of $\boldsymbol{\lambda}_1$ due to a unit-material translation at 2.

The implications of a non-linear formulation of the reciprocity relation (5) have been dealt with in Kienzler and Herrmann (2007). By considering a stressed elastic body subjected sequentially to a material displacement of a defect and the application of a physical force a novel type of coupling of Physical and Material Mechanics by means of a reciprocity theorem has been established in Herrmann and Kienzler (2007b).

It is the intention of the present contribution to explore some applications of the two theorems (5) and (8) in fracture and defect mechanics.

2 Interacting cracks

Interacting cracks have been the object of research over several years (cf., e.g., Erdogan 1962; Panasyuk et al. 1977; Gross 1982). The material forces at crack tips are usually calculated from the path-independent $\boldsymbol{J}$ integral (Rice 1968). Reciprocity relations are, therefore, concerned with the change of the $\boldsymbol{J}$ integral due to the translation of some other defect, e.g., the change of length of a crack 2 in the neighbourhood of the original crack 1. Usually, the solution of crack-interaction problems involve either some advances analytical tools or an extended numerical investigation. Based on the strength-of-materials theories, a simple first estimate for interacting edge cracks in an elastic bar under tension have been given recently by Rohde and Kienzler (2005) and Rohde et al. (2005), and the reciprocity relations are applied within this simplified problem setting. In Rohde and Kienzler (2005) and Rohde et al. (2005) bars are investigated with a set of 2×2 edge cracks symmetrically positioned with respect to its length axis

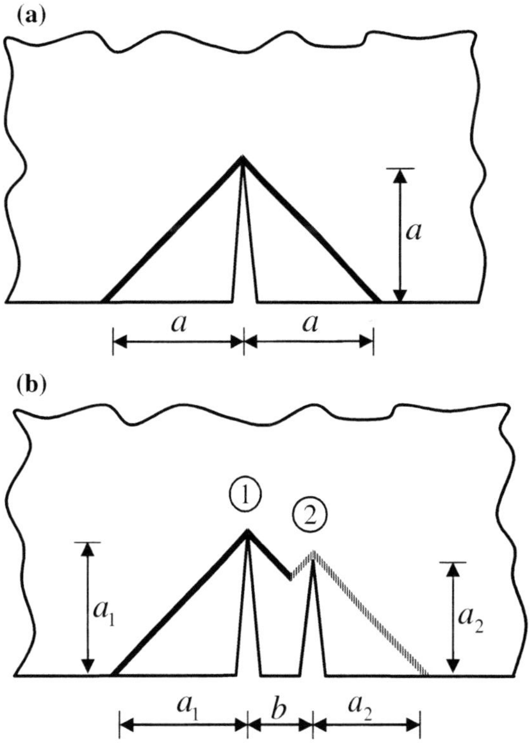

Fig. 4 Single crack (**a**), two neighboured cracks (**b**)

and loaded by an axial force in tension as depicted in Fig. 3.

If a single crack is considered first, the key idea is that the energy-release rate is proportional to the length of the shaded strip in Fig. 4a, i.e.,

$$G = J \propto 2a\sqrt{2}; \quad J = C2a\sqrt{2}. \tag{9}$$

The constant C, say, would be proportional to the square of the applied load, inverse proportional to Young's modulus, and would combine some information about the geometry of the bar.

In the presence of a second crack, cf. Fig. 4b, both strips can not develop completely due to shielding such that the energy-release rate of crack 1 is proportional to the length of the darkly shaded area on the left and the energy release rate of crack 2 is proportional to the lightly shaded area on the right, i.e.,

$$G_1 = J_1 = B_{10} = \frac{1}{2}\sqrt{2}C(3a_1 - a_2 + b), \tag{10a}$$

$$G_2 = J_2 = B_{20} = \frac{1}{2}\sqrt{2}C(3a_2 - a_1 + b), \tag{10b}$$

where a_1 and a_2 are the crack lengths of crack 1 and crack 2, respectively, and b is the distance of both cracks. Eq. 10 is valid as long as

$$|a_1 - a_2| < b < a_1 + a_2. \tag{11}$$

It is observed, of course, that the crack configuration represents a mixed-mode problem and the vector $\boldsymbol{J}$ does not point in the direction of a potential self-similar crack extension. It is sufficient, however, to consider only the component of $\boldsymbol{J}$ in the direction of Δa_1 or Δa_2, corresponding to B_{12}^P or B_{21}^P. Details of the analysis and the validation of the results are given in Rohde and Kienzler (2005) and Rohde et al. (2005).

With the simple relations (10) at hand it is easy to check the reciprocity theorems. A crack extension of crack 1, Δa_1, corresponds to a material translation λ_1. Due to λ_1, the material forces B_{10} and B_{20} change to

$$\begin{aligned} B_{11} &= B_{1\lambda_1} - B_{10} \\ &= \frac{1}{2}\sqrt{2}C\,[3(a_1 + \lambda_1) - a_2 + b - 3a_1 + a_2 - b] \\ &= +\frac{3}{2}\sqrt{2}C\lambda_1, \end{aligned} \tag{12a}$$

$$\begin{aligned} B_{21} &= B_{2\lambda_1} - B_{10} \\ &= \frac{1}{2}\sqrt{2}C\,[3a_2 - (a_1 + \lambda_1) + b - 3a_2 + a_1 - b] \\ &= -\frac{1}{2}\sqrt{2}C\lambda_1, \\ &= \beta_{21}\lambda_1. \end{aligned} \tag{12b}$$

If crack 2 is extended by an amount λ_2 the change of the material forces at crack 1 and 2 is calculated similarly as

$$B_{12} = -\frac{1}{2}\sqrt{2}C\lambda_2 = \beta_{12}\lambda_2, \tag{12c}$$

$$B_{22} = +\frac{3}{2}\sqrt{2}C\lambda_2. \tag{12d}$$

The changes of the various energy-release rates have been made graphic in Fig. 5

Substitution of (12b) and (12c) into relations (5) or (6) and (8) confirms and illustrates the validity of

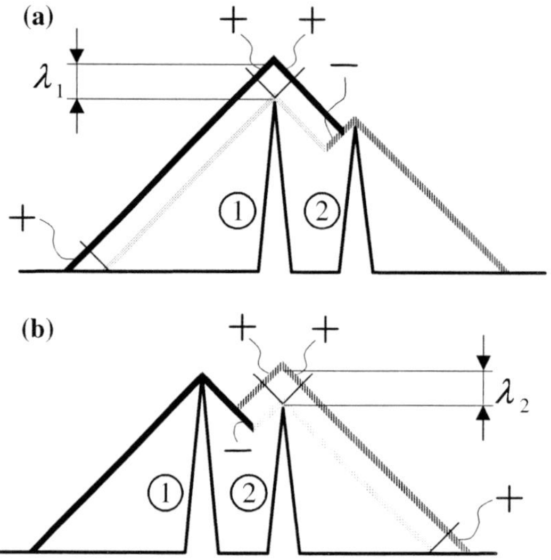

Fig. 5 Change of energy-release rates due to crack extension λ_1 of crack 1 (**a**) and λ_2 of crack 2 (**b**)

reciprocity relations in Material Space. In words: the change of the J integral at crack 1 due to a change of the crack length of crack 2 equals the change of the J integral at crack 2 due to the change of the crack length of crack 1. This relation may also be verified within a more rigorous approach either analytically or numerically and may be used to establish influence surfaces for defects within the stress state of a crack tip. This problem will be further treated in the next section.

3 Interaction of an edge dislocation with a circular hole

As a further example for the application of the material reciprocity relations let us consider the following defect configuration. Within an infinitely extended plane elastic sheet an edge dislocation is situated in the origin of a plane Cartesian coordinate system (x_1, x_2) with Burgers vector pointing (without loss of generality) in x_2-direction. At $x_i = \xi_i (i = 1, 2)$ a circular hole with radius r_0 is centred, or, in polar coordinates, at distance $d > r_0 (\varepsilon = d/r_0 > 1)$ under the angle φ measured from the x_1-axis as depicted in Fig. 6.

Due to a material translation $\xi_i \rightarrow \xi_i + \lambda_i$ the total energy Π of the system changes and the energy-release rate is defined as

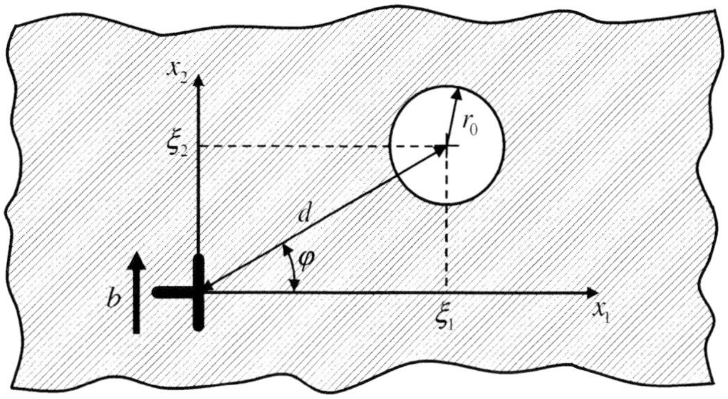

Fig. 6 Edge dislocation and circular hole in a plane elastic plate

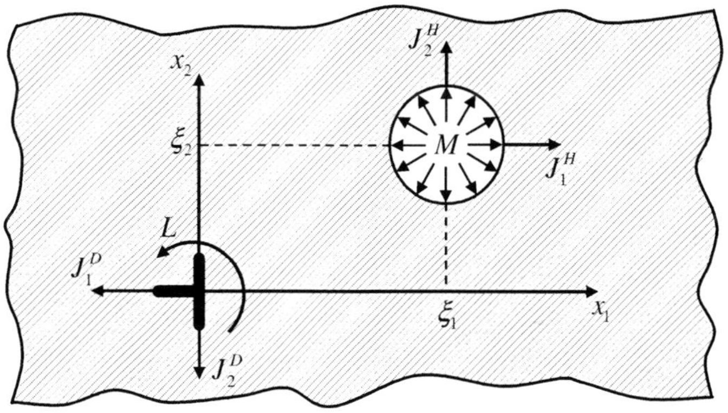

Fig. 7 Free-body diagram for material quantities

$$\lim_{\lambda_i \to 0} \frac{\Pi(\xi_i + \lambda_i) - \Pi(\xi_i)}{\lambda_i} = -J_i. \tag{13}$$

As usual, the energy-release rate is calculated by means of a path-independent contour integral involving the energy-momentum tensor (cf., e.g., Maugin 1993; Gurtin 2000; Kienzler and Herrmann 2000). The contour of integration is arbitrary as long as the same defect, either the hole or the inclusion, is included. In addition, two further path-independent integrals have been introduced (Günther 1962; Knowles and Sternberg 1972; Budiansky and Rice 1973) designated as L and M integrals. The L-integral is a material vector moment ($\boldsymbol{r} \times \boldsymbol{J}$) and calculates the energy-release rate due to a rotation of the defect $\varphi \to \varphi + \omega$. It is path independent if the material is isotropic. The M integral is a material scalar moment ($\boldsymbol{r} \cdot \boldsymbol{J}$) and calculates the energy-release rate due to a self-similar expansion of the defect, here $r_0 \to \alpha r_0$. L and M depend on the choice of the point of reference. Choosing for L the origin of the coordinate system and for M the center of the hole, the path-independent integrals have been evaluated (Kienzler and Herrmann 2000; Kienzler and Kordisch 1990) and are given as

$$J_1 = -\frac{E^* b^2 \cos\varphi}{4\pi r_0 \varepsilon^3} \left(\frac{1}{\varepsilon^2 - 1} + 1 + 2\sin^2\varphi \right), \tag{14a}$$

$$J_2 = -\frac{E^* b^2 \sin\varphi}{4\pi r_0 \varepsilon^3} \left(\frac{1}{\varepsilon^2 - 1} + 2\sin^2\varphi \right), \tag{14b}$$

$$L = -\frac{E^* b^2 \sin\varphi \cos\varphi}{4\pi \varepsilon^2}, \tag{14c}$$

$$M = +\frac{E^* b^2}{4\pi \varepsilon^2} \left(\frac{1}{\varepsilon^2 - 1} + 1 + \sin^2\varphi \right), \tag{14d}$$

with $E^* = E$ for plane stress, $E^* = E/(1 - \nu^2)$ for plane strain, Young's modulus E and Poisson's ratio ν. As in Physical Space, free body diagrams can be sketched in Material Space. This is shown in Fig. 7, where for the virial M, in the sketch a pressure-like symbol and in the equation the symbol $\otimes$ is used.

Likewise, material equilibrium conditions apply, where the expanding-moment condition ("Fliehmoment", cf. Schweins 1849) is normally not used neither in physical nor in material space

$$\rightarrow \; : \; J_1^H - J_1^D = 0 \tag{15a}$$

$$\uparrow \; : \; J_2^H - J_2^D = 0 \tag{15b}$$

$$\cap \; : \; L + \xi_1 J_2^H - \xi_2 J_1^H = 0 \tag{15c}$$

$$\otimes \; : \; M + \xi_1 J_1^D + \xi_2 J_2^D = 0 \tag{15d}$$

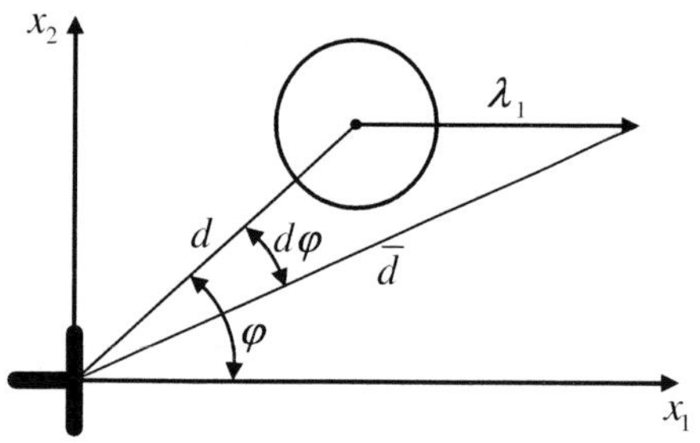

Fig. 8 Material translation λ_1 of the hole

The change of material dynamic quantities J_i, L, M due to the small material kinematic quantities $\lambda_i, \omega, \alpha$ can be calculated by differentiation

$$\Delta_{\lambda_i}(\,) = \frac{\partial(\,)}{\partial \lambda_i}\lambda_i + 0\left(\lambda_i^2\right) \tag{16}$$

and by geometrical considerations.

From Fig. 8, e.g., we read off

$$\bar{d}\cos(\varphi - d\varphi) = d\cos\varphi + \lambda_1, \tag{17a}$$

$$\bar{d}\sin(\varphi - d\varphi) = d\sin\varphi \tag{17b}$$

and after linearization ($d\varphi \ll 1,\ \lambda_1 \ll d$) we find

$$\varepsilon \to \bar{\varepsilon} = \varepsilon + \frac{\lambda_1}{r_0}\cos\varphi \tag{18a}$$

$$\varphi \to \bar{\varphi} = \varphi - \frac{\lambda_1 \sin\varphi}{r_0\varepsilon}. \tag{18b}$$

Inserting in (16) for $i = 1$ and applying the chain role leads to

$$\Delta_{\lambda_1}(\,) = \left(\frac{\partial(\,)}{\partial\varepsilon}\frac{\cos\varphi}{r_0} - \frac{\partial(\,)}{\partial\varphi}\frac{\sin\varphi}{r_0\varepsilon}\right)\lambda_1, \tag{19a}$$

and in a similar way we obtain

$$\Delta_{\lambda_2}(\,) = \left(\frac{\partial(\,)}{\partial\varepsilon}\frac{\sin\varphi}{r_0} + \frac{\partial(\,)}{\partial\varphi}\frac{\cos\varphi}{r_0\varepsilon}\right)\lambda_2. \tag{19b}$$

Due to a rotation of the hole around the dislocation, $\varphi \to \varphi + \omega$, and a self-similar expansion of the hole, $r_0 \to \alpha r_0$ we find

$$\Delta_\omega(\,) = \frac{\partial(\,)}{\partial\varphi}\omega, \tag{19c}$$

$$\Delta_\alpha(\,) = \varepsilon\frac{\partial(\,)}{\partial\varepsilon}\alpha. \tag{19d}$$

Using Eq. 19, the change of the relevant quantities can easily be calculated with the results

- *material translation* λ_1

$$\Delta_{\lambda_1}J_1 = +\lambda_1\frac{E^*b^2}{4\pi r_0^2\varepsilon^4} \times \left(\frac{2\cos^2\varphi}{(\varepsilon^2-1)^2} + \frac{6\cos^2\varphi - 1}{(\varepsilon^2-1)} - 3(4\cos^4\varphi - 6\cos^2\varphi + 1)\right) \tag{20a}$$

$$\Delta_{\lambda_1}J_2 = +\lambda_1\frac{E^*b^2\sin\varphi\cos\varphi}{2\pi r_0^2\varepsilon^4}\left(\frac{1}{(\varepsilon^2-1)^2} + \frac{3}{(\varepsilon^2-1)} + 6\sin^2\varphi\right) \tag{20b}$$

$$\Delta_{\lambda_1}L = -\lambda_1\frac{E^*b^2\sin\varphi}{4\pi r_0\varepsilon^3} \times \left(1 - 4\cos^2\varphi\right) \tag{20c}$$

$$\Delta_{\lambda_1}M = -\lambda_1\frac{E^*b^2\cos\varphi}{2\pi r_0\varepsilon^3} \times \left(\frac{\varepsilon^4}{(\varepsilon^2-1)^2} + 2\sin^2\varphi\right) \tag{20d}$$

- *material translation* λ_2

$$\Delta_{\lambda_2}J_1 = +\lambda_2\frac{E^*b^2\sin\varphi\cos\varphi}{2\pi r_0^2\varepsilon^4}\left(\frac{1}{(\varepsilon^2-1)^2} + \frac{3}{(\varepsilon^2-1)} + 6\sin^2\varphi\right) \tag{21a}$$

$$\Delta_{\lambda_2}J_2 = +\lambda_2\frac{E^*b^2}{4\pi r_0^2\varepsilon^4} \times \left(\frac{2\sin^2\varphi}{(\varepsilon^2-1)^2} + \frac{6\sin^2\varphi - 1}{\varepsilon^2-1} - 6\sin^2\varphi(1 - 2\sin^2\varphi)\right) \tag{21b}$$

$$\Delta_{\lambda_2}L = -\lambda_2\frac{E^*b^2\cos\varphi}{4\pi r_0\varepsilon^3} \times \left(1 - 4\sin^2\varphi\right) \tag{21c}$$

$$\Delta_{\lambda_2}M = -\lambda_2\frac{E^*b^2\sin\varphi}{2\pi r_0\varepsilon^3} \times \left(\frac{\varepsilon^4}{(\varepsilon^2-1)^2} + 1 - 2\cos^2\varphi\right) \tag{21d}$$

- *material rotation* ω

$$\Delta_\omega J_1 = +\omega\frac{E^*b^2\sin\varphi}{4\pi r_0\varepsilon^3} \times\left(\frac{1}{\varepsilon^2-1}+3(1-2\cos^2\varphi)\right) \quad (22a)$$

$$\Delta_\omega J_2 = +\omega\frac{E^*b^2\cos\varphi}{4\pi r_0\varepsilon^3} \times\left(-\frac{1}{\varepsilon^2-1}-6\sin^2\varphi\right) \quad (22b)$$

$$\Delta_\omega L = -\omega\frac{E^*b^2}{4\pi\varepsilon^2}\left(1-2\cos^2\varphi\right) = +\omega\frac{E^*b^2}{4\pi\varepsilon^2}\cos 2\varphi \quad (22c)$$

$$\Delta_\omega M = +\omega\frac{E^*b^2}{2\pi\varepsilon^2}\sin\varphi\cos\varphi = \omega\frac{E^*b^2}{4\pi\varepsilon^2}\sin 2\varphi \quad (22d)$$

- *material self-similar expansion* α

$$\Delta_\alpha J_1 = \alpha\frac{E^*b^2\cos\varphi}{2\pi r_0\varepsilon^3} \times\left(-\frac{\varepsilon^4}{(\varepsilon^2-1)^2}-2\sin^2\varphi\right) \quad (23a)$$

$$\Delta_\alpha J_2 = \alpha\frac{E^*b^2\sin\varphi}{2\pi r_0\varepsilon^3} \times\left(-\frac{\varepsilon^4}{(\varepsilon^2-1)^2}-1+2\cos^2\varphi\right) \quad (23b)$$

$$\Delta_\alpha L = -\alpha\frac{E^*b^2}{2\pi\varepsilon^2}\sin\varphi\cos\varphi = -\alpha\frac{E^*b^2}{4\pi\varepsilon^2}\sin 2\varphi \quad (23c)$$

$$\Delta_\alpha M = +\alpha\frac{E^*b^2}{2\pi\varepsilon^2} \times\left(\frac{\varepsilon^4}{(\varepsilon^2-1)^2}+\sin^2\varphi\right). \quad (23d)$$

Also the modified material quantities have to satisfy material equilibrium conditions. Especially for the vectorial and the scalar moments the change of the lever arms due to the material displacements have to be observed. As an example consider the material rotation depicted in Fig. 9

Equilibrium of moments yields

$$L + \Delta_\omega L + (\xi_1 - \xi_2\omega)(J_2 + \Delta_\omega J_2) -(\xi_2 + \xi_1\omega)(J_1 + \Delta_\omega J_1) = 0. \quad (24)$$

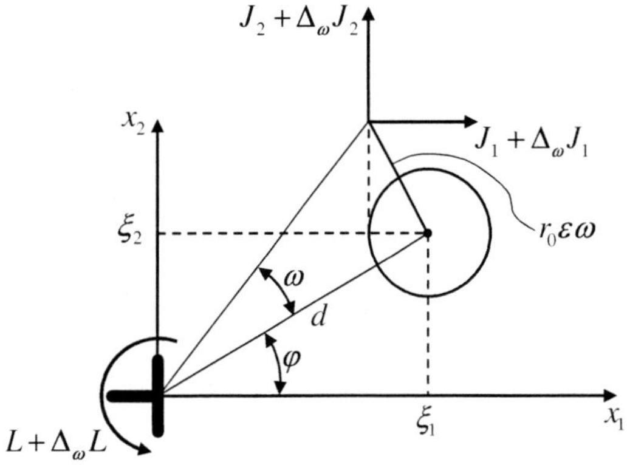

Fig. 9 Free-body diagram after material rotation ω

Due to (15c) and neglection of terms of $0(\omega^2)$ we arrive at

$$\Delta_\omega L + r_0\varepsilon(\cos\varphi\Delta_\omega J_2 - \omega\sin\varphi J_2 - \sin\varphi\Delta_\omega J_1 - \omega\cos\varphi J_1) = 0. \quad (25c)$$

In a similar way we obtain

$$\Delta_{\lambda_1} L + \xi_1\Delta_{\lambda_1} J_2 + \lambda_1 J_2 - \xi_2\Delta_{\lambda_1} J_1 = 0, \quad (25a)$$

$$\Delta_{\lambda_2} L + \xi_1\Delta_{\lambda_2} J_2 - \lambda_2 J_1 - \xi_2\Delta_{\lambda_2} J_1 = 0, \quad (25b)$$

$$\Delta_\alpha L + r_0\varepsilon(\cos\varphi\Delta_\alpha J_2 - \sin\varphi\Delta_\alpha J_1 = 0. \quad (25d)$$

The virial equations are

$$\Delta_{\lambda_1} M + \lambda_1 J_1 + \xi_1\Delta_{\lambda_1} J_1 + \xi_2\Delta_{\lambda_1} J_2 = 0, \quad (26a)$$

$$\Delta_{\lambda_2} M + \lambda_2 J_2 + \xi_1\Delta_{\lambda_2} J_1 + \xi_2\Delta_{\lambda_2} J_2 = 0, \quad (26b)$$

$$\Delta_\omega M + r_0\varepsilon(\cos\varphi\Delta_\omega J_1 - \omega\sin\varphi J_1 + \sin\varphi\Delta_\omega J_2 + \omega\cos\varphi J_2) = 0, \quad (26c)$$

$$\Delta_\alpha M + r_0\varepsilon(\cos\varphi\Delta_\alpha J_1 + \sin\varphi\Delta_\alpha J_2) = 0. \quad (26d)$$

On introducing (20)–(23) into (25) and (26) it is observed that the material equilibrium conditions are satisfied identically.

Let us now turn our attention to reciprocity. As a first application we consider two material displacements λ_1 and λ_2 of the circular hole. The reciprocity theorem states that the work of the change of J_2 due to the material translation λ_1 in the material translation λ_2 is equal to the work of the change of the material force J_1 due to a material translation λ_2 in the material translation λ_1. Thus

$$\lambda_2\Delta_{\lambda_1} J_2 = \lambda_1\Delta_{\lambda_2} J_1. \quad (27a)$$

With (20b) and (21a) it is easily seen that (27a) holds. For the Maxwell-like version we introduce material influence coefficients as

$$\Delta_{\lambda_1} J_2 = \beta_{21}\lambda_1, \tag{28a}$$

$$\Delta_{\lambda_2} J_1 = \beta_{12}\lambda_2, \tag{28b}$$

and it is seen, immediately that

$$\beta_{21} = \beta_{12}. \tag{28c}$$

Another reciprocity relation may be formulated as: the work of the change of L due to a material self-similar expansion α in the material rotation ω is equal to the work of the change of M due to a rotation ω in the material self-similar expansion α, i.e.,

$$-\omega\Delta_\alpha L = \alpha\Delta_\omega M. \tag{27b}$$

The minus sign appears due to the positive definitions of the involved quantities. The Maxwell-like version is obvious. Both relations are confirmed with (23c) and (22d). Likewise, four further reciprocity relations can be established where we restrict ourselves to the Betti-like formulation because the Maxwell-like version is trivially accessible

$$\lambda_1\Delta_\alpha J_1 = \alpha\Delta_{\lambda_1} M, \tag{27c}$$

$$\lambda_2\Delta_\alpha J_2 = \alpha\Delta_{\lambda_2} M, \tag{27d}$$

$$-\omega\Delta_{\lambda_1} L = \lambda_1(\Delta_\omega J_1 + \omega J_2), \tag{27e}$$

$$-\omega\Delta_{\lambda_2} L = \lambda_2(\Delta_\omega J_2 + \omega J_1). \tag{27f}$$

The relations (27e) and (27f) are a little more extensive as mostly when rotations are involved. It might be supporting to give some more ideas on its derivation. Consider (27e) and apply the rotation ω first. J_1 and J_2 would change to $J_1 + \Delta_\omega J_1$ and $J_2 + \Delta_\omega J_2$ the work $W_{0\omega}$ would be

$$W_{0\omega} = J_2 r_0\varepsilon\omega\cos\varphi + \frac{1}{2} r_0\varepsilon\omega\cos\varphi\Delta_\omega J_2 - J_1 r_0\varepsilon\omega\sin\varphi - \frac{1}{2} r_0\varepsilon\omega\sin\varphi\Delta_\omega J_1. \tag{29}$$

L does not contribute to (29) since the hole is rotated around the origin whereas the dislocation is not rotated. An additional translation in x_1-direction produces additional work as

$$W_{\omega\lambda_1} = (J_1 + \Delta_\omega J_1)\lambda_1 + \frac{1}{2}\lambda_1\Delta_{\lambda_1} J_1. \tag{30}$$

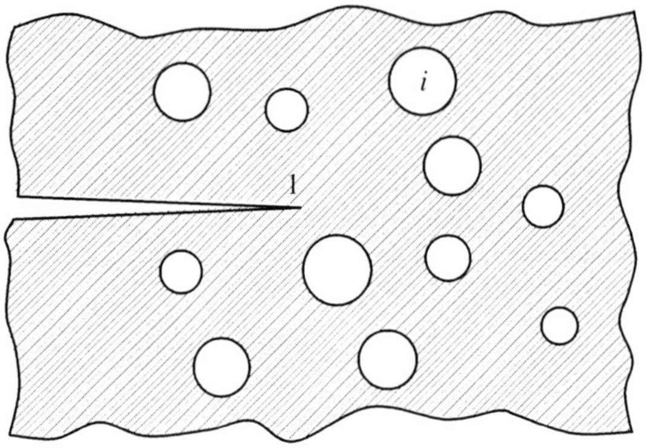

Fig. 10 Crack in a damaged material

If λ_1 is applied first, then the change of work would be

$$W_{0\lambda_1} = J_1\lambda_1 + \frac{1}{2}\lambda_1\Delta_{\lambda_1} J_1. \tag{31}$$

The material forces after this step are $\lambda_1 + \Delta_{\lambda_1} J_1$ and $\lambda_2 + \Delta_{\lambda_1} J_2$. An additional rotation ω causes the change of work to an amount of

$$W_{\lambda_1\omega} = (J_2 + \Delta_{\lambda_1} J_2) r_0\varepsilon\omega\cos\varphi + \frac{1}{2} r_0\varepsilon\omega\cos\varphi\Delta_\omega J_2 - (J_1 + \Delta_{\lambda_1} J_1) r_0\varepsilon\omega\sin\varphi - \frac{1}{2} r_0\varepsilon\omega\sin\varphi\Delta_\omega J_1. \tag{32}$$

Since finally, the energy stored in the system under consideration is the same independently of the sequence of the small material displacements, the work $W_{0\omega} + W_{\omega\lambda_1}$ and the work $W_{0\lambda_1} + W_{\lambda_1\omega}$ must be equal

$$W_{0\omega} + W_{\omega\lambda_1} = W_{0\lambda_1} + W_{\lambda_1\omega}. \tag{33}$$

After cancelling equal terms we find

$$\Delta_\omega J_1\lambda_1 = r_0\varepsilon\omega\cos\varphi\Delta_{\lambda_1} J_2 - r_0\varepsilon\omega\sin\varphi\Delta_{\lambda_1} J_1, \tag{34}$$

and with (25a), Eq. 27e is verified.

As an example of the usefulness of the reciprocity relations given above let us consider a crack surrounded by a damaged material characterized by various holes of different radii as depicted in Fig. 10. the system may support the general idea underlying Gurson's model (Gurson 1977).

Assume that we would be interested in the change of the J integral at the crack tip (defect 1) due to self-similar growth of each void (defect i). We would thus have to calculate $\Delta_{\alpha_i} J_1$. For this purpose we would have to evaluate the original configuration first and, additionally, construct for each void a new FE mesh with an extended radius $r_i \rightarrow \alpha r_i$, perform the FE calculation and calculate the change in J_1 each time. Instead, by using (27c), we translate the crack tip by an amount λ_1 and calculate the change in M of void i, $\Delta_{\lambda_1} M_i$, due to this translation and we have

$$\Delta_{\alpha_i} J_1 = \frac{\alpha_i}{\lambda_1} \Delta_{\lambda_1} M_i . \qquad (35)$$

Note that the change of J_1 due to the growth of any void is obtainable from only one remesh. In this way it is straight forward to construct influence surfaces for the J integral to assess the risk of voids in the neighbourhood of the crack. Of course, the voids could have different forms, they could also be cracks.

4 Conclusion

Reciprocity theorems in Material Space have been applied to problems of crack and defect interactions. Whereas one part of the paper in devoted to the validation of various versions of the theorems, another part and forthcoming research is concerned with the exploration of useful applications of reciprocity relations. One aspect is the construction of influence surfaces to assess the effect of damage (in the widest sense) on a main crack or defect. As in Physical Space, reciprocity relations might be the basis for the establishment of a boundary-integral method in Material Space. The generalization to continuously distributed defects, material forces per unit of volume and material translation fields as $\lambda_i = \lambda_i(x_j)$ $(i, j = 1, 2, 3)$ would result in reciprocity relations involving surface- and volume-integral expressions with probably useful and far-reaching applications.

Acknowledgements The support of G. Herrmann and R. Kienzler by the Research in Pairs Programme of the Mathematisches Forschungsinstitut Oberwolfach is gratefully acknowledged. Thanks are also due to Dipl.-Ing. R. Schröder for carefully drawing the figures.

References

Barber JR (2002) Elasticity, 2nd edn. Kluwer, Dordrecht
Budiansky B, Rice J (1973) Conservation laws and energy-release rates. ASME J Appl Mech 40:201–203
Erdogan F (1962) On the stress distribution in plates with collinear cuts under arbitrary loads. In: Proceedings of the Fourth U.S. National Congress of Applied Mechanics. ASME 1:547–553
Gross D (1982) Spannungsintensitätsfaktoren von Rißsystemen. Ing Arch 51:301–310
Günther W (1962) Über einige Randintegrale der Elastomechanik. Abh Braunschw Wiss Ges 14:53–72
Gurson AL (1977) Continuum theory of ductile rupture by void nucleation and growth. Part I—yield criteria and flow rules for porous ductile media. J Eng Mater Tech 99:2–15
Gurtin ME (2000) Configurational forces as basic concepts of continuum physics. Springer, New York
Herrmann G, Kienzler R (2007a) Reciprocity relations in Eshelbian mechanics. Mech Res Commun 34:338–343
Herrmann G, Kienzler R (2007b) A reciprocity relation couples Newtonian and Eshelbian Mechanics. ASME J Appl Mech (in press)
Kienzler R, Herrmann G (2000) Mechanics in material space. Springer, Berlin
Kienzler R, Herrmann G (2007) Nonlinear and linearized reciprocity relations in structural configurational mechanics. Acta Mech, doi:10.1007/s00707-007-0457-5
Kienzler R, Kordisch H (1990) Calculation of J_1 and J_2, using the L and M integral. Int J Fract 43:213–225
Knowles JK, Sternberg E (1972) On a class of conservation laws in linearized and finite elastostatics. Arch Rat Mech Anal 44:187–211
Marguerre K (1962) Elasticity, basic concepts. In: Flügge W Handbook of engineering mechanics. McGraw-Hill, New York,
Maugin GA (1993) Material inhomogeneities in elasticity. Chapman & Hall, London
Panasyuk MP, Savruk MP, Datsyshyn AP (1977) A general method of solution of two-dimensional problems in the theory of cracks. Eng Fract Mech 9:481–497
Rice JR (1968) A path independent integral and the approximate analysis of strain concentration by notches and cracks. ASME J Appl Mech 27:379–386
Rohde L, Kienzler R (2005) Numerical computation and analytical estimation of stress-intensity factors for strips with multiple edge cracks. Arch Appl Mech 74:846–852
Rohde L, Kienzler R, Herrmann G (2005) On a new method for calculating stress-intensity factors of multiple edge cracks. Phil Mag 85:4231–4244
Schweins G (1849) Fliehmomente oder die Summe $(xX + yY)$ bei Kräften in der Ebene und $(xX + yY + zZ)$ bei Kräften im Raume. Crelles J Reine Angew Math 38:77–88
Timoshenko SP, Goodier JN (1970) Theory of elasticity, 3rd edn. McGraw-Hill, New York

Configurational forces and gauge conditions in electromagnetic bodies

Carmine Trimarco

Abstract Material inhomogeneities, defective materials or fractured solids address the notion of configurational force. In electromagnetic deformable bodies this force also depends upon the electromagnetic potentials, which unfortunately are not uniquely defined. In electrostatics the problem of uniqueness is scarcely relevant, as only the scalar potential plays a role. Thus, dielectrics are adequately described in this context, even in the presence of a crack-line. In electrodynamics also the vector potential plays a prominent role and the lack of uniqueness of the electromagnetic potentials cannot be disregarded. The problem of solving this indeterminacy for the quantities of interest appeals to additional conditions, the gauge conditions. Gauge transformations, which leave invariant the Maxwell equations and the balance of momentum in material form, are here examined. In configurational mechanics of electromagnetic solids, a possible gauge dependent quantity is the material momentum, which seems to be related to some extent to supercurrents in deformable superconductors.

C. Trimarco (✉)
Department of Applied Mathematics,
University of Pisa,
Via F. Buonarroti, 1, 56127 Pisa, Italy
e-mail: trimarco@dma.unipi.it

Keywords Configurational mechanics · Electromagnetism · Variational methods · Gauges · Superconductors

1 Introduction

The description of electromagnetic solids, which undergo finite deformations, is quite involved due to the coupling of the mechanical with the electromagnetic fields. However, the balance laws of mechanics and the equations of electromagnetism can be possibly established in a consistent fashion, through a variational procedure. This procedure is essentially based on a Lagrangian density, which depends on the scalar and the vector potentials, the deformation, the polarisation and the magnetisation in their Lagrangian form and on their space and time derivative (Trimarco and Maugin 2001). Unfortunately, the electromagnetic potentials are not uniquely defined (Jackson 1962). This lack of uniqueness is generally unimportant in electrostatics, as the vector potential is made to identically vanish and the scalar potential is defined apart from a constant function of the domain of interest. Hence, mathematical problems in electrostatics of solid dielectrics, fractured or not, can be satisfactorily formulated in terms of the electric potential, the polarisation and the mechanical quantities of interest (Trimarco 2005a,b). In electrodynamics, the electric field is a time- dependent field, which is

Defect and Material Mechanics. C. Dascalu, G.A. Maugin & C. Stolz (eds.),
doi: 10.1007/978-1-4020-6929-1_3,

unavoidably coupled with the magnetic induction through the vector potential. Due to the indeterminacy of the scalar and vector potentials, additional conditions, which are known as gauge conditions, need to be properly established.

As mentioned, a preliminary issue in deformable solids is to write the whole set of the mechanical and electromagnetic equations. These equations can be shown to lead to new equations, which are instructive and enlightening in the presence of material defects or in crack propagation problems. In fact, the new equations, known as configurational balance laws, though mere identities along the motion, can be viewed as additional balance laws that govern the behaviour of defects or any other material inhomogeneity (Maugin and Trimarco 1992a, b, Gurtin 2000, Kienzler and Herrmann 2000, Trimarco and Maugin 2001). The notion of Eshelby stress tensor and that of material momentum arises in a natural way in this context.

However, the configurational balance laws can be alternatively perceived as primary laws when it is possible to appeal to a nonclassical variational approach, which is based on the inverse motion and inverse deformation (Maugin and Trimarco 1992a, Maugin 1993, Trimarco and Maugin 2001). This approach can be extended to electromagnetic bodies. In this context, the lack of uniqueness of the vector potential affects the unique definition of the Eshelby stress and that of material momentum, which are derived as canonical quantities from a suitable expression of the Lagrangian density. Differently, the physical electromagnetic stress (a Cauchy-like stress) and momentum do not depend explicitly upon the electromagnetic potentials and are thus uniquely defined (Trimarco 2007a).

The set of transformations (the gauge transformations) for the electromagnetic potentials, under which the electromagnetic Maxwell fields are invariant, are re-proposed here in the Lagrangian form and their effect on the physical quantities of interest is examined. The treatment parallels the classical approach to some extent. However, the gauge conditions for the material potentials cannot be a formal replica of the classical Coulomb or Lorenz-Lorentz gauge conditions of electrodynamics (Jackson 2002, Trimarco 2007a, b). The problem has not an obvious solution and seems to represent an open issue in deformable solids.

The gauge transformations for the Lagrangian electromagnetic potentials are also explored and by invariance arguments some interesting results can be shown in evidence. As an example of the role played by the vector potential in the configurational framework, the case of a superconducting material is discussed and a link between the supercurrent and material momentum of configurational mechanics is pointed out.

2 The Lagrangian electromagnetic potentials

The Lagrangian scalar and vector electromagnetic potentials ϕ and $\mathbf{A}$, respectively, are defined through the equations

$$\mathrm{Curl}\,\mathbf{A} = \mathbf{B} \tag{1}$$

$$-\mathbf{A}^{\bullet} - \mathrm{Grad}\,\phi = \mathbf{E}, \tag{2}$$

where $\mathbf{B}$ and $\mathbf{E}$ are the magnetic induction and the electric field, respectively, in the Lagrangian form. Curl and Grad denote the differential operators of curl and gradient, respectively, in the reference configuration V of an electromagnetic body, whose current configuration is denoted by $\mathcal{V}$. V and $\mathcal{V}$ are here assumed to be simply connected open subsets of the Euclidean space E_3. If $\mathbf{x} \in \mathcal{V}$ and $\mathbf{X} \in \mathrm{V}$, $\mathbf{x} = \chi(\mathbf{X}, \mathrm{t})$ represents the motion, $\mathrm{t} \in \mathbb{R}$ being the time. $\mathbf{F}$ denotes the deformation gradient and $\mathbf{v} = \mathbf{x}^{\bullet} \equiv; \frac{\mathrm{d}}{\mathrm{dt}}[\chi(\mathbf{X}, \mathrm{t})]$ the velocity. As usual, det $\mathbf{F} \equiv; \mathrm{J} > 0$ (Truesdell and Noll 1965). With reference to the Eq. 2, $\mathbf{A}^{\bullet} \equiv \frac{\mathrm{d}\mathbf{A}}{\mathrm{dt}} \equiv \partial[\mathbf{A}(\mathbf{X}, \mathrm{t})]/\partial\mathrm{t}|_{\mathrm{X}}$ denotes the material time derivative of $\mathbf{A}$. The potentials ϕ and $\mathbf{A}$ are related to the classical Eulerian electromagnetic potentials φ and $\mathbf{a}$ in $\mathcal{V}$, as follows:

$$\mathbf{A} = \mathbf{F}^{\mathrm{T}}\mathbf{a} \tag{3}$$

$$\phi = \varphi - \mathbf{v} \bullet \mathbf{a}. \tag{4}$$

$\mathbf{F}^{\mathrm{T}}$ denotes the transpose of the deformation gradient. It is worth recalling that

$$\mathrm{curl}\,\mathbf{a} = \mathbf{b} \tag{5}$$

$$-\partial\mathbf{a}/\partial\mathrm{t} \equiv -\mathbf{a}_{,\mathrm{t}} = \mathbf{e} + \mathrm{grad}\,\varphi, \tag{6}$$

$\mathbf{b}$ and $\mathbf{e}$ being the classical Eulerian magnetic induction and electric field, respectively. $(\partial\mathbf{a}/\partial\mathrm{t})$ denotes the Eulerian time derivative of $\mathbf{a}$ (Becker 1964, Jackson 1962). Note that the space and time differential operators that appear in the Eqs. 5 and 6 act in the current configuration $\mathcal{V}$.

The following transformations

$$\mathbf{a}^* = \mathbf{a} - \text{grad g} \tag{7}$$

$$\varphi^* = \varphi + (\mathrm{g}_{,\mathrm{t}}) \tag{8}$$

represent the classical gauge transformations. The arbitrary scalar function g ($\mathbf{x}$, t) represents the classical gauge function and $(\mathrm{g}, _\mathrm{t}) \equiv \partial \mathrm{g}/\partial \mathrm{t}$. With reference to the Eqs. 5 and 6, **e** and **b** are not altered by the transformations (7) and (8).

Similar transformations can be introduced for ϕ and **A** as follows:

$$\mathbf{A}^* = \mathbf{A} - \text{Grad f}, \tag{9}$$

$$\phi^* = \phi + \mathrm{f}^\bullet, \tag{10}$$

where f ($\mathbf{x}$ ($\mathbf{X}$, t), t) is an arbitrary scalar function. Analogously to the classical case and with reference to the Eqs. 1 and 2, the fields **E** and **B** are unaffected by these Lagrangian transformations. f can be viewed as a Lagrangian gauge function (Trimarco 2007a,b).

Along with the transformations (7) and (8) it is worth introducing the following transformations for the time and space derivative differential operators in $\mathcal{V}$, which still involve the electromagnetic potentials φ and **a**:

$$\nabla_\mathbf{a} = (\text{grad}) - i\alpha\, \mathbf{a} \tag{11}$$

$$(\partial/\partial \mathrm{t})_\varphi = (\partial/\partial \mathrm{t}) + i\alpha\, \varphi, \tag{12}$$

where i denotes the imaginary unit and α a real valued positive constant. It can be readily checked that the classical Maxwell equations in the Eulerian form are invariant also with respect to the transformations (11) and (12). These transformations are of quantum-mechanics interest and are known in the literature as gauge transformations of first kind. In quantum mechanics context $\alpha = (\mathrm{q}/\hbar)$, q being an electric charge and ħ the reduced Planck's constant (Becker 1964, Landau and Lifschitz 1974).

Analogous transformations to those expressed by (11) and (12) can be established in the referential configuration V. These transformations, which also do not affect the Maxwell equations in the Lagrangian form, are

$$\nabla_\mathbf{A} = (\text{Grad}) - \beta \mathbf{A} \tag{13}$$

$$(\partial/\partial \mathrm{t})_\phi = (\partial/\partial \mathrm{t}) + \beta\phi, \tag{14}$$

where possibly $\beta = i\alpha$. Note that the transformations (13) and (14) have not been apparently discussed in the literature, although they enter the phenomenological description of complex physical effects such as supercurrents in deformable solids (Maugin 1992, Trimarco 2007b).

3 Inverse variation and configurational forces

A variational approach based on the inverse deformation for elastic solids was introduced and discussed in previous papers and therein extended to the dynamical cases (Maugin and Trimarco 1992a, b). From this approach the configurational stress and momentum and the related balance law are derived by varying the reference configuration of the material body, while the current configuration is kept fixed. In this view, we consider the inverse motion $\mathbf{X} = \chi^{-1}(\mathbf{x}, \mathrm{t})$, the inverse deformation gradient $\mathbf{F}^{-1}$ and introduce the material velocity $\mathbf{V} \equiv \partial \mathbf{X}((\mathbf{x}, \mathrm{t}), \mathrm{t})/\partial \mathrm{t} \equiv \mathbf{X}_{,\mathrm{t}}$. Consider also the following Lagrangian density per unit volume of the current configuration:

$$\mathcal{L} = \mathcal{L}(\phi, (\phi_{,\mathrm{t}}), \text{grad}\,\phi, \mathbf{A}, (\mathbf{A}_{,\mathrm{t}}), \text{grad}\,\mathbf{A}, \mathbb{P}(\mathbb{P}_{,\mathrm{t}}), \text{grad}\,\mathbb{P}, \mathbb{M}, (\mathbb{M}_{,\mathrm{t}}), \text{grad}\,\mathbb{M}, \mathbf{X}, (\mathbf{X}_{,\mathrm{t}}), \mathbf{F}^{-1}, \mathbf{x}), \tag{15}$$

where $\mathbb{P} = \mathbf{J}\mathbf{F}^{-1}\mathbf{P}$ and $\mathbb{M} = \mathbf{F}^{\mathrm{T}}\boldsymbol{M} \equiv \mathbf{F}^{\mathrm{T}}(\mathbf{M} + \mathbf{v}\mathrm{x}\mathbf{P})$ represent the material polarisation and the magnetisation, respectively.

One of the Euler–Lagrange equations that are associated with the Lagrangian density (15) represents the balance equation of the configurational or material momentum:

$$(\partial/\partial \mathrm{t})|\mathrm{x}[\partial\mathcal{L}/\partial\mathbf{V}] - \partial\mathcal{L}/\partial\mathbf{X} + \text{div}\{\partial\mathcal{L}/\partial(\mathbf{F}^{-1}) + [\partial\mathcal{L}/\partial\mathbf{V}] \otimes [\mathbf{F}(\mathbf{V})]\} = 0. \tag{16}$$

From a specific Lagrangian density (Trimarco 2007a, b), the following canonical quantities can be explicitly written:

$$(\partial\mathcal{L}/\partial\mathbf{V}) = \boldsymbol{p} + (\text{Grad}\mathbf{A})\mathbf{D}, \tag{17}$$

$$-\mathrm{J}(\partial\mathcal{L}/\partial\mathbf{F}^{-1})\mathbf{F}^{-\mathrm{T}} = \boldsymbol{b} + (\text{Grad}\,\mathbf{A})[\partial\mathcal{L}/\partial(\text{Grad}\,\mathbf{A})] + \mathbf{A}^\bullet \otimes \mathbf{D}, \tag{18}$$

where

$$\boldsymbol{p} = \rho_0 \mathbf{C}\mathbf{V} + \mathbb{P} \times \mathbf{B} \tag{19}$$

and

$$\boldsymbol{b} = \{W - \mathbb{P}\bullet\mathbf{E} - \mathbf{B}\bullet(\mathbf{V}\mathrm{x}\mathbf{D_0}) - 1/2\rho_0\mathbf{V}\bullet\boldsymbol{C}\nabla\}\mathbf{I} - \mathbf{F}^{\mathrm{T}}(\partial W/\partial \mathrm{F}) + \mathbf{E}\otimes\mathbb{P} - \mathbb{M}\otimes\mathbf{B} + (\mathbf{V}\mathrm{x}\mathbf{D_0})\otimes\mathbf{B} - (\mathbf{B}\mathrm{x}\mathbf{V})\otimes\mathbf{D_0} + (\mathbf{D_0}\mathrm{x}\mathbf{B})\otimes\mathbf{V}, \tag{20}$$

represent the material momentum and the material configurational stress, respectively. $\mathbf{C} \equiv \mathbf{F}^T\mathbf{F}$ is the right Cauchy-Green deformation tensor. **E, B** and **D** are the electric field, the magnetic and the electric induction, respectively, in the Lagrangian form. $\mathbf{D_0} \equiv \mathbf{D} - \mathbb{P} \equiv J\varepsilon_0\mathbf{C}^{-1}(\mathbf{E} + \mathbf{V} \times \mathbf{B})$, ε_0 being the dielectric constant of a vacuum.

With reference to the Eqs. 17 and 18, the canonical quantities $(\partial\mathcal{L}/\partial\mathbf{V})$ and $(\partial\mathcal{L}/\partial\mathbf{F}^{-1})$ depend on the (Grad **A**) and on $\mathbf{A}^\bullet$ explicitly. It is not difficult to see that these quantities are not gauge invariant with none of the introduced transformations. By contrast, it can be shown that the Eq. 16 does enjoy the gauge invariance property with respect to the transformations (9)–(10) and (13)–(14). A unique definition of these canonical quantities addresses gauge conditions for the Lagrangian electromagnetic potentials (Trimarco 2007b).

4 Gauge dependence and gauge invariance

Most of the quantities and of the equations of interest in electromagnetic bodies are invariant with respect to gauge transformations, which have been introduced in the previous section. For instance, the Maxwell equations in the Lagrangian form are invariant under the transformations (9) and (10). So are the Lagrangian electromagnetic fields, the electromagnetic stress and momentum (Nelson 1979, Trimarco 2001). However, there are quantities that clearly depend on the choice of the electromagnetic potentials as in the case of the canonical quantities (17) and (18).

A simple example of gauge dependent quantities in classical electromagnetism is the momentum of a charged particle, which reads (Goldstein 1980, Becker 1966, Landau and Lifschitz 1974)

$$\mathbf{p} = \mathrm{m}\mathbf{v} + \mathrm{q}\mathbf{a}, \tag{21}$$

where m and q represent the mass and the charge of the particle of interest, respectively. It is worth noting that the momentum density for an electromagnetic body can be written as (Nelson 1979, Schoeller and Tellung 1992)

$$\mathbf{p}_\mathrm{d} = \rho\mathbf{v} + \varepsilon_0\mathbf{e} \times \mathbf{b}. \tag{22}$$

This momentum density is clearly invariant with respect to the gauge transformations (7) and (8), differently form the momentum (21). In the expression (22) ρ is the mass density in $\mathcal{V}$. It is worth remarking that the momentum (21) can be recovered from the momentum density (22) by volume integration, in the linear approximation of the second order term $(\mathbf{e} \times \mathbf{b})$, having assumed that **b** coincides with a given external magnetic induction and that **a** is a divergence-free field (transverse or Coulomb gauge condition).

A second instructive example is provided by the London equation (1950) for superconductors. This equation is the simplest equation that governs the electric current in a superconductor and that accounts for the Meixner effect (**b** identically vanishing in the bulk of a superconductor), one of the most peculiar features of superconducting materials (Kittel 1986). The London equation in S.I. units reads

$$\mathbf{j}_\mathrm{s} = -[(\mu_0\lambda^2)^{-1}]\mathbf{a}. \tag{23}$$

$\mathbf{j}_\mathrm{s}$ is the supercurrent density, μ_0 the magnetic permeability of a vacuum and λ a phenomenological positive constant known as the London penetration depth. In order to uniquely define the supercurrent, a restriction on **a** is needed. Similarly to the first example, this restriction is provided by the aforementioned gauge condition

$$\mathrm{div}\,\mathbf{a} = 0. \tag{24}$$

Note that the quantity $(\mathrm{curl}\,\mathbf{j}_\mathrm{s})$ is gauge invariant, differently from $\mathbf{j}_\mathrm{s}$ in the Eq. 23. In fact, by taking into account the Eq. 5, the Eq. 23 can be also written in gauge invariant form as

$$\mathrm{curl}\,\mathbf{j}_\mathrm{s} = -[(\mu_0\lambda^2)^{-1}]\mathbf{b}. \tag{25}$$

Superconductivity is a genuine quantum mechanical effect and a gauge invariant form of $\mathbf{j}_\mathrm{s}$ can be found in a natural way in this context (Kittel 1986, Blatt 1964, von Lau 1952, Fröhlich 1966, Jones and March 1973). However, the phenomenon can be satisfactorily described in the framework of continuum mechanics. In this framework and in the simple case of type 1 superconductors, the following simple model can be proposed. The supercurrent is assumed to be related to the motion of electric supercharges, in analogy with the idea that the electric current is related with the velocity of the free electrons in a metal. Specific assumptions on the density ρ_s of the supercharges (which are not single electrons, but rather group of electrons: in fact the 'electron-pairs' of the Bardeen–Cooper–Schrieffer theory) and on their velocity $\mathbf{v}$ can be proposed. With the additional assumption that the supercurrent density $\mathbf{j}_s$ is proportional to the momentum density and specifically to both, velocity and mass density, one is lead to gauge invariant equations that are reminder of the behaviour of Eulerian fluids (Fröhlich 1966).

The treatment is consistent with the Ginzburg–Landau theory, according to which a free energy density of a superconductor depends upon a macroscopic complex-valued function ψ, this playing the role of an order parameter in the transition between the thermodynamical phases of a superconductor. This assumption is consistent with the microscopic statistical behaviour of the supercharges, which behave as indistinguishable particles that act as 'bosons', contrary to ordinary electrons that act as 'fermions' (Fröhlich 1966, Kittel 1986, Jones March 1973). As a result, a single function ψ describes the behaviour of all supercharges (Fröhlich 1966). Accordingly, the density ρ_s is associated with ψ as follows:

$$\psi(\mathbf{x}, \mathrm{t}) = (\rho_s)^{1/2} \exp[i\theta/u], \tag{26}$$

so that $\psi\bar{\psi} = \rho_s$, $\bar{\psi}$ being the complex conjugate of ψ. Hereafter, it is understood that $\rho_s = \rho_s(\mathbf{x}, \mathrm{t})$ and that $\theta = \theta(\mathbf{x}, \mathrm{t})$, whereas u is a constant whose physical dimensions are defined by the following assumption on the velocity $\mathbf{v}$:

$$\mathbf{v} = \operatorname{grad}\theta - \kappa\mathbf{a}, \tag{27}$$

κ being a positive constant. With reference to the Eq. 27, θ plays the role of a velocity potential. It is worth noting the similarity of this equation with the momentum (21) if ($\mathbf{p}$/m) is replaced by (grad θ). This similarity reproduces in the macroscopic framework one of the fundamental assumption of quantum-mechanics for the canonical momentum, which is viewed as the differential gradient operator acting on the 'wave function' of the particle of interest. A further assumption is

$$\mathbf{j}_s = \kappa\rho_s\mathbf{v}. \tag{28}$$

According to the Eqs. 27 and 28, the quantity (curl $\mathbf{v}$) and thus $\mathbf{j}_s$ are proportional to the magnetic induction $\mathbf{b}$, a result that is still based on the London's idea.

Under conditions that are specified below, the density ρ_s, the velocity $\mathbf{v}$ and the supercurrent $\mathbf{j}_s$ turn out to satisfy the following equations:

$$\rho_s(\partial\mathbf{v}/\partial\mathrm{t}) + \rho_s(\operatorname{grad}\mathbf{v})\mathbf{v} = -\operatorname{grad}\mathrm{p} + \mathbf{f}, \tag{29}$$

$$\operatorname{div}\mathbf{j}_s + (\partial\kappa\rho_s/\partial\mathrm{t}) = 0. \tag{30}$$

The Eq. 29 represents the balance of momentum of an Eulerian fluid in the presence of a pressure term p and a force $\mathbf{f}$ of electromagnetic origin. The Eq. 30 represents the conservation of the supercurrent, which in terms of ψ and $\mathbf{a}$ reads

$$\mathbf{j}_s = (u/2i)[\bar{\psi}(\operatorname{grad}\psi) - \psi(\operatorname{grad}\bar{\psi})] - \kappa\psi\bar{\psi}\mathbf{a}. \tag{31}$$

This expression can be evaluated by taking into account the Eqs. 26, 27 and 28. Note that $\mathbf{j}_s$ is invariant with respect to the gauge transformations (11) and (12), differently from the case expounded in the previous section. It is also invariant with respect to the gauge transformations (7) and (8), provided ψ transforms as follows (Aharonov and Bohm 1959, Jackson 2002):

$$\psi^* = \psi\exp(ig\kappa/u). \tag{32}$$

It is worth remarking that the transformation (32) only affects the phase of $\psi$and thus does not affect the density ρ_s, in accordance with the formula (26). The gauge invariant Eqs. 29 and 30 hold true provided ψ satisfies the equation

$$\begin{aligned}(1/i)(\partial\psi/\partial\mathrm{t}) = {} & (1/2)u[(\operatorname{grad}) - i(\kappa/u)\mathbf{a}]^2\psi \\ & + c\psi - d(\psi\bar{\psi})\psi,\end{aligned} \tag{33}$$

where c and d are positive constants. It is worth remarking that, with reference to the formula (11) the otherwise unspecified macroscopic constants k and u and α can be related to the microscopic ones by assuming that $k = (\mathrm{q/m})$ and $u = (\hbar/\mathrm{m})$, m denoting the mass of the supercharge (Fröhlich 1966).

The gauge invariant Eq. 33 represents an extended version of the Ginzburg–Landau equation, which can be derived from the following free energy density in the presence of magnetic fields

$$\begin{aligned}F(\psi, (\nabla_{\mathbf{a}}\psi), (\operatorname{curl}\mathbf{a}), \mathrm{T}) = {} & (1/2)(\nabla_{\mathbf{a}}\psi)^2 + c\psi\bar{\psi})^2 \\ & -(d/2)(\psi\bar{\psi})^4 \\ & +(1/2)(\mu_o)^{-1}(\operatorname{curl}\mathbf{a})^2.\end{aligned} \tag{34}$$

Here, c and d possibly depend on the temperature T. Invariance of F with respect to ψ and to $\bar{\psi}$

leads to the stationary case of Eq. 33. Invariance with respect to **a** leads to the equation

$$\mathrm{curl}\,(\mathrm{curl}\,\mathbf{a}) = \mu_0(\partial F/\partial \mathbf{a}). \tag{35}$$

In this equation the quantity $(\partial F/\partial \mathbf{a})$ plays the role of a current density, in fact the supercurrent such as defined by the formula (31).

5 The Lagrangian electromagnetic potentials in deformable superconductors

In a deformable electromagnetic body the Lagrangian potentials are more appropriate than the Eulerian ones, although they are related to one another through the formulas (3) and (4). Accordingly, the gauge transformations that are inherent deformable solids are those expressed by the formulas (9)–(10) and (13)–(14). Also, according to the principle of material indifference (Truesdell and Noll 1965), the free energy density $\mathcal{F}$ that generalises the Ginzburg-Landau expression (34) should depend on $(\nabla_{\mathbf{A}}\Psi)$ and on $(\mathrm{Curl}\,\mathbf{A})$, rather than on $(\nabla_{\mathbf{a}}\psi)$ and on $(\mathrm{curl}\,\mathbf{a})$, where

$$\Psi = (\mathrm{J})^{1/2}\psi \tag{36}$$

and $\Psi\bar{\Psi} = \mathrm{J}\rho_{\mathrm{s}}$. With reference to the formulas (3), (11) and (13), the following relationship can be recovered

$$\nabla_{\mathbf{A}} = \mathbf{F}^{\mathrm{T}}\nabla_{\mathbf{a}}, \tag{37}$$

provided that $\beta = i\alpha$.

Invariance arguments like those expounded in Sect. 4 can be re-proposed for the free energy density in the reference configuration $\mathcal{F} = \mathcal{F}(\Psi, (\nabla_{\mathbf{A}}\Psi), (\mathrm{Curl}\,\mathbf{A}), \ldots)$ in order to derive the equation that govern Ψ and the Lagrangian supercurrent $\boldsymbol{J}_{\mathrm{s}}$ (Maugin 1992).

Re-examination of the formulas (27) and (28) in the light of the transformation (37) suggests the following transformation for **v** through the deformation gradient:

$$\mathbf{F}^{\mathrm{T}}\mathbf{v} = \mathrm{Grad}\,\Theta - \kappa\mathbf{A}, \tag{38}$$

Θ being the Lagrangian velocity potential. The transformation (38) is enlightening as the quantity $\mathbf{F}^{\mathrm{T}}\mathbf{v}$ represents the mechanical part of material momentum such as introduced through the Eq. 19. In fact, if $\mathbf{V} \equiv -\mathbf{F}^{-1}\mathbf{v}$ and if the contribution of the electromagnetic field is disregarded in the formula (19), one can write $\boldsymbol{p} = \rho_0\mathbf{C}\mathbf{V} \equiv -\rho_0\mathbf{F}^{\mathrm{T}}\mathbf{v}$. Consistently with this remark, the operator $(\nabla_{\mathbf{A}})$ should be related to the material momentum rather than to the momentum or to the velocity. It is worth recalling that the dependence through $(\nabla_{\mathbf{A}}\Psi)$ is typically quadratic as it accounts for the contribution of the kinetic energy density of particles in the microscopic framework (Becker and Sauter 1964, Landau and Lifschitz 1974).

By taking into account the Eqs. 28 and 38, the Lagrangian supercurrent reads

$$\boldsymbol{J}_{\mathrm{s}} = \kappa\mathrm{J}\rho_{\mathrm{s}}\mathbf{F}^{\mathrm{T}}\mathbf{v}, \tag{39}$$

or, equivalently,

$$\boldsymbol{J}_{\mathrm{s}} = \mathrm{J}\mathbf{F}^{\mathrm{T}}\mathbf{j}_{\mathrm{s}}. \tag{40}$$

Due to the explicit dependence of $\boldsymbol{J}_{\mathrm{s}}$ upon the deformation gradient, it is unlikely that the supercurrent can be treated in this case like an Eulerian fluid as it was shown in the previous section for undeformed bodies, unless the body undergoes specific time-independent homogeneous deformations. With reference to the Eq. 39 or 40, it is worth remarking that supercurrent transforms in the material frame differently than ordinary electric free-current $\boldsymbol{i}$. In fact, the Lagrangian electric current density reads

$$\boldsymbol{j} = \mathrm{J}\mathbf{F}^{-1}(\boldsymbol{i} - \rho_{\mathrm{e}}\mathbf{v}), \tag{41}$$

where ρ_{e} denotes the ordinary electric free charge. The electric current $\boldsymbol{j}$ enters the inherent Maxwell equation in the Lagrangian form, which reads

$$\mathrm{Curl}\,\mathbf{H} - (\mathrm{d}/\mathrm{d}t)\mathbf{D} = \boldsymbol{j}, \tag{42}$$

where **H** represents the Lagrangian magnetic field.

The whole set of the Maxwell equations in the Lagrangian form can be found in Nelson (1979) and in Maugin and Trimarco (1991), though with different notation.

6 Final comments

Variational methods in configurational mechanics introduce the notion of material stress (the Eshelby stress) and material momentum. As is known, variational methods also address the notion of energy-release rate of fracture mechanics (Maugin and Trimarco 1992b, Gurtin 2000, Kienzler and Herrmann 2000). The material stress and momentum, if derived as canonical quantities from the Lagrangian (15), are affected by the indeterminacy that is introduced by the Lagrangian electromagnetic potentials. However, this indeterminacy can be solved by removing the gauge dependent terms that appear in the formulas (17) and (18), as these terms balance each other identically in the Euler-Lagrange equation (Eq. 16). In the most general case, gauge invariant quantities are desirable in order to circumvent the problem of introducing gauge conditions. Here, the interest for superconductors is twofold: first, for the crucial role that the vector potential plays in these materials. Second, for that a mechanical model for the electric current in deformable superconductors addresses the material momentum rather than the momentum or the velocity and thus addresses configurational mechanics.

References

Aharonov Y, Bohm D (1959) Significance of electromagnetic potentials in the quantum theory. Phys Rev 2d Series 115(3):485–491

Becker R, Sauter F (1964) Electromagnetic interactions, vol 2. Blackie & Sons, London

Blatt JM (1964) Theory of superconductivity. Academic Press, London

Fröhlich H (1966) Macroscopic wave functions in superconductors. Proc Phys Soc 87:330–332

Goldstein H (1980) Classical mechanics, 12th edn. Addison-Wesley, Reading Mass

Gurtin ME (2000) Configurational forces as basic concepts of continuum physics. Springer Verlag, New York

Jackson JD (1962) Classical electrodynamics. J Wiley & Sons, New York

Jackson JD (2002) From Lorenz to Coulomb and other explicit gauge transformations. Am J Phys 70(9):917–928

Jones W, March NH (1973) Theoretical solid state physics, vol 2. Wiley-Interscience, Wiley and Sons Ltd., New York. [(1985) unabridged re-publication by Dover Publications Inc., New York]

Kienzler R, Herrmann G (2000) Mechanics in material space. Springer-Verlag, Berlin

Kittel C (1986) Introduction to solid state physics, 6th edn. John Wiley & Sons, New York

Landau L, Lifschitz E (1974) Méchanique quantique, 3rd edn. MIR, Moscou

Laue von M (1952) Theory of superconductivity. Academic Press, New York

London F (1950) Superfluids: macroscopic theory of superconductivity, vol 1. Wiley, New York

Maugin GA (1992) Irreversible thermodynamics of deformable superconductors. CR Acad Sci Paris t 314 Série II:889–894

Maugin GA (1993) Material inhomogeneities in elasticity series applied mathematics and mathematical computation, vol 3. Chapman and Hall, London

Maugin GA, Trimarco C (1991) Pseudomomentum and material forces in electromagnetic solids. Int J Electromagnet Mater 2:207–216

Maugin GA, Trimarco C (1992a) Note on a mixed variational principle in finite elasticity. Rend Mat Acc Naz dei Lincei, Serie IX, v. III:69–74

Maugin GA, Trimarco C (1992b) Pseudomomentum and material forces in nonlinear elasticity. Acta Mechanica 94:1–28

Maugin GA, Trimarco C (2001) Elements of field theory in inhomogeneous and defective materials. Configurational mechanics of materials, CISM courses and lectures, n. 427. Springer Verlag, Wien, pp 55–128

Nelson DF (1979) Electric, optic and acoustic interactions in dielectrics. John Wiley, New York

Schoeller H, Thellung A (1992) Lagrangian formalism and conservation law for electrodynamics in nonlinear elastic dielectrics. Ann Phys 220:18–39

Trimarco C (2001) A Lagrangian approach to electromagnetic bodies. Technische Mechanik 22(3):175–180

Trimarco C (2005a) On the material energy–momentum tensor in electrostatics and in magnetostatics.In: Steinman P, Maugin GA (eds) Advances in mechanics and mathematics, 11, mechanics of material forces. Springer New York, Cptr 16, 161–171

Trimarco C (2005b) The total kinetic energy of an electromagnetic body. Phil Mag 85(33–35):4277–4287

Trimarco C (2007a) Material electromagnetic fields and material forces. Arch Appl Mech 77:177–184 [Online http://maecourses.ucsd.edu/symi/other-papers.html]

Trimarco C (2007b) Configurational forces in dynamics and electrodynamics. Proc Estonian Acad Phys Math 56(2):116–125

Trimarco C, Maugin GA (2001) Material mechanics of electromagnetic bodies. Configurational mechanics of materials, CISM courses and lectures, n. 427, Springer Verlag, Wien, pp 129–172

Truesdell CA, Noll W (1965) The nonlinear field theory of mechanics. Handbuch der Physik, Bd.III/3, S. Flügge(ed). Springer Verlag, Berlin

The anti-symmetry principle for quasi-static crack propagation in Mode III

Gerardo E. Oleaga

Abstract In this note we study a basic propagation criterion for quasi-static crack evolution in Mode III. Using classical techniques of complex analysis, the assumption of stable growth is expressed in terms of the parameters defining the elastic field around the tip. We explore the consequences of the local condition obtained and analyse its role as a crack propagation law. In particular, we herein extend to bounded domains a number of results previously obtained for the whole plane.

Keywords Anti-symmetry principle · Quasistatic crack growth · Linear elasticity · Brittle material · Anti-plane fields · Configurational forces · Crack propagation law

1 Introduction

The present work deals with some fundamental questions in linear elastic fracture mechanics in its simplest framework; namely, out of plane fields, brittle homogeneous materials, quasi-static propagation and Griffith's dissipation model. The starting point is the following basic question: how the shape of a growing crack is determined by the local field around the tip? This problem is intimately related to that of finding a suitable propagation criterion. There is a huge amount of literature about this subject, and we shortly review a few previous contributions to establish a reference for the forthcoming discussion.

G. E. Oleaga (✉)
Departamento de Matemática Aplicada, Facultad de Matemáticas, Universidad Complutense de Madrid, 28040 Madrid, Spain
e-mail: oleaga@mat.ucm.es

To our knowledge, one of the most widely accepted criteria is the so-called *principle of local symmetry*. It was proposed by Goldstein and Salganik in (1974) and later analysed by Cotterell and Rice in (1980). For quasi-static propagation in a two dimensional in-plane elastic field, it can be formulated as follows: the crack grows along the path that cancels the Mode II stress intensity factor, so that:

$$K_{\mathrm{II}} = 0. \tag{1}$$

This equation is added to the energy conservation law, expressed in terms of the stress intensity factors through the celebrated Griffith–Irwin relationship:

$$\frac{1}{Y}\left(K_{\mathrm{I}}^{2} + K_{\mathrm{II}}^{2}\right) = \kappa. \tag{2}$$

Here κ is a material constant defining the amount of energy per unit length that has to be provided to open the crack; Y is the *Young modulus*. Both scalar equations provide conditions linking the path shape and the applied loads through the parameters defining the strength of the field near the tip; a criterion with this property is therefore termed *local*. Notice that the left hand side of (2) represents the so-called *energy release rate* for a crack propagating *smoothly*, i.e., without kinking or branching. The principle completes the

doi: 10.1007/978-1-4020-6929-1_4,

mathematical formulation of the free boundary problem, since we have two scalar equations to find the evolution of a plane curve. Nevertheless, it is not physically complete, because we are imposing the artificial constraint of path smoothness.

Another well-known local criterion is the *maximum energy release rate principle*. In our view, it is the most straightforward way to approach the problem from basic physical principles. Consider a family of *virtual extensions* $\gamma_\alpha(t)$ of the actual crack configuration. In this setting, $t \geq 0$ is a parametrisation of the extended curve, α defines the specific curve of the family and $\gamma_\alpha(t)$ represents the tip of the extended crack at "time" t. Let $\Delta E(t;\alpha)$ denote the amount of mechanical energy released along this virtual extension and $\Delta Q(t;\alpha)$ the amount of *dissipation* involved (i.e., minus the work that must be done to break the atom bonds on crack faces). We can say that the evolution along the path $\Delta\Gamma_\alpha := \{\gamma_\alpha(t), t \geq 0\}$ is possible if the following inequality holds for some small enough $\delta > 0$:

$$-\Delta E(t;\alpha) \geq \Delta Q(t;\alpha) \quad \forall t \leq \delta. \tag{3}$$

If we have a strict inequality in (3), the crack evolution turns to be *unstable*, and other physical ingredients should enter the picture, for instance the kinetic energy flux. On the other hand, we say that a crack configuration is *stable* if the following condition holds: For all virtual extensions $\Delta\Gamma_\alpha$, there exists some $\delta > 0$ such that:

$$-\Delta E(t;\alpha) \leq \Delta Q(t;\alpha) \quad \text{for all } t < \delta. \tag{4}$$

Consider now two cases of this inequality. For a finite extension, $\Delta Q(t;\alpha)$ is always greater than zero (we have to do *negative* mechanical work to break the bonds on crack faces). On the other hand, ΔE is always negative (elastic energy decreases with crack growth). We consider the following limit for a fixed α:

$$L(\alpha) := \lim_{t\to 0}\left(-\frac{\Delta E(t;\alpha)}{\Delta Q(t;\alpha)}\right). \tag{5}$$

If $L(\alpha)$ is strictly greater than 1, according to (3) *unstable* growth along this path will take place. On the other hand, if it is strictly lower than 1 we say that crack growth along this way is not possible. According to this, we say that the curve evolves in a *quasi-static or critical regime* if it satisfies the stability condition (4) and at each configuration there exists some α with $L(\alpha) = 1$ (critical growth).

When Griffith's model enters the picture, the dissipative term is proportional to the length of the extended crack, that is $\Delta Q(t;\alpha) = \kappa l(t)$. Therefore, the limit in (5) is written as:

$$L(\alpha) = \lim_{t\to 0}\left(-\frac{\Delta E(t;\alpha)}{\kappa l(t)}\right) =: \frac{1}{\kappa}G(\alpha), \tag{6}$$

where $G(\alpha)$ is the *energy release rate*. In order to check that the critical growth condition is satisfied, we must ensure first that there are no possible unstable directions along other paths. This leads us to the following inequality:

$$G(\alpha) \leq \kappa. \tag{7}$$

Then, if a crack propagates under a critical or quasi-static regime, the growth directions are defined as follows:

$$\alpha^* = \text{crack growth direction} \Rightarrow G(\alpha^*) = \max_\alpha G(\alpha). \tag{8}$$

This is the so-called *maximum energy release rate criterion*.

In this note we start by reviewing the role of the *configurational force* (6) as a basic object to determine the path of a quasi-static growing crack in Mode III. In Sect. 2 we show that this concept alone *is not enough* to define the shape evolution in an out of plane setting and a more precise study of ΔE should be done. In Sects. 3 and 4 we resume the picture presented in our previous work (Oleaga 2004, 2006), where a basic condition for crack stability called *the anti-symmetry principle* was obtained for an unbounded domain, and give some further insight about the kind of crack propagation law obtained. In Sect. 5 we analyse the case of a bounded domain, not considered in Oleaga (2004, 2006), and the main differences with the former one are pointed out.

2 The role of the *driving force* in crack direction

The discussion in the previous Section indicates that the function $G(\alpha)$ in (6) has to be computed in the first place. This task was performed by several authors and it is worth to review some of these works here. Let us parametrise the crack growing direction by the number $\alpha := \varphi/\pi$, where φ is the kinking angle with respect to the initial crack. Consider first some theoretical formulae for this quantity. In the analysis of the energy released for a virtual crack path, we have the well known Eshelby-Rice-Cherepanov J integral for straight extensions (cf. Rice 1968). Explicitly, for $\alpha = 0$ we have that:

$$G(0) = J_1 \tag{9}$$

where J_1 (formula (68) in Rice 1968) is the component, in the direction parallel to the existing crack, of the vector given by (summation on repeated indices is assumed):

$$J_k := \int_C \left(U\, n_k - \sigma_{ij} u_{i,k} n_j\right) \mathrm{d}s, \tag{10}$$

where $|\vec{n}| = 1$ is normal to C (a Jordan curve surrounding the crack tip), U is the elastic energy density, σ_{ij} are the components of the stress tensor and u_i is the displacement field. In pure Mode III, (10) reduces to the expression:

$$J_k = \int_C \left(U\, n_k - \sigma_{3j} u_{3,k} n_j\right) \mathrm{d}s \quad (j, k = 1, 2).$$

Thus, J_1 gives the energy release rate for a *straight extension of the preexisting crack*. Nevertheless, (9) tells us nothing for other virtual directions of propagation. It should be pointed out that in Rice's article, the quantity J *is not considered as a vectorial force*.

More recently, Gurtin introduced the vector (10) in Gurtin (formula (4.6) in Gurtin 1979) with a different notation. This may be considered as a natural generalization of (9) when the coordinate axes are not parallel to the crack direction. The formula for G turns to be:

$$G = \vec{J} \cdot \vec{e}, \tag{11}$$

where $\vec{e}$ is a unit vector in the direction of (quasi-static) motion. This expression suggests that $\vec{J}$ is a mechanical force yielding the energy released for *arbitrary* directions of motion, *while freezing the actual configuration*. But this is not the case, since (11) is only valid for smooth extensions of the initial crack. Then it is not possible to give $\vec{J}$ the status of a force as a "gradient" of some free energy.

Let us see that (11) gives a wrong answer when $\vec{e}$ is not pointing in the direction of the current crack configuration. For this kind of "kinked" paths in Mode III the stress intensity factor was obtained by Sih (1965) applying uniform loads at infinity. By the well-known Griffith–Irwin relationship, we can apply this result to compute the energy release rate for different directions:

$$G(\alpha) = \frac{K^2}{2\mu}\left(\frac{1-\alpha}{1+\alpha}\right)^{\alpha} \quad \alpha := \varphi/\pi, \tag{12}$$

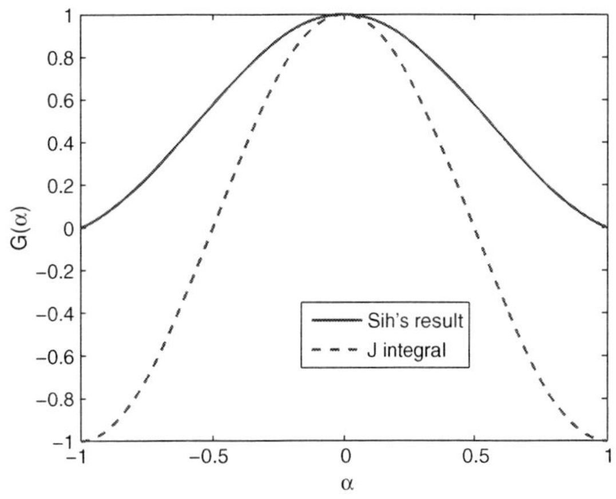

Fig. 1 The energy release rate for different growing directions

K being the stress intensity factor for the *initial configuration*, and μ the shear modulus. In Fig. 1 we see a comparison between the true energy release rate and the one obtained using $\vec{J}$ as a generalised force in (11).

It is to be noted that, even if $\vec{J}$ cannot be used as a good reference for the work done by the elastic forces, it shows correctly that the best direction to release energy *is the same as the one defined by the preexisting crack* (for Mode III). We conclude that the "crack driving force", or more precisely *the first order variation of the energy*, is not enough to determine the crack geometry: any smooth curve is compatible with the condition of maximum G imposed in (8). On the other hand, in addition to the well-known Griffith's balance during critical growth

$$G(0) = \kappa \tag{13}$$

it is certainly useful to establish a necessary condition for the *stability* of the crack configuration.

3 The energy functional for kinks and straight extensions

In the previous Section we showed that in Mode III we should study the energy released functional in more detail to find a complete crack evolution law from first principles. We will now summarize the results obtained in Oleaga (2004) where the energy released for a finite extension of the crack is expanded in terms of crack length and more information about crack direction is obtained. We refer to that article for any technical detailed not included here.

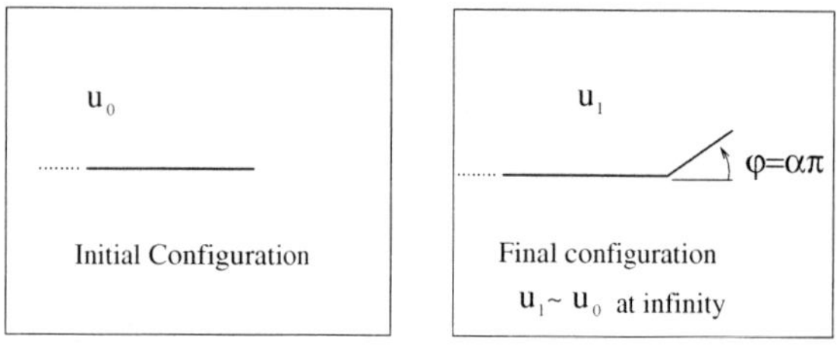

Fig. 2 Problem setting

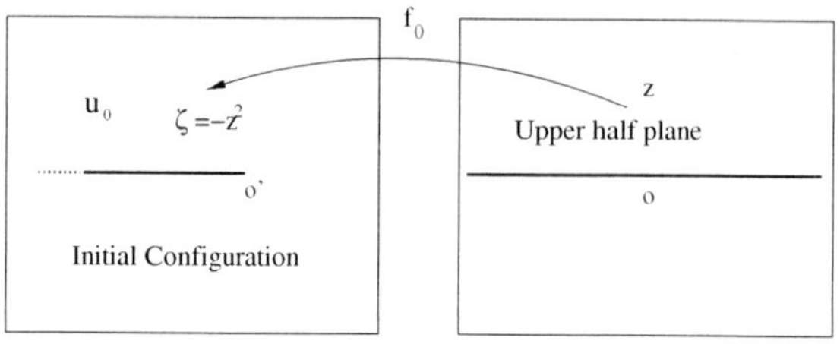

Fig. 3 The map f_0

3.1 The boundary value problem

The energy release rate for different growing directions, shown in (12), was obtained in Sih (1965) for an initially rectilinear crack in the plane, subjected to *uniform loading* at infinity. This boundary conditions define a local field that is probably not rich enough to capture the growing directional preference. To overcome this (possible) drawback, we considered the semi-infinite straight crack as initial configuration with an *arbitrary* displacement field u_0 around it, satisfying the equilibrium equation with no tractions on Γ_0:

$$\Delta u_0 = 0 \quad \text{in } \mathbb{R}^2 \backslash \Gamma_0, \tag{14}$$

$$\partial_n u_0 = 0 \quad \text{on } \Gamma_0. \tag{15}$$

We also add the finite energy condition around the tip, for some $r > 0$:

$$\int_{B_r(\text{tip})} |\nabla u_0|^2 < \infty. \tag{16}$$

Once the crack is extended, similar conditions hold for u_l together with the "matching condition" (Fig. 2):

$$|u_l - u_0| \to 0 \quad \text{uniformly at infinity} \tag{17}$$

Notice that in this case we are using $t \equiv l =$ length of the crack extension to parametrise the displacements. In this way, the elastic field at infinity is fixed during crack advance and there is no mechanical work of the applied loading. The function u_l satisfies a mixed boundary value problem and it is well defined for a given initial field and crack configuration Γ_l (cf. Oleaga 2004 for the details).

Using basic properties of harmonic functions we write the field as the real part of an analytic function η_0:

$$u_0 = \text{Re}\,[\eta_0(\zeta)] \quad \eta_0 \text{ analytic in } \mathbb{C} \backslash \Gamma_0.$$

Taking into account the assumed Neumann homogeneous condition, it follows that η_0' is such that:

$$\eta_0'(\zeta) = \frac{\partial u_0}{\partial x} - i\frac{\partial u_0}{\partial y} \Rightarrow \eta_0' \text{ real on } \Gamma_0.$$

The initial configuration may be carried to the upper half plane by the elementary map

$$f_0(z) = -z^2 \tag{18}$$

Consequently, the complex function given by (Fig. 3)

$$h_0(z) := \eta_0(f_0(z))$$

may be extended analytically by symmetry to the whole plane, and then:

$$h_0(z) = \sum_{n=0}^{\infty} c_n z^n \qquad c_n \in \mathbb{R}.$$

Therefore, u_0 admits the following convergent expansion:

$$\begin{aligned} u_0 &= \text{Re}\left[h_0\left(f_0^{-1}(\zeta)\right)\right] \\ &= c_0 - c_1 r^{1/2} \sin(\theta/2) - c_2 r \cos(\theta) + \cdots \end{aligned} \tag{19}$$

where

$$\zeta = re^{i\theta} \qquad f_0^{-1}(\zeta) = \sqrt{-\zeta}$$

We thus take (19) as the most general initial local field.

3.2 Kinked configurations: conformal mapping

For a given angle φ we compute an expansion of the Energy released:

$$\begin{aligned} f_l(z) &:= -(z - a(l))^{1-\alpha}(z - b(l))^{1+\alpha}, \\ &\alpha := \varphi/\pi \; (-1 \le \alpha \le 1) \end{aligned} \tag{20}$$

where:

$$a(l) := -\sqrt{l}\left(\frac{1-\alpha}{1+\alpha}\right)^{\frac{1+\alpha}{2}},$$

$$b(l) := \sqrt{l}\left(\frac{1+\alpha}{1-\alpha}\right)^{\frac{1-\alpha}{2}}. \tag{21}$$

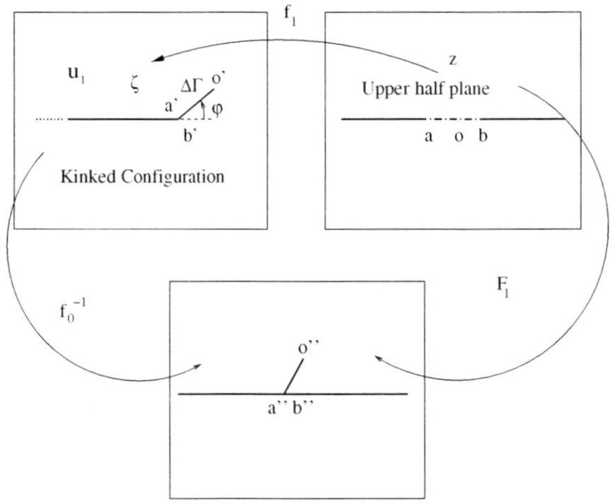

Fig. 4 f_0^{-1}, f_l and F_l

This map sends the upper half plane $\mathbb{H}$ onto the set $\mathbb{C}\setminus\Gamma_l$ as shown in Fig. 4. It can be extended to the whole complex plane outside the interval $[-a(l), b(l)]$ and has the following properties:

$$f_l(\mathbb{R}) = \Gamma_l \quad \text{and} \quad f_l([a,b]) = \Delta\Gamma$$

Using the same arguments as before for $u_l = \mathrm{Re}[\eta_l(\zeta)]$ we can show that the complex function

$$h_l(z) := \eta_l(f_l(z))$$

is analytic in the whole $\mathbb{C}$ and therefore can be expanded in powers of z:

$$h_l(z) = \sum_{n=0}^{\infty} c_n(l)\, z^n, \quad c_n(l) \in \mathbb{R}.$$

We conclude that u_l admits the following expansion in terms of some real coefficients $c_n(l)$:

$$u_l = \mathrm{Re}\left[\sum_{n=0}^{\infty} c_n(l)\left(f_l^{-1}(\zeta)\right)^n\right], \quad \zeta \in \mathbb{C}\setminus\Gamma_l. \tag{22}$$

3.3 Relationship between c_n and $c_n(l)$

Using analyticity, Neumann boundary conditions and the asymptotic matching condition at infinity we can show that

$$c_n(l) = c_n + (n+1)\, c_{n+1} b_0(l) + O(l), \tag{23}$$

where $b_0(l) = -\frac{1}{2}((1+\alpha)b(l) + (1-\alpha)a(l))$ is obtained from the expansion at infinity of the conformal map:

$$F_l(z) = f_0^{-1} \circ f_l(z) = \sqrt{(z-a(l))^{1-\alpha}(z-b(l))^{1+\alpha}}$$
$$= z + b_0(l) + \frac{b_1(l)}{z} + \frac{b_2(l)}{z^2} \cdots \quad z \to \infty. \tag{24}$$

Notice that for $\alpha = 0$ we have that $a(l) = -b(l)$ and then $b_0(l) = 0$.

On the other hand, F_l has the following scale invariance:

$$F_l(z) = \sqrt{l}\, F_1\left(z/\sqrt{l}\right) \Rightarrow b_n(l) = \sqrt{l}^{(n+1)} b_n(1)$$

and may be extended to the lower half plane, being a sectionally holomorphic function in $\mathbb{C}\setminus[a,b]$.

3.4 Complex formula for the energy released

An important tool in Oleaga (2004) is the energy formula

$$\Delta E = \frac{\mu}{4i}\int_C h_l'(z)\, h_0(F_l(z))\, dz, \tag{25}$$

where C is any simple closed curve surrounding the interval $[a,b]$ in the complex plane. By the analytic properties of h_l, h_0, and the expansion for F_l we have that:

$$\Delta E = \frac{\mu}{4i}\int_C \left(\sum_{j=1} l^{\frac{j}{2}}\, j\, c_j(l)\, z^{j-1}\right) \times \left(\sum_{k=0} l^{\frac{k}{2}} c_k \left(F_1(z)\right)^k\right) dz, \tag{26}$$

where C encloses the interval $[a(1), b(1)]$.

The first term of (26) is given by

$$-\Delta E = l\frac{\mu\pi}{4} c_1^2 \left(\frac{1-\alpha}{1+\alpha}\right)^{\alpha} + o\left(l^{3/2}\right).$$

This provides a formula for the energy release rate in terms of the kinking angle:

$$G(\alpha) = \lim_{l\to 0}\left(-\frac{\Delta E}{l}\right) = \frac{\mu\pi}{4} c_1^2 A(\alpha).$$

This is the same behavior as that in (12), already obtained by Sih in (1965) for uniform stresses at infinity. We can see that *even if we take a general equilibrium field around the tip, it is not possible to obtain the crack deviation from the initial configuration by using the first order expansion for the energy released.* Therefore, we must look at the energy functional in more detail.

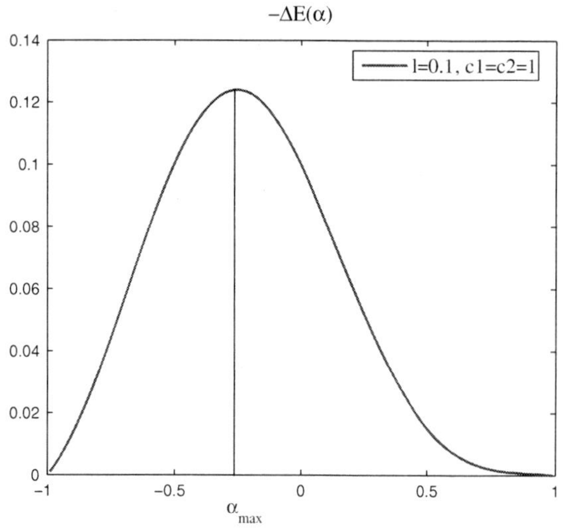

Fig. 5 Two terms in the Energy expansion: the optimal angle

3.5 One more term in the expansion

Using the complex formula for the energy released, the first two terms in the expansion are given by (Fig. 5)

$$-\Delta E(\alpha) = \frac{\mu\pi}{4} c_1^2 A(\alpha)\, l - \frac{4\mu\pi}{3} c_1 c_2 B(\alpha)\, l^{\frac{3}{2}} + O\left(l^2\right) \tag{27}$$

where

$$B(\alpha) := \frac{\alpha A(\alpha)^{\frac{3}{2}}}{\sqrt{1-\alpha^2}}.$$

If we compute the maximum of $-\Delta E$ in terms of α for a fixed l, the optimal angle has the following asymptotic behaviour for $l \to 0$:

$$\alpha_{\max} = -\frac{4}{3}\frac{c_2}{c_1} l^{1/2} + o(l^{1/2}). \tag{28}$$

This suggests the following initial shape for $c_2 \neq 0$:

$$x_2 \approx -\frac{4\pi}{3}\frac{c_2}{c_1}(x_1)^{3/2}, \quad x_1 > 0. \tag{29}$$

The contribution of the extra term introduces a singularity in crack curvature at that point. There is an interesting consequence of the analysis: the term of order $l^{3/2}$ in (27) has no influence on the total amount of energy released by the body for a finite extension of the crack, but *it has a strong influence on crack geometry.*

It has been recently pointed out to me (Buliga Private Communication) that there is an interesting connection between this fact and some properties of the minimisers of the so-called Mumford-Shah functional (Mumford and Shah 1989). The behaviour (29) is heuristically proposed as a singular energy minimiser *when the tip is located at* $x_1 = 0$ (see Sect. 3 Mumford and Shah 1989). The problems are not exactly the same, since for Mumford-Shah this *cuspidal crack tip* appears at the end of the curve and in our problem the singularity occurs between the initial crack tip and the optimal extension. Nevertheless, it seems that there is a close relationship between the optimal shapes for both problems.

3.6 Understanding the outcome for straight kinks

Angle growing preference should be selected in such a way that the body is able to release the maximum amount of energy, provided the cost of bond breaking is the same in all directions. An intuitive way to look at our result is to check that the crack opening is larger in the direction defined by (28). Let us take first $c_1 = 1$, $c_2 = 0$, that is, a purely antisymmetric field. The comparison for different angles in the left of Fig. 6 shows the preference of the $\alpha = 0$ direction when $l = 0.1$. On the other hand, if we add a symmetric contribution to the initial field with $c_2 = 1$, we obtain the pattern for the opening displacement shown on the right side of Fig. 6.

Notice that for $\alpha = -\alpha_{\max}$ the opening is the lower one. On the other hand, the direction given in (28) is optimal among the three for maximum opening. The field was computed analytically, using the conformal map (20), the values $c_1 = c_2 = 1$ for the initial field, $l = 0.1$ and $c_1(l) = c_1 + 2b_0(l)c_2$ for the expanded crack field (cf. (23)).

4 Arbitrary extensions as virtual paths

Kinked configurations are useful to quantify the deviation when we *allow the crack to perform a sudden jump of length l*. Nevertheless, the results show that this kind of abrupt change in direction is not admissible if the crack extends continuously; for vanishing lengths the optimal angle deviation must then go to zero. In this Section we will use the limit shape suggested by (29) to construct suitable *trial paths* in order to derive more precise conditions for crack stability. The technical details can be followed in Oleaga (2006).

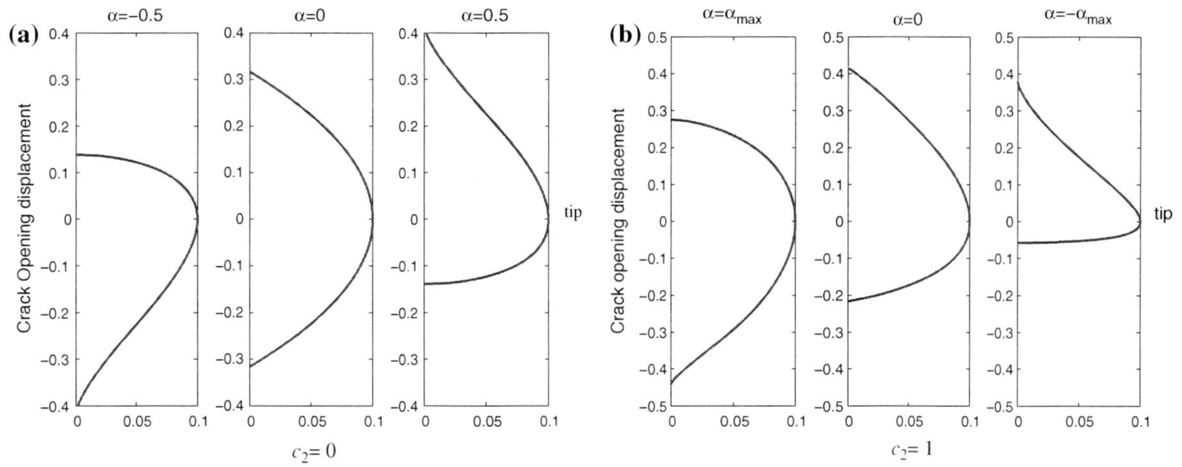

Fig. 6 Crack opening displacement

4.1 Loewner evolutions

It is possible to generate the shape $x^{3/2}$ by means of a suitable conformal transform using the *chordal Loewner equation* for slit maps on the upper half plane (see Marshall and Rohde 2001):

$$\partial_t F_t(z) = -\frac{2\partial_z F_t(z)}{z - \xi_t}, \qquad F_0(z) = z. \tag{30}$$

Let us assume that $f_0^{-1}(\Delta\Gamma)$ is parametrised in a one to one way by $\gamma : t \mapsto \mathbb{H}$. We denote the evolution of the tip as $\gamma(t)$ and the whole set of the crack extension up to time t by γ_t. By (30), it can be shown that each map $F_t : \mathbb{H} \mapsto \mathbb{H} \backslash \gamma_t$ is normalized as follows:

$$F_t(z) = z - \frac{2t}{z} + \frac{b_2(t)}{z^2} + o\left(1/z^3\right).$$

That is, for Loewner evolutions we always have that:

$$b_0(t) = 0, \qquad b_1(t) = -2t. \tag{31}$$

We define $\xi_t := F_t^{-1}(\gamma(t))$ (see Fig. 7). Notice that $\xi_t \in \mathbb{R}$ and the parameter t *is not* in general identified with crack length.

4.2 Generating the virtual path

Recent results of Kager et al. (2004) provide explicit solutions of the Loewner equation for some behaviours of ξ_t. For instance, taking $\xi_t = \lambda t$, (30) takes the form

$$\partial_t F_t(z) = -\frac{2\partial_z F_t(z)}{z - \lambda t}, \qquad F_0(z) = z. \tag{32}$$

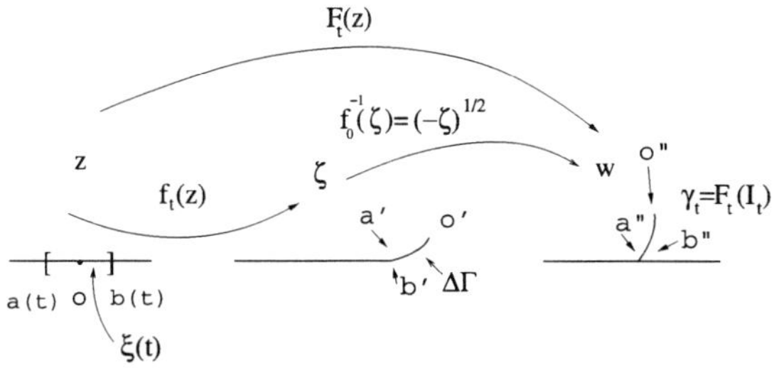

Fig. 7 The conformal map F_t

This case is solved explicitly and it can be shown that generates the following path in the upper half plane:

$$\gamma_{\lambda,t} = 2i\,t^{1/2} + \frac{2}{3}\lambda t - \frac{i}{18}\lambda^2 t^{3/2} + O(t^2). \tag{33}$$

The asymptotic behaviour of the tip on the *physical plane* is given by:

$$x_2 = -\frac{\lambda}{3} x_1^{3/2} + o\left(x_1^{3/2}\right).$$

Moreover, the coefficient $b_2(t)$ of this map is given by:

$$b_2(t) = -\lambda t^2. \tag{34}$$

4.3 The energy released functional and the anti-symmetry principle

It is possible to write the following expansion for the energy released ($\mu = 1$):

$$\Delta E(t) = \frac{1}{4i}\int_C \left(c_1^2 F_t(z) + c_1 c_2 \times \left(F_t(z)^2 + 2F_t(z)(z + b_0(t))\right)\right) dz + O\left(|I_t|^4\right)$$
$$= \frac{\pi}{2}\left(c_1^2 b_1(t) + 4c_1c_2(b_0(t)b_1(t) + b_2(t))\right) + O\left(|I_t|^4\right), \text{ as } |I_t| \to 0. \quad (35)$$

where $|I_t|$ is the length of the interval $I_t := [a(t), b(t)]$ (see Fig. 7 above). Thus, taking into account (31), the asymptotic expression for the energy is given by:

$$\Delta E(t) = \frac{\pi}{2}\left(-2tc_1^2 + 4c_1c_2b_2(t)\right) + O\left(|I_t|^4\right), \quad \text{as } |I_t| \to 0. \quad (36)$$

We have now all the ingredients to compute the energy released by the slit generated by (32). Inserting (34) in (36) we find that an extension evolving up to "time" t will release an amount of energy given by $\Delta E = -\pi c_1^2 t - 2\pi\lambda c_1 c_2 t^2 + O(t^3)$.

According to (4), in order to check the stability we must take into account the dissipation that is required to open this path. For this purpose we need the evolution of the length of the extended curve. The trial path in the physical plane is given by

$$-(\gamma_\lambda(t))^2 = 4t - \frac{8}{3}i\lambda t^{\frac{3}{2}} - \frac{2}{3}\lambda^2 t^2 + O(t^{5/2}).$$

The length of this crack extension is given by the expansion

$$l(t) = 4t + \frac{2\lambda^2}{3}t^2 + O(t^3) \quad t \to 0.$$

We use the last computations in the following equation:

$$\Delta E + \Delta Q = (4\kappa - \pi c_1^2)t + \left(\frac{2\kappa\lambda^2}{3} - 2\pi\lambda c_1 c_2\right) \times t^2 + O\left(t^3\right)$$

The term of order t contains the balance between energy release rate $G(0)$ and κ. It cancels out due to the critical growth equation (13) and we then have that: $|c_1| = \sqrt{\frac{4\kappa}{\pi}}$. To satisfy the stable growth condition (3) the term of order t^2 must be positive or zero. Computing the minimum with respect to the parameter λ we obtain $\lambda = \dfrac{3\pi c_1 c_2}{2\kappa}$, and for this value we have:

$$\Delta E + \Delta Q = -\frac{3}{2}\frac{\pi^2}{\kappa}c_1^2 c_2^2 t^2 + O(t^3) < 0 \quad \text{for } t \to 0, \quad (37)$$

then violating the stability condition (4).

Therefore, since $c_1 \neq 0$, a second necessary condition for a stable configuration is the following:

$$c_2 = 0. \quad (38)$$

This is equivalent to impose $k_2 = 0$ in the typical expansion of the initial displacement field around the tip:

$$u_0 = k_0 + k_1 r^{1/2}\sin(\theta/2) + k_2 r\cos(\theta) + \cdots$$

This condition cancels the symmetric contribution to the displacement near the tip, keeping the first (antisymmetric) term of the expansion. We can state that (38) must hold on every stage of the propagation process, thus providing the second scalar condition to complete the free boundary problem. On the other hand, it has to be satisfied by other models imposing global minimization of the total energy (see for instance Buliga 1999; Francfort and Marigo 1998).

5 The case of a finite domain

The stability of the configuration was studied by imposing a Dirichlet condition at infinity (see (17)). It is natural to ask if the same outcome can be obtained on a *finite domain*. In other words, what happens if we *freeze* the displacements at a finite distance of the tip instead of fixing them at infinity? Bearing this question in mind, we will explain which parts of the previous discussion should be revised. This part of the work was not previously considered in the reviewed articles (Oleaga 2004, 2006).

5.1 A simple setting for the boundary value problem

We start with an initially straight crack configuration with the tip at the center of a circle of unit radius (the results are in fact independent of this particular geometry). The selected field satisfies the equilibrium equation (14), the Neumann boundary condition (15) and the finite energy condition near the tip (16). We now can apply a one-to-one conformal map g_0 transforming the upper half unit disc onto the initial domain. Take $u_0(x, y)$, the real part of an analytic function $\eta_0(x+iy)$ in $D\backslash\Gamma_0$, where

$$D := \{z : |z| < 1\}.$$

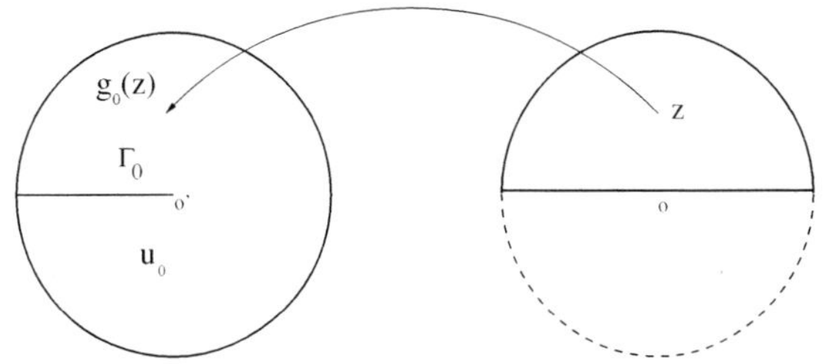

Fig. 8 The basic conformal map g_0

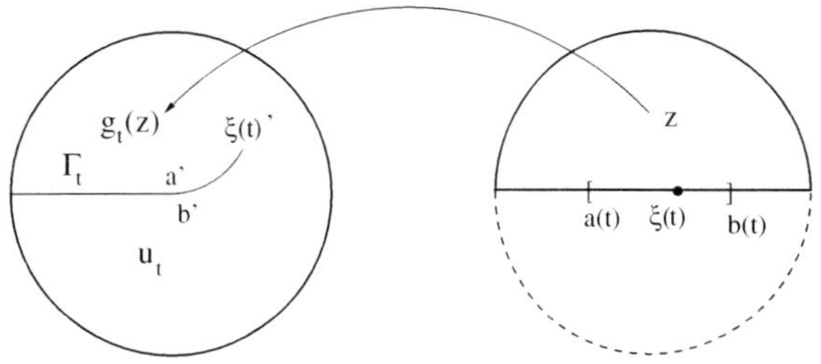

Fig. 9 The basic map g_t

The analytic function $h_0 := \eta_0 \circ g_0$ (defined on the right side of Fig. 8) is extended by symmetry to the lower half disk due to the Neumman homogeneous condition. It then has a Taylor expansion with real coefficients as follows:

$$u_0(g_0(z)) = \mathrm{Re}\,(h_0(z)) = \mathrm{Re}\left(\sum_{n=0}^{n} c_n(0)\, z^n\right)$$

$$c_n \in \mathbb{R}. \tag{39}$$

For simplicity, we will assume that (39) is convergent on some open set containing $\overline{D}$.

We define the field u_t as the one satisfying the equilibrium equation $\Delta u_t = 0$ on $D \setminus \Gamma_t$, the Neumann boundary condition $\partial u / \partial n = 0$ on Γ_t, and the Dirichlet boundary condition on the unit circle:

$$u_t(x, y) = u_0(x, y) \quad \text{for } |x + iy| = 1.$$

There is an analytic function η_t in $D \setminus \Gamma_t$ such that $u_t(x, y) = \mathrm{Re}\ \eta_t(x + iy)$ and the function

$$h_t(z) := \eta_t(g_t(z))$$

is analytic in the upper half unit disk and can be extended to the whole disk by the Neumann homogeneous condition. The map $g_t(z)$ sends the upper half unit disk to $D\backslash\Gamma_t$ as shown in Fig. 9. We have that h_t admits the following expansion:

$$h_t(z) = \sum_{n=0}^{\infty} c_n(t)\, z^n \qquad c_n(t) \in \mathbb{R}. \tag{40}$$

The main difference with the case of an unbounded domain corresponds to the behaviour of the coefficients $c_n(t)$ for $t \to 0$. The Dirichlet condition is applied on a curve at a finite distance of the tip, while in the former case it was imposed on a unique point at infinity.

Assuming for a moment that we know $g_t(z)$, we can write Schwartz integral representation of $h_t(z)$ for the unit disk. The boundary values of the real part of $h_t(z)$ are obtained from the following chain of equalities:

$$u_t\left(g_t\left(e^{i\theta}\right)\right) = u_0\left(g_t\left(e^{i\theta}\right)\right)$$
$$\left(\left|g_t\left(e^{i\theta}\right)\right| = 1\text{cf. Fig. } 9\right),$$
$$u_0\left(g_t\left(e^{i\theta}\right)\right) = u_0\left(g_0\left(g_0^{-1} \circ g_t\left(e^{i\theta}\right)\right)\right)$$
$$u_0\left(g_0\left(G_t\left(e^{i\theta}\right)\right)\right) = \mathrm{Re}\left\{h_0\left(G_t\left(e^{i\theta}\right)\right)\right\},$$

where G_t is defined as

$$G_t := g_0^{-1} \circ g_t.$$

This map is analytic on a circular ring whose interior circle contains the interval $[a(t), b(t)]$ (cf. Fig. 10). It admits a Laurent expansion with real coefficients that we write as follows:

$$G_t(z) = G_t^+(z) + G_t^-(z), \tag{41}$$

with

$$G_t^+(z) := \sum_{k=0}^{\infty} a_k(t)\, z^k \qquad G_t^-(z) := \sum_{k=1}^{\infty} b_k(t)\, z^{-k}. \tag{42}$$

Notice that $G_t(z) \to z$ and $b_{k+1}(t) \ll b_k(t)$ for $t \to 0$. This last assertion is a consequence of the Area Theorem (see the Appendix). Notice that G_t^- is analytic outside of a vanishing circle containing $[a(t), b(t)]$.

To simplify a bit the notation we define the boundary values of the real part of h_t as $U_t(\theta) := \mathrm{Re}\left\{h_0\left(G_t\left(e^{i\theta}\right)\right)\right\}$. Using Schwartz integral in the unit disk, we have the following explicit formula

$$h_t(z) = \frac{1}{2\pi} \int_{-\pi}^{\pi} \frac{e^{i\theta} + z}{e^{i\theta} - z} U_t(\theta)\, d\theta \underset{\zeta := e^{i\theta}}{=} \frac{1}{2\pi i} \times \int_{|\zeta|=1} \frac{\zeta + z}{\zeta - z} \mathrm{Re}\left\{h_0\left(G_t(\zeta)\right)\right\} \frac{d\zeta}{\zeta} \tag{43}$$

We can then write (cf. (40)):

$$h_t(z) = \sum_{n=0}^{\infty} c_n(t) z^n$$
$$= (1 + z) \sum_{n=0}^{\infty} \left(\frac{1}{2\pi} \int_{-\pi}^{\pi} U_t(\theta)\, e^{-in\theta}\, d\theta\right) z^n$$

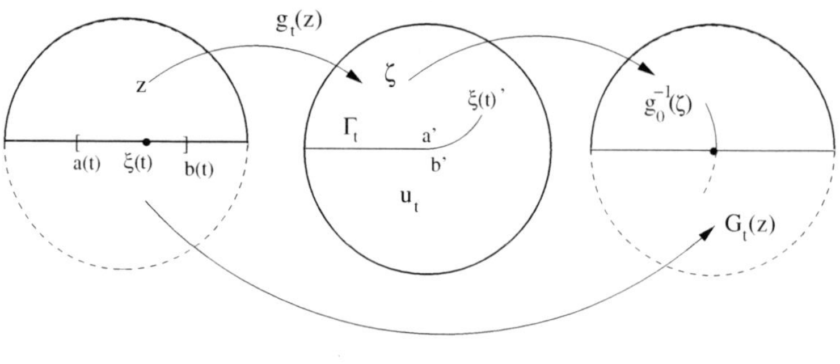

Fig. 10 The mapping properties of the extended G_t

We want now to relate the coefficients $c_n(t)$ with the ones of the initial field $c_n(0)$. Using the expansion for $h_0(z)$ we have that:

$$U_t(\theta) = \mathrm{Re}\left\{h_0\left(G_t\left(e^{i\theta}\right)\right)\right\} = \mathrm{Re}\left\{\sum_{k=0}^{\infty} c_k(0)\left(G_t\left(e^{i\theta}\right)\right)^k\right\}$$

The expression for the first coefficient (notice that $U_t(-\theta) = U_t(\theta)$) should read (we drop for a while the dependence on t of the a_i's and b_i's):

$$c_0(t) = \mathrm{Re}\sum_{j=0}^{\infty} c_k(0)\left\{\frac{1}{\pi}\int_0^{\pi}\left(G_t\left(e^{i\theta}\right)\right)^j\right\} d\theta = c_0(0) + c_1(0)\,a_0 + c_2(0) \times\left(a_0^2 + 2a_1b_1 + 2a_2b_2\right) + \cdots \tag{44}$$

Similarly, we obtain that for $c_1(t)$,

$$c_1(t) = \sum_{j=0}^{\infty} c_j(0)\frac{2}{\pi}\int_0^{\pi} \mathrm{Re}\left(G_t\left(e^{i\theta}\right)\right)^j \cos(\theta) d\theta = c_1(0)\,(a_1 + b_1) + 2c_2(0)\,(a_0\,(a_1 + b_1) + a_1b_2 + a_2\,(b_1 + b_3)) + \cdots \tag{45}$$

For $c_2(t)$ we have:

$$c_2(t) = \sum_{j=0}^{\infty} c_j(0)\frac{2}{\pi}\int_0^{\pi} \mathrm{Re}\left(G_t\left(e^{i\theta}\right)\right)^j \cos(2\theta)\, d\theta = c_1(0)\,(a_2 + b_2) + c_2(0) \times\left(a_1^2 + 2a_0b_2 + 2a_1b_3\right) + \cdots \tag{46}$$

the main contribution being (notice that $a_1(t) \to 1$ for $t \to 0$):

$$c_2(t) = c_2(0)a_1^2(t) + o(1).$$

We can now apply the complex version of the energy release formula (cf. 25):

$$\Delta E = \frac{1}{4i}\int_C h_t'(z)\, h_0\,(G_t(z))\, \mathrm{d}z = \frac{1}{4i}\int_C \left(\sum_{k=1}^{\infty} k\, c_k(t)\, z^{k-1}\right) \times\left(\sum_{j=0}^{\infty} c_j(0)\,(G_t(z))^j\right) \mathrm{d}z,$$

where C is a simple curve enclosing the interval $[a(t), b(t)]$ and contained in D. Taking into account (45, 46) we can approximate the expression for ΔE as follows:

$$\Delta E \approx \frac{\pi}{2}\,\{c_1(t)\,(c_1(0)b_1(t) + 2c_2(0)\,(a_0(t)b_1(t) + a_1(t)\;\; b_2(t))) + 2c_2(t)\,(c_1(0)b_2(t) + 2c_2(0)\;\; (a_0(t)b_2(t) + a_1(t)b_3(t)) + \cdots)\}$$
$$\approx \frac{\pi}{2}\left(c_1^2(0)a_1(t)\,b_1(t) + 2c_1(0)c_2(0) \times\left(a_0(t)a_1(t)b_1(t) + a_1^2(t)b_2(t) + a_0(t)b_1^2(t)\right) \cdots + 2c_1(0)c_2(0)a_1^2(t)b_2(t) + \cdots\right)$$

Summing up:

$$\Delta E \approx \frac{\pi}{2}\left(c_1^2(0)a_1(t)\,b_1(t) + 2c_1(0)c_2(0)(a_0(t)a_1(t) \times b_1(t) + 2a_1^2(t)b_2(t) + a_0(t)b_1^2(t)\right) \tag{47}$$

Notice that for $a_0 = 0$ and $a_1 = 1$ we obtain the same asymptotic expression as that derived for the unbounded domain (cf. (35)).

5.2 The construction of the conformal map

To justify the anti-symmetry principle for a finite domain it only remains to show that a suitable conformal map can be constructed, and that its coefficients

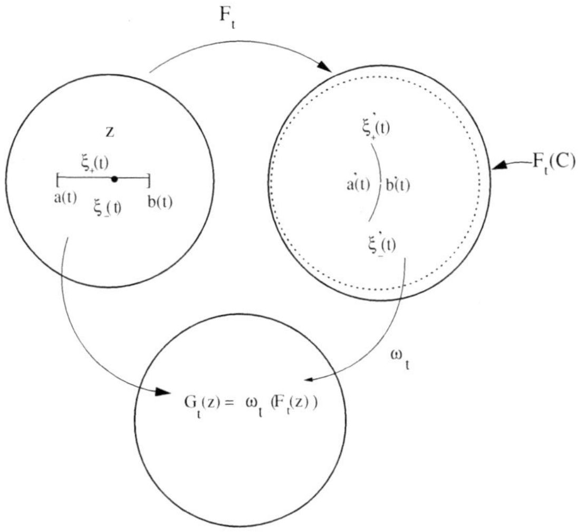

Fig. 11 The map ω_t modifying F_t

$a_0(t)$, $a_1(t)$, $b_1(t)$, $b_2(t)$ produce the instability given in (37). We will proceed by using a modified map of that obtained from the Loewner equation in Sect. 4.2.

Let us take firstly the map F_t for the upper half plane, generated with (32). By restricting the domain, we can see that F_t sends the unit circle with a cut on the real line to a nearly circular domain for small t with a cut inside (cf. Fig. 11). A slight modification of the map F_t should carry us to the desired G_t. This change will be applied as follows:

1. We map the boundary of the unit disk ∂D to the boundary of a nearly circular domain by means of $F_t(z)$.
2. We consider the holomorphic map $\omega_t(z)$ that carries the region bounded by $F_t(\partial D)$ to the unit disc with $\omega_t(0) = 0$ and $\omega_t(z) = \overline{\omega_t(\bar{z})}$. $G_t(z)$ will be given by the composition $\omega_t \circ F_t(z)$.

We summarize in the following Figure the properties of the modified F_t.

We now follow (Nehari 1975) for the technical details and define the real function

$$\rho(\theta) := \lim_{t\to 0} \frac{|F_t(e^{i\theta})| - 1}{t}.$$

Thus, $r = 1 + t\rho(\theta) + o(t)$ is the polar equation for the boundary of the modified domain (the up-right domain in Fig. 11). Notice that $\rho(\theta) = \rho(-\theta)$ due to the symmetry properties of the map F_t. We now have an explicit formula for the first terms of ω_t (cf. p. 264 Nehari 1975):

$$\omega_t(z) = z\left(1 - \frac{t}{2\pi}\int_{-\pi}^{\pi} \frac{e^{i\theta}+z}{e^{i\theta}-z}\,\rho(\theta)\,d\theta + o(t)\right). \tag{48}$$

We may write this as follows:

$$\omega_t(z) = z\,(1 - t\psi(z) + o(t)), \qquad \psi(z) = \sum_{n=0}^{\infty} p_n z^n.$$

We now look at $G_t(z)$:

$$G_t(z) = \omega_t(F_t(z)) = F_t(z)\,(1 - t\psi(F_t(z)) + o(t)),$$

and then we can write

$$G_t(z) = (1 - tp_0)\,F_t(z) - t\left(p_1 F_t^2 + p_2 F_t^3 + \cdots\right)$$

Let us denote by $\tilde{b}_n(t)$ the coefficients of the expansion of $F_t(z)$ (cf. 24) to distinguish them from the ones of $G_t(z)$ (cf. 41). We have that:

$$\begin{aligned}
a_0(t) &= (1 - tp_0)\,\tilde{b}_0(t) - t\left(p_1\left(\tilde{b}_0^2(t) + 2\tilde{b}_1(t)\right)\right)\\
&\quad + o(t\tilde{b}_1(t))\\
a_1(t) &= (1 - tp_0) - t\left(2p_1\tilde{b}_0(t) + 3p_2\tilde{b}_1(t)\right)\\
&\quad + o(t)\\
b_1(t) &= (1 - tp_0)\tilde{b}_1(t) - t\left(2p_1\left(\tilde{b}_0(t)\tilde{b}_1(t) + \tilde{b}_2(t)\right)\right.\\
&\quad \left. + 3p_2\left(\tilde{b}_0^2\tilde{b}_1 + \tilde{b}_1^2 + \tilde{b}_0\tilde{b}_2\right)\right) + O(t\tilde{b}_1)\\
b_2(t) &= (1 - tp_0)\,\tilde{b}_2(t) - t\left(2p_1(t)\tilde{b}_3(t)\right) + o(t^2)
\end{aligned}$$

For the linear forcing path generated by Loewner's equation we have that

$$\tilde{b}_0(t) = 0, \quad \tilde{b}_1(t) = -2t, \quad \tilde{b}_2 = -\lambda t^2.$$

Then there holds:

$$\begin{aligned}
a_0(t) &= 4p_1 t^2 + o(t^2),\\
a_1(t) &= (1 - tp_0) + o(t),\\
b_1(t) &= -2t\,(1 - tp_0) + o(t^2),\\
b_2(t) &= -\lambda t^2 + o(t^2).
\end{aligned}$$

The energy released can be written now as (cf. 47):

$$\Delta E \approx -\frac{\pi}{2}\,(1 - tp_0)^2\left(2c_1^2(0)t + 4\lambda c_1(0)c_2(0)t^2\right) + o(t^2) \tag{49}$$

Notice that this expression is (up to order t^2) the one obtained for the unbounded domain *multiplied by the factor* $(1 - tp_0)^2$. This factor is *non-universal* in the sense that p_0 depends on the geometric properties of the body boundary:

$$p_0 = \frac{1}{\pi}\int_0^{\pi} \rho(\theta)\,d\theta$$

This is consistent with the asymptotics of the stress intensity factor obtained by Leblond in (1989), but a thorough discussion of this subject would carry us far from our main point here. We mention by pass that the physical length of the crack is (up to order t) $l \sim 4t$, and then the energy release rate is consistent with the expression obtained before:

$$\lim_{t\to 0} -\frac{\Delta E}{4t} = \frac{\pi}{4} c_1^2(0).$$

This is also in accordance to Leblond's paper, since we should not have an explicit dependence of the body geometry in this order (of course, there is an *implicit* information encoded in c_1).

To conclude, it remains to find the expression for the length in terms of t, up to order t^2. We have to take into account that $\omega_t(z) \approx (1 - t\rho_0)z$. The path in the (non physical) lower domain in Fig. 11 is given by (cf. 33):

$$\gamma_{\lambda,t} = (1 - t\rho_0)\left(2i\, t^{1/2} + \frac{2}{3}\lambda t - \frac{i}{18}\lambda^2 t^{3/2}\right) + O(t^2).$$

In the physical domain we should have, recalling that $G_t(z) = g_0^{-1} \circ g_t(z)$ and $g_0(z) = -z^2$:

$$g_0(\gamma_{\lambda,t}) = (1 - t\rho_0)^2 \left(4t - \frac{8}{3} i\lambda t^{\frac{3}{2}} - \frac{2}{3}\lambda^2 t^2 + O\left(t^{5/2}\right)\right)$$

Therefore, the length of the path is given by:

$$l(t) = (1 - t\rho_0)^2 \left(4t + \frac{2\lambda^2}{3} t^2\right) + O(t^3) \quad t \to 0$$

We conclude that the dissipation term obtained in the unbounded case is to be multiplied by the positive non-universal factor $(1 - t\rho_0)^2$, the same as that multiplying the relevant contribution to the released energy in (49). Therefore, selecting $\lambda = \dfrac{3\pi c_1 c_2}{2\kappa}$ as in the unbounded case, the constructed virtual path shows again the unstable character of an initial field with $c_2(0) \neq 0$ (cf. (37)):

$$\Delta E + \Delta Q = -(1 - t\rho_0)^2 \left(\frac{3}{2}\frac{\pi^2}{\kappa} c_1^2 c_2^2 t^2\right) + O(t^3) < 0 \quad \text{for } t \to 0.$$

This path violates the stability assumption (4) and provides a necessary condition for the initial field, namely $c_2 = 0$.

Acknowledgements This research was supported by Spanish research project BFM 2004-05634. Part of this work was presented at the International Symposium on Defects and Material Mechanics held in Aussois, 25-29 March 2007. I wish to acknowledge the organizers C. Dascalu and G. Maugin for the kind invitation to participate in this meeting.

Appendix A: The area theorem for the Laurent expansion

Consider a family of univalent functions indexed by t, on a circular ring whose interior circle is of radius $\rho(t)$, with $\rho(t) \to 0$ as $t \to 0$. Moreover, assume that each function admits a Laurent expansion with real coefficients written as (41). It is possible to show that the area inside the image circle C_r of radius r inside the circular ring, is given by:

$$\text{Area} = \frac{1}{2}\text{Im} \int_{C_r} \overline{G_t(z)}\, G_t'(z)\, dz.$$

Taking into account the Laurent expansion for G_t in (41), we can write:

$$\begin{aligned} \text{Area} &= \frac{1}{2}\text{Im} \int_{C_r} \overline{\left(\sum_{n=0}^{\infty} a_n z^n + \sum_{n=1}^{\infty} b_n z^{-n}\right)} \\ &\quad \times \left(\sum_{n=1}^{\infty} n a_n z^{n-1} - \sum_{n=1}^{\infty} n b_n z^{-n-1}\right) dz \\ &= \pi \left(\sum_{n=1}^{\infty} n r^n a_n^2 - \sum_{n=1}^{\infty} n r^{-n} b_n^2\right) \end{aligned}$$

where we avoided the dependence on t for the coefficients. Since Area is non-negative, and taking $r \to \rho(t)$ we finally obtain the inequality:

$$\sum_{n=1}^{\infty} n\rho(t)^{-n} b_n^2 \le \sum_{n=1}^{\infty} n\rho(t)^n a_n^2. \tag{50}$$

The right hand side is convergent by construction; in fact, it goes to zero for $t \to 0$. Notice that

$$\frac{1}{2}\text{Im} \int_{C_r} \overline{G_t^+(z)}\; G_t^{+\prime}(z)\, dz = \pi \sum_{n=1}^{\infty} n\rho(t)^n a_n^2.$$

We can then take the limit for $r \to \rho(t)$ since G_t^+ is analytic inside the outer circle of the ring (cf. (42)). This shows that the coefficients of the left hand series in (50) must be, at least, bounded:

$$n\rho(t)^{-n} b_n^2(t) \le M \Rightarrow b_n^2(t) \le \frac{M}{n}\rho(t)^n, \quad \text{for } n \ge 1,$$

for a suitable positive constant M. This provides the asymptotic relative magnitude of the different b_n's for $t \to 0$.

References

Buliga M (1999) Energy minimizing brittle crack propagation. J Elast 52:201–238

Cotterell B, Rice JR (1980) Slightly curved or kinked cracks. Int J Frac 16(2):155–169

Duren PL (1983) Univalent functions, a series of comprehensive studies in mathematics. Springer-Verlag, New York

Francfort GA, Marigo JJ (1998) Revisiting brittle fracture as an energy minimization problem. J Mech Phys Solids 46(8): 1319–1342

Goldstein RV, Salganik RL (1974) Brittle fracture of solids with arbitrary cracks. Int J Frac 10:507

Gurtin M (1979) On the energy release rate in quasi-static elastic crack propagation. J Elast 9(2):187–195

Irwin GR (1957) Analysis of stresses and strains near the end of a crack transversing a plate. J Appl Mech 24:361–364

Kager W, Nienhuis B, Kadanoff L (2004) Exact Solutions for Loewner Evolutions. J Stat Phys 115(3/4):805–822

Leblond JB (1989) Crack paths in plane situations I: General form of the expansion of the stress intensity factors. Int J Solids Struct 25(11):1311–1325

Marshall DE, Rohde S (2001) The Loewner differential equation and slit mappings. J Am Math Soc 18:763–768

Mumford D, Shah J (1989) Optimal Approximations by Piecewise Smooth Functions and Associated Variational Problems. Commun Pure Appl Math XLII:577–685

Nehari Z (1975) Conformal mapping. Dover Publications, New York

Oleaga GE (2004) On the path of a quasi-static crack in Mode III. J Elast 76(2):163–189

Oleaga GE (2006) The classical theory of univalent functions and quasi-static crack propagation. Eur J Appl Math 17:233–255

Rice JR (1968) Mathematical analysis in the mechanics of fracture. In: Liebowitz H (ed) Fracture. Academic, New York, pp 191–311

Sih GC (1965) Stress distribution near internal crack tips for longitudinal shear problems. J Appl Mech 51–58

Configurational balance and entropy sinks

Marcelo Epstein

Abstract For evolutionary processes of material remodelling and growth, a comparison is drawn between a conventional formulation and one that postulates the existence of additional balance laws for the configurational forces.

Keywords Growth · Remodelling · Plasticity · Configurational forces · Eshelby stress · Material evolution

1 Introduction

One of the recent trends in Continuum Mechanics is the unified description of processes that alter not only the spatial distribution of the body particles but also their material arrangement. These changes taking place in the material manifold are known as processes of *material evolution*. They encompass such diverse phenomena as crack propagation, plasticity, viscoelasticity, and biological growth and remodelling. Their description is the task of a branch of Continuum Mechanics that can alternatively be called Material Mechanics or Configurational Mechanics. The terminology Eshelbian Mechanics is also used in deference to John D. Eshelby (1916–1981), one of the first scientists to clearly identify the nature of the forces responsible for the evolution of material inhomogeneities (Eshelby 1951). In biological applications, the term remodelling is applied to those situations in which the mass density of the body remains unaffected, as opposed to processes of growth or resorption. Following this clear biological distinction, we will extend the terminology to all processes, regardless of the physical context in which they appear. Thus, for example, metal plasticity will be regarded as a remodelling phenomenon, whereas erosion as a process of (negative) growth (or: resorption).

The equations that govern processes of remodelling and growth can be regarded as the result of focusing attention on a single component of a chemically reacting mixture of several substances. Mass is, therefore, not necessarily conserved and transfers of momentum, angular momentum, energy and entropy appear in the equations as extra contributors to the total balance. In a process of growth, these extra terms may be, at least in part, attributed to the very addition or subtraction of mass. At best, we may have a case of "compliant" growth, whereby the new mass happens to enter at the same velocity, specific energy and specific entropy as the local substratum. At worst, not only will the entering quantities be at a different state than the substratum, but the process of growth (or even just remodelling) itself may entail other sources of discrepancy.

Remark 1.1 In (Cowin and Hegedus 1976) the "non-compliant" source of internal energy is lumped as a clearly indicated extra term in the balance equation,

M. Epstein (✉)
Department of Mechanical and Manufacturing Engineering, The University of Calgary, Calgary, Alberta, Canada T2N 1N4
e-mail: mepstein@ucalgary.ca

doi: 10.1007/978-1-4020-6929-1_5,

so that the compliant case can be easily recovered by eliminating this extra term. This practice, which will be followed here, was also adopted for the totality of balance laws in (Epstein and Maugin 2000), where the compliant contributions were called "reversible". To avoid unnecessary confusion with the use of this terminology in the strictest thermodynamical sense, we prefer to call them "compliant".

A physical interpretation of the non-compliant sources can be inferred from the fact that certain evolution processes cannot take place spontaneously, but need the participation of some external agents. From the mixture point of view alluded to above, these agents can be perhaps traced back to the excluded components of the mixture. In certain cases (Epstein 2005), one may imagine the presence of some microactuators which, after detecting the present state of affairs as far as the presence of certain material defects, are programmed to make the defect pattern evolve in a prescribed way. Since to the unaware observer these mechanisms may appear to act against the natural tendency of an isolated system to increase its entropy, we refer to the corresponding non-compliant terms as *entropy sinks*.

An alternative point of view advocates the postulation of additional balance laws to be satisfied by the material or configurational forces, thus rendering the entropy sinks unnecessary. The question as to whether these extra balance laws are to be admitted into the analysis of remodelling processes is still the subject of some controversy. The purpose of this paper is to show that, at least in the context of the class of problems discussed, there exists a precise relation between the two approaches and that they are in fact equivalent.

2 The field equations of remodelling and bulk growth

By bulk, or volumetric, growth we understand a process of addition or removal of mass while the body particles retain their identity. It is only the mass density that changes with time. In contradistinction with this situation it is possible, and certainly meaningful and practical, to consider growth by addition of mass at the boundaries of the body. Thus, for example, holes may be closed or created which change the topology of the original body. These more complicated processes are excluded from the present analysis. For some indication of the difficulties involved in modelling these processes, see Segev and Epstein (1996).

2.1 Balance equations

While at some conceivable level of analysis (such as that of chemically reacting mixtures) the appearance or disappearance of mass of one species may be accounted for by concomitant losses or gains in other species, in a bulk growth model we accept the existence of sources and sinks of mass as part of the theory.[1] On this basis, we obtain the following referential form for the balance of mass:

$$\frac{\partial \rho_R}{\partial t} = \Pi + Div\mathbf{M}. \tag{2.1}$$

where t denotes time, ρ_R is the possibly time-varying mass density in the reference configuration, Π is the (smooth) volumetric source of mass, Div is the referential divergence operator and $\mathbf{M}$ is a vector of possible mass flux through the boundary with exterior unit normal $\mathbf{N}$. The mass flowing through the boundary per unit time is given by:

$$M = \mathbf{M} \cdot \mathbf{N} \tag{2.2}$$

where $\mathbf{N}$ is the unit normal. In what follows, to simplify the analysis, we will assume the mass flux to vanish identically, so that only the volumetric source of mass remains.

By the standard procedure, we obtain the following equation for the balance of linear momentum:

$$\rho_R \frac{\partial \mathbf{v}}{\partial t} = \mathbf{f}_R + \bar{\mathbf{p}}_R + Div\mathbf{T}. \tag{2.3}$$

In this equation $\mathbf{v}$ is the velocity, $\mathbf{f}_R$ is the body force per unit referential volume and $\mathbf{T}$ is the Piola stress. The term $\bar{\mathbf{p}}_R$ represents the non-compliant source of momentum.

In the absence of body and surface couples, the balance of angular momentum yields the classical symmetry of the Piola stress, namely:

$$\mathbf{TF}^T = \mathbf{FT}^T \tag{2.4}$$

We note that if mass flux had not been neglected, this equation would no longer hold.

[1] The author wishes to thank an anonymous reviewer for bringing to his attention two recent papers by Garikipati et al. (2004, 2006) where some of these issues are discussed in greater depth. Our purpose here, however, is to deal with the simplest possible setting in which the comparison of approaches is meaningful.

Denoting by u_ρ the internal energy per unit mass, we can write the local form of the energy balance as:

$$\rho_R \frac{\partial u_\rho}{\partial t} = \rho_R r_\rho + \bar{U} + tr[\mathbf{T}(\nabla \mathbf{v})^T] - Div\mathbf{Q}, \quad (2.5)$$

where ∇ is the referential gradient operator, r_ρ and $-\mathbf{Q}.\mathbf{N}$ represent, respectively, the rate of non-mechanical energy supply per unit mass ("radiation") and per unit area ("conduction") and where $\bar{U}$ is the non-compliant volumetric contribution to the internal energy. In principle, this term could have been absorbed into r_ρ, with the understanding that it may eventually be specified constitutively, rather than just externally.

2.2 The Clausius-Duhem inequality

What does the Second Law of Thermodynamics have to say when one has come to terms with the assumption that, at least for modelling purposes, matter is continuous? A possible answer to this question is the one embodied in the Clausius-Duhem inequality. It is not our intention to either contest or defend the validity of this, or any other, particular form of the Second Law. We simply do not know how this fundamental law of nature can be rendered compatible with such a bold, albeit manifestly useful, assumption. To further complicate matters, there is the issue of the use of internal state variables (Coleman and Gurtin 1967), which can be regarded as implicated in any model of material evolution. Is the formulation of the Second Law for a continuous medium to be modified, perhaps augmented, in the presence of internal variables? Or must we recalcitrantly cling to the original form and live with the consequences? We choose to proceed to the formulation of the Clausius-Duhem inequality adding terms that are consistent with the philosophy used for the balance equations formulated so far.

Defining the entropy content per unit mass, s_ρ, and introducing the Helmholtz free energy per unit mass as:

$$\psi_\rho \equiv u_\rho - \theta s_\rho, \quad (2.6)$$

θ being the absolute temperature, the Clausius-Duhem inequality states that:

$$\rho_R \dot{\psi}_\rho + \rho_R \dot{\theta} s_\rho - \bar{H} - tr[\mathbf{T}(\nabla \mathbf{v})^T] + \frac{1}{\theta}\mathbf{Q}\nabla\theta \leq 0. \quad (2.7)$$

We remark that, following Cowin and Hegedus (1976), we have added a further volumetric sink $\bar{H}$ of non-compliant entropy, which should be specified constitutively. Notice the subtle distinctions between this form of the Clausius-Duhem inequality, which does not presume mass conservation, and its usual counterpart, namely:

$$\dot{\psi}_R + s_R \dot{\theta} - tr[\mathbf{T}(\nabla \mathbf{v})^T] + \frac{1}{\theta}\mathbf{Q}\nabla\theta \leq 0, \quad (2.8)$$

where ψ_R and s_R are measured per unit referential *volume*. Notice also that the entropy sink $\bar{H}$ may be present even if growth does not occur. It may, indeed, be responsible for the process of remodelling.

3 Material evolution

3.1 The material archetype, its implants and evolution

A theory of growth and remodelling is inextricably intertwined with the concept of *internal state variables*. From the point of view of a putative mixture theory, this fact can be seen as a consequence of not including all the components of the mixture. Be that as it may, what are the appropriate internal variables to be used? Clearly, this is a matter of definition of the model being used. Here, we will be following the *anelastic model*, without claiming that it is the only possible one.[2] In an anelastic-like theory, the internal variables can be motivated as follows (Epstein and Maugin 1990):

Let us introduce the notion of a *hyperelastic material archetype* as a material point endowed with a particular (orthonormal) frame in which its material response is expressible in terms of a single scalar function:

$$\bar{\psi} = \bar{\psi}(\mathbf{F}), \quad (3.1)$$

representing a free energy per unit volume in terms of a linear map $\mathbf{F}$. This map can be interpreted physically as the deformation of the standard unit cube into an arbitrary parallelepiped.

The role of this archetype is double. In the first instance, it can be used to model the material response of a point $\mathbf{X}$ belonging to a given material body $\mathcal{B}$. Let the body be in some global reference configuration, in which we adopt a Cartesian coordinate system. Then,

[2] In fact, in some theories of bone remodelling, it is customary to use a different model, whereby the change of density is accompanied by a change in elastic properties, rather than by a mere re-accommodation of the "relaxed configuration".

if $\mathbf{X}$ is actually made of the archetypal material, there must exist a linear map $\mathbf{P}(\mathbf{X})$ from the archetype to the tangent space $T_X\mathcal{B}$, called an *implant of the archetype into* $\mathbf{X}$, such that the response of this point in the given reference configuration is given precisely by:

$$\psi_R(\mathbf{F}, \mathbf{X}) = J_P^{-1}\bar{\psi}(\mathbf{FP}(\mathbf{X})). \tag{3.2}$$

Otherwise, what right would we have to claim that both the archetype and the body point are made of the same material? Here, J denotes the determinant of the subscripted tensor. The only difference that may exist between the responses is one of "scaling" (via some linear map) and, possibly, of zero energy level (via an additive constant). In the case of a solid material this last degree of freedom can be easily disposed of (as we have assumed in the above equation) by assigning a value of zero energy to the natural states. Given a material body, it may so happen that each and every point has a response given by the above equation. In that case we say that the body is *materially uniform* (according to Noll's terminology (Noll 1967)).

The second role that an archetype may play, whether or not the body is materially uniform, is nothing but the temporal counterpart of the spatial notion of uniformity just introduced. What we mean by this is that, although a material point, as time goes on, may retain the "chemical identity" (as it were) provided by the archetype, it may so happen that the implant $\mathbf{P}(\mathbf{X})$ evolves in time, thus becoming:

$$\mathbf{P} = \mathbf{P}(\mathbf{X}, t). \tag{3.3}$$

If this is the case, we say that the material point $\mathbf{X}$ exhibits an *anelastic* behaviour, or that its constitution undergoes an *anelastic time evolution*.

Assuming now, for the sake of simplicity, that the body is *both* uniform and anelastic, we may say that such a body is characterized by a constitutive law that includes, in addition to the deformation gradient, a collection of internal variables encapsulated in the matrix $\mathbf{P}$:

$$\psi_R = \psi_R(\mathbf{F}, \mathbf{P}, \mathbf{X}, t). \tag{3.4}$$

But these internal variables are not completely arbitrary, as one might think by a cursory glance at the preceding equation. Indeed, they enter the constitutive law in a *right-multiplicative way* only, namely:

$$\psi_R = \psi_R(\mathbf{F}, \mathbf{P}, \mathbf{X}, t) = J_P^{-1}\bar{\psi}(\mathbf{FP}(\mathbf{X}, t)). \tag{3.5}$$

This particular detail makes the Eshelby stress emerge as the natural consequence of calculating the forces associated with the internal variables. Moreover, the Eshelby stress acquires precisely the meaning attributed to it by Eshelby himself, namely, the energy expenditure associated with the change in the inhomogeneity pattern of a material. More graphically, the Eshelby stress is the energy expended per "unit" remodelling. To see that this is indeed the case, let us calculate first the derivative of Eq. 3.5 with respect to the deformation gradient, thus obtaining the Piola stress as:

$$\mathbf{T} = \frac{\partial \psi_R}{\partial \mathbf{F}} = J_P^{-1}\frac{\partial \bar{\psi}}{\partial (\mathbf{FP})}\mathbf{P}^T. \tag{3.6}$$

The remodelling counterpart of this calculation should be obtained by taking the derivative with respect to the implant, namely:

$$\frac{\partial \psi_R}{\partial \mathbf{P}} = -J_P^{-1}\mathbf{P}^{-T}\bar{\psi} + J_P^{-1}\mathbf{F}^T\frac{\partial \bar{\psi}}{\partial (\mathbf{FP})}. \tag{3.7}$$

Combining the last three equations yields the following result:

$$\frac{\partial \psi_R}{\partial \mathbf{P}} = -\left[\psi_R\mathbf{I} - \mathbf{F}^T\mathbf{T}\right](\mathbf{P}^{-T}), \tag{3.8}$$

where $\mathbf{I}$ is the identity tensor. The quantity within the square brackets, namely:

$$\mathbf{b} = \psi_R\mathbf{I} - \mathbf{F}^T\mathbf{T}, \tag{3.9}$$

is precisely the classical expression of the Eshelby stress.

If we denote the constant density of the archetype by $\bar{\rho}_R$, the density at the reference configuration is given by:

$$\rho_R = \bar{\rho}_R J_P^{-1}, \tag{3.10}$$

whence:

$$\dot{\rho}_R = -\rho_R tr(\mathbf{P}^{-1}\dot{\mathbf{P}}), \tag{3.11}$$

or:

$$tr(\bar{\mathbf{L}}_P) = -\frac{\dot{\rho}_R}{\rho_R}, \tag{3.12}$$

where we have denoted:

$$\bar{\mathbf{L}}_P = \mathbf{P}^{-1}\dot{\mathbf{P}}. \tag{3.13}$$

Comparing this result with Eq. 2.1, we conclude that:

$$\Pi = -\left(\rho_R tr\bar{\mathbf{L}}_P + Div\mathbf{M}\right), \tag{3.14}$$

which in the present analysis reduces to:

$$\Pi = -\rho_R tr\bar{\mathbf{L}}_P. \tag{3.15}$$

This equation implies that no separate constitutive law needs to be given for the volumetric mass source Π, since it will be dictated by the evolution of the implants. It is worthwhile noting that the time evolution of the determinant of **P** tells us whether there is instantaneous growth (negative time-derivative) or resorption (positive time-derivative). In other words, growth is taking place if $tr\bar{\mathbf{L}}_P$ is negative, while the reduced (or increased) value of the determinant of **P** is a measure of accumulated growth (or resorption) from a given time origin. If **P** is non-spherical, there may also be rotations and distortions in addition to growth or resorption. Thus, the only difference between a theory of growth and a purely remodelling theory (such as the case of visco-elastoplasticity) resides in the fact that in the latter we require that the evolution function be traceless. In a theory of growth and remodelling it is precisely the trace of the inhomogeneity velocity gradient that will carry the information about the volumetric growth.

3.2 Thermodynamic restrictions

The Clausius-Duhem inequality (2.7) is not to be violated in any conceivable thermomechanical process, thus leading to restrictions on the possible constitutive and evolution laws of a given class of materials. A class of materials is defined by a choice of a class of constitutive functionals and of a list of independent arguments therein. In the case of bodies undergoing material evolution in the sense just defined, the list of independent variables must include the time-dependent implant field **P**, which acts, therefore, as a special kind of internal state variable. What is special about this internal state variable is that in enters the constitutive law mainly in a multiplicative way, like some kind of "change of scale", only. The evolution itself is assumed to be governed by a first-order ODE, such as:

$$\dot{\mathbf{P}} = \mathbf{f}(\mathbf{P}, \mathbf{b}, \ldots), \tag{3.16}$$

where **f** is a constitutive function of material evolution.

Assuming that the archetype consists of a simple thermoelastic heat conductor, the list of arguments of its constitutive law consists of **F**, θ and $\nabla\theta$. Consequently,

$$\psi_R = J_P^{-1}\bar{\psi}(\mathbf{FP}, \theta, (\nabla\theta)\mathbf{P}). \tag{3.17}$$

Following the standard procedure (which in this case is somewhat more involved than in the non-evolving case) we obtain:

$$\frac{\partial\psi_R}{\partial\mathbf{F}} = \mathbf{T}, \tag{3.18}$$

$$\frac{\partial\psi_R}{\partial\theta} = -s_R, \tag{3.19}$$

and

$$\frac{\partial\psi_R}{\partial(\nabla\theta)} = 0. \tag{3.20}$$

These restrictions coincide, not surprisingly, with the constitutive restrictions of thermoelasticity. It is only the residual inequality which turns out to be augmented in the case of an evolving material. Notice that the fact that the term involving $\dot{\mathbf{P}}$ does not vanish is due to the relation imposed by the evolution law between $\dot{\mathbf{P}}$ and at least some of the variables appearing in its bracketed coefficient. Using Eqs. (3.18) and (3.20) as well as the definition of the Eshelby stress as given in Eq. 3.9, we obtain the residual inequality:

$$tr(\mathbf{L}_P\boldsymbol{\beta}^T) - \bar{H} + \frac{1}{\theta}\mathbf{Q}\nabla\theta \leq 0. \tag{3.21}$$

where:

$$\mathbf{L}_P = \dot{\mathbf{P}}\mathbf{P}^{-1}, \tag{3.22}$$

and where $\boldsymbol{\beta}$ is the Mandel stress, namely, (minus) the Eshelby stress devoid of its spherical part (proportional to the free energy). Notice the natural way in which the Mandel stress makes its appearance as the thermodynamic dual of the inhomogeneity velocity gradient. The contraction of the product of these two quantities provides the dissipation associated with the material evolution (such as motion of dislocations or growth). Although it is conceivable to satisfy the residual inequality (3.21) by some delicate compensation between the mechanisms of dislocation motion and heat transfer, it seems more reasonable to assume the stronger condition that the conduction term $\mathbf{Q}\ \nabla\theta$ and the inhomogeneity part are independently non-positive, namely:

$$tr(\mathbf{L}_P\boldsymbol{\beta}^T) - \bar{H} \leq 0. \tag{3.23}$$

The generic evolution law (3.16) can be shown to be restricted to a dependence on **P** which can be expressed in the form:

$$\bar{\mathbf{L}}_P = \bar{\mathbf{f}}(\bar{\mathbf{b}}, \ldots), \tag{3.24}$$

where superimposed bars denote quantities pulled back to the archetype. For the particular case in which the argument of this evolution law is just the Mandel stress $\boldsymbol{\beta}$, this restriction boils down to the following constraint on the evolution function

$$tr[\bar{\mathbf{f}}(\mathbf{z})\,\mathbf{z}^T] - \bar{H} \leq 0, \tag{3.25}$$

for all values of the argument $\mathbf{z}$. If one sets $\bar{H} = 0$, a trivial way to satisfy this inequality is by choosing $\bar{f}(\mathbf{z}) = -k\mathbf{z}$, where k is a positive material constant (analogous to a viscosity or a heat conduction coefficient). On the other hand, the presence of the extra term $\bar{H}$ makes, in principle, possible to sustain any evolution law by means of a finely adjusted control mechanism that systematically removes entropy from the system. In living organisms, it is quite possible that equilibrium is actively maintained in this way, just as one might hold a broom upside-down on the tip of a finger. The preceding calculations indicate that the thermodynamic dual of the inhomogeneity velocity gradient is not exactly the Eshelby stress, but rather the Mandel stress. In some sense this is a sobering and, at the same time, comforting remark, since it seems to eliminate the apparent paradox implied by the arbitrariness of the zero level of the free energy.

4 Configurational balance

Our presentation of the fundamental equations of the mechanics and thermodynamics of evolving materials has been based on the consideration of the implant maps $\mathbf{P}$ as a set of passive internal state variables. This point of view, however minimalist it may be, leads to whole new vistas in terms of our ability to encompass a great variety of anelastic behaviors, including viscoelasticity, growth and remodelling. Many authors (e.g., Di Carlo 2005; Di Carlo and Quiligotti 2002; Gurtin 2000; Podio-Guidugli 2002), on the other hand, have suggested that this framework can be considerably enlarged by advancing the assumption of extra balance laws to be satisfied by the material forces (such as the Eshelby stress). It seems somewhat futile to argue as to which is the right approach, since, as convincingly explained by Clifford Truesdell, Continuum Mechanics is, by its very nature, a *theory of models*. The philosophical boundary between balance laws (supposed to apply to a very large universe of models) and constitutive laws (restricted to smaller classes) is at best fuzzy. The main requirement of any model is ultimately one of mathematical consistency. The value of a model is determined by its ability to withstand the vicissitudes of time, an ability determined to a great extent by its suitability to describe a large number of phenomena accurately, its elegance and simplicity and, in some measure, the authority of those propounding it. The present coexistence of several points of view (which, in our view, are not antagonistic but complementary) is a testimony of the richness of the subject and the vibrancy of Continuum Mechanics, both as a scientific discipline and as a veritable Weltanschauung, a vibrancy that seems to gainsay, time and time again, all the prophesies of its imminent doom and to cause it to rise cyclically from its proverbial ashes.

Forces are bounded linear operators on vector spaces of velocities (or virtual displacements). What this means is that once a kinematic substratum has been established for a continuum theory, the most general notion of force (as an entity producing virtual power on the space of virtual velocities) is completely determined and it encompasses both internal and external forces. This general approach can be pursued to the point of formulating completely global notions of stresses and of configurational (Eshelby-like) forces (Epstein and Śniatycki 2005). In the case of evolving materials of the kind we have been considering in this paper, we may regard the implant fields $\mathbf{P}$ as added kinematic degrees of freedom on the same footing with the deformation gradient $\mathbf{F}$. At a given body point and at a given value of $\mathbf{F}$, we consider the collection of associated virtual "velocities" $\dot{\mathbf{F}}$ (or virtual "displacements" $\delta\mathbf{F}$). These are mixed body-spatial quantities. A local linear operator on the collection of virtual velocities at a point is known as a Piola (or first Piola-Kirchhoff) stress $\mathbf{T}$. The result of the linear operation, namely:

$$tr(\mathbf{T}\dot{\mathbf{F}}^T) = T_i^I \dot{F}_I^i, \tag{4.1}$$

is known as the *virtual power* of the "force" (or stress) $\mathbf{T}$ on the virtual velocity $\dot{\mathbf{F}}$. A similar statement can be made in a purely static context in terms of virtual displacements, but we will continue using the language of velocities.

Having enlarged our kinematic outlook to include the time-dependent implant fields $\mathbf{P}$, we consider at a given body point and at a given value of the implant the collection of virtual implant velocities $\dot{\mathbf{P}}$, and we call a linear operator thereon a *material or configurational force* $\tilde{\mathbf{b}}$. Note that when so introduced this is a mixed tensor with one leg standing on the archetype and the other on the reference configuration (just as a Piola stress stands between the reference configuration and physical space). Moreover, although we are using the suggestive notation $\tilde{\mathbf{b}}$, this configurational force is yet to be connected with our previous notion of Eshelby stress. The evaluation of the linear operator $\tilde{\mathbf{b}}$ on a virtual implant velocity, namely:

$$tr(\tilde{\mathbf{b}}\dot{\mathbf{P}}^T) = \tilde{b}_I^\alpha \dot{P}_\alpha^I, \tag{4.2}$$

is called the *virtual power of the configurational force* on the given virtual implant velocity.

So far, we have just formulated definitions. Consider for a moment the classical case of a non-evolving material. We know that the mechanical field equations can be derived from the postulation of a *principle of virtual power*. Recall that to achieve this aim one introduces the virtual power (EVP) of the external forces as:

$$EVP \equiv \int_\Omega \mathbf{f}_R \mathbf{v} d\Omega + \int_{\partial\Omega} \mathbf{t}_R \mathbf{v} dS, \tag{4.3}$$

where $\mathbf{t}_R$ is the surface traction per unit referential area of the unsupported part of the boundary. We are not including inertia effects, for the sake of simplicity. We now define the internal virtual power (IVP) as the integral of the expression (4.1), viz:

$$IVP \equiv \int_\Omega tr(\mathbf{T}\dot{\mathbf{F}}^T) d\Omega. \tag{4.4}$$

The principle of virtual power stipulates the satisfaction of the identity:

$$IVP \equiv EVP, \tag{4.5}$$

for all virtual velocities. Now, in this classical case there is an understood extra compatibility condition between the virtual ordinary velocities ($\mathbf{v}$) and the virtual velocities of the deformation gradient ($\dot{\mathbf{F}}$). This condition establishes that the deformation gradient velocities must be derived from the (globally smooth) ordinary velocity field. That is, the identity (4.5) is enforced only under the condition that:

$$\dot{\mathbf{F}} = \nabla \mathbf{v}. \tag{4.6}$$

In this case, integration by parts and the arbitrariness of the virtual velocity field $\mathbf{v}$ are immediately seen to imply the field equation:

$$Div\mathbf{T} + \mathbf{f}_R = \mathbf{0}, \tag{4.7}$$

and the natural boundary condition:

$$\mathbf{TN} = \mathbf{t}_R, \tag{4.8}$$

where $\mathbf{N}$ is the unit exterior normal to the (unsupported part of the) boundary in the reference configuration. Thus, the "weak formulation" (4.5) delivers the same differential equations and boundary conditions as the usual "strong" formulation and, in fact, generalizes the latter for weaker conditions on the space of admissible functions.

While the consistency between the weak and strong points of view in the classical case is well grounded on a tradition that goes as far back as d'Alembert, there is no a-priori reason to suppose that the principle of virtual power can be extended meaningfully to include remodelling phenomena or, for that matter, any case where internal state variables enter the physical picture. But it certainly can be done formally, as the following treatment shows. The first thing to be done is straightforward: the internal virtual power is to be supplemented with the integral of the local virtual power of the configurational force. Instead of Eq. 4.4, we have now the augmented expression:

$$IVP \equiv \int_\Omega \left(tr(\mathbf{T}\dot{\mathbf{F}}^T) + tr(\tilde{\mathbf{b}}\dot{\mathbf{P}}^T) \right) d\Omega. \tag{4.9}$$

The second modification entails, as one would expect, a generalization of the external virtual power expression. It is here that an important difference between the classical case and the new formulation arises. For, whereas in the classical case we have at our disposal an ordinary velocity field (whose referential gradient delivers the velocity of $\mathbf{F}$), in the augmented formulation the implant field "velocity" $\dot{\mathbf{P}}$ is not necessarily integrable. In other words, there doesn't in general exist a vector field whose referential gradient is $\dot{\mathbf{P}}$. One way to come out of this analogical impasse is to postulate the existence of an external material body force $\mathbf{B}$ which performs virtual power on *the same* virtual field $\dot{\mathbf{P}}$ as the internal material force $\tilde{\mathbf{b}}$. The nature of this new external force is left to be specified as part of each particular theory. It may very well vanish altogether or it may be meaningfully stipulated from physical considerations. At any rate, as an external body force, $\mathbf{B}$ is supposed to be given directly, rather than determined by any constitutive equation. The augmented external virtual power becomes:

$$EVP \equiv \int_\Omega \left(\mathbf{f}_R \mathbf{v} + tr(\mathbf{B}\dot{\mathbf{P}}^T) \right) d\Omega + \int_{\partial\Omega} \mathbf{t}_R \mathbf{v} dS. \tag{4.10}$$

Assuming the independence of the spatial and material virtual velocity fields, we obtain now the extra balance equation:

$$\tilde{\mathbf{b}} - \mathbf{B} = \mathbf{0}, \tag{4.11}$$

which doesn't seem much of a balance equation, but which can be intelligently exploited, as we shall presently see.

To complete the theory, a *dissipation principle* is postulated in the form of the following inequality:

$$\frac{D}{Dt}\int_{\Omega} \psi_R d\Omega \leq \int_{\Omega} \left(tr(\mathbf{T}\dot{\mathbf{F}}^T) + tr(\tilde{\mathbf{b}}\dot{\mathbf{P}}^T)\right) d\Omega, \tag{4.12}$$

where we have omitted the non-mechanical (heating) terms. Assuming the body to be uniform with a constitutive law given by:

$$\psi_R(\mathbf{F}, \mathbf{X}) = J_P^{-1}\bar{\psi}(\mathbf{FP}(\mathbf{X})), \tag{4.13}$$

we obtain the local form of (4.12) as:

$$tr\left[\left(-J_P^{-1}\,\mathbf{P}^{-T}\bar{\psi} + J_P^{-1}\,\mathbf{F}^T\frac{\partial\bar{\psi}}{\partial(\mathbf{FP})} - \tilde{\mathbf{b}}\right)\dot{\mathbf{P}}^T\right] + tr\left[\left(J_P^{-1}\frac{\partial\bar{\psi}}{\partial(\mathbf{FP})}\,\mathbf{P}^T - \mathbf{T}\right)\dot{\mathbf{F}}^T\right] \leq 0, \tag{4.14}$$

which, in view of the identity:

$$\frac{\partial\psi_R}{\partial\mathbf{F}} = J_P^{-1}\frac{\partial\bar{\psi}}{\partial(\mathbf{FP})}\,\mathbf{P}^T, \tag{4.15}$$

can be written as:

$$-tr\left[\left(\psi_R\mathbf{I} - \mathbf{F}^T\frac{\partial\psi_R}{\partial\mathbf{F}} - \tilde{\mathbf{b}}_R\right)\mathbf{L}_P^T\right] + tr\left[\left(\frac{\partial\psi_R}{\partial\mathbf{F}} - \mathbf{T}\right)\dot{\mathbf{F}}^T\right] \leq 0, \tag{4.16}$$

with

$$\tilde{\mathbf{b}}_R \equiv -\tilde{\mathbf{b}}\,\mathbf{P}^T. \tag{4.17}$$

Rather than hastily interpreting this as the identical vanishing of the quantities within the square brackets, one now introduces (in the terminology of Di Carlo 2005) the *extra energetic responses* $\hat{\mathbf{T}}$ and $\hat{\mathbf{b}}$ satisfying the residual inequality:

$$-\,tr[\hat{\mathbf{T}}\dot{\mathbf{F}}^T] + tr[\hat{\mathbf{b}}\mathbf{L}_P^T] \leq 0. \tag{4.18}$$

while making the identifications:

$$\psi_R\mathbf{I} - \mathbf{F}^T\frac{\partial\psi_R}{\partial\mathbf{F}} - \tilde{\mathbf{b}}_R = -\hat{\mathbf{b}}, \tag{4.19}$$

and

$$\frac{\partial\psi_R}{\partial\mathbf{F}} - \mathbf{T} = -\hat{\mathbf{T}}. \tag{4.20}$$

We are still left with the task of specifying the extra energetic quantities $\hat{\mathbf{b}}$ and $\hat{\mathbf{T}}$. Assume, for simplicity, that the latter vanishes (so that we recover the usual formula for the Piola stress in terms of the derivative of the referential free-energy density), but that the former does not. We see, then, that in this approach, the quantity $\tilde{\mathbf{b}}_R$ is not quite the Eshelby stress, but rather the sum of the Eshelby stress and the extra energetic term $\hat{\mathbf{b}}$. To satisfy the inequality (4.18), an evolution equation will have to be given in terms of the pull-back of $\hat{\mathbf{b}}$ (rather than that of $\tilde{\mathbf{b}}_R$) to the archetype.

Returning now to the extra balance Eq. 4.11, we see that, even if the external material body force $\mathbf{B}$ were to vanish, the repercussion will not be the vanishing of the Eshelby stress, but the vanishing of $\tilde{\mathbf{b}}_R$. In that case, the extra energetic term $\hat{\mathbf{b}}$ will boil down to the classical Eshelby stress, and the evolution equation will coincide with that of the previous formulation, whereby no new balance law was postulated.

At the other extreme, the dissipation inequality $tr[\hat{\mathbf{b}}\mathbf{L}_P^T] \leq 0$ can always be trivially satisfied by setting $\hat{\mathbf{b}} = \mathbf{0}$ identically, thus rendering all remodelling processes apparently "reversible". In this case, it follows from Eq. 4.19 that the material force $\tilde{\mathbf{b}}_R$ coincides with the classical Eshelby stress. The extra balance Eq. 4.11, however, requires now that the external material body force $\mathbf{B}$ be also equal to the Eshelby stress. In this way, one can specify any evolution law whatsoever (for example, one of those we call of the "self-driven" type (Epstein 2005)) and always satisfy the dissipation inequality at the price of an external agent ($\mathbf{B}$) doing the job of carrying out the prescription of the evolution law by means of external sources of power. More to the point, if we substitute the extra balance law (4.11) into the dissipation principle (4.16), we obtain in general:

$$-tr\left[\left(\psi_R\mathbf{I} - \mathbf{F}^T\frac{\partial\psi_R}{\partial\mathbf{F}} - \mathbf{B}_R\right)\mathbf{L}_P^T\right] + tr\left[\left(\frac{\partial\psi_R}{\partial\mathbf{F}} - \mathbf{T}\right)\dot{\mathbf{F}}^T\right] \leq 0, \tag{4.21}$$

with

$$\mathbf{B}_R \equiv -\mathbf{B}\mathbf{P}^T. \tag{4.22}$$

Focusing attention on the evolutionary part only, namely:

$$-\,tr\left[\left(\psi_R\mathbf{I} - \mathbf{F}^T\frac{\partial\psi_R}{\partial\mathbf{F}} - \mathbf{B}_R\right)\mathbf{L}_P^T\right] \leq 0, \tag{4.23}$$

and comparing with the standard expression (3.23), we conclude that, from the point of view of the standard formulation (without the extra balance law), what we have here is an entropy sink:

$$\bar{H} = tr\left[(\psi_R \mathbf{I} - \mathbf{B}_R)\,\mathbf{L}_P^T\right] \tag{4.24}$$

regulated from outside the system. In conclusion, whether one opts for the interpretation of these extraneous agents as external material body forces or as entropy sinks in the standard dissipation inequality (3.23), there doesn't seem to be (at least in the situations just considered) a great deal of practical difference in the final equations, since the extra configurational balance ultimately delivers entropy sinks of a particular form.

Acknowledgements This work has been partially supported by the Natural Sciences and Engineering Research Council of Canada.

References

Coleman BD, Gurtin ME (1967) Thermodynamics with internal state variables. J Chem Phys 47(2):597–613

Cowin SC, Hegedus DH (1976) Bone remodeling I: theory of adaptive elasticity. J Elast 6:313–326

Di Carlo A (2005) Surface and bulk growth unified. In: Steinmann P, Maugin GA (eds) Mechanics of material forces. Advances in mechanics and mathematics, vol 11. Springer, pp 53–64

Di Carlo A, Quiligotti S (2002) Growth and balance. Mech Res Commun 29:449–456

Epstein M (2005) Self-driven continuous dislocations and growth. In: Steinmann P, Maugin GA (eds) Mechanics of material forces. Springer, pp 129–139

Epstein M, Maugin GA (1990) The energy-momentum tensor and material uniformity in finite elasticity. Acta Mech 83:127–133

Epstein M, Maugin GA (2000) Thermomechanics of volumetric growth in uniform bodies. Int J Plast 16:951–978

Epstein M, Śniatycki J (2005) Non-local inhomogeneity and Eshelby entities. Phil Mag 85(33–35):3939–3955

Eshelby JD (1951) The force on an elastic singularity. Phil Trans Roy Soc London A 244:87–112

Garikipati K, Arruda EM, Grosh K, Narayanan H, Calve S (2004) A continuum treatment of growth in biological tissue: the coupling of mass transport and mechanics. J Mech Phys Solids 52(7):1595–1625

Garikipati K, Olberding JE, Narayanan H, Arruda EM, Grosh K, Calve S (2006) Biological remodelling: stationary energy, configurational change, internal variables and dissipation. J Mech Phys Solids 54(7):1493–1515

Gurtin ME (2000) Configurational forces as basic concepts of continuum physics. Springer-Verlag

Noll W (1967) Materially uniform simple bodies with inhomogeneities. Arch Rational Mech Anal 27:1–32

Podio-Guidugli P (2002) Configurational forces: are they needed? Mech Res Commun 29:513–519

Segev R, Epstein M (1996) On theories of growing bodies. In: Batra RC, Beatty MF (eds) Contemporary research in the mechanics and mathematics of materials. CIMNE, Barcelona, pp 119–130

Application of invariant integrals to the problems of defect identification

Robert V. Goldstein · Efim I. Shifrin · Pavel S. Shushpannikov

Abstract A problem of parameters identification for embedded defects in a linear elastic body using results of static tests is considered. A method, based on the use of invariant integrals is developed for solving this problem. A problem on identification the spherical inclusion parameters is considered as an example of the proposed approach application. It is shown that the radius, elastic moduli and coordinates of a spherical inclusion center are determined from one uniaxial tension (compression) test. The explicit formulae expressing the spherical inclusion parameters by means of the values of corresponding invariant integrals are obtained for the case when a spherical defect is located in an infinite elastic solid. If the defect is located in a bounded elastic body, the formulae can be considered as approximate ones. The values of the invariant integrals can be calculated from the experimental data if both applied loads and displacements are measured on the surface of the body in the static test. A numerical analysis of the obtained explicit formulae is fulfilled. It is shown that the formulae give a good approximation of the spherical inclusion parameters even in the case when the inclusion is located close enough to the surface of the body.

R. V. Goldstein · E. I. Shifrin (✉) · P. S. Shushpannikov
Institute for Problems in Mechanics RAS, Prosp. Vernadskogo 101-1, Moscow, 119526 Russia
e-mail: shifrin@ipmnet.ru

R. V. Goldstein
e-mail: goldst@ipmnet.ru

P. S. Shushpannikov
e-mail: shushpan@ipmnet.ru

Keywords Defect identification · Invariant integrals · Spherical inclusion · Explicit formulae

1 Introduction

The problems of defect, mainly cracks and cavities, identification were considered in a number of publications. Most methods of defect identification use the surface measurements for bodies subjected to dynamic forces (see, for example, Bostrom and Wirdelius (1995); Alves and Ha-Duong (1999); Glushkov et al. (2002); Guzina et al. (2003); Vatuliyan (2004)). The data of static tests are also often used for the defect detection (see, Keat et al. (1998); Andrieux et al. (1999); Ammari et al. (2002); Engelhardt et al. (2006)). A review of different approaches for solving elastostatic and elastodynamic inverse problems is presented by Bonnet and Constantinescu (2005). Usually inverse problems are solved as follows:

A defect and its location are described by some parameters;
A direct problem is solved by one of the numerical methods for the prescribed parameters of the defect and its location;
An error function, describing the difference between calculated and experimental data, is constructed;
One of the optimization methods is used for the determination of the unknown defect parameters, giving an extremum for the error function.

doi: 10.1007/978-1-4020-6929-1_6,

Since the error function can have several extrema, the realization of optimization methods becomes a difficult problem. In this connection the methods, which help determine some defect parameters without using the error function optimization, are of great interest (see, Andrieux et al. (1999); Ammari et al. (2002)). In particular, a reciprocity gap principle was used in Andrieux et al. (1999) for a plane crack identification.

The aims of the present publication are as follows:

To supplement the reciprocity gap principle approach with other types of invariant integrals;

To develop an approach for obtaining explicit formulae for the defect parameters in the case when the sizes of a defect are small as compared to the distance between the defect and the body boundary.

2 Statement of the problem

Let V be a simply connected domain in a three-dimensional space R^3, $G \subset V$ is an embedded subdomain, $\Omega = V \backslash G$. Let us suppose that Ω is an isotropic linear elastic body with a shear modulus μ_M and Poisson ratio ν_M. The defect G can be a cavity, a crack or an isotropic linear elastic inclusion. Let us introduce Cartesian coordinates $OX_1X_2X_3$. The stress-strain state in the matrix Ω we'll mark with the superscript (f) : $\sigma_{ij}^{(f)}$ is the stress tensor, $e_{ij}^{(f)}$ is the strain tensor and $u^{(f)} = \left(u_1^{(f)}, u_2^{(f)}, u_3^{(f)}\right)$ is the displacement vector. According to our suppositions the following equalities are valid

$$e_{ij}^{(f)}(X) = \left(u_{i,j}^{(f)}(X) + u_{j,i}^{(f)}(X)\right)/2, \quad (i = 1, 2, 3;\ j = 1, 2, 3)$$

$$\sigma_{ij}^{(f)}(X) = 2\mu_M\left[\frac{\nu_M}{1-2\nu_M}\theta^{(f)}(X)\delta_{ij} + e_{ij}^{(f)}(X)\right],$$

$$\theta^{(f)}(X) = \sum_{k=1}^{3} e_{kk}^{(f)}(X) \tag{2.1}$$

$$\sigma_{ij,j}^{(f)}(X) = 0,\ X = (X_1, X_2, X_3) \in \Omega$$

where the convention of summation for repeated indices is used, δ_{ij} is the Kronecker delta.

It is supposed that the loads $t^{(f)} = \left(t_1^{(f)}, t_2^{(f)}, t_3^{(f)}\right)$ are applied to the external boundary of Ω, coinciding with the boundary of the domain $V - \partial V$

$$\sigma_{ij}^{(f)}(X)n_j(X) = t_i^{(f)}(X),\ X \in \partial V \tag{2.2}$$

where $n(X) = (n_1(X), n_2(X), n_3(X))$ is a unit outward normal to the boundary ∂V at the point X.

The applied loads are self-equilibrated

$$\int_{\partial V} t_i^{(f)}(X)dS = 0,\ \int_{\partial V} X \wedge t^{(f)}(X)dS = 0 \tag{2.3}$$

where $\wedge$ is a vector product.

If the defect G is a cavity or a crack we suppose that the boundary ∂G is unloaded

$$\sigma_{ij}^{(f)}(X)N_j(X) = 0,\ X \in \partial G \tag{2.4}$$

where $N(X) = (N_1(X), N_2(X), N_3(X))$ is a unit normal to ∂G at the point X.

If the defect G is an inclusion, we suppose that G is an isotropic and linear elastic body with unknown shear modulus μ_I and Poisson ratio ν_I. It is supposed also complete bonding between the matrix and inclusion. Let us denote by σ_{ij}^I, e_{ij}^I and $u^I = \left(u_1^I, u_2^I, u_3^I\right)$ the stresses, strains and displacements of the inclusion. The mentioned suppositions lead to the following equations

$$e_{ij}^{(I)}(X) = \left(u_{i,j}^{(I)}(X) + u_{j,i}^{(I)}(X)\right)/2, \quad X \in G, (i = 1, 2, 3;\ j = 1, 2, 3)$$

$$\sigma_{ij}^{(I)}(X) = 2\mu_I\left[\frac{\nu_I}{1-2\nu_I}\theta^{(I)}(X)\delta_{ij} + e_{ij}^{(I)}(X)\right],$$

$$\theta^{(I)}(X) = \sum_{k=1}^{3} e_{kk}^{(I)}(X) \tag{2.5}$$

$$\sigma_{ij,j}^{(I)}(X) = 0$$

The bonding conditions have the following form

$$u^I(X) = u^{(f)}(X),\ \sigma_{ij}^I(X)N_j(X) = \sigma_{ij}^{(f)}(X)N_j(X), \quad X \in \partial G \tag{2.6}$$

We suppose that overdetermined boundary data (the applied loads $t^{(f)}(X)$ and displacements $u^{(f)}(X)$) are available on the whole boundary ∂V. The problem consists in searching for the shape, location and elastic moduli (in the case of inclusion) of the defect G using the available data.

3 Invariant integrals and their use in elastostatic inverse problems

The idea of the reciprocity gap principle, used in Andrieux et al. (1999) for the plane crack identification, is as follows. Let us suppose that isotropic linear elastic body with shear modulus μ_M and Poisson ratio

v_M occupies the domain V. A regular elastic field in the body we'll mark by a superscript $(r)(\sigma_{ij}^{(r)}, e_{ij}^{(r)}, u^{(r)} = (u_1^{(r)}, u_2^{(r)}, u_3^{(r)}))$. Consider an integral

$$RG^{(f)}(r) = \int_S \left(t_i^{(f)} u_i^{(r)} - t_i^{(r)} u_i^{(f)} \right) dS \qquad (3.1)$$

where $S \subset \Omega$ is a closed surface, $t_i^{(r)}(X) = \sigma_{ij}^{(r)}(X) n_j(X)$, $n(X) = (n_1(X), n_2(X), n_3(X))$ is a unit outward normal to S.

If the surface S doesn't contain the domain G inside then $RG^{(f)}(r) = 0$, otherwise the values $RG^{(f)}(r)$ can differ from zero and give some information about the defect G. In the case when the loads $t^{(f)}$ and displacements $u^{(f)}$ are available on the surface of the body ∂V, it is possible to take $S = \partial V$ and for all known regular fields the values $RG^{(f)}(r)$ can be calculated. It was shown in Andrieux et al. (1999) that it is possible to reconstruct a plane crack using the appropriate regular fields.

It is well-known that for isotropic linear elastic solids the following invariant integrals are valid (see, Knowles and Sternberg (1972)):

$$J_i = \int_S (W n_i - t_j u_{j,i}) dS, \ i = 1, 2, 3$$

$$L_i = \int_S \varepsilon_{ijk} \left(W X_k n_j + t_j u_k - t_p u_{p,j} X_k \right) dS, \quad i = 1, 2, 3 \qquad (3.2)$$

$$M = \int_S \left(W X_i n_i - t_j u_{j,i} X_i - \frac{1}{2} t_i u_i \right) dS$$

where S as above is a closed surface; σ_{ij}, e_{ij} and $u = (u_1, u_2, u_3)$ are stress and strain tensors and displacement vector corresponding to some stress-strain state of elastic body; $W = \sigma_{kl} e_{kl}/2$; $t_i = \sigma_{ij} n_j$; ε_{ijk} is the alternating tensor.

All these integrals are equal to zero if there are no defects inside S. If a defect is located inside S then the integrals can differ from zero and the values of the integrals give some information about the defect. Due to this property all invariant integrals (3.2) can be used for the defect detection analogously to the reciprocity gap principle given by Eq. (3.1).

Let us mark the invariant integrals (3.2) for the elastic field $u^{(f)}$ by the superscript (f) : $J_i^{(f)}, L_i^{(f)}, M^{(f)}$. Consider invariant integrals for the sum of the applied and regular elastic fields and mark these integrals by the superscript $(f) + (r)$. Because the invariant integrals for the regular elastic fields are equal to zero the following equalities are valid

$$\begin{aligned} J_i^{(f)+(r)} &= J_i^{(f)} + J_{i\text{int}}^{(f)}(r) \\ L_i^{(f)+(r)} &= L_i^{(f)} + L_{i\text{int}}^{(f)}(r) \\ M^{(f)+(r)} &= M^{(f)} + M_{\text{int}}^{(f)}(r) \end{aligned} \qquad (3.3)$$

where the integrals describing the interaction between the applied and regular elastic fields have the following form

$$J_{i\text{int}}^{(f)}(r) = \int_S \left(\sigma_{kl}^{(f)} e_{kl}^{(r)} n_i - t_j^{(f)} u_{j,i}^{(r)} - t_j^{(r)} u_{j,i}^{(f)} \right) dS$$

$$L_{i\text{int}}^{(f)}(r) = \int_S \varepsilon_{ijk} \left(\sigma_{mn}^{(f)} e_{mn}^{(r)} X_k n_j + t_j^{(f)} u_k^{(r)} + t_j^{(r)} u_k^{(f)} - t_p^{(f)} u_{p,j}^{(r)} X_k - t_p^{(r)} u_{p,j}^{(f)} X_k \right) dS \qquad (3.4)$$

$$M_{\text{int}}^{(f)}(r) = \int_S \left(\sigma_{kl}^{(f)} e_{kl}^{(r)} X_i n_i - t_j^{(f)} u_{j,i}^{(r)} X_i - t_j^{(r)} u_{j,i}^{(f)} X_i - \frac{1}{2} t_i^{(f)} u_i^{(r)} - \frac{1}{2} t_i^{(r)} u_i^{(f)} \right) dS$$

Integrals (3.4) are also invariant.

If we suppose that applied loads $t^{(f)}$ and displacements $u^{(f)}$ are known on the boundary ∂V then all stresses $\sigma_{ij}^{(f)}$, strains $e_{ij}^{(f)}$ and distortion tensor $u_{j,i}^{(f)}$ can be calculated on ∂V. So for $S = \partial V$ and known regular fields invariant integrals (3.4) can be calculated.

In the case when the sizes of the defect are small as compared to the distance between the defect and the boundary ∂V, the values of the integrals (3.1) and (3.4) only slightly differ from the integrals for the defect G, located in an infinite elastic solid. Integrals (3.1) and (3.4) for the infinite elastic body with a defect G can be expressed by means of the defect parameters and coordinates of its location. Equating the values of integrals in Eqs. (3.1) and (3.4) calculated by using the experimental data and their expressions by means of the defect parameters and coordinates one obtains a system of equations relative to the defect parameters and coordinates.

4 Identification of a spherical inclusion using one static uniaxial tension test

Let us suppose that applied loads are related to uniform uniaxial tension (compression) in the direction of the axis $X_3, t^{(f)} = (0, 0, \sigma n_3), X \in \partial V$. To emphasize the form of applied loads we'll mark below the stress-strain state of the body outside the defect G by superscript (3) instead of the superscript (f).

Let a defect G be a spherical inclusion of a radius a. Its center is located at the point $M^0\left(x_1^0, x_2^0, x_3^0\right)$. Consider Cartesian coordinates $M^0 x_1 x_2 x_3$ with the origin at M^0

$$X_i = x_i + x_i^0, \; i = 1, 2, 3 \tag{4.1}$$

Introduce the spherical coordinates with the origin at M^0

$$x_1 = r \sin\theta \cos\varphi, \; x_2 = r \sin\theta \sin\varphi, \; x_3 = r\cos\theta \tag{4.2}$$

Solution of the problem for a spherical inclusion in an infinite elastic solid under uniaxial tension (compression) in the direction of the axis x_3 was obtained by Goodier (1933). According to Goodier (1933) the solution of the problem outside the inclusion has the following form

$$\begin{aligned} u_r^{(3)} &= \frac{\sigma a^3}{r^2}\left\{-A - \frac{3a^2 B}{r^2} + \left[5\left(5 - 4\nu_M\right) - \frac{9a^2}{r^2}\right]\right. \\ &\left.\cos 2\theta \cdot B\right\} + \frac{\sigma r}{4\mu_M\left(1+\nu_M\right)}\left[\left(1-\nu_M\right)\right. \\ &\left. + \left(1+\nu_M\right)\cos 2\theta\right] \end{aligned} \tag{4.3}$$

$$\begin{aligned} u_\theta^{(3)} &= \frac{-2B\sigma a^3 \sin 2\theta}{r^2}\left[5\left(1-2\nu_M\right) + \frac{3a^2}{r^2}\right] \\ &- \frac{\sigma r}{4\mu_M}\sin 2\theta, \; u_\varphi^{(3)} = 0 \\ \sigma_{rr}^{(3)} &= \frac{2\mu_M \sigma a^3}{r^3}\left\{2A + \left(-10\nu_M + \frac{12a^2}{r^2}\right)B\right. \\ &\left. + \left[10\left(\nu_M - 5\right) + \frac{36a^2}{r^2}\right] B\cos 2\theta\right\} \\ &+ \frac{\sigma}{2}\left(1 + \cos 2\theta\right) \end{aligned}$$

$$\begin{aligned} \sigma_{\theta\theta}^{(3)} &= \frac{2\mu_M \sigma a^3}{r^3}\left\{-A - \left(10\nu_M + \frac{3a^2}{r^2}\right)B\right. \\ &\left. + \left[5\left(1-2\nu_M\right) - \frac{21a^2}{r^2}\right] B\cos 2\theta\right\} \\ &+ \sigma \sin^2\theta \\ \sigma_{\varphi\varphi}^{(3)} &= \frac{2\mu_M \sigma a^3}{r^3}\left\{-A - \left(10\left(1-\nu_M\right) + \frac{9a^2}{r^2}\right)B\right. \\ &\left. + 15\left[\left(1-2\nu_M\right) - \frac{a^2}{r^2}\right] B\cos 2\theta\right\} \\ \sigma_{r\theta}^{(3)} &= \frac{2\mu_M \sigma a^3 B}{r^3}\left[-10\left(1+\nu_M\right) + \frac{24a^2}{r^2}\right]\sin 2\theta \\ \sigma_{r\varphi}^{(3)} &= 0, \; \sigma_{\theta\varphi}^{(3)} = 0 \end{aligned} \tag{4.4}$$

where

$$B = \frac{\mu_M - \mu_I}{8\mu_M\left[\left(7 - 5\nu_M\right)\mu_M + \left(8 - 10\nu_M\right)\mu_I\right]} \tag{4.5}$$

$$\begin{aligned} A &= -B\frac{\left(1-2\nu_I\right)\left(6-5\nu_M\right)2\mu_M + \left(3+19\nu_I - 20\nu_I\nu_M\right)\mu_I}{2\left(1-2\nu_I\right)\mu_M + \left(1+\nu_I\right)\mu_I} + D \\ D &= \frac{\left(1-\nu_M - 2\nu_I\nu_M\right)\mu_I - \left(1-2\nu_I\right)\left(1+\nu_M\right)\mu_M}{4\mu_M\left(1+\nu_M\right)\left[2\left(1-2\nu_I\right)\mu_M + \left(1+\nu_I\right)\mu_I\right]} \end{aligned} \tag{4.6}$$

For the spherical inclusion identification we'll use the following regular elastic fields with constant, linear and quadratic stresses. Below the stress tensors and displacements vectors are presented in the initial Cartesian coordinates $OX_1X_2X_3$

$$\begin{aligned} \sigma^{(C1)} &= \begin{pmatrix} \sigma & 0 & 0 \\ 0 & 0 & 0 \\ 0 & 0 & 0 \end{pmatrix}, \; u^{(C1)} = \frac{\sigma}{2\mu_M\left(1+\nu_M\right)}\begin{pmatrix} X_1 \\ -\nu_M X_2 \\ -\nu_M X_3 \end{pmatrix} \\ \sigma^{(C2)} &= \begin{pmatrix} 0 & 0 & 0 \\ 0 & \sigma & 0 \\ 0 & 0 & 0 \end{pmatrix}, \; u^{(C2)} = \frac{\sigma}{2\mu_M\left(1+\nu_M\right)}\begin{pmatrix} -\nu_M X_1 \\ X_2 \\ -\nu_M X_3 \end{pmatrix} \\ \sigma^{(C3)} &= \begin{pmatrix} 0 & 0 & 0 \\ 0 & 0 & 0 \\ 0 & 0 & \sigma \end{pmatrix}, \; u^{(C3)} = \frac{\sigma}{2\mu_M\left(1+\nu_M\right)}\begin{pmatrix} -\nu_M X_1 \\ -\nu_M X_2 \\ X_3 \end{pmatrix} \end{aligned} \tag{4.7}$$

$$\sigma^{(L1)} = \begin{pmatrix} \frac{\sigma X_1}{L} & \frac{-\sigma X_2}{L} & 0 \\ \frac{-\sigma X_2}{L} & 0 & 0 \\ 0 & 0 & 0 \end{pmatrix}, u^{(L1)} = \frac{\sigma}{4L\mu_M(1+\nu_M)} \begin{pmatrix} X_1^2 - (2+\nu_M)X_2^2 + \nu_M X_3^2 \\ -2\nu_M X_1 X_2 \\ -2\nu_M X_1 X_3 \end{pmatrix}$$

$$\sigma^{(L2)} = \begin{pmatrix} 0 & \frac{-\sigma X_1}{L} & 0 \\ \frac{-\sigma X_1}{L} & \frac{\sigma X_2}{L} & 0 \\ 0 & 0 & 0 \end{pmatrix}, u^{(L2)} = \frac{\sigma}{4L\mu_M(1+\nu_M)} \begin{pmatrix} -2\nu_M X_1 X_2 \\ -(2+\nu_M)X_1^2 + X_2^2 + \nu_M X_3^2 \\ -2\nu_M X_2 X_3 \end{pmatrix} \quad (4.8)$$

$$\sigma^{(L3)} = \begin{pmatrix} 0 & 0 & \frac{-\sigma X_1}{L} \\ 0 & 0 & 0 \\ \frac{-\sigma X_1}{L} & 0 & \frac{\sigma X_3}{L} \end{pmatrix}, u^{(L3)} = \frac{\sigma}{4L\mu_M(1+\nu_M)} \begin{pmatrix} -2\nu_M X_1 X_3 \\ -2\nu_M X_2 X_3 \\ -(2+\nu_M)X_1^2 + \nu_M X_2^2 + X_3^2 \end{pmatrix}$$

$$\sigma^{(Q)} = \begin{pmatrix} \frac{\sigma X_3^2}{L^2} & 0 & 0 \\ 0 & \frac{\nu_M \sigma (X_3^2 - X_1^2)}{L^2} & 0 \\ 0 & 0 & \frac{-\sigma X_1^2}{L^2} \end{pmatrix}, u^{(Q)} = \frac{\sigma}{2\mu_M L^2} \begin{pmatrix} (1-\nu_M)X_1 X_3^2 + \nu_M X_1^3/3 \\ 0 \\ -\left[(1-\nu_M)X_1^2 X_3 + \nu_M X_3^3/3\right] \end{pmatrix} \quad (4.9)$$

where L is a typical linear size of the domain V.

It is necessary to note that the integrals $J_i^{(f)}$, $J_{i\text{int}}^{(f)}(r)$ and $RG^{(f)}(r)$ do not depend on the location of the origin of coordinates. At the same time the values of $L_i^{(f)}$, $L_{i\text{int}}^{(f)}(r)$, $M^{(f)}$and $M_{\text{int}}^{(f)}(r)$ depend on the location of the origin of coordinates. In this connection we'll denote below the integrals (3.4) in coordinates $OX_1X_2X_3$ by $L_{i\text{int}}^{(f)}(r, O)$ and $M_{\text{int}}^{(f)}(r, O)$. The integrals in coordinates $M^0x_1x_2x_3$ we'll denote by $L_{i\text{int}}^{(f)}(r, M^0)$ and $M_{\text{int}}^{(f)}(r, M^0)$

$$L_{i\text{int}}^{(f)}\left(r, M^0\right) = \int_S \varepsilon_{ijk}\left(\sigma_{mn}^{(f)} e_{mn}^{(r)} x_k n_j + t_j^{(f)} u_k^{(r)} + t_j^{(r)} \times u_k^{(f)} - t_p^{(f)} u_{p,j}^{(r)} x_k - t_p^{(r)} u_{p,j}^{(f)} x_k\right) dS \quad (4.10)$$

$$M_{\text{int}}^{(f)}\left(r, M^0\right) = \int_S \left(\sigma_{kl}^{(f)} e_{kl}^{(r)} x_i n_i - t_j^{(f)} u_{j,i}^{(r)} x_i - t_j^{(r)} u_{j,i}^{(f)} x_i - \frac{1}{2} t_i^{(f)} u_i^{(r)} - \frac{1}{2} t_i^{(r)} u_i^{(f)}\right) dS$$

There is a simple connection between the integrals in different coordinates. From (3.4) and (4.10) one has

$$L_{i\text{int}}^{(f)}(r, O) = L_{i\text{int}}^{(f)}\left(r, M^0\right) + \varepsilon_{ijk} x_k^0 J_{j\text{int}}^{(f)}(r) \quad (4.11)$$

$$M_{\text{int}}^{(f)}(r, O) = M_{\text{int}}^{(f)}\left(r, M^0\right) + x_i^0 J_{i\text{int}}^{(f)}(r) \quad (4.12)$$

All the interaction integrals (3.1), (3.4) and (4.10) for the applied elastic field (3) and regular elastic fields (Ci), (Li) and (Q) can be calculated analytically. For the calculation of the integrals they are written in the coordinates $M^0x_1x_2x_3$ and the sphere ∂G is taken as a surface S. In the expressions (4.7)–(4.9) the coordinates X_i are replaced by $x_i + x_i^0$ according to (4.1) and the applied elastic field (4.3), (4.4) is transformed from the spherical to Cartesian coordinates. The analytical expressions for some of interaction integrals (3.1), (3.4) and (4.10) are as follows

$$J_{i\text{int}}^{(3)}(Ck) = 0 \quad (4.13)$$

It follows from (4.11)–(4.13)

$$L_{i\text{int}}^{(3)}(Ck, O) = L_{i\text{int}}^{(3)}\left(Ck, M^0\right), M_{\text{int}}^{(3)}(Ck, O) = M_{\text{int}}^{(3)}\left(Ck, M^0\right) \quad (4.14)$$

$$RG^{(3)}(C1) = \frac{4\pi(1-\nu_M)a^3\sigma^2(A+15B)}{1+\nu_M},$$
$$RG^{(3)}(C2) = RG^{(3)}(C1),$$
$$RG^{(3)}(C3) = \frac{4\pi(1-\nu_M)a^3\sigma^2[A-5(1+4\nu_M)B]}{1+\nu_M} \quad (4.15)$$

$$RG^{(3)}(L1) = \frac{4\pi(1-\nu_M)a^3\sigma^2(A+15B)}{1+\nu_M}\left(\frac{x_1^0}{L}\right)$$

$$RG^{(3)}(L2) = \frac{4\pi(1-\nu_M)a^3\sigma^2(A+15B)}{1+\nu_M}\left(\frac{x_2^0}{L}\right) \tag{4.16}$$

$$RG^{(3)}(L3) = \frac{4\pi(1-\nu_M)a^3\sigma^2[A-5(1+4\nu_M)B]}{1+\nu_M}\left(\frac{x_3^0}{L}\right)$$

It follows from (4.15) and (4.16) that in the case when $A+15B \neq 0$ and $A-5(1+4\nu_M)B \neq 0$ the coordinates of the center M^0 of the ball G can be expressed by means of invariant integrals

$$\frac{x_1^0}{L} = \frac{RG^{(3)}(L1)}{RG^{(3)}(C1)}, \frac{x_2^0}{L} = \frac{RG^{(3)}(L2)}{RG^{(3)}(C2)}, \frac{x_3^0}{L} = \frac{RG^{(3)}(L3)}{RG^{(3)}(C3)} \tag{4.17}$$

Let us note that displacements in the regular fields (4.7)–(4.9) are homogeneous vector-functions in the coordinates $OX_1X_2X_3$. It is possible to prove the following proposition.

If $u^{(r)}(X)$ is a homogeneous vector-function of order m $\left(u^{(r)}(kX) = k^m u^{(r)}(X)\right)$ then the following equality is valid

$$M_{\text{int}}^{(f)}(r, O) = -\left(m+\frac{1}{2}\right)RG^{(f)}(r) \tag{4.18}$$

In particular

$$M_{\text{int}}^{(3)}(r, O) = -\left(m+\frac{1}{2}\right)RG^{(3)}(r) \tag{4.19}$$

From (4.17) and (4.19) it follows that coordinates of the defect center can be expressed also in terms of M-integrals

$$\frac{x_1^0}{L} = \frac{3M_{\text{int}}^{(3)}(L1, O)}{5M_{\text{int}}^{(3)}(C1, O)}, \frac{x_2^0}{L} = \frac{3M_{\text{int}}^{(3)}(L2, O)}{5M_{\text{int}}^{(3)}(C2, O)}, \frac{x_3^0}{L} = \frac{3M_{\text{int}}^{(3)}(L3, O)}{5M_{\text{int}}^{(3)}(C3, O)} \tag{4.20}$$

The coordinates can be also calculated by other types of integrals, for example

$$x_1^0 = \frac{-RG^{(3)}(L1)}{J_{1\text{int}}^{(3)}(L1)}, x_2^0 = \frac{-RG^{(3)}(L2)}{J_{2\text{int}}^{(3)}(L2)}, x_3^0 = \frac{-RG^{(3)}(L3)}{J_{3\text{int}}^{(3)}(L3)} \tag{4.21}$$

Let us also note that

$$M_{\text{int}}^{(3)}\left(Lk, M^0\right) = -\frac{3}{2}RG^{(3)}(Lk); \quad k = 1, 2, 3 \tag{4.22}$$

To determine a radius and elastic moduli of the inclusion we'll use the regular elastic field (4.9) with quadratic stresses. Consider some interaction integrals for the applied elastic field (3) and the regular field (Q) given by Eq. (4.9)

$$M_{\text{int}}^{(3)}\left(Q, M^0\right) = \frac{2(1-\nu_M)\pi a^3\sigma^2}{L^2} \times \left[-28Ba^2 + 3(A-5B)\left(x_1^0\right)^2 - 3(A+15B)\left(x_3^0\right)^2\right] \tag{4.23}$$

$$RG^{(3)}(Q) = \frac{-4(1-\nu_M)\pi a^3\sigma^2}{L^2}\left[-4Ba^2 + (A-5B)\left(x_1^0\right)^2 - (A+15B)\left(x_3^0\right)^2\right] \tag{4.24}$$

It follows from (4.23) and (4.24)

$$2M_{\text{int}}^{(3)}\left(Q, M^0\right) + 3RG^{(3)}(Q) = \frac{-64(1-\nu_M)\pi a^5\sigma^2 B}{L^2} \tag{4.25}$$

From (4.15) and (4.19) one has

$$M_{\text{int}}^{(3)}(C3, O) - M_{\text{int}}^{(3)}(C1, O) = 120(1-\nu_M)\pi a^3\sigma^2 B \tag{4.26}$$

It is interesting to note that the sign of the expression $M_{\text{int}}^{(3)}(C3, O) - M_{\text{int}}^{(3)}(C1, O)$ given by Eq. (4.26), coinciding with the sign of the constant B (Eq. (4.5)), indicates the matrix or the inclusion is more stiff. If $M_{\text{int}}^{(3)}(C3, O) - M_{\text{int}}^{(3)}(C1, O) > 0$ then $\mu_M > \mu_I$, if $M_{\text{int}}^{(3)}(C3, O) - M_{\text{int}}^{(3)}(C1, O) < 0$ then $\mu_M < \mu_I$.

Let us suppose that $M_{\text{int}}^{(3)}(C3, O) - M_{\text{int}}^{(3)}(C1, O) \neq 0$. In this case from (4.23) and (4.24) one has

$$\frac{a^2}{L^2} = -\frac{15}{8}\frac{\left(2M_{\text{int}}^{(3)}(Q, M^0) + 3RG^{(3)}(Q)\right)}{\left(M_{\text{int}}^{(3)}(C3, O) - M_{\text{int}}^{(3)}(C1, O)\right)} \tag{4.27}$$

It follows from (4.12), (4.19) and (4.27)

$$\frac{a^2}{L^2} = \frac{15}{4}\frac{\left(2RG^{(3)}(Q) + x_i^0 J_{i\text{int}}^{(3)}(Q)\right)}{\left(M_{\text{int}}^{(3)}(C3, O) - M_{\text{int}}^{(3)}(C1, O)\right)} \tag{4.28}$$

Using (4.28) in (4.26) one obtains an expression for the constant B

$$B = \frac{M_{\text{int}}^{(3)}(C3, O) - M_{\text{int}}^{(3)}(C1, O)}{120(1-\nu_M)\pi a^3\sigma^2} \tag{4.29}$$

After the calculation of the value B one can calculate the shear modulus μ_I using (4.5)

$$\frac{\mu_I}{\mu_M} = \frac{1 - 8\mu_M(7-5\nu_M)B}{16\mu_M(4-5\nu_M)B+1} \tag{4.30}$$

It follows from (4.15) and (4.19)

$$(1+4\nu_M)M_{\text{int}}^{(3)}(C1, O) + 3M_{\text{int}}^{(3)}(C3, O) = -24\pi(1-\nu_M)a^3\sigma^2 A \tag{4.31}$$

From (4.29) and (4.31) one has

$$A = \frac{5B\left[(1+4\nu_M)M_{\text{int}}^{(3)}(C1, O) + 3M_{\text{int}}^{(3)}(C3, O)\right]}{M_{\text{int}}^{(3)}(C1, O) - M_{\text{int}}^{(3)}(C3, O)} \tag{4.32}$$

Finally one obtains an expression for the Poisson ratio ν_I of the inclusion

$$\nu_I = R/S \tag{4.33}$$

where

$$\begin{aligned} R &= 4\mu_M(1+\nu_M)\{(2\mu_M+\mu_I)A \\ &\quad + [2\mu_M(6-5\nu_M)+3\mu_I]B\} \\ &\quad + (1+\nu_M)\mu_M - (1-\nu_M)\mu_I \\ S &= -4\mu_M(1+\nu_M)\{(\mu_I - 4\mu_M)A \\ &\quad + [-4\mu_M(6-5\nu_M) + \mu_I(19-20\nu_M)]B\} \\ &\quad + 2[(1+\nu_M)\mu_M - \nu_M\mu_I] \end{aligned}$$

4.1 Some special cases

Consider some special cases of the considered problem. Let us suppose that G is a spherical cavity ($\mu_I \to 0$). In this case the formulae (4.5) and (4.6) have the following form

$$B = \frac{1}{8\mu_M(7-5\nu_M)}, \quad A = \frac{10\nu_M - 13}{8\mu_M(7-5\nu_M)} \tag{4.34}$$

Using (4.34), (4.15) and (4.18) one has

$$M_{\text{int}}^{(3)}(C3, O) = \frac{3\pi(1-\nu_M)(9+5\nu_M)a^3\sigma^2}{2\mu_M(1+\nu_M)(7-5\nu_M)} \tag{4.35}$$

$$M_{\text{int}}^{(3)}(C1, O) = M_{\text{int}}^{(3)}(C2, O) = \frac{-3\pi(1-\nu_M)(1+5\nu_M)a^3\sigma^2}{2\mu_M(1+\nu_M)(7-5\nu_M)} \tag{4.36}$$

It follows from (4.35) and (4.36) that if $a \neq 0$ then $M_{\text{int}}^{(3)}(Ci, O) \neq 0, i = 1, 2, 3$. In this case the radius of the defect can be calculated from (4.35)

$$a^3 = \frac{2\mu_M(1+\nu_M)(7-5\nu_M)M_{\text{int}}^{(3)}(C3, O)}{3\pi(1-\nu_M)(9+5\nu_M)\sigma^2} \tag{4.37}$$

The coordinates of the cavity center are calculated by means of the formulae (4.17), (4.20) or (4.21) and the problem of the spherical cavity identification is completely solved.

In the case of a rigid inclusion ($\mu_I \to +\infty$) the values A and B according to (4.5) and (4.6) have the form

$$B = \frac{-1}{16\mu_M(4-5\nu_M)}, \quad A = \frac{19 - 33\nu_M + 20\nu_M^2}{16\mu_M(4-5\nu_M)(1+\nu_M)} \tag{4.38}$$

It follows from (4.15), (4.18) and (4.38)

$$M_{\text{int}}^{(3)}(C1, O) = M_{\text{int}}^{(3)}(C2, O) = \frac{-3\pi(1-\nu_M)(5\nu_M^2 - 12\nu_M + 1)a^3\sigma^2}{2\mu_M(4-5\nu_M)(1+\nu_M)^2} \tag{4.39}$$

$$M_{\text{int}}^{(3)}(C3, O) = \frac{-3\pi(1-\nu_M)(5\nu_M^2 - \nu_M + 3)a^3\sigma^2}{\mu_M(4-5\nu_M)(1+\nu_M)^2} \tag{4.40}$$

The radius of the rigid inclusion is determined from (4.40)

$$a^3 = \frac{-\mu_M(4-5\nu_M)(1+\nu_M)^2 M_{\text{int}}^{(3)}(C3, O)}{3\pi(1-\nu_M)(5\nu_M^2 - \nu_M + 3)\sigma^2} \tag{4.41}$$

Let us note that according to (4.39) the values $M_{\text{int}}^{(3)}(C1, O)$ and $M_{\text{int}}^{(3)}(C2, O)$ can become zero when ν_M is a root of the quadratic equation $5\nu_M^2 - 12\nu_M + 1 = 0$ in the interval $(0, 1/2)$, more precisely $\nu_M = \nu_M^0 \approx 0.0864$. So if $\nu_M \neq \nu_M^0$ then the coordinates of the rigid inclusion center are calculated by formulae (4.17), (4.20) or (4.21). If $\nu_M = \nu_M^0$ then the value x_3^0/L can be calculated by the formulae (4.17), (4.20) or (4.21) as before. For the calculation of the values x_1^0/L and x_2^0/L it is necessary to use some other regular elastic fields with linear stresses. Let us take for an example

$$\sigma^{(L4)} = \begin{pmatrix} \frac{\sigma X_1}{L} & \frac{-\sigma X_2}{L} & 0 \\ \frac{-\sigma X_2}{L} & 0 & 0 \\ 0 & 0 & \frac{\nu_M \sigma X_1}{L} \end{pmatrix}, u^{(L4)} = \frac{\sigma}{4\mu_M L} \begin{pmatrix} (1-\nu_M) X_1^2 + (\nu_M - 2) X_2^2 \\ -2\nu_M X_1 X_2 \\ 0 \end{pmatrix} \tag{4.42}$$

$$\sigma^{(L5)} = \begin{pmatrix} 0 & \frac{-\sigma X_1}{L} & 0 \\ \frac{-\sigma X_1}{L} & \frac{\sigma X_2}{L} & 0 \\ 0 & 0 & \frac{\nu_M \sigma X_2}{L} \end{pmatrix}, u^{(L5)} = \frac{\sigma}{4\mu_M L} \begin{pmatrix} -2\nu_M X_1 X_2 \\ (\nu_M - 2) X_1^2 + (1-\nu_M) X_2^2 \\ 0 \end{pmatrix}$$

The calculations lead to the following expressions

$$RG^{(3)}(L4) = \frac{\pi(1-\nu_M)(1-2\nu_M)(1-5\nu_M)a^3\sigma^2}{\mu_M(1+\nu_M)(4-5\nu_M)} \left(\frac{x_1^0}{L}\right)$$

$$RG^{(3)}(L5) = \frac{\pi(1-\nu_M)(1-2\nu_M)(1-5\nu_M)a^3\sigma^2}{\mu_M(1+\nu_M)(4-5\nu_M)} \left(\frac{x_2^0}{L}\right) \tag{4.43}$$

It follows from (4.15), (4.38) and (4.43) that in the case when ν_M is equal or close to ν_M^0 the coordinates x_1^0 and x_2^0 can be calculated from the following formulae

$$\frac{x_1^0}{L} = \frac{2(5\nu_M^2 - \nu_M + 3)}{(1-2\nu_M)(1-5\nu_M)(1+\nu_M)} \frac{RG^{(3)}(L4)}{RG^{(3)}(C3)} \tag{4.44}$$

$$\frac{x_2^0}{L} = \frac{2(5\nu_M^2 - \nu_M + 3)}{(1-2\nu_M)(1-5\nu_M)(1+\nu_M)} \frac{RG^{(3)}(L5)}{RG^{(3)}(C3)}$$

5 Numerical analysis of the obtained formulae

The obtained formulae (4.17), (4.20), (4.21), (4.37) and others are exact for a spherical defect in an infinite elastic solid, but in the case of a bounded domain they can be considered only as approximate ones. It is clear that the formulae give a good approximation in the case when the sizes of the defect are small as compared to the distance between the defect and the boundary of the body. The aim of the numerical analysis is to determine how close to the boundary of the body can be a defect so that the formulae will still be applicable.

Let $OX_1X_2X_3$ is the Cartesian coordinate system. As an example consider the cube domain $V : |X_i| \leq 10,\ i = 1, 2, 3$. The Poisson ratio of the matrix is $\nu_M = 0.25$. We consider below two types of the defects G: (1) G is a spherical cavity of the radius 1; (2) G is a rigid inclusion of the radius 1. The applied loads correspond to the uniform uniaxial tension in the direction of the axis $X_3 : t^{(3)}(X_1, X_2, 10) = (0, 0, \sigma),\ t^{(3)}(X_1, X_2, -10) = (0, 0, -\sigma),\ t^{(3)}(\pm 10, X_2, X_3) = t^{(3)}(X_1, \pm 10, X_3) = (0, 0, 0)$. For different locations of the defect center the direct problem was solved by the FEM and the elastic field on the surface ∂V was calculated. After that the invariant integrals were calculated and the defect parameters were determined by the formulae (4.17) and (4.37) for a cavity and by the formulae (4.17) and (4.41) for a rigid inclusion. We took $L = 10$ for the linear regular elastic fields.

The results of the calculations are presented in the Tables 1 and 2, respectively. The numerical results show that obtained explicit formulae give a good approximation to the inverse problem solution for a spherical cavity and a rigid inclusion even when the defect is located close enough to the boundary of the body.

Besides the limiting cases of a cavity and a rigid inclusion we considered also the case of elastic spherical inclusion. The calculations show that formulae (4.17) also give very accurate results for the coordinates of a defect center in the case of elastic inclusion. At the same time formulae (4.28)–(4.33) used for the calculation of the radius and elastic moduli of the inclusion are much more sensitive to the errors in the boundary data and accuracy of the invariant integrals calculations. To improve the accuracy of the calculations one can use other formulae containing other types of invariant integrals and regular elastic fields. In this connection the problem arises to find the formulae which are less sensitive to the boundary data errors and accuracy of invariant integrals calculations.

6 Conclusion

A method, based on the use of invariant integrals, is proposed for the defect identification. The method extends the possibilities of the reciprocity gap principle.

Explicit formulae for determination of the parameters of the spherical cavities and inclusions by means of

Table 1 Identification of a spherical cavity

Coordinates of the cavity center	The values obtained by approximate formulae			
(x_1^0, x_2^0, x_3^0)	a	x_1^0	x_2^0	x_3^0
(0,0,0)	0.9995	−0.0028	−0.0015	0.0019
(0,3,0)	0.9995	−0.0076	3.0047	0.0006
(0,6,0)	1.0028	−0.0119	6.0705	0.0071
(0,8,0)	1.0162	0.0005	8.5417	−0.0005
(0,0,3)	0.9991	0.0092	−0.0006	2.9968
(0,0,6)	1.0102	−0.0193	0.0035	5.9759
(0,0,8)	1.0077	0.0006	−0.0019	8.6567
(3,3,3)	1.0010	3.0067	3.0004	2.9963
(6,6,6)	1.0010	6.1185	6.1197	6.0664
(8,8,8)	1.1205	8.4437	8.4729	8.0198

Table 2 Identification of a spherical rigid inclusion

Coordinates of the rigid inclusion center	The values obtained by approximate formulae			
(x_1^0, x_2^0, x_3^0)	a	x_1^0	x_2^0	x_3^0
(0,0,0)	0.9999	0.0027	−0.0033	0.0050
(0,3,0)	0.9994	0.0109	3.0112	0.0094
(0,6,0)	1.0013	−0.0037	5.9933	−0.0166
(0,8,0)	1.0054	−0.0095	8.0358	0.0015
(0,0,3)	0.9996	0.0054	−0.0120	3.0067
(0,0,6)	1.0035	−0.0005	−0.0140	5.9949
(0,0,8)	1.0275	−0.0081	0.0015	8.0275
(3,3,3)	1.0004	2.9841	3.0074	2.9977
(6,6,6)	1.0050	5.9610	5.9753	5.9666
(8,8,8)	1.0387	7.9138	7.9201	8.0081

the results of one uniaxial tension (compression) static test are obtained. These formulae are exact for the infinite elastic solids and approximate for the bounded elastic bodies.

Numerical analysis of the formulae is fulfilled. The results of the analysis show that the formulae give a good approximation of the spherical cavity and spherical rigid inclusion parameters even in the case when a defect is close enough to the body boundary. If a defect is a spherical elastic inclusion then its location is also determined very accurately. It is required to find the formulae for determination of the radius and elastic moduli of the inclusion less sensitive to the errors in the boundary data and accuracy of the invariant integrals calculation.

Acknowledgements E. Sh and P. Sh. gratefully acknowledge the support of RFBR grant 07-01-00448.

References

Alves CJS, Ha-Duong T (1999) Inverse scattering for elastic plane cracks. Inverse Probl 15:91–97

Ammari H, Kang H, Nakamura G, Tanuma K (2002) Complete asymptotic expansions of solutions of the system of elastostatics in the presence of an inclusion of small diameter and detection of an inclusion. J Elast 67:97–129

Andrieux S, Ben Abda A, Bui H (1999) Reciprocity principle and crack identification. Inverse Probl 15:59–65

Bonnet M, Constantinescu A (2005) Inverse problems in elasticity. Inverse Probl 21:R1–R50

Bostrom A, Wirdelius H (1995) Ultrasonic probe modeling and nondestructive crack detection. J Acoust Soc Am 97(5 Pt.1):2836–2848

Engelhardt M, Schanz M, Stavroulakis G, Antes H (2006) Defect identification in 3-D elastostatics using a genetic algorithm. Optim Eng 7:63–79

Glushkov EV, Glushkova NV, Ehlakov AV (2002) Mathematical model of the ultrasonic defectoscopy of spatial cracks. Appl Math Mech (PMM) 66(1):147-156 (in Russian)

Goodier JN (1933) Concentration of stress around spherical and cylindrical inclusions and flaws. J Appl Mech APM-55-7 39–44

Guzina B, Nintcu Fata S, Bonnet M (2003) On the stress-wave imaging of cavities in a semi-infinite solid. Int J Solids Struct 40:1505–1523

Keat W, Larson M, Verges M (1998) Inverse method of identification for three-dimensional subsurface cracks in a half-space. Int J Fract 92:253–270

Knowles JK, Sternberg E (1972) On a class of conservation laws in linearized and finite elastostatics. Arch Rational Mech Anal 44(3):187–211

Vatuliyan AO (2004) On the determination of crack configuration in anisotropic medium. Appl Math Mech (PMM) 68(1): 180-188 (in Russian)

On application of classical Eshelby approach to calculating effective elastic moduli of dispersed composites

K. B. Ustinov · R. V. Goldstein

Abstract The problem of finding effective elastic moduli of media with spheroid inclusions in case of small concentration of these inclusions is addressed. A number of particular solutions, both known and new, were obtained as limit transitions and asymptotical expansion of the general solution, based on Eshelby's approach. A special attention was paid to determining the ranges of applicability of the obtained asymptotical solutions. It was shown that for spheroid inclusions the areas of applicability of the asymptotic solutions are determined by two parameters: the ratio of elastic moduli of the inclusion and the matrix and aspect ratio of the inclusions.

Keywords Effective properties · Eshelby's tensor · Inclusion

1 Introduction: general relations

The problem of finding the effective elastic characteristics of heterogeneous media was addressed in numerous papers. The current work is restricted to considering the case of dilute suspension of one phase (inclusions) within the material of the other phase (matrix), both phases supposed to be isotropic, the inclusions in the forms of spheroid or three-axial flat ellipsoid supposed to be distributed uniformly and isotropic in space. The main goal of the paper is to demonstrate the influence of two parameters—ratio of shear moduli of the inclusions and aspect ratio (ratio of the major and minor axes)—on the effective elastic properties of the composite. A special attention is devoted to limiting cases, when both parameters became simultaneously extremely small (large).

The general algorithm of calculating the effective elastic characteristics of media with inclusions of another material is known (Eshelby 1957; Mura 1982). The modern state of the problem may be found e.g., in the following works: Tucker and Liang (1999), Odegard et al. (2002), Ustinov (2003a). Below the brief account is given. Let the Hooke law is satisfied for each point, r, of the media

$$\sigma\,(r) = \Lambda\,(r)\,\varepsilon\,(r) \tag{1}$$

Where σ and ε are stress and strain tensors, respectively. Λ is the 4th rank tensor of elasticity, which can be expressed as follows

$$\Lambda(r) = \Lambda_0 + (\Lambda_* - \Lambda_0)\,\theta(r), \tag{2}$$

where $\theta(r) = 1$ inside the inclusions, and $\theta(r) = 0$ outside the inclusions. All the values corresponding to the matrix are marked with index 0, the values corresponding to the inclusions are marked with asterisk. The direct (indexless) notation is used whenever it is possible; the averaged values are denoted with the bar over the corresponding symbol.

K. B. Ustinov (✉) · R. V. Goldstein
Institute for Problems in Mechanics, Prospect Vernadskogo 101 Moscow RF, Moscow, 119526 Russian Federation
e-mail: ustinov@ipmnet.ru

Defect and Material Mechanics. C. Dascalu, G.A. Maugin & C. Stolz (eds.),
doi: 10.1007/978-1-4020-6929-1_7, © Springer Science+Business Media B.V. 2007

Substituting (2) into (1) and averaging over the volume, one has

$$\overline{\sigma} = \Lambda_0 \overline{\varepsilon} + (\Lambda_* - \Lambda_0)\, \overline{\varepsilon_{incl}}\, \Omega, \tag{3}$$

where $\overline{\varepsilon_{incl}}$ are the stains, averaged over the volume of inclusions; Ω is the volumetric concentration of the inclusions. On the other hand

$$\overline{\sigma} = \Lambda_{eff} \overline{\varepsilon}, \tag{4}$$

where Λ_{eff} is the tensor of effective moduli to be found. Therefore, on expressing $\overline{\varepsilon_{incl}}$ in terms of $\overline{\varepsilon}$ the problem may be regarded as solved.

That may be done for ellipsoid inclusions. Thus, equate the stresses within a single inclusion in the matrix subjected to uniform field, ε_0, with the stresses within an equivalent inclusion (the same as in the material of matrix) subjected to eigenstrains, ε_*, (Eshelby 1957, 1961; see also Mura 1982),

$$\Lambda_* \varepsilon_{incl} = \Lambda_0 (\varepsilon_{incl} - \varepsilon_*), \quad \varepsilon_{incl} = \varepsilon_0 + \varepsilon_{add} \tag{5}$$

where ε_{add} is the additional, or constrained, strain in the equivalent inclusion, which is related to Eshelby's tensor as follows

$$\varepsilon_{add} = S\varepsilon_* \tag{6}$$

The expression for the components of Eshelby's tensor in case of isotropic matrix is known (Eshelby 1957, 1961; see also Mura 1982). For the case of spheroid inclusions they are given in Appendix I. Solving the system of (5), (6) yields

$$\varepsilon_{incl} = \left[I + S\Lambda_0^{-1}(\Lambda_* - \Lambda_0)\right]^{-1} \varepsilon_0 \tag{7}$$

Here I stands for the unit 4th rank tensor $I_{ijkl} = \frac{1}{2}(\delta_{ik}\delta_{jl} + \delta_{il}\delta_{jk})$. The inverse of any 4th rank tensor A is understood in the sense that $A^{-1}A = AA^{-1} = I$ (e.g. Walpole 1969).

In case of a small concentration of the aligned inclusions, externally applied strain, ε_0, may be considered to be equal to $\overline{\varepsilon}$, and accounting this, substituting (7) into (3) gives

$$\overline{\sigma} = \Lambda_0 \overline{\varepsilon} + \left((\Lambda_* - \Lambda_0)^{-1} + S\Lambda_0^{-1}\right)^{-1} \overline{\varepsilon}\, \Omega \tag{8}$$

The comparison of (8) and (4) finally yields

$$\Lambda_{eff} = \Lambda_0 + D\Omega, \quad D = \left((\Lambda_* - \Lambda_0)^{-1} + S\Lambda_0^{-1}\right)^{-1} \tag{9}$$

In case of spheroid inclusion with semi-axes a_1, $a_2 = a_1, a_3, \Omega = \frac{4}{3}\pi a_1^2 a_3 N$, where N is the number of inclusions per unit of volume. It is worth to note that expression (9) remains valid for arbitrary anisotropic matrix and aligned inclusions (also anisotropic), the expression for Eshelby's tensor in case of general anisotropy, however, has to be obtained. In case of anisotropic inclusions in isotropic matrix no additional work is required.

In case of arbitrary distribution of the inclusions over their orientation in space, their influence remains additive due to linearity of the case of small concentration, and the effective properties may be calculated by integration of (9) over all possible orientations with the appropriate weight functions. However more elegant way may be used. Consider the isotropic space distribution of the inclusions in isotropic matrix; the effective medium has to remain isotropic with its tensor of elasticity expressed in terms of Lame constants by well-known formula

$$\Lambda_{ijkl} = \lambda\delta_{ij}\delta_{kl} + \mu\left(\delta_{ik}\delta_{jl} + \delta_{il}\delta_{jk}\right) \tag{10}$$

On the other hand, since each symmetric 4th rank tensor has 2 linear invariants, the elastic constants may be expressed in terms of these invariants

$$k = \lambda + \frac{2}{3}\mu = \frac{1}{9}\Lambda_{iijj} \quad \mu = \frac{1}{30}\left(3\Lambda_{ijij} - \Lambda_{iijj}\right) \tag{11}$$

Thus for the case under consideration

$$k_{eff} = k_0 + \frac{\Omega}{9} D_{iijj}$$
$$\mu_{eff} = \mu_0 + \frac{\Omega}{30}\left(3D_{ijij} - D_{iijj}\right) \tag{12}$$

The idea of using such invariants for calculating the effective characteristics has been known at least since Kroner's work (1958).

Remark 1
As it has been noted, due to linear approximation, that appears naturally in case of small concentration of the inclusions, the influence of each inclusion is additive. Obviously, the influence of each inclusion on the invariants of the tensor of elasticity is independent of its orientation in space. Hence, the invariants of elastic tensor remain the same for an arbitrary distribution of the inclusions in space.

Remark 2
The considered case of a small concentration has both the inherent significance and may serve as a basis for more advanced approaches, e.g. differential self-consistent method (Salganik 1973; Roscoe 1973).

Remark 3
The direct (indexless) notation used in the paper is also convenient because if we represent the involved 4th rank tensors by 6×6 matrixes, all formulae may be considered as formulae of matrix algebra. Moreover, if the classes of symmetry of the involved tensors are not lower then transversal, the formalism of Walpole may be successfully applied, i.e. the tensors may be represented by 6-dimensioned vector-like objects for which operations of multiplication and inversion are defined (Walpole 1969).

2 Asymptotical solution for spheroid inclusions

2.1 General, parameters: limit transitions

The procedure of calculating the effective moduli with the help of (9), (12) and (A1), (A2) is rather tedious, therefore, solutions for particular cases are of importance. A number of such solutions are known, which describe the following cases:

- thin disk-like inclusions of arbitrary rigidity (Wu 1966);
- inclusions in the form of sphere of arbitrary rigidity (Krivoglas and Cherevko 1959);
- thin needle-like inclusions of arbitrary rigidity (Wu 1966);
- thin disk-like inclusions of zero rigidity (cracks) (Walsh 1965);
- thin disk-like inclusions of infinite (very high) rigidity (Kovalenko and Salganik 1977).

However, some of these solutions contradict each other, and neither of them can be obtained from another by limit transition. Below, these solutions will be obtained as the limit transitions and asymptotic expansions of the general solution, because all the cases may be treated as the media with spheroid inclusions. The problem is reduced to finding coefficients at concentration, Ω, in (12), which may be represented as multiplications of elastic modulus (shear or compressive) and dimensionless functions depending on four dimensionless parameters, $\frac{\mu_*}{\mu_0}, \frac{a_3}{a_1}, \nu_0, \nu_*$, ratio of shear moduli of inclusions and matrix, aspect ratio of the inclusions, Poisson's ratios of the matrix and inclusions, respectively.

2.2 Flat and long spheroids

2.2.1 *Flat spheroids*

Substituting (A1), (A2) into (9), and then result into (12), and performing limit transition $a_3/a_1 \to 0$ yields

$$k_{eff} = k_0 \left[1 + \frac{(k_* - k_0)(3k_0 + 4\mu_*)}{k_0 (3k_* + 4\mu_*)} \Omega \right] \tag{13}$$

$$\mu_{eff} = \mu_0 \left[1 + \frac{(\mu_* - \mu_0)\left(9k_*\mu_* + 8\mu_*^2 + 6k_*\mu_0 + 12\mu_*\mu_0\right)}{5\mu_*\mu_0 (3k_* + 4\mu_*)} \Omega \right] \tag{14}$$

The other elastic coefficients may be found by means of well-known formulae of the theory of elasticity. This result coincides with the known results by Wu (1966). However, it is seen that the above formulae are no longer valid for the unlimited increase or decrease of elastic moduli of the inclusions. On performing limit transition $a_3/a_1 \to 0$ one may not be allowed to consider any parameters to be smaller then this ratio. Hence these formulae do not obey to (but still may) cover the range of parameters exceeding the following

$$\frac{a_3}{a_1} << \frac{\mu_*}{\mu_0} << \frac{a_1}{a_3}; \ \frac{a_3}{a_1} << 1 \tag{15}$$

2.2.2 *Long spheroids*

Substituting (A1), (A2) into (9), and then result into (12), and performing limit transition $a_3/a_1 \to \infty$ yields

$$k_{eff} = k_0 \left[1 + \frac{(k_* - k_0)(3k_0 + \mu_* + 3\mu_0)}{k_0 (3k_* + \mu_* + 3\mu_0)} \Omega \right] \tag{16}$$

$$\begin{aligned} \mu_{eff} = \mu_0 \Bigg[1 + & \frac{3k_0 (\mu_* - \mu_0)}{5\mu_0 (\mu_* + \mu_0)(3k_* + \mu_* + 3\mu_0)\left(3k_0\mu_* + 3k_0\mu_0 + 7\mu_*\mu_0 + \mu_0^2\right)} \times \\ & \times \Big[(\mu_* + \mu_0)(4\mu_0 (3\mu_* + 7\mu_0) + 3k_* (\mu_* + 9\mu_0)) \\ & + \mu_0 \left(8\mu_0 \left(9\mu_*^2 + 23\mu_*\mu_0 + 8\mu_0^2 \right) + 3k_* \left(7\mu_*^2 + 52\mu_*\mu_0 + 21\mu_0^2 \right) \right) \Big] \Bigg] \end{aligned} \tag{17}$$

This result coincides with the known results by Wu (1966). However, similar to the case of disk-like inclusions, it should be kept in mind that the area of applicability of this solution is also restricted, although remaining wider

$$\frac{\mu_*}{\mu_0} << \frac{a_3}{a_1};\ \frac{a_3}{a_1} >> 1, \tag{18}$$

i.e. these formulae remain valid for soft needle-like inclusions. Mathematically that means that in case of soft needle-like inclusions the limit transition is uniform.

2.3 Soft and rigid inclusions

2.3.1 Penny-shaped cracks

The classical solution for the effective elastic moduli of solids with cracks (e.g. Walsh 1965) can not be obtained from (13), (14) by limit transition $\frac{\mu_*}{\mu_0} \to 0$. It may be obtained, however, from the general solution (9), (12), by limit transition $\frac{\mu_*}{\mu_0} \to 0$ (or setting $\mu_* = 0$) followed by limit transition $\frac{a_3}{a_1} \to 0$.

$$k_{eff} = k_0\left[1 - \frac{4}{3\pi}\frac{(1-\nu_0^2)}{(1-2\nu_0)}\frac{a_1}{a_3}\Omega\right] = k_0\left[1 - \frac{4}{3\pi}\frac{(1-\nu_0^2)}{(1-2\nu_0)}\Omega'\right] \tag{19}$$

$$\mu_{eff} = \mu_0\left[1 - \frac{8}{15\pi}\frac{(1-\nu_0)(5-\nu_0)}{(2-\nu_0)}\frac{a_1}{a_3}\Omega\right] = \mu_0\left[1 - \frac{8}{15\pi}\frac{(1-\nu_0)(5-\nu_0)}{(2-\nu_0)}\Omega'\right] \tag{20}$$

where

$$\Omega' = \Omega\frac{a_1}{a_3} = \frac{4\pi}{3}a_1^3 N. \tag{21}$$

Such a sequence of limit transitions implies that in that case zero rigidity is of more importance then zero thickness, and that this solution may be considered as the asymptotic solution for

$$\frac{\mu_*}{\mu_0} << \frac{a_3}{a_1} << 1 \tag{22}$$

2.3.2 Penny-shaped extremely rigid inclusions

Similar to the case of crack, there exists a solution for flat rigid inclusions that can not be obtained by limit transition $\frac{\mu_*}{\mu_0} \to \infty$ in (13), (14), but may be obtained from the general solution (9), (12), by limit transition $\frac{\mu_*}{\mu_0} \to \infty$ followed by limit transition $\frac{b}{a} \to 0$ (Ustinov 2003a).

$$k_{eff} = k_0\left[1 + \frac{16}{3\pi}\frac{(1-\nu_0)(1-2\nu_0)}{(3-4\nu_0)(1+\nu_0)}\Omega'\right] \tag{23}$$

$$\mu_{eff} = \mu_0\left[1 + \frac{8}{15\pi}\frac{(1-\nu_0)(43-56\nu_0)}{(3-4\nu_0)(7-8\nu_0)}\Omega'\right] \tag{24}$$

Such a sequence of limit transitions implies that the range of applicability of this solution is

$$\frac{\mu_*}{\mu_0} >> \frac{a_1}{a_3} >> 1 \tag{25}$$

This situation was considered in the work by Kovalenko and Salganik (1977), where three cases of relative values of parameters $\frac{\mu_*}{\mu_0}$, $\frac{a_1}{a_3}$ were pointed out, and the role of the sequence of the limit transitions was stressed (the similar problem for effective properties described by 2nd rank tensors were considered by Salganik 1974). However, the final result in the above paper contained a mistake, appeared in calculation of the average of the contribution to the elastic moduli due to the inclusions over all the orientations, and does not coincide with formulae (23), (24).

2.3.3 Needle-like extremely rigid inclusions

Let us consider the range of extremely rigid long inclusions

$$\frac{\mu_*}{\mu_0} >> \frac{a_3}{a_1} >> 1 \tag{26}$$

The solution corresponding to this case may be obtained from the general solution (9), (12) by limit transition $\frac{\mu_*}{\mu_0} \to \infty$ followed by limit transition $\frac{a_3}{a_1} \to \infty$ (Ustinov 2003a).

$$k_{eff} = k_0\left[1 + \frac{2(1-\nu_0)(1-2\nu_0)}{3(1+\nu_0)\left[4(1-\nu_0)\left(\ln\frac{2a_3}{a_1}-1\right)-1\right]}\Omega''\right] \tag{27}$$

$$\mu_{eff} = \mu_0\left[1 + \frac{4(1-\nu_0)}{15\left[4(1-\nu_0)\left(\ln\frac{2a_3}{a_1}-1\right)-1\right]}\Omega''\right] \tag{28}$$

where $\Omega'' = \Omega\left(\frac{a_3}{a_1}\right)^2 = \frac{4\pi}{3}a_3^3 N$.

Similarly to the previously considered cases, these formulae can not be obtained from the solution (16), (17) by limit transition $\frac{\mu_*}{\mu_0} \to \infty$. Hence, in this case there is also a non-uniform limit transition.

2.4 Non-uniform limit transitions and areas of applicability of the asymptotics

In the previous sections asymptotic formulae for calculating effective elastic moduli of media with inclusions were obtained for some particular important cases. However, the question arises: what are the numerical values of the parameters $\frac{a_3}{a_1}$, and $\frac{\mu_*}{\mu_0}$, for which the considered asymptotics are valid with acceptable accuracy. In order to answer this question, the diagrams of isolines of relative errors of asymptotical solutions were plotted. The diagrams were presented in Fig. 1 for $\nu_0 = 0.45$, $\nu_* = 0.25$ for compressive and shear moduli, respectively (Ustinov 2003b). The abscises correspond to logarithms of ratio of elastic (compressive and shear) moduli of the inclusions and matrix; the ordinates correspond to logarithms of aspect ratio of the inclusions. The upper zones in the diagrams correspond to disk-like inclusions, the lower zones do to needle-like ones; the left zones correspond to soft inclusions, the right zones do to rigid ones. It is seen that the numerical results are in accordance with the performed theoretical estimations.

2.5 Combined asymptotics

In all considered cases there were two main parameters: ratio of moduli and aspect ratio. Therefore four cases of their combinations may be distinguished for which both of them became simultaneously large or small, three of them being aggravated with non-uniform limit transitions. Besides, as it is seen from the Fig. 1, that for the cases

$$\frac{\mu_*}{\mu_0} \sim \frac{a_3}{a_1} << 1 \tag{29}$$

corresponding to the soft disk-like inclusions, or

$$\frac{\mu_0}{\mu_*} \sim \frac{a_3}{a_1} << 1 \tag{30}$$

corresponding to the rigid disk-like inclusions, or

$$\frac{\mu_0}{\mu_*} \sim \left(\frac{a_1}{a_3}\right)^n << 1 \tag{31}$$

corresponding to the rigid needle-like inclusions, the obtained asymptotics fail to produce acceptable results. Consider these cases in more details.

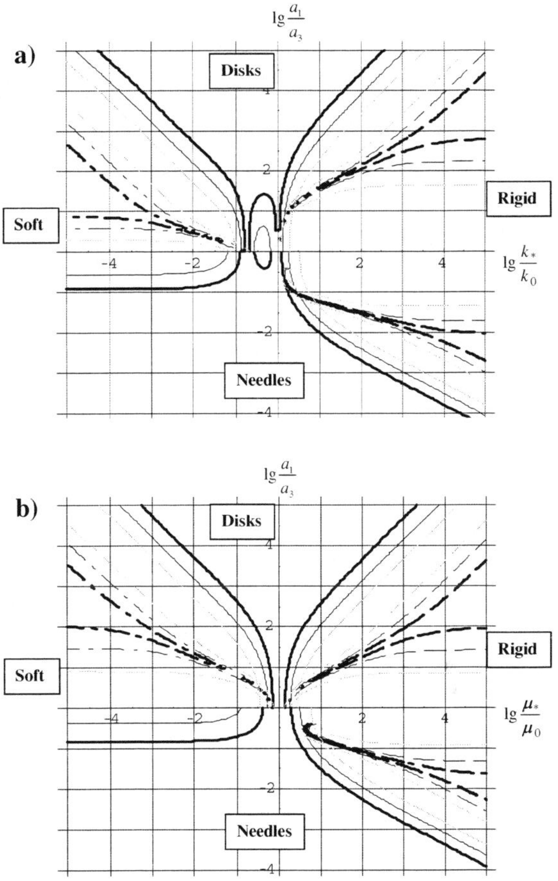

Fig. 1 Ranges of applicability of asimptotics for effective compressive (**a**) and shear (**b**) moduli of media with spheroid inclusions ($\nu_0 = 0.45$, $\nu_* = 0.25$). The abscises correspond to logarithms of ratio of elastic moduli of the inclusions and matrix; the ordinates correspond to logarithms of aspect ratio of the inclusions. Solid lines correspond to asymptotics for disk-like inclusions (13), (14) for upper half-plane and needle-like inclusions (16), (17) for lower half-plane; dashed lines do for extremely rigid disk-like inclusions (23), (24) for upper half-plane, and for extremely rigid needle-like inclusions (27), (28) for lower half-plane; dashed-dotted lines do for extremely soft (crack-like) disk-like inclusions (19), (20). Bold lines border the areas, where the relative errors do not exceed 3%, thin lines do for 10%, gray lines do for 30%

2.5.1 *Flat soft inclusions*

Let us introduce variable, d, interrelating the introduced small parameters so that

$$\frac{\mu_*}{\mu_0} = d\frac{a_3}{a_1} \tag{32}$$

Substituting (A1), (A2) into (9), and then result into (12), accounting for (32) and performing limit transition $a_3/a_1 = \rho \to 0$ one, after the back substitution of d by means of (32) in terms of the introduced small parameters, obtains the following expressions

$$k_{eff} = k_0 \left[1 - \frac{4k_0\left(1-\nu_0^2\right)(1+\nu_*)}{3\left[4k_*\left(1-\nu_0^2\right)(1-\nu^*)+\pi\frac{a_3}{a_1}k_0(1-2\nu_0)(1+\nu_*)\right]}\Omega \right] \tag{33}$$

$$\mu_{eff} = \mu_0 \left[1 - \frac{16(1-\nu_0)\left[4\mu_*(1-\nu_0)(4-5\nu_*)+\mu_0(5-\nu_0)(1-2\nu_*)\pi\frac{a_3}{a_1}\right]}{30\left[4\mu_*(1-\nu_0)+\mu_0(2-\nu_0)\pi\frac{a_3}{a_1}\right]\left[\mu_*(1-\nu_0)(1-\nu_*)+\mu_0(1-2\nu_0)\pi\frac{a_3}{a_1}\right]}\Omega \right] \tag{34}$$

This limit transition corresponds to simultaneous approach $\frac{\mu_*}{\mu_0} \to 0$ and $\frac{a_3}{a_1} \to 0$ so that $\frac{\mu_*}{\mu_0}\frac{a_1}{a_3} = d = const$. These formulae are valid for the range of parameters (29).

The formulae for cracks (19), (20) may be obtained from here by limit transition $\frac{\mu_*}{\mu_0} \to 0$. On the other hand, the result of limit transition $\frac{a_3}{a_1} \to 0$ of these formulae coincides with the result of limit transition $\frac{\mu_*}{\mu_0} \to 0$ of formulae for flat spheroids (13), (14). This means that the corresponding limit transitions are uniform.

2.5.2 *Flat rigid inclusions*

For this case variable, d, interrelating the main small parameters are introduced as follows

$$\frac{\mu_0}{\mu_*} = d\frac{a_3}{a_1} \tag{35}$$

Substituting (A1), (A2) into (9), and then result into (12), accounting for (35) and performing limit transition $a_3/a_1 = \rho \to 0$ one, after the back substitution of d by means of (35) in terms of the introduced small parameters, obtains the following expressions

$$k_{eff} = k_0 \left[1 + \frac{16k_*(1-\nu_0)(1-2\nu_0)(1-2\nu_*)}{8k_0(1-\nu_*)(1-\nu_0)(1-2\nu_0)+\pi k_*\frac{a_3}{a_1}(1-2\nu_*)(1+\nu_0)(3-4\nu_0)}\frac{\Omega}{3} \right] \tag{36}$$

$$\mu_{eff} = \mu_0 \left[1 + \frac{8\mu_*(1-\nu_0)\left[16(1-\nu_0)(7-5\nu_*)+\pi\frac{\mu_*}{\mu_0}\frac{a_3}{a_1}(43-56\nu_0)(1+\nu_*)\right]}{15\left[16(1-\nu_0)+\pi\frac{\mu_*}{\mu_0}\frac{a_3}{a_1}(7-8\nu_0)\right]\left[8(1-\nu_0)(1-\nu_*)+\pi\frac{\mu_*}{\mu_0}\frac{a_3}{a_1}(3-4\nu_0)(1+\nu_*)\right]}\Omega \right] \tag{37}$$

The formulae for extremely rigid flat inclusions (23), (24) may be obtained from here by limit transition $\frac{\mu_*}{\mu_0} \to \infty$. On the other hand, the result of limit transition $\frac{a_3}{a_1} \to 0$ of these formulae coincides with the result of limit transition $\frac{\mu_*}{\mu_0} \to \infty$ of formulae for flat spheroids (13), (14). This means that the corresponding limit transitions are uniform.

2.5.3 *Needle-like rigid inclusions*

As it is seen from Fig. 1, the range of applicability of the asimptotics for intermediate and small rigidity of needle-like inclusions is wider then in the other cases. This suggests another parameter, d, interrelating the main small parameters, namely

$$\frac{\mu_0}{\mu_*} = d\left(\frac{a_3}{a_1}\right)^n \tag{38}$$

where n is the power to be determined from the condition of the uniform limit transition of the formulae to be obtained into (27), (28) for $d = 0$, and into (16), (17) for $d \to \infty$.

Substituting (A1), (A2) into (9), and then result into (12), accounting for (38) and performing limit transition $\frac{a_3}{a_1} \to \infty$ one, after the back substitution of d by means of (38) in terms of the introduced small parameters, obtains the following expressions (for $n = 2$)

$$k_{eff} = k_0 \left[1 + \frac{2k_* (1-\nu_0)(1-2\nu_0)(1-2\nu_*)}{6k_0 (1-\nu_0)(1-2\nu_0) + 3k_* \left(\frac{a_1}{a_3}\right)^2 (1-2\nu_*)(1+\nu_0) \left[4(1-\nu_0)\left(\ln\frac{2a_3}{a_1} - 1\right) - 1\right]} \Omega \right] \quad (39)$$

$$\mu_{eff} = \mu_0 \left[1 + \frac{4\mu_* (1-\nu_0)(1+\nu_*)}{30\mu_0 (1-\nu_0) + 15\mu_* \left(\frac{a_1}{a_3}\right)^2 (1+\nu_*) \left[4(1-\nu_0)\left(\ln\frac{2a_3}{a_1} - 1\right) - 1\right]} \Omega \right] \quad (40)$$

The formulae for extremely rigid needle-like inclusions (27), (28) may be obtained from here by limit transition $\frac{\mu_*}{\mu_0} \to \infty$. On the other hand, the result of limit transition $\frac{a_3}{a_1} \to \infty$ of these formulae coincides with the result of limit transition $\frac{\mu_*}{\mu_0} \to \infty$ of formulae for needle-like inclusions (16), (17). This means that the corresponding limit transitions are uniform.

2.5.4 Soft and rigid inclusions

Besides the above considered cases one may outline cases of rigid (absolutely rigid) and soft (voids) inclusions. Such solutions may be obtained by limit transitions $\frac{\mu_*}{\mu_0} \to \infty$ and $\frac{\mu_*}{\mu_0} \to 0$, respectively, in the result of substitution of (A1), (A2) into (9), and then into (12). The final formulae are not presented here because of its awkwardness. Distinguishing these cases appears to have theoretical value only, because for the numerical calculation one can use successfully the exact solution (A1), (A2), (9), (12), and for analytical manipulation the simplifications yielding by them are not significant to make the results more transparent comparing to the general case.

2.6 Diagram of the areas of applicability of the combined asymptotics

Using the similar procedure as before, the diagram showing the range of applicability of the new asymptotics is plotted and given in Fig. 2. It represents the ranges for which the relative errors yielding by the obtained asymptotics for compressive modulus for $\nu_0 = 0.25$, $\nu_* = 0.25$ do not exceed 3%. The abscises correspond to logarithms of ratio of compressive modulus of the inclusions and matrix; the ordinates correspond to logarithms of aspect ratio of the inclusions.

3 Asymptotical solution for inclusions in the form of flat 3-axial ellipsoids: direct asymptotis

Foı the extremely flat inclusions with semi-axes a_1, a_2, a_3

$$a_1 \geq a_2 >> a_3 \quad (41)$$

the components of Eshelby's tensor are represented in Appendix 2 (A9). The expressions for volumetric and shear moduli for the case of isotropic distribution of the inclusion in space are obtained by substituting (A9) into (9), and then result into (12). In the following sections some particular cases are considered. Here, one again faces with the problem of non-uniform limit transitions.

3.1 Inclusions of intermediate rigidity

The result of substitution of (A9) into (9) and then into (12) on limit transition $a_3 \to 0$ coincides with (13), (14). This case corresponds obviously to the range of

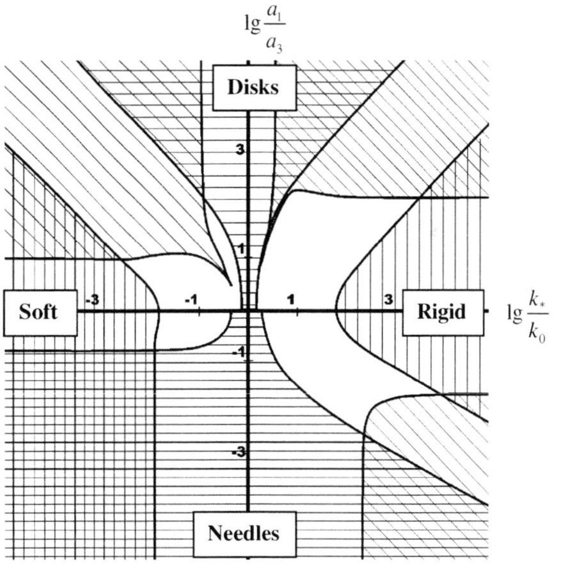

Fig. 2 Range of applicability of asimptotics for effective compressive modulus, errors do not exceed 3% ($\nu_0 = 0.25$, $\nu_* = 0.25$). The abscises correspond to logarithms of ratio of elastic moduli of the inclusions and matrix; the ordinates correspond to logarithms of aspect ratio of the inclusions. The ranges of applicability of asymptotics of disk-like and needle-like inclusions are shaded with horizontal lines; the ones of rigid and soft inclusions are shaded with vertical lines; the ones of combined asymptotics are shaded with inclined lines

parameters determined by (15). Hence, for the case in question, the particular details of the shape in plane of the inclusions are not essential, at least in the frame of considered asymptotics (Ustinov 2003b). Therefore, the applicability of the formulae by (Wu 1966) (13), (14) on the one hand is restricted to the range determined by (15), and on the other hand is extended, at least with the accuracy of $O\left(\frac{a_3}{a_1}\right)$ to the arbitrary flat inclusions. At the same time it has to be remembered, that these formulae do not describe the case of needle-like inclusions, $a_1 >> a_2 \approx a_3$ because the conditions (41) were used essentially while deriving the formulae.

3.2 Extremely soft inclusions

Let us generalize the case of extremely soft inclusions (22) for the elliptical in plane inclusions. The effective moduli for this case may be obtained by substituting (A9) into (9) and then into (12) followed by limit transition $\mu_* \to 0$ and then $a_3 \to 0$. The result coincides with the result obtained by Hashin (1988). Taking into account that the volume and the maximal area of the cross-section of the considered ellipsoid are $V_e = \frac{4}{3}\pi a_1 a_2 a_3$, $S = \pi a_1 a_2$, the expressions for the effective elastic moduli may be written as

$$k^{eff} = k^0 \left[1 - \frac{\Omega'}{3}\sqrt{\frac{a_2}{a_1}}\frac{2(1-\nu_0^2)}{(1-2\nu_0)E(t)}\right] \tag{42}$$

$$\mu^{eff} = \mu^0 - \mu^0 \frac{\Omega'}{5}\sqrt{\frac{a_2}{a_1}}(1-\nu_0) \times \left(\frac{4}{3E(t)} - \frac{a_1^2 - a_2^2}{(a_2^2 - a_1^2(1-\nu_0))E(t) - a_2^2\nu_0 K(t)} + \frac{a_1^2 - a_2^2}{(a_1^2 - a_2^2(1-\nu_0))E(t) - a_2^2\nu_0 K(t)}\right) \tag{43}$$

where $E(t)$, and $K(t)$ are elliptical integrals (A6) and

$$t = 1 - \frac{a_2^2}{a_1^2} \tag{44}$$

$$\Omega' \equiv \Omega\frac{\sqrt{a_1 a_2}}{a_3} = \frac{4}{3}\pi\,(a_1 a_2)^{3/2}\,N = \frac{4}{3\sqrt{\pi}}S^{3/2}N \tag{45}$$

This quantity is reduced to (21) for the circular in plane inclusions $a_1 = a_2$. Formulae (42), (43) reduce to (19), (20). Using the well-known formulae of the theory of elasticity the expression for the Young modulus may be written

$$E^{eff} = E^0 - E^0\Omega'\sqrt{\frac{a_2}{a_1}}\frac{2(1-\nu_0^2)}{15E(t)} \times \left\{\frac{3}{E(t)} + \frac{a_1^2 - a_2^2}{\left[(1-\nu_0)a_1^2 - a_2^2\right]E(t) + \nu_0 a_2^2 K(t)} + \frac{a_1^2 - a_2^2}{\left[a_1^2 - (1-\nu_0)a_2^2\right]E(t) - \nu_0 a_2^2 K(t)}\right\} \tag{46}$$

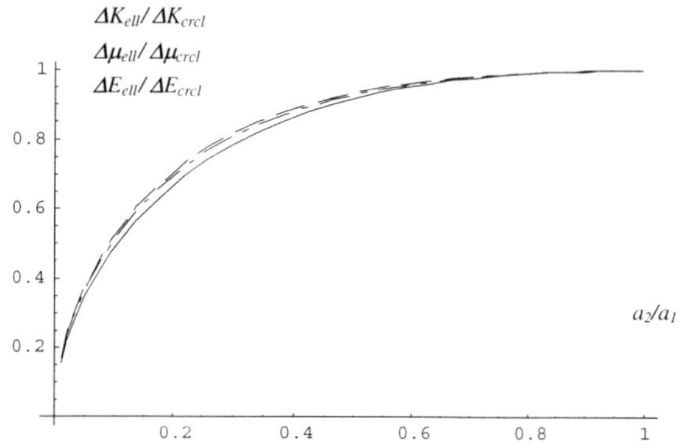

Fig. 3 Influence of the ratio of major and intermediate axes of the cracks on the relative change of effective elastic moduli. Solid line corresponds to volumetric modulus, dashed line does to shear modulus, dotted-dashed line does to Young modulus

Figure 3 represents the influence of the shape of the ellipsoids in plane on the changes in the effective moduli. The abscise axis corresponds to the ratio of major and intermediate axes of the ellipsoid (the minor axis corresponds to the thickness and does not affect the result). The ordinate corresponds to the proportional to concentration change in elastic moduli, divided by the corresponding changes due to the circular cracks. The lines for shear and Young moduli correspond $\nu = 1/2$. It is seen from (42), (43), (46) that the influence of the Poisson's ratio is weak. Thus for $\nu = 0$ all three dependences coincide (the absolute value varies in coefficients). It is seen that reduction of ratio a_2/a_1 from 1 to 0.6 leads only to 5% reduce of the moduli.

3.3 Extremely rigid inclusions

Similarly, for the extremely rigid elliptical in plane flat inclusion, substituting (A9) into (9) and then into (12), and then performing limit transition $\mu^* \to \infty$ and then $a_3 \to 0$, one obtains (Ustinov 2003b).

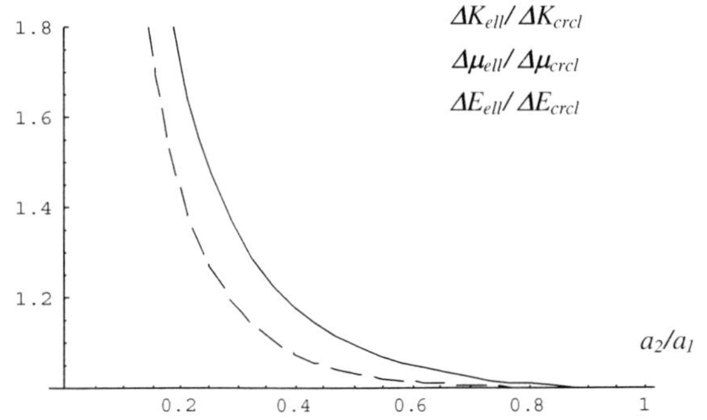

Fig. 4 Influence of the ratio of major and intermediate axes of the rigid inclusions to the relative change of effective elastic moduli. Solid line corresponds to volumetric modulus, dashed line does to shear modulus, dotted-dashed line does to Young modulus

Similar to what was done for spheroid inclusion, the combined asymptotics may be obtained for simultaneous approach of both relative rigidity and aspect ratio to their extreme values, however the corresponding formulae are not represented here due to their awkwardness.

4 Conclusions

The number of exact analytical asymptotic formulae, both known and new, for effective elastic characteristics of media with spheroid inclusions of another material was obtained by means of limit transitions of the general solutions for effective elastic characteristics of media with spheroid inclusions. The problem

$$k^{eff} = k^0 + \frac{k^0}{a_2^2}\frac{\Omega'}{3}\sqrt{\frac{a_2}{a_1}}\frac{(1-\nu_0)}{(1+\nu_0)(3-4\nu_0)} \times \frac{2\left[2a_1^2a_2^2(7-18\nu_0+8\nu_0^2)-(a_1^4+a_2^4)(3-10\nu_0+8\nu_0^2)\right]E(t)-8a_2^2(a_1^2+a_2^2)(1-2\nu_0)K(t)}{a_1^2(5-4\nu_0)E^2(t)-4(a_1^2+a_2^2)(1-\nu_0)E(t)K(t)+a_2^2(3-4\nu_0)K^2(t)} \tag{47}$$

$$\mu^{eff} = \mu^0 + \mu^0\Omega'\sqrt{\frac{a_2}{a_1}}(1-\nu_0)\left[\frac{1}{5}\frac{(a_1^2-a_2^2)^2}{\left[\left(a_1^4+a_2^4\right)(1-\nu_0)+2a_1^2a_2^2\nu_0\right]E(t)-a_2^2(a_1^2+a_2^2)K(t)} - \frac{4\left[\left(a_1^4+a_2^4\right)(3-4\nu_0)-8a_1^2a_2^2(1-\nu_0)\right]E(t)+4a_2^2(a_1^2+a_2^2)K(t)}{15a_2^2(3-4\nu_0)\left[a_1^2(5-4\nu_0)E^2(t)-4(a_1^2+a_2^2)(1-\nu_0)E(t)K(t)+a_2^2(3-4\nu_0)K^2(t)\right]}\right] \tag{48}$$

Similar to the case of flat voids (cracks) the effective moduli are determined by their shape and concentration and are independent of their thickness.

The Fig. 4 represents the influence of the ratio of major and intermediate axes of the rigid inclusions on the relative change of effective elastic moduli. The abscise axis corresponds to the ratio of major and intermediate axes of the ellipsoid (the minor axis corresponds to the thickness and does not affect the result). The ordinate corresponds to the proportional to concentration changes in elastic moduli, divided by the corresponding changes due to circular rigid inclusions. All the lines correspond to $\nu = 1/2$. It is seen that the influence of the rigid particle increases with the divergence of the inclusions shape from the circular one.

For circular inclusions formulae (47), (48) are reduced to (23), (24). However formulae for needle-like inclusions (27), (28) may not be obtained from (47), (48) by a limit transition, because formulae (47), (48) were obtained in the frame of the assumption (25) that contradict assumption (26) used in deriving (27), (28).

was considered in the frame of dilute concentration of the inclusions, for which the influence of each inclusion into effective moduli may be treated additively. The parameters determining the additive, proportional to the inclusions concentration, part of moduli are Poisson's ratios of matrix and inclusions, the ratio of elastic (compressive, or shear) moduli of the inclusions and matrix and the aspect ratio of the inclusions, the last two being of primary importance. In case of the large deviation of these parameters from unity, seven overlapping regions were distinguished, for which asymptotics are valid corresponding to approaching to zero (infinity) of various combinations of these two parameters. These regions correspond to

1. Rigid inclusions.
2. Thin rigid inclusions. This is a combined asymptotics, the effective elastic moduli are determined by (36), (37), the range of applicability does for (30).
3. Thin inclusions of moderate rigidity. The effective elastic moduli are determined by (13), (14), the range of applicability does for (15).

4. Soft thin inclusions. This is a combined asymptotics, the effective elastic moduli are determined by (33), (34), the range of applicability does for (29).
5. Soft inclusions.
6. Needle-like inclusions of low and moderate rigidity. The effective elastic moduli are determined by (16), (17), the range of applicability does for (18).
7. Rigid needle-like inclusions. This is a combined asymptotics, the effective elastic moduli are determined by (39), (40), the range of applicability does for (31).

Let us mention once more that the case of soft needle-like inclusion is absent since the ranges described above as regions 5 and 6 are overlapping (Fig. 2).

Each zone of intersection of the distinguished ranges corresponds to succeeding limit transition (to zero, or infinity) of the parameters $\frac{a_3}{a_1}$, and $\frac{\mu_*}{\mu_0}$.

The case of the inclusions in the form of flat three-axial ellipsoids is also considered. Depending on the values of the same parameters (the ratio of elastic moduli of the inclusions and matrix and the ratio of the highest and lowest axes of the inclusions) three particular cases were distinguished and their areas of applicability were estimated. These cases are: extremely soft (crack-like) inclusions, inclusions of intermediate rigidity, and extremely rigid inclusions.

Note in conclusion that the obtained results enable to calculate local stress-deformation field near the non-homogeneities and, hence, to search for the conditions of local strength violation according to the appropriate criteria. The examples will be considered separately.

The study was performed within the framework of the Program for Fundamental Research of the Department of Energetics, Machinery, Mechanics and Control Processes, No. 12 and project 05-02-17542-a of RFFI.

Appendix 1

The nonzero components of Eshelby's tensor in case of spheroid inclusions with semi-axes, $a_1, a_2 = a_1, a_3$, may be expressed as follows (e.g., Chow 1977):

$$S_{1111} = S_{2222} = \left(\frac{3P}{4} + R\right) I - \pi P \rho^2$$

$$S_{1122} = S_{2211} = \left(\frac{P}{4} - R\right) I - \frac{\pi}{3} P \rho^2$$

$$S_{1133} = S_{2233} = \frac{4\pi}{3} P \rho^2 - \left(P\rho^2 + R\right) I$$

$$S_{3311} = S_{3322} = 4\pi \left(\frac{P}{3} - R\right) - (P - 2R)\, I$$

$$S_{3333} = 4\pi \left(\frac{P}{3}(1 - 3\rho^2) + R\right) + 2\left(P\rho^2 - R\right) I$$

$$2S_{2323} = 2S_{1313} = 2S_{3232} = 2S_{3131} = 4\pi \left(\frac{P}{3}(1+\rho^2) + R\right) - \left(P(1 + \rho^2) + R\right) I$$

$$2S_{1212} = 2S_{2121} = S_{1111} - S_{1122} \quad \text{(A1)}$$

where

$$P = \frac{3}{8\pi} \frac{1}{(1 - \nu_0)\left(1 - \rho^2\right)}$$

$$R = \frac{1}{8\pi} \frac{1 - 2\nu_0}{(1 - \nu_0)}$$

$$I = \frac{2\pi\rho}{\left(1 - \rho^2\right)^{3/2}} \left[\arccos \rho - \rho\sqrt{1 - \rho^2}\right] \quad for\ \rho < 1$$

$$I = \frac{2\pi\rho}{\left(\rho^2 - 1\right)^{3/2}} \left[\rho\sqrt{\rho^2 - 1} - \cosh^{-1} \rho\right] \quad for\ \rho > 1$$

$$\rho = \frac{a_3}{a_1} \quad \text{(A2)}$$

Appendix 2

For the extremely flat inclusions with semi-axes $a_1 \geq a_2 >> a_3$ the components of Eshelby's tensor are presented in (Mura 1982) as follows

$$S_{1111} = \frac{3a_1^2 \mathrm{I}_{11}}{8\pi(1 - \nu)} + \frac{(1 - 2\nu)\mathrm{I}_1}{8\pi(1 - \nu)}$$

$$S_{1122} = \frac{a_2^2 \mathrm{I}_{12}}{8\pi(1 - \nu)} - \frac{(1 - 2\nu)\mathrm{I}_1}{8\pi(1 - \nu)}$$

$$S_{1133} = \frac{a_3^2 \mathrm{I}_{13}}{8\pi(1 - \nu)} - \frac{(1 - 2\nu)\mathrm{I}_1}{8\pi(1 - \nu)}$$

$$S_{1212} = \frac{(a_1^2 + a_2^2)\mathrm{I}_{12}}{16\pi(1 - \nu)} + \frac{(1 - 2\nu)(\mathrm{I}_1 + \mathrm{I}_2)}{16\pi(1 - \nu)} \quad \text{(A3)}$$

where $\nu = \nu_0$, and

$$\mathrm{I}_1 = \frac{4\pi a_2 a_3 \left(K(t) - E(t)\right)}{a_1^2 - a_2^2}$$

$$\mathrm{I}_2 = \frac{4\pi a_3 E(t)}{a_2} - \frac{4\pi a_2 a_3 \left(K(t) - E(t)\right)}{a_1^2 - a_2^2}$$

$$\mathrm{I}_3 = 4\pi - \frac{4\pi a_3 E(t)}{a_2} \quad \text{(A4)}$$

$$
\begin{aligned}
\mathrm{I}_{12} &= \frac{\mathrm{I}_2 - \mathrm{I}_1}{a_1^2 - a_2^2} \\
\mathrm{I}_{23} &= \frac{\mathrm{I}_3 - \mathrm{I}_2}{a_2^2 - a_3^2} \\
\mathrm{I}_{31} &= \frac{\mathrm{I}_1 - \mathrm{I}_3}{a_3^2 - a_1^2}
\end{aligned} \tag{A5}
$$

Here $E(t)$ and $K(t)$ denotes the elliptical integrals

$$
\begin{aligned}
E(t) &= \int_0^{\pi/2} (1 - t\sin^2\phi)^{1/2} d\phi \\
K(t) &= \int_0^{\pi/2} (1 - t\sin^2\phi)^{-1/2} d\phi
\end{aligned} \tag{A6}
$$

where

$$
t = \frac{a_1^2 - a_2^2}{a_1^2} \tag{A7}
$$

The values $\mathrm{I}_i, \mathrm{I}_{ij}$ are satisfied with the following condition (Mura 1982):

$$
\begin{aligned}
&\mathrm{I}_1 + \mathrm{I}_2 + \mathrm{I}_3 = 4\pi \\
&3\mathrm{I}_{11} + \mathrm{I}_{12} + \mathrm{I}_{13} = 4\pi/a_1^2 \\
&3a_1^2\mathrm{I}_{11} + a_2^2\mathrm{I}_{12} + a_3^2\mathrm{I}_{13} = 3\mathrm{I}_1 \\
&\mathrm{I}_{12} = (\mathrm{I}_2 - \mathrm{I}_1)/(a_1^2 - a_2^2)
\end{aligned} \tag{A8}
$$

All the other non-zero components of Eshelby's tensor may be obtained by cycle permutation.

It has to be remembered that the above expressions has asymptotic character and are valid for small a_3 only. Therefore the expressions, obtained by the pointed substitutions and permutations, should be expanded into series with respect to this small parameter keeping the leading term only. On performing these one obtains

$$
\begin{aligned}
S_{1111} &= a_2a_3\frac{2\left[a_1^2(2-\nu_0) - a_2^2(1-\nu_0)\right]K(t) - \left[a_1^2(5-2\nu_0) - a_2^2(3-2\nu_0)\right]E(t)}{2(a_1^2-a_2^2)^2(1-\nu_0)} \\
S_{1122} &= a_2a_3\frac{2\left[a_1^2(1-\nu_0) + \nu_0 a_2^2\right]E(t) - \left[a_1^2(1-2\nu_0) + a_2^2(1+2\nu_0)\right]K(t)}{2(a_1^2-a_2^2)^2(1-\nu_0)} \\
S_{1133} &= \frac{a_2a_3(1-2\nu_0)\left(E(t) - K(t)\right)}{2(a_1^2-a_2^2)(1-\nu_0)} \\
S_{2211} &= a_3\frac{2a_1^2\left[a_1^2\nu_0 + a_2^2(1-\nu_0)\right]E(t) - a_2^2\left[a_1^2(1+2\nu_0) + a_2^2(1-2\nu_0)\right]K(t)}{2a_2(a_1^2-a_2^2)^2(1-\nu_0)} \\
S_{2222} &= a_3\frac{a_1^2\left[a_1^2(3-2\nu_0) - a_2^2(5-2\nu_0)\right]E(t) - 2a_2^2\left[a_1^2(1-\nu_0) - a_2^2(2-\nu_0)\right]K(t)}{2a_2(a_1^2-a_2^2)^2(1-\nu_0)} \\
S_{2233} &= -\frac{a_3(1-2\nu_0)\left(a_1^2E(t) - a_2^2K(t)\right)}{2a_2(a_1^2-a_2^2)(1-\nu_0)} \\
S_{3311} &= \frac{\nu_0}{1-\nu_0} - a_3\frac{\left[2a_1^2\nu_0 - a_2^2(1+2\nu_0)\right]E(t) + a_2^2K(t)}{2a_2(a_1^2-a_2^2)(1-\nu_0)} \\
S_{3322} &= \frac{\nu_0}{1-\nu_0} - a_3\frac{\left[a_1^2(1+2\nu_0) - 2a_2^2\nu_0\right]E(t) - a_2^2K(t)}{2a_2(a_1^2-a_2^2)(1-\nu_0)} \\
S_{3333} &= 1 - a_3\frac{(1-2\nu_0)E(t)}{2a_2(1-\nu_0)} \\
S_{1212} &= S_{2121} = S_{1221} = S_{2112} = a_3\frac{\left[\left(a_1^4 + a_2^4\right)(1-\nu_0) + 2a_1^2a_2^2\nu_0\right]E(t) - a_2^2(a_1^2+a_2^2)K(t)}{2a_2(a_1^2-a_2^2)^2(1-\nu_0)} \\
S_{1313} &= S_{3131} = S_{1331} = S_{3113} = \frac{1}{2} + a_3\frac{\left[a_2^2 - a_1^2(1-\nu_0)\right]E(t) - a_2^2\nu_0K(t)}{2a_2(a_1^2-a_2^2)(1-\nu_0)} \\
S_{2323} &= S_{3232} = S_{2332} = S_{3223} = \frac{1}{2} + a_3\frac{\left[-a_1^2 + a_2^2(1-\nu_0)\right]E(t) + a_2^2\nu_0K(t)}{2a_2(a_1^2-a_2^2)(1-\nu_0)}
\end{aligned} \tag{A9}
$$

References

Chow TS (1977) Elastic moduli of filled polymers: The effects of particle shape. J Appl Phys 48:4072–4075

Eshelby JD (1957) The determination of the elastic field of an ellipsoidal inclusion, and related problems. Proc R Soc Lon A 241:376–396

Eshelby JD (1961) Elastic inclusions and inhomogeneities. In: Sneddon IN, Hill R (eds) Progress in Solid Mechanics, vol 2. North-Holland, Amsterdam, pp 89–140

Hashin Z (1988) The differential scheme and its application to cracked materials. J Mech Phys Solids 36:719–733

Kovalenko YF, Salganik RL (1977) Treshinovatye neodnorodnosti i ih vlianie na effektivnye mehanicheskie harakteristiki (Crack-like Inhomogeneities and their Influence on Effective Mechanical Characteristics). Izv AN SSSR Mech Tv Tela No 5:76–86

Krivoglas MA, Cherevko AS (1959) Ob uprugih modulah tverdoi smesi (On elastic moduli of solid mixture). Fizika metallov i metalovedenie 8(2):161–164

Kroner E (1958) Berechnung der Elastischen Konstanten des Vielkristalls aus den Konstanten der Einkristalls. Z Phys 151:504–518

Mura T (1982) Micromechanics of defects in solids. Martinus Nijhoff Publishers, The Hague-Boston

Odegard GM, Gates TS, Wise KE, Park C, Siochi EJ (2002) Constitutive modeling of nanotube-reinforced polymer composites. NASA/CR-2002-211760 ICASE Report No/2002-27

Roscoe RA (1973) Isotropic composites with elastic and viscoelastic phases: General bounds for the moduli and solutions for special geometries. Rheol Acta 12:404–411

Salganik RL (1973) Mehanika tel s bolshim chislom treshin (Mechanics of Bodies with Many Cracks). Izv AN SSSR Mech Tv Tela No 4:149–158

Salganik RL (1974) Protsessy perenosa v telah s bolshim chislom treshin (Transition Processes in Bodies with Many Cracks). Inzhenerno-fizicheskii zhurnal Tom 27(6):1069–1075

Tucker CLIII, Liang E (1999) Stiffness predictions for unidirectional short-fiber composites: review and evaluation. Composites Sci Technol 59:655–671

Ustinov KB (2003a) Ob opredelenii effektivnyh uprugih harakteristik dvuhfaznyh sred. Sluchai izolirovannyh neodnorodnostei v forme ellipsoidov vrashenia (On determining the effective characteristics of elastic two phases media. The case of spheroid inclusions). Uspehi Mehaniki 2:126–168

Ustinov KB (2003b) Ob opredelenii effektivnyh uprugih harakteristik dvuhfaznyh sred. Sluchai izolirovannyh uploshennyh neodnorodnostei i oblasti primenimosti asimptotik. (On determining effective elastic characteristics of two phases media Flat inclusions and ranges of applicability of asymptotics). Preprint IPM RAS N735, 24 pp

Walpole LJ (1969) On the overall elastic moduli of composite materials. J Mech Phys Solids 17:235–251

Walsh JB (1965) The effect of cracks on the compressibility of rocks. J Geophys Res 70(2):381–389

Wu TT (1966) The effect of inclusion shape on the elastic moduli of a two-phase material. Int J Solids Struct 2:1–8

Material forces in finite elasto-plasticity with continuously distributed dislocations

Sanda Cleja-Ţigoiu

Abstract In this paper we propose a thermodynamically consistent model for elasto-plastic material with structural inhomogeneities such as dislocations, subjected to large deformations, in isothermal processes. The plastic measure of deformation is represented by a pair of plastic distortion, and plastic connection with non-zero torsion (in order to have the non-zero Burgers vector). The developments are focused on the balance equations (for material forces and for physical force system), derived from an appropriate principle of the virtual power formulated within the constitutive framework of finite elasto-plasticity and on constitutive restrictions imposed by the free energy imbalance. The presence of the material forces (microforce and microstress momentum) is a key point in the exposure, and viscoplastic (generally rate dependent) constitutive representation are derived.

Keywords Finite deformation · Plastic distortion · Configuration with torsion · Plastic connection · Stress momentum · Material forces · Free energy imbalance · Principle of virtual power · Dislocations

AMS Subject Classifications (2000) 74C99 · 74A20

S. Cleja-Ţigoiu (✉)
Faculty of Mathematics and Computer Science, University of Bucharest, str. Academiei nr. 14, Bucharest 010014, Romania
e-mail: stigoiu@yahoo.com

1 Introduction

The aim of the paper is to propose a *thermodynamically consistent model* for material with structural inhomogeneities such as dislocations, subjected to large deformations, in isothermal processes. *Continuously distributed dislocations* are modeled in Teodosiu (1970), Steinmann (2002, 1997), Cleja-Tigoiu (2002a), Gurtin (2004), by the non-zero Burgers vector, which is related to the non-zero *curl* of the plastic deformation component $\mathbf{F}^p$. In our model the plastic measure of deformation is represented by a pair $(\mathbf{F}^p, \overset{(p)}{\mathbf{\Gamma}})$. $\mathbf{F}^p$ is an invertible second order tensor, called *plastic distortion*, and $\overset{(p)}{\mathbf{\Gamma}}$ is an affine *plastic connection* with non-zero torsion, represented by a third order tensor. On the background of differential geometry concepts we introduce the *multiplicative decomposition of the deformation gradient* into elastic and plastic distortions, and the *decomposition of the motion connection* in terms of the elastic and plastic connections. Unlike $\mathbf{F}$—the deformation gradient that is derived from the potential χ, which represents the motion function, the plastic distortion *cannot be derived from a certain potential.* The pair $(\mathbf{F}^p, \overset{(p)}{\mathbf{\Gamma}})$ defines an *anholonomic configuration* (see Acharya 2004; Bilby 1960; Schouten 1954), or a so called *configuration with torsion* $\mathcal{K}$.

In this paper a new appropriate *principle of the virtual power*, that can be applicable to the constitutive framework of finite elasto-plasticity, has been

Defect and Material Mechanics. C. Dascalu, G.A. Maugin & C. Stolz (eds.),
doi: 10.1007/978-1-4020-6929-1_8, © Springer Science+Business Media B.V. 2007

formulated. No Cosserat type kinematics is assumed in the model, and non-symmetric Cauchy stress $\mathbf{T}$ and the *stresses momentum* $\boldsymbol{\mu}$ (described by a third order tensorial field) are involved in the internal power $\mathcal{P}_{int}$. The pair $(\mathbf{T}, \boldsymbol{\mu})$ is power conjugated to the appropriate rate of the elastic deformation measure.

The principle of virtual power allows to formulate the *macroscopic balance laws* of macromomentum and angular momentum for ($\mathbf{T}$ and $\boldsymbol{\mu}$), which are similar to those derived in Fleck et al. (1994), and to derive the *micro-balance equations* (i.e. the microstructural material field equations) which are specific for the material force system $(\boldsymbol{\Upsilon}^p_{\mathcal{K}}, \boldsymbol{\mu}^p_{\mathcal{K}})$, like Gurtin (2003).

To complete the constitutive equations within the constitutive framework of finite elasto-plasticity, the *second law for isothermal processes* is formalized similar to Gurtin (2003, 2004). The second law leads to *local free energy inequality* or to *free energy imbalance* $\dot{\psi} - \mathcal{P}_{int} \leq 0$, where ψ is the free energy density. We determine the thermodynamic restrictions imposed by the requirement to have the free energy imbalance satisfied in *any virtual process*, for a given deformation state.

The *admissible virtual process* is defined based on the *kinematical relationships*, being consistent with them. The time derivatives of the appropriate distortions and connections put into evidence the presence of the velocity gradient $\mathbf{L}$ and of its gradient $\nabla_{\chi}\mathbf{L}$ in the deformed configuration, as well as of the rate of plastic distortion $\mathbf{L}^p$ and of its gradient $\nabla_{\mathcal{K}}\mathbf{L}^p$, written in the configuration with torsion, respectively.

The *constitutive hypotheses* concerning *microforces and micromomentum* (called the material forces) are motivated by *the dissipation inequality*. The material forces are represented by a non-dissipative part *energetic microforces*, derived from the free energy and a *dissipative part*, defined in a such way that the dissipation be positive. We emphasized the role played in the theory by the micro-stress $\boldsymbol{\Upsilon}^p_{\mathcal{K}}$, which is force conjugated to the rate of plastic distortion $\mathbf{L}^p$, and by the micro-stress momentum $\boldsymbol{\mu}^p_{\mathcal{K}}$, which generates work in conjunction with $\nabla_{\mathcal{K}}\mathbf{L}^p$.

We briefly recall different issues, involving continuously distributed dislocations results, which are closely related to the model proposed here.

A continuum theory for material with continuously distributed dislocations has been developed by Kondo and Yuki (1958), Bilby (1960), Kröner in (1963, 1992), Kröner and Lagoudas in (1992) (for elastic models), and mathematically founded by Noll (1967) and Wang (1967) (within the constitutive framework of simple materials), using the differential geometry concepts. The decomposition theorem of the connection with metric property into a Levy-Civita connection and contortion was studied in Schouten (1954), Kondo and Yuki (1958), and applied in finite elasto-plasticity by Steinmann (1994), Cleja-Tigoiu (2002a), see also Beju et al. (1983), Le and Stumpf (1996c)

A Cosserat theory for elasto-(visco)plastic single crystals, at finite deformations and based on the crystallographic slip mechanism of plastic deformation, was elaborated in Naghdi and Srinivasa (1994), Le and Stumpf (1996a), Steinmann (1994). In Forest et al. (1997) the natural Cosserat strains are considered for the development of the constitutive equations and evolution laws are proposed also for lattice torsion-curvature (second order) tensor.

Elasto-plastic model with dislocations was developed in Le and Stumpf (1996b) based on an appropriate principle of virtual work, and a thermodynamically consistent analysis of the anisotropic damage evolution was been performed by Stumpf and Hackle in (2003). Gurtin (2000) developed a gradient theory of single crystal plasticity that accounts for geometrically necessary dislocations.

Within the framework of finite elasto-plasticity the evolution equations to describing the irreversible behavior plays a fundamental role. Certain compatibility conditions concerning the evolution equation for the measure of continuously distributed dislocation could arrive, see for instance Acharia (2004), Cleja-Tigoiu (2002a), Gupta et al (2006), Cleja-Tigoiu et al. (in press). The compatibility conditions are viewed in Cleja-Tigoiu et al. (in press), say for the given plastic metric tensor, as partial differential equations for the torsion. We also mention here a second order theory, which allows for growth diffusion, developed in Epstein and Maugin (2000), especially for the evolution equation of the second order gradients that has been considered.

The boundary conditions (appropriate to plasticity that account for the dislocation) are discussed for instance by Gurtin and Needlemen (2005) and Gupta et al. (2006).

Maugin in (1999), combining the energy and momentum balance derives a so-called *pseudo-momentum* equation to derive material forces. Well-known examples of the material forces are driving forces on defects and the J-Integral in fracture mechanics.

We mention the proposed model in our paper Cleja-Ţigoiu (2002a), where the results from Cleja-Ţigoiu (1990), Cleja-Ţigoiu and Soós (1990), and Cleja-Ţigoiu (2001) have been extended to elasto-plastic materials with continuous distributed dislocations, based on the hypotheses listed below in an heuristic manner: (i) The crystalline body is not *homogeneous* and it has no relaxed (natural) global configuration. (ii) The local relaxed state is characterized by non-Euclidean and non-Riemannian metric space. (iii) Dynamical balance equations involve *non-symmetric Cauchy stress tensor* and *couple stresses*. (iv) The crystalline body behaves as an *elastic material element*, which means that the stress and the stress momentum are functions of the elastic distortion and elastic connection. (v) The *irreversible behavior* of the material is described by the evolution equations (of the rate independent type) for plastic distortion as well as for the gradient of plastic distortion.

In the present paper the developments are focused on the balance equations (micro and macro), derived from an appropriate principle of the virtual power formulated within the constitutive framework of finite elasto-plasticity and on the restriction on the constitutive equations imposed by the imbalanced free energy (i.e. *second law for isothermal processes*). The presence of the *material forces* is a key point in the exposure, and viscoplastic (generally rate dependent) constitutive representation are derived.

List of notations. Further the following notations will be used:

$\mathcal{E}$—the three dimensional Euclidean space, with the vector space of translations $\mathcal{V}$;
Lin—the set of the linear mappings from $\mathcal{V}$ to $\mathcal{V}$, i.e the set of second order tensor, $Skew \subset Lin$ the set of all skew-symmetric second order tensors;
$\mathbf{u} \times \mathbf{v}$ is the cross product, $\mathbf{u} \otimes \mathbf{v}$ and $\mathbf{u} \cdot \mathbf{v}$ denote the tensorial product and the scalar product of the vectors $\mathbf{u}, \mathbf{v} \in \mathcal{V}$.
$\mathbf{A}^s$ and $\mathbf{A}^a$ are the symmetrical and skew-symmetrical parts of the tensor $\mathbf{A}$, here $\mathbf{A}^T$ denotes the transpose of $\mathbf{A}$;
$\partial_{\mathbf{A}}\phi(x)$ denotes the partial differential of the function ϕ with respect to the field $\mathbf{A}$.
Curl of a second order tensor field $\mathbf{A}$ is defined by the second order tensor field

$$(\mathrm{curl}\mathbf{A})(\mathbf{u} \times \mathbf{v}) := (\nabla\mathbf{A}(\mathbf{u}))\mathbf{v} - (\nabla\mathbf{A}(\mathbf{v}))\mathbf{u} \quad \forall \mathbf{u}, \mathbf{v} \in \mathcal{V}. \tag{1}$$

The component representation of the *curl* is given in a Cartesian basis by $(\mathrm{curl}\,\mathbf{A})_{pi} = \epsilon_{ijk}\dfrac{\partial A_{pk}}{\partial x^j}$, while the third order tensor field $\nabla\mathbf{A}$ is characterized by $\nabla\mathbf{A} = \dfrac{\partial A_{ij}}{\partial x^k}\mathbf{i}^i \otimes \mathbf{i}^j \otimes \mathbf{i}^k$. Thus we have the formulae $\nabla_\chi \mathbf{L} \equiv \dfrac{\partial}{\partial x^k}\left(\dfrac{\partial v_i}{\partial x^j}\right)\mathbf{i}^i \otimes \mathbf{i}^j \otimes \mathbf{i}^k$.
$Lin(\mathcal{V}, Lin) = \{\mathbf{N} : \mathcal{V} \longrightarrow Lin \;\; \text{linear}\}$– defines the third order tensors and it is given by $\mathbf{N} = N_{ijk}\mathbf{i}^i \otimes \mathbf{i}^j \otimes \mathbf{i}^k$. The scalar product of two third order tensors is given by $\mathbf{N} \cdot \mathbf{M} = N_{ijk}M_{ijk}$.
Three configurations will be considered: k be a fixed reference configuration of the body $\mathcal{B}$, $k(\mathcal{B}) \subset \mathcal{E}$ with the vector space $\mathcal{V}_k$, $\chi(\cdot, t)$ the deformed configuration at time t, for any motion of the body $\mathcal{B}$, $\chi : \mathcal{B} \times R \longrightarrow \mathcal{E}$, there exists $\mathcal{K}$, time dependent (non-local) configuration with torsion, defined by the pair $(\mathbf{F}^p, \overset{(p)}{\mathbf{\Gamma}}_k)$, $\mathbf{F}^p$—plastic distortion and $\overset{(p)}{\mathbf{\Gamma}}_k$—plastic connection.
$\mathbf{F}$– the deformation gradient is defined by

$$\mathbf{F}(\mathbf{Z}, t) = \nabla(\chi(\cdot, t) \circ k^{-1})(\mathbf{Z}), \quad \forall \;\; \mathbf{Z} \in k(\mathcal{B}). \tag{2}$$

We recall here the *integrability theorem*: Let $\mathbf{F}$ be a function defined on a arcwise connected domain $\mathcal{U}$. $\mathbf{F}$ is a gradient of χ, i.e. $\mathbf{F}(\mathbf{x}) = \nabla\chi(\mathbf{x})$ for $\mathbf{x} \in \mathcal{U}$ if and only if

$$(\nabla\mathbf{F}(\mathbf{x})\mathbf{u})\mathbf{v} - (\nabla\mathbf{F}(\mathbf{x})\mathbf{v})\mathbf{u} = 0 \iff (\mathrm{curl}\mathbf{F}(\mathbf{x}))(\mathbf{u} \times \mathbf{v}) = 0, \quad \forall \;\; \mathbf{x} \in \mathcal{U}, \quad \forall \mathbf{u}, \mathbf{v} \in \mathcal{V}_k. \tag{3}$$

Dislocation means (1) non-zero curl or (2) non-zero plastic torsion

(1) $\mathrm{curl}(\mathbf{F}^P) \neq 0$ or
$\exists \; \mathbf{u}, \mathbf{v}$ such that $(\nabla\mathbf{F}^p(\mathbf{u}))\mathbf{v} - (\nabla\mathbf{F}^p(\mathbf{v}))\mathbf{u} \neq 0$

(2) $\exists \; \mathbf{u}, \mathbf{v}$ such that $(\mathbf{S}^p\mathbf{u})\mathbf{v} = (\mathbf{\Gamma}^p\mathbf{u})\mathbf{v} - (\mathbf{\Gamma}^p\mathbf{v})\mathbf{u} \neq 0$.

Let us introduce the following notation for a third order field generated by a connection, say $\mathbf{\Gamma}$, and by second order tensors, for instance $\mathbf{F}_1$, $\mathbf{F}_2$,

$$(\mathbf{\Gamma}[\mathbf{F}_1, \mathbf{F}_2]\mathbf{u})\mathbf{v} = (\mathbf{\Gamma}(\mathbf{F}_1\mathbf{u}))\,\mathbf{F}_2\mathbf{v}, \quad \forall \mathbf{u}, \mathbf{v} \in \mathcal{V}_k. \tag{4}$$

2 Second order elastic and plastic pairs of deformations

Let us consider the body $\mathcal{B}$, k the initial configuration of the body (which will be not explicitly mentioned further) and $\chi(\cdot, t)$ the actual configuration attached to the function χ which defines the motion of the body.

Ax.1 (the existence of the second order pair of plastic deformations) For any motion χ of the body $\mathcal{B}$, at any material particle **X** and at any time t, there exists a pair $(\mathbf{F}^p, \overset{(p)}{\mathbf{\Gamma}})$ with $\mathbf{F}^p$ an invertible second order tensor, called plastic distortion and $\overset{(p)}{\mathbf{\Gamma}}$ a third order field, which represent an affine connection, called plastic connection. The pair $(\mathbf{F}^p, \overset{(p)}{\mathbf{\Gamma}})$ is invariant with respect to a change of frame in the actual configuration.

Definition The connection of the motion χ with respect to the reference configuration can be introduced by

$$\mathbf{\Gamma u} = \mathbf{F}^{-1}\nabla \mathbf{F u}, \quad \forall\, \mathbf{u} \in \mathcal{V}. \tag{5}$$

Ax.2 For any pair $(\mathbf{F}, \mathbf{\Gamma})$ of the deformation gradient and motion connection, there exists a *second order pair of elastic deformation*, where the *elastic distortion* is defined by

$$\mathbf{F}^e = \mathbf{F}(\mathbf{F}^p)^{-1}, \tag{6}$$

and the *elastic connection* is introduced in terms of the motion and plastic connections, both of them being related to the initial configuration, through the formula

$$\overset{(e)}{\mathbf{\Gamma}}_{\mathcal{K}}\, \tilde{\mathbf{u}} = \mathbf{F}^p((\mathbf{\Gamma} - \overset{(p)}{\mathbf{\Gamma}})(\mathbf{F}^p)^{-1}\tilde{\mathbf{u}})(\mathbf{F}^p)^{-1}, \quad \forall\ \tilde{\mathbf{u}} \in \mathcal{V}_{\mathcal{K}}. \tag{7}$$

Here $\mathcal{V}_{\mathcal{K}} := \mathbf{F}^p(\mathcal{V}_k)$.

The pull back to the reference configuration k leads to the following relationship between connections:

$$\overset{(e)}{\mathbf{\Gamma}}_{back} := (\mathbf{F}^p)^{-1}\, \overset{(e)}{\mathbf{\Gamma}}_{\mathcal{K}}\, [\mathbf{F}^p, \mathbf{F}^p] = \mathbf{\Gamma} - \overset{(p)}{\mathbf{\Gamma}}, \tag{8}$$

derived from (7).

Definition The differential of any smooth tensor field $\bar{\mathbf{F}}$, defined on $k(\mathcal{B})$, with respect to the configuration with torsion $\mathcal{K}$ is given by

$$(\nabla_{\mathcal{K}}\bar{\mathbf{F}})\tilde{\mathbf{u}} = (\nabla\bar{\mathbf{F}})(\mathbf{F}^p)^{-1}\tilde{\mathbf{u}}, \quad \forall\ \tilde{\mathbf{u}} \in \mathcal{V}_{\mathcal{K}} \tag{9}$$

Proposition 1

1. *The* multiplicative decomposition *of the deformation gradient* **F** *into the elastic and plastic distortions* $\mathbf{F}^e$, $\mathbf{F}^p$ *follows*

$$\mathbf{F} = \mathbf{F}^e\mathbf{F}^p. \tag{10}$$

2. The connections $\mathbf{\Gamma}$, $\overset{(e)}{\mathbf{\Gamma}}_{\mathcal{K}}$ and $\overset{(p)}{\mathbf{\Gamma}}$ *are related by*

$$\mathbf{\Gamma u} = (\mathbf{F}^p)^{-1}(\overset{(e)}{\mathbf{\Gamma}}_{\mathcal{K}}\, (\mathbf{F}^p\mathbf{u}))\mathbf{F}^p + \overset{(p)}{\mathbf{\Gamma}}\, \mathbf{u}, \quad \text{with}$$
$$\mathbf{\Gamma} = (\mathbf{F})^{-1}\nabla\mathbf{F} \tag{11}$$

$\forall\, \mathbf{u} \in \mathcal{V}$, *or using* (4)

$$\mathbf{\Gamma} = (\mathbf{F}^p)^{-1}(\overset{(e)}{\mathbf{\Gamma}}_{\mathcal{K}}\, [\mathbf{F}^p, \mathbf{F}^p]) + \overset{(p)}{\mathbf{\Gamma}}\, . \tag{12}$$

We put into evidence the **rules of calculus** for the expressions of the differential of a smooth tensor field $\bar{\mathbf{F}}$, in various configurations, k, χ and $\mathcal{K}$, when we pass from one configuration to the other one:

$$\begin{aligned}
&(\nabla\bar{\mathbf{F}})\mathbf{u} = (\nabla_\chi\bar{\mathbf{F}})\nabla(\chi \circ k^{-1})\mathbf{u} \equiv (\nabla_\chi\bar{\mathbf{F}})\mathbf{F}\mathbf{u},\\
&\quad(\nabla\bar{\mathbf{F}})\mathbf{u} = (\nabla_{\mathcal{K}}\bar{\mathbf{F}})\mathbf{F}^p\mathbf{u},\\
&(\nabla_\chi\bar{\mathbf{F}})\bar{\mathbf{u}} = (\nabla\bar{\mathbf{F}})\mathbf{F}^{-1}\bar{\mathbf{u}} = (\nabla_{\mathcal{K}}\bar{\mathbf{F}})\mathbf{F}^p\mathbf{F}^{-1}\bar{\mathbf{u}}\\
&\quad\equiv (\nabla_{\mathcal{K}}\bar{\mathbf{F}})(\mathbf{F}^e)^{-1}\bar{\mathbf{u}}, \quad \forall \mathbf{u} \in \mathcal{V}, \bar{\mathbf{u}} \in \mathcal{V}_\chi.
\end{aligned} \tag{13}$$

Ax.3 The plastic connection $\overset{(p)}{\mathbf{\Gamma}}$ has the non-zero *Cartan torsion*, defined by the skew-symmetric part of the connection, as it follows

$$(\mathbf{S}^p\mathbf{v})\mathbf{u} \equiv (\overset{(p)}{\mathbf{\Gamma}}\ \mathbf{v})\mathbf{u} - (\overset{(p)}{\mathbf{\Gamma}}\ \mathbf{u})\mathbf{v}. \tag{14}$$

Ax.4 The plastic distortion and the plastic connection are compatible each other, in the sense that the *Frobenius integrability condition* is satisfied

$$\overset{(p)}{\mathbf{\Gamma}} = (\mathbf{F}^p)^{-1}\nabla\, \mathbf{F}^p. \tag{15}$$

2.1 Relationships between the connections attached to the plastic and elastic distortions

When we pass from one configuration, say from the initial configuration to another one, say $\mathcal{K}$, as a consequence of rule of calculus formulae (13), from (15) the relationship between the plastic connections follows

$$\overset{(p)}{\mathbf{\Gamma}} = -(\mathbf{F}^p)^{-1}\, \overset{(p)}{\mathbf{\Gamma}}_{\mathcal{K}}\ [\mathbf{F}^p, \mathbf{F}^p], \tag{16}$$

where the plastic connection with respect to $\mathcal{K}$ is introduced by

$$\overset{(p)}{\mathbf{\Gamma}}_{\mathcal{K}} = \mathbf{F}^p(\nabla_{\mathcal{K}}(\mathbf{F}^p)^{-1}). \tag{17}$$

Remark Consequently we defined two pairs $(\mathbf{F}^p, \overset{(p)}{\mathbf{\Gamma}})$ and $((\mathbf{F}^p)^{-1}, \overset{(p)}{\mathbf{\Gamma}}_{\mathcal{K}})$ of the appropriate plastic distortions, $\mathbf{F}^p$ and $(\mathbf{F}^p)^{-1}$, and the connections $\overset{(p)}{\mathbf{\Gamma}}$ and $\overset{(p)}{\mathbf{\Gamma}}_{\mathcal{K}}$, which are compatible.

The appropriate torsion (14), for the plastic connection defined in (15), becomes

$$(\mathbf{S}^p\mathbf{v})\mathbf{u} = (\mathbf{F}^p)^{-1}[((\nabla\,\mathbf{F}^p)\mathbf{v})\mathbf{u} - ((\nabla\,\mathbf{F}^p)\mathbf{u})\mathbf{v}]. \quad (18)$$

Let us introduce the *elastic connections* associated with respect to the specified configurations

$$\overset{(e)}{\boldsymbol{\Gamma}}_{\mathcal{K}} = (\mathbf{F}^e)^{-1}\nabla_{\mathcal{K}}\,\mathbf{F}^e, \quad \overset{(e)}{\boldsymbol{\Gamma}}_{\chi} = \mathbf{F}^e\nabla_{\chi}\,(\mathbf{F}^e)^{-1}. \quad (19)$$

From the above definitions together with (9) the following relationship can be put into evidence

$$\overset{(e)}{\boldsymbol{\Gamma}}_{\chi} = -\mathbf{F}^e\,\overset{(e)}{\boldsymbol{\Gamma}}_{\mathcal{K}}\,[(\mathbf{F}^e)^{-1}, (\mathbf{F}^e)^{-1}]. \quad (20)$$

On the other hand for the elastic connection, say $\overset{(e)}{\boldsymbol{\Gamma}}_{\mathcal{K}}$, the *torsion* can be similarly defined as the skew-symmetric part of the connection

$$(\mathbf{S}^e_{\mathcal{K}}\mathbf{v})\mathbf{u} \equiv (\overset{(e)}{\boldsymbol{\Gamma}}_{\mathcal{K}}\,\mathbf{v})\mathbf{u} - (\overset{(e)}{\boldsymbol{\Gamma}}_{\mathcal{K}}\,\mathbf{u})\mathbf{v}. \quad (21)$$

As a consequence of the multiplicative decomposition (10) and of the adopted definition for the connections, the relationships between appropriate elastic and plastic connections follows.

Due to the symmetry of the motion connection introduced in (5), provided by the fact that at any time t $\mathbf{F}$ is the gradient of an appropriate application (2),

$$(\boldsymbol{\Gamma}\mathbf{v})\mathbf{u} - (\boldsymbol{\Gamma}\mathbf{u})\mathbf{v} = 0. \quad (22)$$

As a consequence of (11) the torsion of the plastic connection with respect to the reference configuration and the torsion of the elastic connection, with respect to the so called configuration with torsion $\mathcal{K}$, are related one to another by

$$\overset{(p)}{\mathbf{S}} = -(\mathbf{F}^p)^{-1}(\overset{(e)}{\mathbf{S}}_{\mathcal{K}}\,[\mathbf{F}^p, \mathbf{F}^p]). \quad (23)$$

On the other hand, from (16) it follows

$$\overset{(p)}{\mathbf{S}} = -(\mathbf{F}^p)^{-1}(\overset{(p)}{\mathbf{S}}_{\mathcal{K}}\,[\mathbf{F}^p, \mathbf{F}^p]). \quad (24)$$

Thus the equality between the elastic torsion and plastic torsion is derived

$$\overset{(e)}{\mathbf{S}}_{\mathcal{K}} = \overset{(p)}{\mathbf{S}}_{\mathcal{K}} \equiv \mathbf{S}_{\mathcal{K}}. \quad (25)$$

Proposition 2 *As a consequence of the axioms* **Ax.3**, **Ax.4**, *of the definitions* (14), (7), *as well as of the relationship* (22), *it follows* $\mathbf{F}^p$, $\mathbf{F}^e$ *are not the gradients of certain mappings.*

2.2 Time-derivatives of the elastic connection with respect to relaxed configuration

When we take the time derivative of the motion connection (5), the rate of the total connection is expressed in term of the second order velocity gradient

$$\frac{d}{dt}(\boldsymbol{\Gamma}) = \mathbf{F}^{-1}(\nabla_{\chi}\mathbf{L})[\mathbf{F}, \mathbf{F}] \quad (26)$$

where the velocity gradient in the actual configuration is characterized by

$$\mathbf{L} := \nabla_{\chi}\mathbf{v}(\mathbf{x}, t), \quad \mathbf{L} = \dot{\mathbf{F}}(\mathbf{F})^{-1}, \quad (27)$$

where $\mathbf{v}$ is the vector field in the actual configuration.

As a consequence of the multiplicative decomposition (10) the kinematics relationships

$$\begin{aligned} &\mathbf{L} = \dot{\mathbf{F}}^e(\mathbf{F}^e)^{-1} + \mathbf{F}^e\mathbf{L}^p(\mathbf{F}^e)^{-1}, \\ &\mathbf{L}^e = \dot{\mathbf{F}}^e(\mathbf{F}^e)^{-1}, \quad \mathbf{L}^p = \dot{\mathbf{F}}^p(\mathbf{F}^p)^{-1} \end{aligned} \quad (28)$$

follow. $\mathbf{L}^e$, $\mathbf{L}^p$ are the rates of elastic and plastic distortions in the deformed configuration and the configuration with torsion, respectively.

The rate of the plastic connection, relative to the reference configuration, can be derived from (15) under the form similar to (26)

$$\frac{d}{dt}(\overset{(p)}{\boldsymbol{\Gamma}}) = (\mathbf{F}^p)^{-1}(\nabla_{\mathcal{K}}\mathbf{L}^p)[\mathbf{F}^p, \mathbf{F}^p]. \quad (29)$$

Taking into account the relationship between the plastic connection when we pass from one configuration to the other one, the time derivative applied to (16) leads to

$$\begin{aligned} \frac{d}{dt}(\overset{(p)}{\boldsymbol{\Gamma}}_{\mathcal{K}})\tilde{\mathbf{u}} &= \mathbf{L}^p(\overset{(p)}{\boldsymbol{\Gamma}}_{\mathcal{K}})\tilde{\mathbf{u}} \\ &\quad -\mathbf{F}^p\frac{d}{dt}(\overset{(p)}{\boldsymbol{\Gamma}})((\mathbf{F}^p)^{-1}\tilde{\mathbf{u}})(\mathbf{F}^p)^{-1} \\ &\quad -\overset{(p)}{\boldsymbol{\Gamma}}_{\mathcal{K}}\,(\mathbf{L}^p\tilde{\mathbf{u}}) - (\overset{(p)}{\boldsymbol{\Gamma}}_{\mathcal{K}}\,\tilde{\mathbf{u}})\mathbf{L}^p. \end{aligned} \quad (30)$$

When we replace (29) into the above relation the rate of plastic connection is calculated through

$$\begin{aligned} \frac{d}{dt}(\overset{(p)}{\boldsymbol{\Gamma}}_{\mathcal{K}})\tilde{\mathbf{u}} &= \mathbf{L}^p(\overset{(p)}{\boldsymbol{\Gamma}}_{\mathcal{K}})\tilde{\mathbf{u}} - (\nabla_{\mathcal{K}}\mathbf{L}^p)\tilde{\mathbf{u}} - \overset{(p)}{\boldsymbol{\Gamma}}_{\mathcal{K}}\,(\mathbf{L}^p\tilde{\mathbf{u}}) \\ &\quad -(\overset{(p)}{\boldsymbol{\Gamma}}_{\mathcal{K}}\,\tilde{\mathbf{u}})\mathbf{L}^p. \end{aligned} \quad (31)$$

When we take the time derivative in (12) we get

$$\frac{d}{dt}(\overset{(e)}{\boldsymbol{\Gamma}}_{\mathcal{K}})(\tilde{\mathbf{u}}) = \mathbf{F}^p\left(\frac{d}{dt}(\boldsymbol{\Gamma}-\overset{(p)}{\boldsymbol{\Gamma}})\right)((\mathbf{F}^p)^{-1}\tilde{\mathbf{u}})(\mathbf{F}^p)^{-1}$$
$$+\mathbf{L}^p\mathbf{F}^p((\boldsymbol{\Gamma}-\overset{(p)}{\boldsymbol{\Gamma}})(\mathbf{F}^p)^{-1}\tilde{\mathbf{u}})(\mathbf{F}^p)^{-1}$$
$$-\mathbf{F}^p((\boldsymbol{\Gamma}-\overset{(p)}{\boldsymbol{\Gamma}})(\mathbf{F}^p)^{-1}\mathbf{L}^p\tilde{\mathbf{u}})(\mathbf{F}^p)^{-1}$$
$$-\mathbf{F}^p((\boldsymbol{\Gamma}-\overset{(p)}{\boldsymbol{\Gamma}})(\mathbf{F}^p)^{-1}\tilde{\mathbf{u}})((\mathbf{F}^p)^{-1}\mathbf{L}^p)$$
$$\forall\ \tilde{\mathbf{u}} \in \mathcal{V}_{\mathcal{K}}. \tag{32}$$

Using again (12) and the multiplicative decomposition (10) in relationship (32), the time derivative of the appropriate connections are related by

$$\frac{d}{dt}(\overset{(e)}{\boldsymbol{\Gamma}}_{\mathcal{K}})(\tilde{\mathbf{u}})-\mathbf{L}^p(\overset{(e)}{\boldsymbol{\Gamma}}_{\mathcal{K}}\,\tilde{\mathbf{u}})+\overset{(e)}{\boldsymbol{\Gamma}}_{\mathcal{K}}\,(\mathbf{L}^p\tilde{\mathbf{u}})+(\overset{(e)}{\boldsymbol{\Gamma}}_{\mathcal{K}}\,\tilde{\mathbf{u}})\mathbf{L}^p$$
$$\equiv \mathbf{F}^p\left(\frac{d}{dt}(\boldsymbol{\Gamma}-\overset{(p)}{\boldsymbol{\Gamma}})\right)((\mathbf{F}^p)^{-1}\tilde{\mathbf{u}})(\mathbf{F}^p)^{-1}. \tag{33}$$

Let us introduce a linear operator applied to the elastic connection $\overset{(e)}{\boldsymbol{\Gamma}}_{\mathcal{K}}$, dependent on the rate of plastic distortion $\mathbf{L}^p$, by

$$(\mathcal{L}_{\mathbf{L}^p}[\overset{(e)}{\boldsymbol{\Gamma}}_{\mathcal{K}}])\tilde{\mathbf{u}} := \frac{d}{dt}(\overset{(e)}{\boldsymbol{\Gamma}}_{\mathcal{K}})\tilde{\mathbf{u}} - \mathbf{L}^p(\overset{(e)}{\boldsymbol{\Gamma}}_{\mathcal{K}}\,\tilde{\mathbf{u}})$$
$$+\ \overset{(e)}{\boldsymbol{\Gamma}}_{\mathcal{K}}\,(\mathbf{L}^p\tilde{\mathbf{u}}) + (\overset{(e)}{\boldsymbol{\Gamma}}_{\mathcal{K}}\,\tilde{\mathbf{u}})\mathbf{L}^p \tag{34}$$

for all $\tilde{\mathbf{u}}$. The expression of the above operator (34), introduced in the left hand side of (33), leads to the following equivalent formula

$$(\mathbf{F}^p)^{-1}((\mathcal{L}_{\mathbf{L}^p}[\overset{(e)}{\boldsymbol{\Gamma}}_{\mathcal{K}}])[\mathbf{F}^p,\mathbf{F}^p]) = \frac{d}{dt}(\boldsymbol{\Gamma}) - \frac{d}{dt}(\overset{(p)}{\boldsymbol{\Gamma}}). \tag{35}$$

Using (26), (29) and the multiplicative decomposition (10), (35) becomes

$$(\mathcal{L}_{\mathbf{L}^p}[\overset{(e)}{\boldsymbol{\Gamma}}_{\mathcal{K}}]) = (\mathbf{F}^e)^{-1}(\nabla_\chi \mathbf{L})[\mathbf{F}^e,\mathbf{F}^e] - \nabla_{\mathcal{K}}\mathbf{L}^p. \tag{36}$$

Definition Starting from the kinematic relationships derived above, for a given deformation state (i.e. $\mathbf{F}$, $\mathbf{F}^e$, $\mathbf{F}^p$, $\overset{(e)}{\boldsymbol{\Gamma}}_{\mathcal{K}}$, $\overset{(p)}{\boldsymbol{\Gamma}}$ are considered to be given), we characterize a *virtual process* by

$\widetilde{\mathbf{v}}$—the virtual velocity, $\widetilde{\mathbf{L}}$—the virtual velocity gradient,

$\widetilde{\mathbf{L}}^e$ and $\widetilde{\mathbf{L}}^p$ the virtual rate of the elastic and plastic distortion, compatible with the kinematical relationships (27), (28), which means

$$\widetilde{\mathbf{L}} := \nabla_\chi\widetilde{\mathbf{v}}, \quad \text{and} \quad \widetilde{\mathbf{L}} = \widetilde{\mathbf{L}}^e + \mathbf{F}^e\widetilde{\mathbf{L}}^p(\mathbf{F}^e)^{-1}. \tag{37}$$

Consequently, taking into account (34) and (36) the *virtual time-derivative of the elastic connection* with respect to the plastically deformed configuration can be introduced by the **Definition:**

$$\text{virt}\frac{d}{dt}(\overset{(e)}{\boldsymbol{\Gamma}}_{\mathcal{K}})(\tilde{\mathbf{u}}) = \mathbf{F}^{e-1}((\nabla_\chi\widetilde{\mathbf{L}})[\mathbf{F}^e,\mathbf{F}^e])\tilde{\mathbf{u}} - (\nabla_{\mathcal{K}}\widetilde{\mathbf{L}}^p)\tilde{\mathbf{u}}$$
$$+\widetilde{\mathbf{L}}^p(\overset{(e)}{\boldsymbol{\Gamma}}_{\mathcal{K}}\,\tilde{\mathbf{u}}) - (\overset{(e)}{\boldsymbol{\Gamma}}_{\mathcal{K}}\,(\widetilde{\mathbf{L}}^p\tilde{\mathbf{u}}))$$
$$-(\overset{(e)}{\boldsymbol{\Gamma}}\,\tilde{\mathbf{u}})\widetilde{\mathbf{L}}^p, \quad \forall\ \tilde{\mathbf{u}} \in \mathcal{V}_{\mathcal{K}}. \tag{38}$$

3 The macro and micro balance equations

The principle of the virtual power at any arbitrary fixed moment of the time t is built starting from the principle of the virtual power derived from Fleck et al. (1994) and using the result already proved by Cleja-Tigoiu in (2002a), relative to the expressions of the power expanded by an elasto-plastic material (without any relation with a principle of the virtual power).

First we recall the **definitions** for Piola-Kirchhoff stress tensor and the stress momentum as pulled back to the configuration with torsion, and Mandel's non-symmetric stress measure, all of them being expressed relative to $\mathcal{K}$

$$\begin{aligned}
&\boldsymbol{\Pi}_{\mathcal{K}} \equiv \boldsymbol{\Pi} = \det(\mathbf{F}^e)(\mathbf{F}^e)^{-1}\mathbf{T}^s(\mathbf{F}^e)^{-T}, \quad \det\mathbf{F}^e = \frac{\rho_{\mathcal{K}}}{\rho}\\
&\boldsymbol{\mu}_{\mathcal{K}} = (\det\mathbf{F}^e)(\mathbf{F}^e)^T\boldsymbol{\mu}[(\mathbf{F}^e)^{-T},(\mathbf{F}^e)^{-T}],\\
&\frac{1}{\rho_{\mathcal{K}}}\boldsymbol{\Sigma}_{\mathcal{K}} = (\mathbf{F}^e)^T\frac{\mathbf{T}}{\rho}(\mathbf{F}^e)^{-T}, \qquad \mathbf{C}^e = (\mathbf{F}^e)^T\mathbf{F}^e,
\end{aligned} \tag{39}$$

associated to the non-symmetric Cauchy stress $\mathbf{T}$.

The virtual power at a fixed moment of time is written for any part $\mathcal{P} \subset \mathcal{B}$ and we accept that a surface element in the actual configuration, with unit normal $\mathbf{n}$, may transmit both force vector and couple vector. We start from the virtual power principle, **VPP-I**, formulated in Continuum Mechanics with couple stresses (see Fleck et al. (1994)).

PVP-I. In the deformed configuration, $\forall\ \ \mathcal{P} \subset \mathcal{B}$ bounded by a smooth surface $\partial\mathcal{P}$, the virtual power at a fixed moment of time

$$\begin{aligned}
&\int_{\mathcal{P}_t}\rho\mathbf{a}\cdot\mathbf{w}d\mathbf{x} + \int_{\mathcal{P}_t}\mathbf{T}\cdot\nabla\mathbf{w}d\mathbf{x} + \int_{\mathcal{P}_t}2\overset{\times}{\mathbf{T}}\cdot\boldsymbol{\theta}d\mathbf{x}\\
&\quad+\int_{\mathcal{P}_t}\mathbf{M}\cdot\nabla\boldsymbol{\theta}d\mathbf{x} = +\int_{\partial\mathcal{P}_t}\mathbf{T}\mathbf{n}\cdot\mathbf{w}d\sigma\\
&\quad+\int_{\partial\mathcal{P}_t}\mathbf{M}\mathbf{n}\cdot\boldsymbol{\theta}d\sigma + \int_{\mathcal{P}_t}\rho\mathbf{b}_f\cdot\mathbf{w}d\mathbf{x}\\
&\quad+\int_{\mathcal{P}_t}\rho\mathbf{b}_m\cdot\boldsymbol{\theta}d\mathbf{x}
\end{aligned} \tag{40}$$

holds for $\forall$ virtual velocity $\mathbf{w}$, and $\boldsymbol{\theta} = \frac{1}{2}\text{curl }\mathbf{w}$. Here $\mathbf{a}$ is the acceleration vector and the vector field $\overset{\times}{\mathbf{T}}$ is the coaxial vector associated with the skew-symmetric part of the Cauchy stress tensor $\mathbf{T}$

$$\mathbf{u}\times \overset{\times}{\mathbf{T}} = \mathbf{T}^a\mathbf{u}, \quad \forall\, \mathbf{u} \in \mathcal{V}_\chi. \tag{41}$$

$\mathbf{b}_f$, $\mathbf{b}_m$ are densities of the body forces and body couples, the vector force $\mathbf{Tn}$ and couple vector $\mathbf{Mn}$, acting on the surface of the normal characterized by the unit vector $\mathbf{n}$, $\mathbf{M}$ is the couple stress tensor. $\mathbf{T}$ and $\mathbf{M}$ are generally non-symmetric second order tensors.

Let us consider some simple properties, concerning the second order tensor fields with non-zero associated *curl*, which will be useful in defining an appropriate form of the virtual power principle.

Proposition 3 $\forall \mathbf{L} = \mathbf{W} \in Skew$ such that $curl\mathbf{L} = 0$, then

$$\begin{aligned} &(\nabla\mathbf{Lu})\mathbf{v} - (\nabla\mathbf{Lv})\mathbf{u} = 0 \quad \text{and} \\ &\exists\ \omega,\ \ \psi \in \mathcal{V}\ :\ \ \mathbf{Wu} = \omega \times \mathbf{u}, \\ &\forall \mathbf{u} \in \mathcal{V}\ \ \omega = curl\psi. \end{aligned} \tag{42}$$

Proposition 4 $\forall \mathbf{L} = \mathbf{W} \in Skew$ such that $curl\mathbf{L} \neq 0$, then

$$\begin{aligned} &(\nabla\mathbf{L})\mathbf{u} + ((\nabla\mathbf{L})\mathbf{u})^T = 0 \iff \\ &\quad \exists\ \Lambda \in Lin\ \ \nabla\mathbf{Lu} = \Lambda\mathbf{u} \times \mathbf{I} \quad \text{and} \\ &\quad curl\mathbf{W} \equiv det\Lambda\mathbf{I} - \Lambda^T. \end{aligned} \tag{43}$$

Based on the two propositions derived above we propose a virtual power principle appropriate to finite elasto-plasticity with continuously distributed dislocation.

We pass to the virtual power principle, **PVP-II**, that can be derived from the previous one, when $\frac{1}{2}\text{curl }\mathbf{w}$—vector field replaced by $\{\nabla\mathbf{w}\}^a$—tensor field.

PVP-II. $\forall\ \ \mathcal{P}$ and $\forall$ virtual velocity $\mathbf{w}$

$$\begin{aligned} &\int_{\mathcal{P}_t} \rho\mathbf{a}\cdot\mathbf{w}dx + \int_{\mathcal{P}_t} \mathbf{T}\cdot\nabla\mathbf{w}dx + \int_{\mathcal{P}_t} \mathbf{T}^*\cdot\{\nabla\mathbf{w}\}^a dx \\ &\quad + \int_{\mathcal{P}_t} \boldsymbol{\mu}\cdot\nabla_\chi\{\nabla\mathbf{w}\}^a dx = \int_{\mathcal{P}_t} \rho\mathbf{b}_f\cdot\mathbf{w}dx \\ &\quad + \int_{\partial\mathcal{P}_t} \mathbf{Tn}\cdot\mathbf{w}d\sigma + \int_{\partial\mathcal{P}_t} \boldsymbol{\mu}\mathbf{n}\cdot\{\nabla\mathbf{w}\}^a d\sigma \\ &\quad + \int_{\mathcal{P}_t} \rho\mathbf{B}_m\cdot\{\nabla\mathbf{w}\}^a dx \end{aligned} \tag{44}$$

with $$\{\nabla\mathbf{w}\}^a = \frac{1}{2}(\nabla\mathbf{w} - \nabla\mathbf{w}^T). \tag{45}$$

In the above formulae $\mathbf{B}_m$ denotes the second order body couple density and $\boldsymbol{\mu}$ is a third order field associated with a second order tensor field $\mathbf{M}$ as it follows

$$\boldsymbol{\mu}\cdot\nabla_\chi\{\nabla\mathbf{w}\}^a = \frac{1}{2}\mathbf{M}\cdot\nabla\text{curl}\mathbf{w}. \tag{46}$$

The principle of the virtual power **PVP-III** follows from **PVP-II**, when $\{\nabla\mathbf{w}\}^a$ replaced by $\overline{\mathbf{W}} \in Skew$.

PVP-III $\forall\ \ \mathcal{P}$ and $\forall$ virtual fields $\mathbf{w}$ and $\overline{\mathbf{W}}$ the equality

$$\begin{aligned} &\int_{\mathcal{P}_t} \rho\mathbf{a}\cdot\mathbf{w}dx + \int_{\mathcal{P}_t} \mathbf{T}\cdot\nabla\mathbf{w}dx + \int_{\mathcal{P}_t} \mathbf{T}^*\cdot\overline{\mathbf{W}}dx \\ &\quad + \int_{\mathcal{P}_t} \boldsymbol{\mu}\cdot\nabla\overline{\mathbf{W}}dx = \int_{\mathcal{P}_t} \rho\mathbf{b}_f\cdot\mathbf{w}dx \\ &\quad + \int_{\partial\mathcal{P}_t} \mathbf{Tn}\cdot\mathbf{w}d\sigma + \int_{\partial\mathcal{P}_t} \boldsymbol{\mu}\mathbf{n}\cdot\overline{\mathbf{W}}d\sigma \\ &\quad + \int_{\mathcal{P}_t} \rho\mathbf{B}_m\cdot\overline{\mathbf{W}}dx \end{aligned} \tag{47}$$

holds.

Finally a general form for the principle of the virtual power principle, **PVP-general**, can be put into evidence, when $\overline{\mathbf{W}} \in Skew$ in **PVP-III** is replaced by $\overline{\mathbf{L}}$, restricted to the conditions either $\overline{\mathbf{L}} = 0$ or $curl\overline{\mathbf{L}} \neq 0$.

Let us introduce the micro-stress $\boldsymbol{\Upsilon}^p_{\mathcal{K}}$ in the configuration with torsion, and the micro stress momentum $\boldsymbol{\mu}^p_{\mathcal{K}}$.

Ax.5 The **principle of the virtual power in finite elasto-plastic**, formulated $\forall$ part $\mathcal{P} \subset \mathcal{B}$ bounded by a smooth surface $\partial\mathcal{P}$

$$\begin{aligned} &\int_{\chi(\mathcal{P},t)} \rho\mathbf{a}\cdot\tilde{\mathbf{v}}dV + \int_{\chi(\mathcal{P},t)} \{\mathbf{T}\cdot\nabla_\chi\tilde{\mathbf{v}} + \mathbf{T}^*\cdot\widetilde{\mathbf{L}}^e\}dV \\ &\quad + \int_{\chi(\mathcal{P},t)} \boldsymbol{\mu}\cdot\nabla_\chi\widetilde{\mathbf{L}}^e dV \\ &\quad + \int_{\mathcal{K}_t(\mathcal{P})} \{\boldsymbol{\Upsilon}^p_{\mathcal{K}}\cdot\widetilde{\mathbf{L}}^p + \boldsymbol{\mu}^p_{\mathcal{K}}\cdot\nabla_{\mathcal{K}}\widetilde{\mathbf{L}}^p\}dV_{\mathcal{K}} \\ &\quad = \int_{\chi(\mathcal{P},t)} \rho\mathbf{b}_f\cdot\tilde{\mathbf{v}}dV \\ &\qquad + \int_{\chi(\mathcal{P},t)} \rho\mathbf{B}_m\cdot\widetilde{\mathbf{L}}^e dV + \int_{\mathcal{K}_t} \tilde{\rho}\mathbf{B}^p_m\cdot\widetilde{\mathbf{L}}^p dV_{\mathcal{K}} \\ &\qquad + \int_{\partial\chi(\mathcal{P},t)} \mathbf{t}(\mathbf{n})\cdot\tilde{\mathbf{v}}dV + \int_{\partial\chi(\mathcal{P},t)} \mathbf{M}(\mathbf{n})\cdot\widetilde{\mathbf{L}}^e dA \\ &\qquad + \int_{\partial\mathcal{K}_t(\mathcal{P})} \mathbf{M}^p(\mathbf{n})\cdot\widetilde{\mathbf{L}}^p dA_{\mathcal{K}}, \end{aligned} \tag{48}$$

holds for any all generalized virtual velocities and virtual rates $\tilde{\mathbf{v}}$, $\nabla_\chi\tilde{\mathbf{v}}$, $\widetilde{\mathbf{L}}^p$, $\nabla_{\mathcal{K}}\widetilde{\mathbf{L}}^p$, for all $\widetilde{\mathbf{L}}^e$, $\nabla_\chi\widetilde{\mathbf{L}}^e$.

Now we justify the postulated form for the principle of the virtual power (48), using the expression of the mechanical internal power as it follows from Cleja-Ţigoiu (2002a).

Proposition 5

1. *The density of the internal mechanical power produced by the non-symmetric Cauchy stress tensor can be written under the form*

$$\frac{1}{\rho}\mathbf{T}\cdot\mathbf{L} = \frac{1}{\rho}\mathbf{T}\cdot\mathbf{L}^e + \frac{1}{\rho_{\mathcal{K}}}\mathbf{\Sigma}_{\mathcal{K}}\cdot\mathbf{L}^p. \tag{49}$$

2. *The density of the internal mechanical power produced by the couple stresses can be expressed with the aid of the third order tensor of stress momentum defined in the actual configuration by $\boldsymbol{\mu}$ or in terms of the stress momentum pulled back to the configuration with torsion $\boldsymbol{\mu}_{\mathcal{K}}$*

$$\begin{aligned}\frac{1}{\rho}\boldsymbol{\mu}\cdot\nabla_{\chi}\mathbf{L} &= \frac{1}{\rho_{\mathcal{K}}}\boldsymbol{\mu}_{\mathcal{K}}\cdot(\mathbf{F}^e)^{-1}\nabla_{\chi}\mathbf{L}[\mathbf{F}^e,\mathbf{F}^e] \\ &= \frac{1}{\rho_{\mathcal{K}}}\boldsymbol{\mu}_{\mathcal{K}}\cdot(\mathcal{L}_{\mathbf{L}^p}[\overset{(e)}{\mathbf{\Gamma}}_{\mathcal{K}}]) \\ &\quad+\frac{1}{\rho_{\mathcal{K}}}\boldsymbol{\mu}_{\mathcal{K}}\cdot\nabla_{\mathcal{K}}\mathbf{L}^p.\end{aligned} \tag{50}$$

The last equality, that put in evidence the elastic and plastic part of the appropriate internal power, has been reformulated as a consequence of (36).

Remark We use Gurtin's argument, since there are no motivation to suppose that the terms which enter the previous formulae (49) and (50) refer to the same stress and the same momentum, and we replace in (49) $\mathbf{\Sigma}_{\mathcal{K}}$ by the microstress $\mathbf{\Upsilon}^p_{\mathcal{K}}$ and in (50) $\boldsymbol{\mu}_{\mathcal{K}}$, which is power conjugated with $\nabla_{\mathcal{K}}\mathbf{L}^p$, by $\boldsymbol{\mu}^p_{\mathcal{K}}$.

The *macro-balance equation* at any time t can be derived from (48) if we take $\widetilde{\mathbf{L}}^p = 0$ and $\nabla_{\mathcal{K}}\widetilde{\mathbf{L}}^p = 0$. For any $\widetilde{\mathbf{v}}$, a virtual velocity and for any second order tensor field $\widetilde{\mathbf{L}}^e$ with $curl(\widetilde{\mathbf{L}}^e) \neq 0$, i.e non-reducible to a gradient of an appropriate vector field, the *macro-balance equations* are derived. The *micro-balance equation* at any time t can be derived from (48) if $\widetilde{\mathbf{L}}^p$ and $\nabla_{\mathcal{K}}\widetilde{\mathbf{L}}^p$ are non-zero.

Theorem 1

1. *The impulse and momentum local balance equations can be written under the form*

$$\begin{aligned}\rho\mathbf{a} &= \operatorname{div}\mathbf{T} + \rho\mathbf{b}_f \\ \mathbf{T}^* &= \operatorname{div}\boldsymbol{\mu} + \rho\mathbf{B}_m, \quad \text{on } \mathcal{P}_t\end{aligned} \tag{51}$$

with the appropriate boundary conditions on $\partial\mathcal{P}_t$

$$\mathbf{Tn} = \mathbf{t}(\mathbf{n}), \quad \textit{and} \quad \boldsymbol{\mu}\mathbf{n} = \mathbf{M}(\mathbf{n}). \tag{52}$$

2. *The micro balance equation is expressed by*

$$\begin{aligned}&\mathbf{\Upsilon}^p_{\mathcal{K}} - div\,\boldsymbol{\mu}^p_{\mathcal{K}} = \tilde{\rho}\mathbf{B}^p_m, \quad in\ \mathcal{K}(\mathcal{P},t), \\ &\boldsymbol{\mu}^p_{\mathcal{K}}\mathbf{n} = \mathbf{M}^p\mathbf{n} \quad on \quad \partial\mathcal{K}(\mathcal{P},t),\end{aligned} \tag{53}$$

micro-traction condition.

We remark that (51), $(53)_1$ together with the boundary conditions (52) and $(53)_2$ follow directly from the principle of the virtual power, stipulated in (48), without any additional assumptions.

Finally we **conclude** that the theory can be based either on the appropriate postulate of the variational principle for physical and material space or on the postulates of the physical and material balance laws.

4 Free energy imbalance

Ax.6 There exists a free energy density function ψ, invariant with respect to a change of frame in the actual configuration

$$\psi = \psi_{\mathcal{K}}(\mathbf{C}^e, \overset{(e)}{\mathbf{\Gamma}}_{\mathcal{K}}, (\mathbf{F}^p)^{-1}, \overset{(p)}{\mathbf{\Gamma}}_{\mathcal{K}}), \tag{54}$$

represented in the configuration with torsion $\mathcal{K}$ by a function of the second order elastic pair $(\mathbf{C}^e, \overset{(e)}{\mathbf{\Gamma}}_{\mathcal{K}})$, and dependent on the plastic measure of deformation $((\mathbf{F}^p)^{-1}, \overset{(p)}{\mathbf{\Gamma}}_{\mathcal{K}})$.

Ax.7 The elasto-plastic behavior of the material is restricted to satisfy in $\mathcal{K}$ the *free energy imbalance*

$$-\dot{\psi}_{\mathcal{K}} + \frac{1}{\rho_{\mathcal{K}}}(\mathcal{P}_{int})_{\mathcal{K}} \geq 0 \tag{55}$$

for any *virtual (isothermic) processes.*

Proposition 6 In (55) *the internal power in the configuration with torsion can be calculated starting from the expression*

$$\begin{aligned}(\mathcal{P}_{int})_{\mathcal{K}} &= \frac{1}{\rho}(\mathbf{T}+\mathbf{T}^*)\cdot\mathbf{L}^e + \mathbf{\Upsilon}^p_{\mathcal{K}}\cdot\mathbf{L}^p \\ &\quad+\frac{1}{\rho_{\mathcal{K}}}\boldsymbol{\mu}_{\mathcal{K}}\cdot\mathcal{L}_{\mathbf{L}^p}[\overset{(e)}{\mathbf{\Gamma}}_{\mathcal{K}}] \\ &\quad+\frac{1}{\rho_{\mathcal{K}}}\boldsymbol{\mu}^p_{\mathcal{K}}\cdot\nabla_{\mathcal{K}}\mathbf{L}^p,\end{aligned} \tag{56}$$

while *the time derivative of the free energy is expressed through*

$$\dot{\psi}_{\mathcal{K}} = \partial_{\mathbf{C}^e}\psi_{\mathcal{K}} \cdot \dot{\mathbf{C}}^e - (\mathbf{F}^p)^{-T}\partial_{(\mathbf{F}^p)^{-1}}\psi_{\mathcal{K}} \cdot \mathbf{L}^p + \partial_{\overset{(e)}{\boldsymbol{\Gamma}}_{\mathcal{K}}}\psi_{\mathcal{K}} \cdot \left(\frac{d}{dt}\overset{(e)}{\boldsymbol{\Gamma}}_{\mathcal{K}}\right) + \partial_{\overset{(p)}{\boldsymbol{\Gamma}}_{\mathcal{K}}}\psi_{\mathcal{K}} \cdot \left(\frac{d}{dt}\overset{(p)}{\boldsymbol{\Gamma}}_{\mathcal{K}}\right). \tag{57}$$

The formula (56) is derived from macro and micro balance equations, multiplied by $\mathbf{L}^e$ and by $\mathbf{L}^p$, respectively.

In order to pursuit the calculus

1. We eliminate the rate of the elastic distortion, which enters the expression (56) via formula (28), as well the gradient of the rate of elastic ditorsion via the formula (36). Then only $\mathbf{L}$ and $\mathbf{L}^p$ and their appropriate differentials, $\nabla_\chi \mathbf{L}$ and $\nabla_{\mathcal{K}}\mathbf{L}^p$, enter the internal power.
2. In (57) the rate of the elastic strain is replaced by

$$\dot{\mathbf{C}}^e = 2\,(\mathbf{F}^e)^T\{\mathbf{L}\}^s\mathbf{F}^e - 2\,\{\mathbf{C}^e\mathbf{L}^p\}^s, \tag{58}$$

using the elastic strain $\mathbf{C}^e$ expressed as a consequence of the multiplicative decomposition under the form

$$\mathbf{C}^e := (\mathbf{F}^e)^T\mathbf{F}^e = (\mathbf{F}^p)^{-T}\mathbf{C}(\mathbf{F}^p)^{-1}, \quad \text{where} \quad \mathbf{C} = \mathbf{F}^T\mathbf{F}. \tag{59}$$

In order to obtain the imbalanced energy condition, we pass to the virtual kinematic process, as it follows:

$$\begin{aligned}
&\mathrm{virt}(\dot{\mathbf{C}}^e) = 2\,(\mathbf{F}^e)^T\{\widetilde{\mathbf{L}}\}^s\mathbf{F}^e - 2\,\{\mathbf{C}^e \cdot \widetilde{\mathbf{L}}^p\}^s, \quad \text{related to (58)}\\
&\mathrm{virt}(\mathcal{L}_{\mathbf{L}^p}[\overset{(e)}{\boldsymbol{\Gamma}}_{\mathcal{K}}]) = (\mathbf{F}^e)^{-1}(\nabla_\chi\widetilde{\mathbf{L}})[\mathbf{F}^e, \mathbf{F}^e] - \nabla_{\mathcal{K}}\widetilde{\mathbf{L}}^p, \quad \text{related to (36).}
\end{aligned} \tag{60}$$

Everywhere we replace $\mathbf{L}$, and $\nabla_\chi\mathbf{L}$, by the virtual $\widetilde{\mathbf{L}}$ $\nabla_\chi\widetilde{\mathbf{L}}$, $\mathbf{L}^p$, and $\nabla_{\mathcal{K}}\mathbf{L}^p$ are replaced by $\widetilde{\mathbf{L}}^p$ and $\nabla_\chi\widetilde{\mathbf{L}}^p$.

Proposition 7 *The* free energy imbalance *is satisfied for any* virtual process, *if the inequality written below*

$$\begin{aligned}
&\left\{\frac{1}{\rho}(\mathbf{F}^e)^{-1}\{\mathbf{T}+\mathbf{T}^*\}^s(\mathbf{F}^e)^{-T} - 2\partial_{\mathbf{C}^e}\psi_{\mathcal{K}}\right\} \cdot [(\mathbf{F}^e)^T\{\widetilde{\mathbf{L}}\}^s\mathbf{F}^e - \{\mathbf{C}^e \cdot \widetilde{\mathbf{L}}^p\}^s]\\
&+\frac{1}{\rho}\{\mathbf{T}+\mathbf{T}^*\}^a \cdot (\widetilde{\mathbf{L}} - \mathbf{F}^e\widetilde{\mathbf{L}}^p(\mathbf{F}^e)^{-1})\\
&+\left\{\frac{1}{\rho_{\mathcal{K}}}\boldsymbol{\Upsilon}^p_{\mathcal{K}} + (\mathbf{F}^p)^{-T}\partial_{(\mathbf{F}^p)^{-1}}\psi_{\mathcal{K}}\right\} \cdot \widetilde{\mathbf{L}}^p\\
&+\frac{1}{\rho_{\mathcal{K}}}\boldsymbol{\mu}_{\mathcal{K}} \cdot [(\mathbf{F}^e)^{-1}(\nabla_\chi\widetilde{\mathbf{L}})[\mathbf{F}^e, \mathbf{F}^e] - \nabla_{\mathcal{K}}\widetilde{\mathbf{L}}^p]\\
&+\frac{1}{\rho_{\mathcal{K}}}\boldsymbol{\mu}^p_{\mathcal{K}} \cdot \nabla_{\mathcal{K}}\widetilde{\mathbf{L}}^p\\
&-\partial_{\boldsymbol{\Gamma}^e{}_{\mathcal{K}}}\psi_{\mathcal{K}} \cdot \mathrm{virt}\left(\frac{d}{dt}\boldsymbol{\Gamma}^e{}_{\mathcal{K}}\right) - \partial_{\overset{(p)}{\boldsymbol{\Gamma}}_{\mathcal{K}}}\psi_{\mathcal{K}} \cdot \mathrm{virt}\left(\frac{d}{dt}\overset{(p)}{\boldsymbol{\Gamma}}_{\mathcal{K}}\right) \geq 0
\end{aligned} \tag{61}$$

holds for any $\widetilde{\mathbf{L}} \equiv \nabla_\chi\widetilde{\mathbf{v}}$, $\nabla_\chi\widetilde{\mathbf{L}}$, and for arbitrarily given $\widetilde{\mathbf{L}}^p$, $\nabla_{\mathcal{K}}\widetilde{\mathbf{L}}^p$.

The *virtual kinematic processes* are also characterized by the virtual variations of the fields via the formulae

$$\mathrm{virt}\left(\frac{d}{dt}(\overset{(e)}{\boldsymbol{\Gamma}}_{\mathcal{K}})\right) = \widetilde{\mathbf{L}}^p\,\overset{(e)}{\boldsymbol{\Gamma}}_{\mathcal{K}} - \overset{(e)}{\boldsymbol{\Gamma}}_{\mathcal{K}}\,\widetilde{\mathbf{L}}^p - \overset{(e)}{\boldsymbol{\Gamma}}_{\mathcal{K}}\,[\mathbf{I}, \widetilde{\mathbf{L}}^p] + (\mathbf{F}^e)^{-1}(\nabla_\chi\widetilde{\mathbf{L}})[\mathbf{F}^e, \mathbf{F}^e] - \nabla_{\mathcal{K}}\widetilde{\mathbf{L}}^p \quad \text{related to (38)} \tag{62}$$

and related to (31)

$$\mathrm{virt}\left(\frac{d}{dt}(\overset{(p)}{\boldsymbol{\Gamma}}_{\mathcal{K}})\right) = \widetilde{\mathbf{L}}^p\,\overset{(p)}{\boldsymbol{\Gamma}}_{\mathcal{K}} - \nabla_{\mathcal{K}}\widetilde{\mathbf{L}}^p - \overset{(p)}{\boldsymbol{\Gamma}}\,\widetilde{\mathbf{L}}^p - \overset{(p)}{\boldsymbol{\Gamma}}\,[\mathbf{I}, \widetilde{\mathbf{L}}^p]. \tag{63}$$

5 Thermodynamic restrictions

We provide the thermomechanic restrictions on the constitutive description of elasto-plastic material, based on the *imbalanced free energy condition.* We require the *imbalanced condition* written in (61) to be satisfied for any *virtual process*, defined by the formulae (60)–(63), when $\widetilde{\mathbf{L}}$, $\nabla_\chi\widetilde{\mathbf{L}}$ are arbitrary, and for the given $\widetilde{\mathbf{L}}^p$, $\nabla_{\mathcal{K}}\widetilde{\mathbf{L}}^p$.

I. First step: we consider a virtual process in a such way to have $\widetilde{\mathbf{L}}^p = 0$, $\nabla_{\mathcal{K}}\widetilde{\mathbf{L}}^p = 0$.

Thus (61) holds for any $\widetilde{\mathbf{L}}$, $\nabla_\chi \widetilde{\mathbf{L}}$, if and only if the following *constitutive restrictions*

$$\begin{aligned}&(\mathbf{F}^e)^{-1}\{\mathbf{T}+\mathbf{T}^*\}^s(\mathbf{F}^e)^{-T} = 2\rho\partial_{\mathbf{C}^e}\psi_{\mathcal{K}},\\ &\{\mathbf{T}+\mathbf{T}^*\}^a = 0\\ &\frac{1}{\rho_{\mathcal{K}}}\boldsymbol{\mu}_{\mathcal{K}} = \partial_{\boldsymbol{\Gamma}^e{}_{\mathcal{K}}}\psi_{\mathcal{K}},\end{aligned} \tag{64}$$

are satisfied.

II. Second step: we introduce the thermodynamic restriction (64) into *imbalanced free energy condition* (61) and we get the *dissipation inequality*

$$\begin{aligned}&\left\{\frac{1}{\rho_{\mathcal{K}}}\boldsymbol{\Upsilon}^p_{\mathcal{K}} + (\mathbf{F}^p)^{-T}\partial_{(\mathbf{F}^p)^{-1}}\psi_{\mathcal{K}}\right\}\cdot\widetilde{\mathbf{L}}^p\\ &+\left\{\frac{1}{\rho_{\mathcal{K}}}\boldsymbol{\mu}^p_{\mathcal{K}} + \partial_{\overset{(p)}{\boldsymbol{\Gamma}}_{\mathcal{K}}}\psi_{\mathcal{K}}\right\}\cdot\nabla_{\mathcal{K}}\widetilde{\mathbf{L}}^p - \partial_{\overset{(e)}{\boldsymbol{\Gamma}}_{\mathcal{K}}}\psi_{\mathcal{K}}\\ &\cdot\{\widetilde{\mathbf{L}}^p\,\overset{(e)}{\boldsymbol{\Gamma}}_{\mathcal{K}} - \overset{(e)}{\boldsymbol{\Gamma}}_{\mathcal{K}}\,\widetilde{\mathbf{L}}^p - \overset{(e)}{\boldsymbol{\Gamma}}_{\mathcal{K}}\,[\mathbf{I},\widetilde{\mathbf{L}}^p]\} - \partial_{\overset{(p)}{\boldsymbol{\Gamma}}_{\mathcal{K}}}\psi_{\mathcal{K}}\\ &\cdot\{\widetilde{\mathbf{L}}^p\,\overset{(p)}{\boldsymbol{\Gamma}}_{\mathcal{K}} - \overset{(p)}{\boldsymbol{\Gamma}}_{\mathcal{K}}\,\widetilde{\mathbf{L}}^p - \overset{(p)}{\boldsymbol{\Gamma}}_{\mathcal{K}}\,[\mathbf{I},\widetilde{\mathbf{L}}^p]\} \geq 0.\end{aligned} \tag{65}$$

Let us introduce the free energy in the reference configuration

$$\begin{aligned}\psi &= \psi_{\mathcal{K}}(\mathbf{C}^e, \overset{(e)}{\boldsymbol{\Gamma}}_{\mathcal{K}}, (\mathbf{F}^p)^{-1}, \overset{(p)}{\boldsymbol{\Gamma}}_{\mathcal{K}})\\ &\equiv \bar{\psi}(\mathbf{C}, \overset{(e)}{\boldsymbol{\Gamma}}_{back}, \mathbf{F}^p, \overset{(p)}{\boldsymbol{\Gamma}}),\end{aligned} \tag{66}$$

taking into account the relationships (59), (8) and (16).

Proposition 8 *When we pass to the free energy density in the reference configuration k, the dissipation inequality becomes*

$$\begin{aligned}&\left\{\frac{1}{\rho_{\mathcal{K}}}\boldsymbol{\Upsilon}^p_{\mathcal{K}} - 2\mathbf{C}^e(\mathbf{F}^p)\partial_{\mathbf{C}}\bar{\psi}(\mathbf{F}^p)^T - \partial_{\mathbf{F}^p}\bar{\psi}(\mathbf{F}^p)^T\right\}\cdot\widetilde{\mathbf{L}}^p\\ &+\left\{\frac{1}{\rho_{\mathcal{K}}}\boldsymbol{\mu}^p_{\mathcal{K}} - (\mathbf{F}^p)^{-T}\partial_{\overset{(p)}{\boldsymbol{\Gamma}}}\bar{\psi}[(\mathbf{F}^p)^T,(\mathbf{F}^p)^T]\right\}\\ &\cdot\nabla_{\mathcal{K}}\widetilde{\mathbf{L}}^p \geq 0, \quad \forall\ \widetilde{\mathbf{L}}^p, \nabla_{\mathcal{K}}\widetilde{\mathbf{L}}^p.\end{aligned} \tag{67}$$

In order to **prove** the above formulae we take into account the relationships between the partial derivatives of the free energy expressed relative to the initial configuration and to the configuration with torsion, derived from (66) together with (59), (8) and (16), under the form

$$\begin{aligned}&\partial_{\mathbf{C}}\bar{\psi} = (\mathbf{F}^p)^{-1}\partial_{\mathbf{C}^e}\psi_{\mathcal{K}}(\mathbf{F}^p)^{-T}\\ &\partial_{\overset{(p)}{\boldsymbol{\Gamma}}}\bar{\psi} = -(\mathbf{F}^p)^T\partial_{\overset{(p)}{\boldsymbol{\Gamma}}_{\mathcal{K}}}\psi_{\mathcal{K}}[(\mathbf{F}^p)^{-T},(\mathbf{F}^p)^{-T}],\\ &\partial_{\overset{(e)}{\boldsymbol{\Gamma}}_{\mathcal{K}}}\psi_{\mathcal{K}} = (\mathbf{F}^p)^{-T}\partial_{\overset{(e)}{\boldsymbol{\Gamma}}_{back}}\bar{\psi}[(\mathbf{F}^p)^T,(\mathbf{F}^p)^T],\\ &\partial_{\mathbf{F}^p}\bar{\psi}(\mathbf{F}^p)^T\cdot\widetilde{\mathbf{L}}^p = -2\mathbf{C}^e\partial_{\mathbf{C}^e}\psi_{\mathcal{K}}\cdot\widetilde{\mathbf{L}}^p\\ &\quad-(\mathbf{F}^p)^{-T}\partial_{(\mathbf{F}^p)^{-1}}\psi_{\mathcal{K}}\cdot\widetilde{\mathbf{L}}^p\\ &\quad+\partial_{\overset{(e)}{\boldsymbol{\Gamma}}_{\mathcal{K}}}\psi_{\mathcal{K}}\cdot\{\widetilde{\mathbf{L}}^p\,\overset{(e)}{\boldsymbol{\Gamma}}_{\mathcal{K}} - \overset{(e)}{\boldsymbol{\Gamma}}_{\mathcal{K}}\,\widetilde{\mathbf{L}}^p - \overset{(e)}{\boldsymbol{\Gamma}}_{\mathcal{K}}\,[\mathbf{I},\widetilde{\mathbf{L}}^p]\}\\ &\quad+\partial_{\overset{(p)}{\boldsymbol{\Gamma}}_{\mathcal{K}}}\psi_{\mathcal{K}}\cdot\{\widetilde{\mathbf{L}}^p\,\overset{(p)}{\boldsymbol{\Gamma}}_{\mathcal{K}} - \overset{(p)}{\boldsymbol{\Gamma}}_{\mathcal{K}}\,\widetilde{\mathbf{L}}^p - \overset{(p)}{\boldsymbol{\Gamma}}_{\mathcal{K}}\,[\mathbf{I},\widetilde{\mathbf{L}}^p]\}\end{aligned} \tag{68}$$

The constitutive form of $\partial_{\mathbf{F}^p}\bar{\psi}$ following from (68) can be expressed through

$$\begin{aligned}\partial_{\mathbf{F}^p}\bar{\psi} &= -2\mathbf{C}^e\partial_{\mathbf{C}^e}\psi_{\mathcal{K}} - (\mathbf{F}^p)^{-T}\partial_{(\mathbf{F}^p)^{-1}}\psi_{\mathcal{K}}\\ &\quad+\mathcal{D}\psi_{\mathcal{K}}(\overset{(e)}{\boldsymbol{\Gamma}})[\overset{(e)}{\boldsymbol{\Gamma}}] + \mathcal{D}\psi_{\mathcal{K}}(\overset{(p)}{\boldsymbol{\Gamma}})[\overset{(p)}{\boldsymbol{\Gamma}}].\end{aligned} \tag{69}$$

Here the second order tensor denoted by $\mathcal{D}\psi_{\mathcal{K}}(\overset{(e)}{\boldsymbol{\Gamma}})[\overset{(e)}{\boldsymbol{\Gamma}}]$ is defined for any third order tensor field $\overset{(e)}{\boldsymbol{\Gamma}} \equiv \mathcal{X}$ as it follows

$$\begin{aligned}\mathcal{D}\psi_{\mathcal{K}}(\mathcal{X})[\mathcal{X}] := \Bigg[&\frac{\partial\psi_{\mathcal{K}}}{\partial\mathcal{X}_{pjk}}\mathcal{X}_{sjk} - \mathcal{X}_{ijp}\frac{\partial\psi_{\mathcal{K}}}{\partial\mathcal{X}_{ijs}}\\ &-\mathcal{X}_{ipk}\frac{\partial\psi_{\mathcal{K}}}{\partial\mathcal{X}_{isk}}\Bigg]\mathbf{i}_p\otimes\mathbf{i}_s.\end{aligned} \tag{70}$$

Based on the dissipation inequality, we formulate the *constitutive hypotheses*:

Ax.8 The microforces contain:

(1) a *dissipative part*,
(2) a non-dissipative part, which are derived from the free energy, the so-called *energetic microforces*,

and they are represented through

$$\begin{aligned}&\frac{1}{\rho_{\mathcal{K}}}\boldsymbol{\Upsilon}^p_{\mathcal{K}} = 2\mathbf{C}^e(\mathbf{F}^p)\partial_{\mathbf{C}}\bar{\psi}(\mathbf{F}^p)^T + \partial_{\mathbf{F}^p}\bar{\psi}(\mathbf{F}^p)^T + Y_\gamma\widetilde{\mathbf{L}}^p\\ &\frac{1}{\rho_{\mathcal{K}}}\boldsymbol{\mu}^p_{\mathcal{K}} = (\mathbf{F}^p)^{-T}\partial_{\overset{(p)}{\boldsymbol{\Gamma}}}\bar{\psi}[(\mathbf{F}^p)^T,(\mathbf{F}^p)^T] + Y_\mu\nabla_{\mathcal{K}}\widetilde{\mathbf{L}}^p.\end{aligned} \tag{71}$$

We **remark** that the non-dissipative parts of the microforces were derived from the free energy through (71), but only the last terms of the right sides of (71) are dissipative, because the free energy cannot depend on the rates.

Ax.9 The scalar constitutive functions Y_γ, Y_μ are defined in such a way to be compatible with the dissipation inequality

$$Y_\mu\nabla_{\mathcal{K}}\widetilde{\mathbf{L}}^p\cdot\nabla_{\mathcal{K}}\widetilde{\mathbf{L}}^p + Y_\gamma\widetilde{\mathbf{L}}^p\cdot\widetilde{\mathbf{L}}^p \geq 0. \tag{72}$$

Remark The presence of the non-dissipative part (the first term in the right-hand side, which has the significance of the Mandel's type stress) in the formula $(71)_1$ couples the macroscopic and microscopic forces.

Following Gurtin (2004), we introduce the *intensity for the accumulated effect* of the rate of plastic distortion and of the gradient of plastic distortion, through

$$d^p := \sqrt{\mathbf{L}^p \cdot \mathbf{L}^p + h^2 \nabla_{\mathcal{K}} \mathbf{L}^p \cdot \nabla_{\mathcal{K}} \mathbf{L}^p}, \tag{73}$$

and we define

$$Y_\gamma := Y(d^p), \quad Y_\mu := h^2 Y(d^p), \tag{74}$$

with h a length scale.

We resume the constitutive equations for macrostress and macrostress momentum, when $\mathbf{T}^* = -\{\mathbf{T}\}^a$,

$$\{\mathbf{T}\}^s = 2\rho \mathbf{F}^e \partial_{\mathbf{C}^e} \psi_{\mathcal{K}} (\mathbf{F}^e)^T, \quad \frac{1}{\rho_{\mathcal{K}}} \boldsymbol{\mu}_{\mathcal{K}} = \partial_{\boldsymbol{\Gamma}^e{}_{\mathcal{K}}} \psi_{\mathcal{K}}, \tag{75}$$

which can be equivalently represented as a consequence of the relationships derived in (68) through

$$\{\mathbf{T}\}^s = 2\rho \mathbf{F} \partial_{\mathbf{C}} \bar{\psi} \mathbf{F}^T, \quad \frac{1}{\rho} \boldsymbol{\mu} = \mathbf{F}^{-1} \partial_{\boldsymbol{\Gamma}^e_{\text{back}}} \bar{\psi} [\mathbf{F}^T, \mathbf{F}^T]. \tag{76}$$

The microforces and microstress momentum have been represented under the form of viscoplastic constitutive equations

$$\begin{aligned} \frac{1}{\rho_{\mathcal{K}}} \boldsymbol{\Upsilon}^p_{\mathcal{K}} &= 2\mathbf{C}^e (\mathbf{F}^p) \partial_{\mathbf{C}} \bar{\psi} (\mathbf{F}^p)^T + \partial_{\mathbf{F}^p} \bar{\psi} (\mathbf{F}^p)^T \\ &\quad + Y(d^p) \mathbf{L}^p \\ \frac{1}{\rho_{\mathcal{K}}} \boldsymbol{\mu}^p_{\mathcal{K}} &= (\mathbf{F}^p)^{-T} \partial_{\underset{\boldsymbol{\Gamma}}{(\mathrm{p})}} \bar{\psi} [(\mathbf{F}^p)^T, (\mathbf{F}^p)^T] \\ &\quad + h^2 Y(d^p) \nabla_{\mathcal{K}} \mathbf{L}^p, \end{aligned} \tag{77}$$

with $d^p := \sqrt{\mathbf{L}^p \cdot \mathbf{L}^p + h^2 \nabla_{\mathcal{K}} \mathbf{L}^p \cdot \nabla_{\mathcal{K}} \mathbf{L}^p}$. Thus the dissipation inequality (72) becomes

$$\begin{aligned} & Y(d^p) \mathbf{L}^p \cdot \mathbf{L}^p + h^2 Y(d^p) \nabla_{\mathcal{K}} \mathbf{L}^p \cdot \nabla_{\mathcal{K}} \mathbf{L}^p \\ & \equiv Y(d^p)(d^p)^2 \geq 0, \end{aligned} \tag{78}$$

under the supposition that $Y(d^p) \geq 0$.

Let us remark that the stress

$$\overline{\boldsymbol{\Upsilon}}^p_{\mathcal{K}} := \boldsymbol{\Upsilon}^p_{\mathcal{K}} - \boldsymbol{\Sigma}^p_{\mathcal{K}} \tag{79}$$

leads to the *viscoplastic constitutive equations*

$$\begin{aligned} \frac{1}{\rho_{\mathcal{K}}} \overline{\boldsymbol{\Upsilon}}^p_{\mathcal{K}} &= \partial_{\mathbf{F}^p} \bar{\psi} (\mathbf{F}^p)^T + Y(d^p) \mathbf{L}^p \\ \frac{1}{\rho_{\mathcal{K}}} \boldsymbol{\mu}^p_{\mathcal{K}} &= (\mathbf{F}^p)^{-T} \partial_{\underset{\boldsymbol{\Gamma}}{(\mathrm{p})}} \bar{\psi} [(\mathbf{F}^p)^T, (\mathbf{F}^p)^T] \\ &\quad + h^2 Y(d^p) \nabla_{\mathcal{K}} \widetilde{\mathbf{L}}^p. \end{aligned} \tag{80}$$

The micro balance Equation $(53)_1$ involves the *material forces* and finally it is written under the form

$$\overline{\boldsymbol{\Upsilon}}^p_{\mathcal{K}} = \boldsymbol{\Sigma}_{\mathcal{K}} + \operatorname{div} \boldsymbol{\mu}^p_{\mathcal{K}} + \tilde{\rho} \mathbf{B}^p_m, \quad \text{in } \mathcal{K}(\mathcal{P}, t). \tag{81}$$

6 Screw dislocations

We recall here the definitions of the *Burgers vector* $\mathbf{b}_{\mathcal{K}}$ adopted in our context of finite elasto-plasticity, first in terms of plastic distortion $\mathbf{F}^p$.

Definition For a given $\mathcal{A}_0$ surface with normal $\mathbf{N}$ bounded by C_0, a closed curve in the reference configuration, we introduce the definition

$$\begin{aligned} \mathbf{b}_{\mathcal{K}} &= \int_{C_0} \mathbf{F}^p \, d\mathbf{X} = \int_{\mathcal{A}_0} \text{curl } (\mathbf{F}^p) \mathbf{N} dA \\ &= \int_{\mathcal{A}_{\mathcal{K}}} \boldsymbol{\alpha}_{\mathcal{K}} \mathbf{n}_{\mathcal{K}} dA_{\mathcal{K}}, \end{aligned} \tag{82}$$

where $\mathbf{N}$ is the unit normal on the surface in the reference configuration and Noll's dislocation (second order) tensor is given in terms of the plastic distortion

$$\boldsymbol{\alpha}_{\mathcal{K}} \equiv \frac{1}{\det \mathbf{F}^p} \text{curl}(\mathbf{F}^p)(\mathbf{F}^p)^T. \tag{83}$$

The meaning of Burgers vector clearly appears from the approximate formula, derived from (82)

$$\mathbf{b}_{\mathcal{K}} \simeq \text{curl } (\mathbf{F}^p) \mathbf{N} \text{ area}(\mathcal{A}_0), \tag{84}$$

with the functions calculated in an appropriate point.

A similar **definition** for the *Burgers vector* $\mathbf{b}_{\mathcal{K}}$ can be also done in terms of elastic distortion $\mathbf{F}^e$ and the approximate formula can be derived under the form $\mathbf{b}_{\mathcal{K}} \simeq \text{curl } (\mathbf{F}^e)^{-1} \mathbf{n} \text{ area } (A_t)$, with the functions calculated in a fixed point inside A_t—an appropriate surface in the actual configuration.

The geometric dislocation tensor $\mathbf{G}$, which represents the Noll's second order dislocation tensor, $\mathbf{G} = \bar{\boldsymbol{\alpha}}_{\mathcal{K}} = \boldsymbol{\alpha}_{\mathcal{K}}$, is decomposed by Gurtin (2002) in pure *screw* and *edge dislocations*, corresponding to different slip system within the constitutive framework of Crystal plasticity.

Remark As it follows from the characteristics attributed to the *edge dislocation* (Hirth and Lothe 1982), the Burgers vector is defined by

$$\begin{aligned} & \mathbf{b} = (\text{curl}(\mathbf{F}^p)) \, \mathbf{N}(\text{area } \mathcal{A}_0) \quad \text{with} \\ & \quad \text{curl } (\mathbf{F}^p) = a^p (\mathbf{e}_3 \otimes \mathbf{e}_1) \end{aligned} \tag{85}$$

where the unit vectors $\mathbf{e}$ are chosen in a such way to have $\mathbf{e}_1 = \mathbf{N}$, $\mathbf{e}_3$ is the Burgers direction, $\mathbf{e}_3 = \dfrac{\mathbf{b}}{|\, \mathbf{b} \,|}$, $\mathbf{e}_2 \perp \mathbf{b}$.

Consequently the edge dislocation is characterized by the *curl of plastic distortion* of the form

$$\text{curl } (\mathbf{F}^p) = \left(\frac{\partial F^p_{32}}{\partial x^3} - \frac{\partial F^p_{33}}{\partial x^2} \right) (\mathbf{e}_3 \otimes \mathbf{e}_1) \tag{86}$$

Let us **remark** that if there exists *a potential for plastic deformation*

$$F^p_{32} = \frac{\partial u^3}{\partial x^2}, \quad F^p_{33} = \frac{\partial u^3}{\partial x^3} \quad \text{with} \quad u^3 = u^3(x^2, x^3)$$
$$\equiv \mathbf{u} \cdot \frac{\mathbf{b}}{|\,\mathbf{b}\,|}, \tag{87}$$

which can be associated to a certain *plane motion* of dislocation, then in order to have non-zero *curl* it is necessary to accept a non-simply arc wise connected physical domain in a certain neighborhood of the given material point.

Following again (Hirth and Lothe 1982) the *screw dislocation* is characterized by a Burgers vector with the property listed below

$$\mathbf{b} = (\text{curl }(\mathbf{F}^p))\,\mathbf{N}\ (\text{area}\mathcal{A}_0) \quad \text{for}$$
$$\text{curl }(\mathbf{F}^p) = a^p(\mathbf{b} \otimes \mathbf{b}), \tag{88}$$

$$\mathbf{N} = \mathbf{e}_3, \quad \mathbf{e}_3 = \frac{\mathbf{b}}{|\,\mathbf{b}\,|}.$$

Let us characterize the *plastic curl* in this case

$$\text{curl }(\mathbf{F}^p) = \left(\frac{\partial F^p_{31}}{\partial x^2} - \frac{\partial F^p_{32}}{\partial x^1}\right)(\mathbf{e}_3 \otimes \mathbf{e}_3) \tag{89}$$

If there exists a potential for plastic deformation, defining the only two components F^p_{31}, F^p_{32} of the plastic distortion supposed to be non-zero, then

$$F^p_{31} = \frac{\partial u^3}{\partial x^1}, \quad F^p_{32} = \frac{\partial u^3}{\partial x^2} \quad \text{with}$$
$$u^3 = u^3(x^1, x^2) \equiv \mathbf{u} \cdot \frac{\mathbf{b}}{|\,\mathbf{b}\,|}, \tag{90}$$

which means that an anti-plane motion of dislocation has been considered. Again, in order to have non-vanishing *curl*, the physical space in a certain neighborhood of the material point is locally a simply arc wise connected domain.

6.1 Characteristics of the plastic distortion in the case of screw dislocation

The *simplest form* of the plastic distortion, compatible with the characterization given for the *screw dislocation* can be represented by the matrix

$$\mathbf{F}^p = \begin{pmatrix} 1 & 0 & 0 \\ 0 & 1 & 0 \\ F^p_{31} & F^p_{32} & 1 \end{pmatrix}, \tag{91}$$

in the Cartesian basis denoted by $\{\mathbf{e}_j\}_{\{j=\overline{1,3}\}}$, where F^p_{31}, F^p_{32} are functions of (x^1, x^2), and

$$\frac{\partial F^p_{31}}{\partial x^2} - \frac{\partial F^p_{32}}{\partial x^1} \neq 0, \quad \det \mathbf{F}^p = 1. \tag{92}$$

We kept the same notation for the tensor and its matrix representation, in a certain mentioned basis.

The plastic metric tensor $\mathbf{C}^p$ associated to the plastic distortion (91) is given by

$$\mathbf{C}^p = \begin{pmatrix} 1 + (\gamma_1)^2 & \gamma_1\gamma_2 & \gamma_1 \\ \gamma_1\gamma_2 & 1 + \gamma_2^2 & \gamma_2 \\ \gamma_1 & \gamma_2 & 1 \end{pmatrix}, \tag{93}$$

with the notation $F^p_{31} = \gamma_1, \quad F^p_{32} = \gamma_2$.

Let us introduce the function $\boldsymbol{\gamma}$

$$\boldsymbol{\gamma} = \gamma_1\mathbf{e}_1 + \gamma_2\mathbf{e}_2 \quad \text{with}$$
$$\gamma_1 = \gamma_1(x^1, x^2), \quad \gamma_2 = \gamma_2(x^1, x^2). \tag{94}$$

From (91) together with (94) we get

$$\mathbf{C}^p = \mathbf{I} + \boldsymbol{\gamma} \otimes \boldsymbol{\gamma} + \boldsymbol{\gamma} \otimes \mathbf{e}_3 + \mathbf{e}_3 \otimes \boldsymbol{\gamma}. \tag{95}$$

Let us introduce a local basis $(\boldsymbol{\mu}, \boldsymbol{\nu}, \mathbf{e}_3)$

$$\boldsymbol{\nu} = \frac{\boldsymbol{\gamma}}{|\,\boldsymbol{\gamma}\,|}, \quad \text{for} \quad |\,\boldsymbol{\gamma}\,| \equiv \sqrt{\gamma_1^2 + \gamma_2^2},$$
$$\boldsymbol{\mu} \in (\mathbf{e}_1, \mathbf{e}_2), \quad \text{such that} \quad \boldsymbol{\mu} \cdot \boldsymbol{\nu} = 0. \tag{96}$$

From (91) together with (94) we get

$$\mathbf{C}^p = \boldsymbol{\mu} \otimes \boldsymbol{\mu} + \mathbf{A}^p \quad \text{with}$$
$$\mathbf{A}^p = (1 + |\,\boldsymbol{\gamma}\,|^2)\boldsymbol{\nu} \otimes \boldsymbol{\nu} + |\,\boldsymbol{\gamma}\,|\,(\boldsymbol{\nu} \otimes \mathbf{e}_3$$
$$+ \mathbf{e}_3 \otimes \boldsymbol{\nu}) + \mathbf{e}_3 \otimes \mathbf{e}_3. \tag{97}$$

The second order tensor $\mathbf{A}^p$ has the matrix representation in the basis $\{\boldsymbol{\nu}, \mathbf{e}_3\}$

$$\mathbf{A}^p = \begin{pmatrix} 1 + |\,\boldsymbol{\gamma}\,|^2 & |\,\boldsymbol{\gamma}\,| \\ |\,\boldsymbol{\gamma}\,| & 1 \end{pmatrix}. \tag{98}$$

We define the positive square root tensor $(\mathbf{A}^p)^{1/2}$ and the symmetric and positive definite tensor $\mathbf{U}^p$.

Proposition 9 $(\mathbf{A}^p)^{1/2}$ *is defined by*

$$(\mathbf{A}^p)^{1/2} = \frac{1}{\sqrt{4 + |\,\boldsymbol{\gamma}\,|^2}}\left(\mathbf{A}^p + \hat{\mathbf{I}}_2\right),$$
$$\hat{\mathbf{I}}_2 = \boldsymbol{\nu} \otimes \boldsymbol{\nu} + \mathbf{e}_3 \otimes \mathbf{e}_3 \tag{99}$$

or in a matrix representation

$$(\mathbf{A}^p)^{1/2} = \frac{1}{\sqrt{4 + |\,\boldsymbol{\gamma}\,|^2}}\begin{pmatrix} 2 + |\,\boldsymbol{\gamma}\,|^2 & |\,\boldsymbol{\gamma}\,| \\ |\,\boldsymbol{\gamma}\,| & 2 \end{pmatrix}. \tag{100}$$

The symmetric and positive definite tensor $\mathbf{U}^p$ *can be expressed under the form*

$$\mathbf{U}^p = \boldsymbol{\mu} \otimes \boldsymbol{\mu} + (\mathbf{A}^p)^{1/2}. \tag{101}$$

Finally, we prove the formula

$$\begin{aligned}\mathbf{U}^p &\equiv (\mathbf{C}^p)^{1/2} = \boldsymbol{\mu}\otimes\boldsymbol{\mu} \\ &+\frac{1}{\sqrt{4+\mid\boldsymbol{\gamma}\mid^2}}(2+\mid\boldsymbol{\gamma}\mid^2)\boldsymbol{\nu}\otimes\boldsymbol{\nu} \\ &+\frac{\mid\gamma\mid}{\sqrt{4+\mid\boldsymbol{\gamma}\mid^2}}(\boldsymbol{\nu}\otimes\mathbf{e}_3+\mathbf{e}_3\otimes\boldsymbol{\nu}) \\ &+\frac{2}{\sqrt{4+\mid\boldsymbol{\gamma}\mid^2}}\mathbf{e}_3\otimes\mathbf{e}_3 \\ &\text{where}\quad \boldsymbol{\nu}=\frac{\boldsymbol{\gamma}}{\mid\boldsymbol{\gamma}\mid},\quad \boldsymbol{\mu}\cdot\boldsymbol{\nu}=0,\quad \boldsymbol{\mu}\cdot\mathbf{e}_3=0.\end{aligned} \tag{102}$$

Proof We write the Hamilton-Caley theorem for the tensor fields $(\mathbf{A}^p)^{1/2}$

$$\mathbf{A}^p-(\mathbf{A}^p)^{1/2}(\mathrm{tr}(\mathbf{A}^p)^{1/2})+\sqrt{(\det\mathbf{A}^p)}\hat{\mathbf{I}}_2=0 \tag{103}$$

When we apply the trace operator in (103) we get

$$(\mathrm{tr}(\mathbf{A}^p)^{1/2})^2=(\mathrm{tr}\mathbf{A}^p)+2\sqrt{(\det\mathbf{A}^p)} \tag{104}$$

and consequently

$$(\mathbf{A}^p)^{1/2}=\frac{1}{\mathrm{tr}(\mathbf{A}^p)^{1/2}}\left(\mathbf{A}^p+\sqrt{(\det\mathbf{A}^p)}\hat{\mathbf{I}}_2\right). \tag{105}$$

Using the explicit expression of the trace, the formula (99) follows at once.

Proposition 10

1. *The torsion of the plastic connection, attached to the plastic connection in the reference configuration , can be expressed in term of* ω, *by*

$$\mathbf{S}=\omega\mathbf{e}_3\otimes[\mathbf{e}_2\otimes\mathbf{e}_1-\mathbf{e}_1\otimes\mathbf{e}_2]. \tag{106}$$

2. *The second order torsion tensor is derived under the form*

$$\mathbf{N}=\omega\mathbf{e}_3\otimes\mathbf{e}_3,\quad\text{with}\quad\omega=\frac{\partial\gamma_2}{\partial x_1}-\frac{\partial\gamma_1}{\partial x_2}. \tag{107}$$

Proof By direct calculus we derive the expression of the plastic connection

$$\begin{aligned}\boldsymbol{\Gamma}^p &\equiv(\mathbf{F}^p)^{-1}(\nabla\mathbf{F}^p)=\mathbf{e}_3\otimes\mathbf{e}_1\otimes(\nabla\gamma_1) \\ &+\mathbf{e}_3\otimes\mathbf{e}_2\otimes(\nabla\gamma_2)\end{aligned} \tag{108}$$

A full characterization of the edge dislocation as well as the appropriate compatibility condition can be found in Cleja-Ţigoiu et al. (2007, in press).

Concluding remarks.

In order to compare the presented here results with the existing in the literature results in the field, we put into evidence different internal power expressions, proposed by Gurtin.

a. The additive decomposition of the displacement vector $\nabla\mathbf{u}=\mathbf{H}^e+\mathbf{H}^p$ into elastic and plastic parts is accepted in Gurtin (2003), within the framework of small deformation theory. The power expended in terms of an appropriate force system is given under the form $\int_{\mathcal{P}}(\mathbf{T}\cdot\dot{\mathbf{H}}^e+\mathbf{T}^p\cdot\dot{\mathbf{H}}^p+\mathcal{S}^p\cdot\nabla\dot{\mathbf{H}}^p)d\mathbf{x}$. Here $\mathbf{T}^p$ a second-order microstress and $\mathcal{S}^p$ a polar (third-order) microstress that together perform work in the evolution of the defects through this structure, for any part $\mathcal{P}$ of the body.
b. In Gurtin (2004) the power expended within any part $\mathcal{P}$ has the form $\int_{\mathcal{P}}(\mathbf{T}\cdot\dot{\mathbf{E}}^e+\mathbf{T}^p\cdot\dot{\mathbf{H}}^p+\mathcal{S}^p\cdot\mathrm{curl}\dot{\mathbf{H}}^p)d\mathbf{x}$. Here $\mathbf{E}^e=\{\mathbf{H}^e\}^S$, and the microscopic stress performs work in conjunction with the rate of Burgers vector, which is characterized by $\mathbf{G}=\mathrm{curl}(\mathbf{H}^p)$
c. Within the constitutive framework of Crystal plasticity, based on the multiplicative decomposition of the deformation gradient, the tensor field $\mathbf{G}=1/(\det\mathbf{F}^p)\mathbf{F}^p\mathrm{curl}(\mathbf{F}^p)$ (i.e just Noll's dislocation density, in Noll (1967), here in terms of the plastic distortion $\mathbf{F}^p$) is considered by Gurtin (2000) to be a measure of geometrically necessary dislocations. In this case the internal power is written in the form $\int_{\mathcal{P}}(\mathbf{T}\cdot\mathbf{L}^e+\sum_{\alpha}(\pi^\alpha v^\alpha+\xi^\alpha\cdot\mathrm{grad}v^\alpha))d\mathbf{x}$, where the sum is espanded to the slip systems and for each α, π^α—internal microforces and ξ^α—microstresses are introduced as forces conjugated to slip and they produce work by slip and by slip gradient respectively.

We conclude

1. As a peculiar aspect of the models proposed by Gurtin in the mentioned papers, the microbalance equation *generates the viscoplastic yield conditions* or the viscoplastic flow rules. On the other hand *no yield criteria* has been introduced, and the irreversible behavior can develop at the very beginning.
2. Let us remark that the micro balance equation together with the viscopalstic constitutive equation for the microforces and microstress momentum generate an appropriate *flow rule* (see also the non-local yield condition in Gurtin (2000))

$$\begin{aligned}&2\mathbf{C}^e(\mathbf{F}^p)\partial_{\mathbf{C}}\bar{\psi}(\mathbf{F}^p)^T+\partial_{\mathbf{F}^p}\bar{\psi}(\mathbf{F}^p)^T+Y(d^p)\mathbf{L}^p \\ &\quad-\mathrm{div}_{\mathcal{K}}((\mathbf{F}^p)^{-T}\partial_{\underset{\boldsymbol{\Gamma}}{(\mathrm{p})}}\bar{\psi}[(\mathbf{F}^p)^T,(\mathbf{F}^p)^T] \\ &\quad+h^2Y(d^p)\nabla_{\mathcal{K}}\widetilde{\mathbf{L}}^p)=\tilde{\rho}\mathbf{B}^p_m,\end{aligned} \tag{109}$$

with $d^p := \sqrt{\mathbf{L}^p \cdot \mathbf{L}^p + h^2 \nabla_{\mathcal{K}} \mathbf{L}^p \cdot \nabla_{\mathcal{K}} \mathbf{L}^p}$, for the plastic incompressible case $\rho_{\mathcal{K}} = \rho_0$.

3. Following the methodology developed in Cleja-Ţigoiu (2002b), we can derive from the proposed here model the behavior of the elasto-plastic material, in the case of small elastic strains but large elastic rotation $\mathbf{R^e}$. In this case $\mathbf{C}^e \simeq \mathbf{I} + 2\boldsymbol{\epsilon}$ with the elastic strain $| \boldsymbol{\epsilon} | << 1$, and the connection is generated by the gradient of the elastic rotation as well as the $\nabla\boldsymbol{\epsilon}$. Moreover, when we restrict ourselves to small elastic and plastic deformations, i.e. the small rotation and small strains, model within the constitutive framework adopted by Gurtin (2000) follows, but the microstress momentum are still presented.
4. The *plasticity and the damage defects* localize over the narrow region of the material. The plastic and the damage evolution process are inhomogeneous at the macroscale. The macroscopic inelastic deformation are sensitive to the structural defects within the volume element. Hence, *nonlocal theories are necessary* to adequately take into account mechanisms which take place in the neighborhood of the considered material points (see the comments by Brünig and Ricci (2005)). A nonlocal theory of an isotropically damaged materials can be developed based on our proposed model, in order to characterize the damage state configuration (identified with a configuration with torsion), independently of the current elastic deformation, i.e. for $\mathbf{C}^e = \mathbf{I}$. The damage produced by the microdefects can be identified with the presence of screw and/or edge dislocations.
5. The proposed description of the edge and screw dislocations (corresponding to deformation fields, which remain *incompatible* due to the non-zero $curl\mathbf{F}^p$) can be utilized in order to describe the appropriate fracture mode, by plane deformation and anti-plane deformation. The non-zero Burgers vector can be utilized in order to describe the cracks motion.
6. The quantities with the dimension of length appears as additional material parameters, in the expression of the free energy density, in the *intensity for the accumulated effect* of the rate of plastic distortion $\mathbf{L}^p$ and of the gradient of $\mathbf{L}^p$.
7. In the model (although at the general framework presented here there is not necessary) we can assume the existence of the viscoplastic (or an yield) function defined in the physical force system $(\mathbf{T}, \boldsymbol{\mu})$, such that the plastic (viscoplastic) behavior can develope if and only if during the deformation process the physical force system lays on the yield surface or it is situated outward the surface $f_{\mathcal{K}}(\mathbf{T}, \boldsymbol{\mu}) \geq 0$. Consequently the behavior would be elastic if the physical force system remains inside the yield surface.

Acknowledgments The author acknowledges support from the Romanian Ministry of Education and Research through CEEX programm (Contract No. CERES-11-12/25.07.2006).

References

1. Acharya A (2004) Constitutive analysis of finite deformation field dislocation mechanics. J Mech Phys Solid 52:301–316
2. Bilby BA (1960) Continuous distribution of dislocations. In: Sneddon IN, Hill R (eds) Progress in solid mechanics. North-Holland, Amsterdam, pp 329–398
3. Beju I, Soós E, Teodorescu PP (1983) Spinor and non-Euclidean tensor calculus with applications. Ed. Tehnica, Bucureşti Romania, Abacus Press, Tunbridge Wells, Kent, England (romanian version 1979)
4. Brüning M, Ricci S (2005) Nonlocal continuum theory of an isotropically damaged metals. Int J Plast 21:1346–1382
5. Cleja-Ţigoiu S (1990) Large elasto-plastic deformations of materials with relaxed configurations – I. Constitutive assumptions, II. Role of the complementary plastic factor. Int J Eng Sci 28:171–180, 273–284
6. Cleja-Ţigoiu S (2001) A model of crystalline materials with dislocations. In: Cleja-Ţigoiu S, Ţigoiu V (eds) Proceedings of 5th international Seminar geometry continua and microstructures. Sinaia, Romania, pp 25–36
7. Cleja-Ţigoiu S (2002a) Couple stresses and non-Riemannian plastic connection in finite elasto-plasticity. ZAMP 53:996–1013
8. Cleja-Ţigoiu S (2002b) Small elastic strains in finite elasto-plastic materials with continuously distributed dislocations. Theor Appl Mech 28(29):93–112
9. Cleja-Ţigoiu S, Maugin GA (2000) Eshelby's stress tensors in finite elastoplasticity. Acta Mechanica 139:231–249
10. Cleja-Ţigoiu S, Soós E (1990) Elastoplastic models with relaxed configurations and internal state variables. Appl Mech Rev 43:131–151
11. Cleja-Ţigoiu S, Fortunée D, Vallée C (2007) Torsion equation in anisotropic elasto-plastic materials with continuously distributed dislocations. Math Mech Solids doi:10.1177/1081286507079157
12. Epstein M, Maugin GA (2000) Thermomechanics of volumetric growth in uniform bodies. Int J Plast 16:951–978
13. Fleck NA, Muller GM, Ashby MF, Hutchinson JW (1994) Strain gradient plasticity: theory and experiment. Acta Metall Mater 42:475–487

14. Forest S, Cailletand G, Sievert R (1997) A Cosserat theory for elastoviscoplastic single crystals at finite deformation. Arch Mech 49(4):705–736
15. Gupta A, Steigmann D, Stölken JS (2006) On the evolution of plasticity and incompatibility. Math Mech Solids. online: doi:10.1177/1081286506064721
16. Gurtin ME (2000) On the plasticity of single crystal: free energy, microforces, plastic-strain gradients. J Mech Phys Solids 48:989–1036
17. Gurtin ME (2002) A gradient theory of single-crystal viscoplasticity that accounts for geometrically necessary dislocations. J Mech Phys Solids 50:5–32
18. Gurtin ME (2003) On a framework for small-deformation viscoplasticity: free energy, microforces, strain gradients. Int J Plast 19:47–90
19. Gurtin ME (2004) A gradient theory of small-deformation isotropic plasticity that accounts for the Burgers and for dissipation due to plastic spin. J Mech Phys Solids 52:2545–2568
20. Gurtin ME, Needleman A (2005) Boundary conditions in small-deformation, single-crystal plasticity that account for the Burgers. J Mech Phys Solids 53:1–31
21. Hirth J, Lothe JP (1982) Theory of dislocations. Krieger Publishing, Malabar, Florida
22. Kondo K, Yuki M (1958) On the current viewpoints of non-Riemannian plasticity theory. In: RAAG memoirs of the unifying study of basic problems in engng and physical sciences by means of geometry II (D), Tokyo, pp 202–226
23. Kröner E (1963) On the physical reality of torque stresses in continuum mechanics. Gauge theory with disclinations. Int J Eng Sci 1:261–278
24. Kröner E (1992) The internal mechanical state of solids with defects. Int J Solids Struct 29:1849–1857
25. Kröner E, Lagoudas DC (1992) Gauge theory with disclinations. Int J Eng Sci 30:1849–1857
26. Le KC, Stumpf H (1996a) A model of elastoplastic bodies with continuously distributed dislocations. Int J Plast 12((5):611–627
27. Le KC, Stumpf H (1996b) Nonlinear continuum with dislocations. Int J Eng Sci 34:339–358
28. Le KC, Stumpf H (1996c) On the determination of the crystal reference in nonlinear continuum theory of dislocation. Proc Roy Soc London A 452:359–371
29. Maugin GA (1999) The thermomechanics of nonlinear irreversible behaviors. World Scientific
30. Naghdi PM, Srinivasa AR (1994) Characterization of dislocations and their influence on plastic deformation in single crystal. Int J Eng Sci 32(7):1157–1182
31. Noll W (1967) Materially uniform simple bodies with inhomogeneities. Arch Rat Mech Anal 27:1–32
32. Schouten JA (1954) Ricci-Calculus. Springer-Verlag, Berlin
33. Steinmann P (1994) A micropolar theory of finite deformation and finite rotation multiplicative elastoplasticity. Int J Solids Struct 31:1063–1084
34. Steinmann P (1997) Continuum theory of dislocations: impact to single cristal plasticity. In: Owen DRJ, Onãte E, Hinton EComputational plasticity, fundamental and applications. CIME, Barcelona,
35. Steinmann P (2002) On spatial and material settings of hyperelastostatic crystal defects. J Mech Phys Solids 50:1743–1766
36. Stumpf H, Hackle K (2003) Micromechanical concept for analysis of damage evolution in thermo-viscoplastic and quasi-brittle materials. Int J Solids Struct 40:1567–1584
37. Teodosiu C (1970) A dynamic theory of dislocations and its applications to the theory of the elastic-plastic continuum. In: Simmons JA, de Witt R, Bullough R (eds) Fundamental aspects of dislocation theory, Nat Bur Stand (U.S.), Spec. Publ. 317, II, pp 837–876
38. Wang CC (1967) On the geometric structure of simple bodies, a Mathematical foundation for the theory of continuous distributions of dislocations. Arch Rat Mech Anal 27:33–94

Distributed dislocation approach for cracks in couple-stress elasticity: shear modes

P. A. Gourgiotis · H. G. Georgiadis

Abstract The distributed dislocation technique proved to be in the past an effective approach in studying crack problems within classical elasticity. The present work aims at extending this technique in studying crack problems within couple-stress elasticity, i.e. within a theory accounting for effects of microstructure. As a first step, the technique is introduced to study finite-length cracks under remotely applied shear loadings (mode II and mode III cases). The mode II and mode III cracks are modeled by a continuous distribution of glide and screw dislocations, respectively, that create both standard stresses and couple stresses in the body. In particular, it is shown that the mode II case is governed by a singular integral equation with a more complicated kernel than that in classical elasticity. The numerical solution of this equation shows that a cracked material governed by couple-stress elasticity behaves in a more rigid way (having increased stiffness) as compared to a material governed by classical elasticity. Also, the stress level at the crack-tip region is appreciably higher than the one predicted by classical elasticity. Finally, in the mode III case the corresponding governing integral equation is hypersingular with a cubic singularity. A new mechanical quadrature is introduced here for the numerical solution of this equation. The results in the mode III case for the crack-face displacement and the near-tip stress show significant departure from the predictions of classical fracture mechanics.

P. A. Gourgiotis · H. G. Georgiadis (✉)
Mechanics Division, National Technical University of Athens, Zographou Campus, Zographou, 15773, Greece
e-mail: georgiad@central.ntua.gr

Keywords Distributed dislocations · Cracks · Couple-stress elasticity · Integral equations

1 Introduction

The present work is concerned with the study of mode II and mode III *finite-length* cracks in a material with microstructure. We assume that the response of the material is governed by couple-stress elasticity. This theory falls into the category of generalized continuum theories and is a particular case of the general approaches of Toupin (1962), Mindlin (1964), and Green and Rivlin (1964). As is well-known, ideas underlying couple-stress elasticity were advanced first by Voigt (1887) and the Cosserat brothers (1909), but the subject was generalized and reached maturity only with the works of Toupin (1962), Mindlin and Tiersten (1962), Mindlin (1964), and Koiter (1964).

Earlier application of the couple-stress elasticity, mainly on stress-concentration problems, met with some success providing solutions physically more adequate than solutions based on classical elasticity (see e.g. Mindlin and Tiersten 1962; Weitsman 1965; Bogy and Sternberg 1967a, b). Work employing couple-stress theories on elasticity and plasticity problems is also continued in recent years (see e.g. Vardoulakis and Sulem 1995; Huang et al. 1997; Chen et al. 1998;

doi: 10.1007/978-1-4020-6929-1_9,

Anthoine 2000; Lubarda and Markenscoff 2000; Bardet and Vardoulakis 2001; Georgiadis and Velgaki 2003; Grentzelou and Georgiadis 2005).

Nevertheless, there is only a limited number of studies concerning the effects of couple-stresses in crack problems. One of the earlier works in this subject is that of Sternberg and Muki (1967) who considered the mode I finite-length crack by employing the method of dual integral equations. They provided only asymptotic results and showed that both the stress and couple-stress fields exhibit a square-root singularity while the rotation field is bounded at the crack-tip. The same method was adopted by Ejike (1969) for a circular (penny-shaped) crack in couple-stress elasticity and by Paul and Sridharan (1980, 1981) for a finite-length crack in micropolar elasticity. Using the Wiener-Hopf technique, Atkinson and Leppington (1977) studied the problem of a semi-infinite crack with exponentially decayed normal tractions on the crack faces. More recently, Huang et al. (1997) provided near-tip asymptotic fields for the mode I and mode II crack problems, in couple-stress elasticity, by using the method of eigenfunction expansions. Also, Zhang et al. (1998) by employing the Wiener-Hopf technique investigated the mode III semi-infinite crack in couple-stress elasticity in the special case where the second couple-stress moduli is set equal to zero. Moreover, using a similar approach, Huang et al. (1999) obtained full-field solutions for semi-infinite cracks under mode I and mode II loadings in elastic-plastic materials with strain-gradient effects.

Here, we aim at providing full-field solutions to the mode II and mode III *finite-length* crack problems within couple-stress elasticity by introducing an approach based on *distributed dislocations*. Since the pioneering work of Bilby et al. (1963), Bilby and Eshelby (1968) the distributed-dislocation technique has been employed to analyze various crack problems in classical elasticity. A thorough exposition of the technique can be found in the treatise by Hills et al. (1996). The strength of this analytical/numerical technique lies in the fact that it gives detailed *full-field* solutions for crack problems at the expense of relatively little analytical demands as compared to the elaborate technique of dual integral equations and, also, of relatively little computational demands as compared to the Finite Element and Boundary Element methods. Although the technique has proven to be very successful in studying crack problems within classical elasticity, it appears that there is no work at all in modeling cracks with distribution of dislocations in materials with microstructure. Therefore, the present work aims at extending the technique in couple-stress elasticity. In another recent work by the present authors (Gourgiotis and Georgiadis 2007) the mode I crack problem was also considered within the same framework. A comparison between the mode II case studied here and the mode I case leads to the conclusion that the opening mode is mathematically more involved than the shear mode. This is in some contrast with situations of classical elasticity where the two plane-strain crack modes involve equivalent mathematical effort.

As in analogous situations of classical elasticity, a superposition scheme will be followed. Thus, the solution to the basic problem (body with a traction-free crack under remote shear field) will be obtained by the superposition of the stress field arising in the un-cracked body (of the same geometry) to the 'corrective' stresses and couple-stresses induced by a continuous distribution of dislocations chosen so that the crack-faces become traction-free. The stress field for a discrete glide and screw dislocation in couple-stress elasticity will serve, respectively, as the Green's function for the mode II and mode III problem. However, we note that deriving the stress field of a discrete dislocation within generalized continua is by no means a straightforward task. Within the framework of couple-stress elasticity a lot of research has been devoted to dislocations. Representative references include work by Kroner (1963), Misicu (1965), Teodosiu (1965), Cohen (1966), Anthony (1970), Knesl and Semela (1972) and Nowacki (1974). Finally, it is shown that due to the nature of the above Green's functions and the boundary conditions that arise in couple-stress elasticity, the aforementioned procedure results for the mode II case in a singular integral equation (SIE), whereas for the mode III case in a hypersingular integral equation (IE) with a cubic singularity. In order to solve this hypersingular IE, a new mechanical quadrature is constructed.

2 Basic concepts and equations of couple-stress elasticity

In this Section, we briefly present the basic ideas and equations of couple-stress elasticity. The theory employed here is a particular case of form III in the general Mindlin's (1964) approach. Nevertheless, we chose to present an alternative approach to Mindlin's variational approach. Indeed, our derivation of basic results relies on the *momentum balance laws*, which—in our opinion—provide more physical insight. It should also be mentioned that versions of the quasi-static couple-stress theory were given by, among others, Aero and Kuvshinskii (1960), Mindlin and Tiersten (1962), Koiter (1964), Palmov (1964), and Muki and Sternberg (1965). The basic equations of dynamical couple-stress theory (including the effects of micro-inertia) were given by Georgiadis and Velgaki (2003).

In the absence of inertia effects, for a control volume CV with bounding surface S, the balance laws for the linear and angular momentum read

$$\int_S T_i^{(n)} dS + \int_{CV} F_i d(CV) = 0, \tag{1}$$

$$\int_S \left(x_j T_k^{(n)} e_{ijk} + M_i^{(n)}\right) dS + \int_{CV} (x_j F_k e_{ijk} + C_i) d(CV) = 0, \tag{2}$$

where $T_i^{(n)}$ is the surface force per unit area (force traction), F_i is the body force per unit volume, $M_i^{(n)}$ is the surface moment per unit area (couple traction), and C_i is the body moment per unit volume.

Next, pertinent *force-stress* and *couple-stress* tensors are introduced by considering the equilibrium of the elementary material tetrahedron and enforcing (1) and (2), respectively. The force-stress tensor σ_{ij} (which is asymmetric) is defined by

$$T_i^{(n)} = \sigma_{ji} n_j, \tag{3}$$

and the couple-stress tensor μ_{ij} (which is also asymmetric) by

$$M_i^{(n)} = \mu_{ji} n_j, \tag{4}$$

where n_j are the direction cosines of the outward unit vector $\mathbf{n}$, which is normal to the surface. In addition just like the third Newton's law $\mathbf{T}^{(\mathbf{n})} = -\mathbf{T}^{(-\mathbf{n})}$ is proved to hold by considering the equilibrium of a material 'slice', it can also be proved that $\mathbf{M}^{(\mathbf{n})} = -\mathbf{M}^{(-\mathbf{n})}$. The couple-stresses μ_{ij} are expressed in dimensions of [force][length]$^{-1}$. Further, σ_{ij} can be decomposed into a symmetric and anti-symmetric part

$$\sigma_{ij} = \tau_{ij} + \alpha_{ij}, \tag{5}$$

with $\tau_{ij} = \tau_{ji}$ and $\alpha_{ij} = -\alpha_{ji}$, whereas it is advantageous to decompose μ_{ij} into its deviatoric $\mu_{ij}^{(D)}$ and spherical $\mu_{ij}^{(S)}$ part in the following manner

$$\mu_{ij} = m_{ij} + \frac{1}{3}\delta_{ij}\mu_{kk}, \tag{6}$$

where $m_{ij} = \mu_{ij}^{(D)}$, $\mu_{ij}^{(S)} = (1/3)\,\delta_{ij}\mu_{kk}$ and δ_{ij} is the Kronecker delta. Now, with the above definitions in hand and with the help of the divergence theorem, one may obtain the equations of equilibrium. Thus, Eq. 2 leads to the following moment equation

$$\partial_i \mu_{ij} + \sigma_{ki} e_{ijk} + C_j = 0, \tag{7}$$

which can also be written as

$$\frac{1}{2}\partial_i \mu_{il} e_{jkl} + \alpha_{jk} + \frac{1}{2} C_l e_{jkl} = 0, \tag{8}$$

since by its definition the anti-symmetric part of stress is written as $\boldsymbol{\alpha} \equiv -(1/2)\,\mathbf{I} \times (\boldsymbol{\sigma} \times \mathbf{I})$, where $\mathbf{I}$ is the idemfactor. Also, Eq. 1 leads to the following force equation

$$\partial_j \sigma_{jk} + F_k = 0, \tag{9}$$

or, by virtue of (5), to the equation

$$\partial_j \tau_{jk} + \partial_j \alpha_{jk} + F_k = 0. \tag{10}$$

Further, combining (8) and (10) yields the *single* equation

$$\partial_j \tau_{jk} - \frac{1}{2}\partial_j \partial_i \mu_{il} e_{jkl} + F_k - \frac{1}{2}\partial_j C_l e_{jkl} = 0. \tag{11}$$

Finally, in view of Eq.6 and by taking into account that curl $\left(\text{div}\left((1/3)\,\delta_{ij}\mu_{kk}\right)\right) = 0$, we write (11) as

$$\partial_j \tau_{jk} - \frac{1}{2}\partial_j \partial_i m_{il} e_{jkl} + F_k - \frac{1}{2}\partial_j C_l e_{jkl} = 0. \tag{12}$$

Equation 12 is therefore the single equation of equilibrium.

As for the kinematical description of the continuum, the following quantities are defined within the geometrically linear theory

$$\varepsilon_{ij} = \frac{1}{2}\left(\partial_j u_i + \partial_i u_j\right), \tag{13}$$

$$\omega_{ij} = \frac{1}{2}\left(\partial_j u_i - \partial_i u_j\right), \tag{14}$$

$$\omega_i = \frac{1}{2} e_{ijk} \partial_j u_k, \tag{15}$$

$$\kappa_{ij} = \partial_i \omega_j, \tag{16}$$

where ε_{ij} is the strain tensor, ω_{ij} is the rotation tensor, ω_i is the rotation vector, and κ_{ij} is the curvature tensor (i.e. the gradient of rotation or the curl of the strain) expressed in dimensions of $[\text{length}]^{-1}$. Notice also that Eq. 16 can alternatively be written as

$$\kappa_{ij} = \frac{1}{2} e_{jkl} \partial_i \partial_k u_l = e_{jkl} \partial_k \varepsilon_{il}. \tag{17}$$

Equation 17 expresses compatibility for curvature and strain fields. In addition, there is an identity, i.e. $\partial_k \kappa_{ij} = \partial_k \partial_i \omega_j = \partial_i \partial_k \omega_j = \partial_i \kappa_{kj}$, which expresses compatibility for the curvature components. The compatibility equations for the strain components are the usual Saint Venant's compatibility equations. We notice also that $\kappa_{ii} = 0$ because $\kappa_{ii} = \partial_i \omega_i = (1/2)\, e_{ijk} u_{k,ji} = 0$ and, therefore, that κ_{ij} has only eight independent components. The tensor κ_{ij} is obviously an *asymmetric* tensor.

Now, regarding the traction boundary conditions, we note that at first sight, it might seem plausible that the surface tractions (i.e. the force-traction and the couple-traction) can be prescribed arbitrarily on the external surface of the body through relations (3) and (4), which stem from the equilibrium of the material tetrahedron. However, as Koiter (1964) pointed out, the resulting number of six traction boundary conditions (three force-tractions and three couple-tractions) would be in contrast with the *five* geometric boundary conditions that can be imposed. Indeed, since the rotation vector ω_i in couple-stress elasticity is not independent of the displacement vector u_i (cf. (15)), the normal component of the rotation is fully specified by the distribution of tangential displacements over the boundary. Therefore, only the three displacement and the two tangential rotation components can be prescribed independently. As a consequence, only *five* surface tractions (i.e. the work conjugates of the above five independent kinematical quantities) can be specified at a point of the bounding surface of the body. These are three *reduced* force-tractions and two tangential couple-tractions (Mindlin and Tiersten 1962; Koiter 1964)

$$P_i^{(n)} = \sigma_{ji} n_j - \frac{1}{2} e_{ijk} n_j \partial_k m_{(nn)}, \tag{18}$$

$$R_i^{(n)} = m_{ji} n_j - m_{(nn)} n_i, \tag{19}$$

where $m_{(nn)} = n_i n_j m_{ij}$ is the normal component of the deviatoric couple-stress tensor m_{ij}. Finally, it is worth noting that in the micropolar (Cosserat) theory of elasticity (see e.g. Nowacki 1972), the traction boundary conditions are six since the rotation is fully independent of the displacement vector. In this case the tractions can directly be derived from the equilibrium of the material tetrahedron, i.e. the relations between tractions and stresses are given by (3) and (4).

Introducing the constitutive equations of the theory is now in order. We assume a linear and isotropic material response, in which case the potential-energy density takes the form

$$W \equiv W\left(\varepsilon_{ij}, \kappa_{ij}\right) = \frac{1}{2} \lambda \varepsilon_{ii} \varepsilon_{jj} + \mu \varepsilon_{ij} \varepsilon_{ij} + 2\eta \kappa_{ij} \kappa_{ij} + 2\eta' \kappa_{ij} \kappa_{ji}, \tag{20}$$

where $\left(\lambda, \mu, \eta, \eta'\right)$ are material constants. Then, Eq. 20 leads, through the standard variational manner, to the following constitutive equations

$$\tau_{ij} \equiv \sigma_{(ij)} = \frac{\partial W}{\partial \varepsilon_{ij}} = \lambda \delta_{ij} \varepsilon_{kk} + 2\mu \varepsilon_{ij}, \tag{21}$$

$$m_{ij} = \frac{\partial W}{\partial \kappa_{ij}} = 4\eta \kappa_{ij} + 4\eta' \kappa_{ji}. \tag{22}$$

In view of (21) and (22), the moduli (λ, μ) have the same meaning as the Lamé constants of classical elasticity theory, whereas the moduli $\left(\eta, \eta'\right)$ account for couple-stress effects.

Finally, the following points are of notice: (i) The couple-stress moduli $\left(\eta, \eta'\right)$ are expressed in dimensions of [force]. (ii) Since $\kappa_{ii} = 0$, $m_{ii} = 0$ is also valid and therefore the tensor m_{ij} has only eight independent components. (iii) The scalar $(1/3)\, \mu_{kk}$ of the couple-stress tensor does not appear in the final equation of equilibrium, nor in the reduced boundary conditions and the constitutive equations. Consequently, $(1/3)\, \mu_{kk}$ is left indeterminate within the couple-stress theory. (iv) The following restrictions for the material constants should prevail on the basis of a positive definite potential-energy density (Mindlin and Tiersten 1962)

$$3\lambda + 2\mu > 0, \quad \mu > 0, \quad \eta > 0, \quad -1 < \frac{\eta'}{\eta} < 1. \tag{23a,b,c,d}$$

3 Plane problems of couple-stress elasticity

The cases of plane strain and anti-plane strain are examined here and the basic equations are given. In what follows, vanishing body forces and body couples are assumed.

3.1 Plane-strain

For a body that occupies a domain in the (x, y)-plane under conditions of plane strain, the displacement field takes the general form

$$u_x \equiv u_x(x, y) \neq 0, \quad u_y \equiv u_y(x, y) \neq 0, \quad u_z \equiv 0. \tag{24a,b,c}$$

By virtue of (13)–(16), the non-vanishing components of strain, rotation and curvature are given as

$$\varepsilon_{xx} = \frac{\partial u_x}{\partial x}, \quad \varepsilon_{yy} = \frac{\partial u_y}{\partial y}, \quad \varepsilon_{xy} = \varepsilon_{yx} = \frac{1}{2}\left(\frac{\partial u_y}{\partial x} + \frac{\partial u_x}{\partial y}\right), \tag{25a,b,c}$$

$$\omega_z = \omega_{xy} = \frac{1}{2}\left(\frac{\partial u_y}{\partial x} - \frac{\partial u_x}{\partial y}\right), \tag{26}$$

$$\kappa_{xz} = \frac{\partial \omega_z}{\partial x}, \quad \kappa_{yz} = \frac{\partial \omega_z}{\partial y}. \tag{27a,b}$$

Also, from the constitutive Eqs. 21 and 22, the following relations are derived between stress and strain and between couple-stress and curvature

$$\tau_{xx} = (2\mu + \lambda)\,\varepsilon_{xx} + \lambda\varepsilon_{yy}, \quad \tau_{yy} = (2\mu + \lambda)\,\varepsilon_{yy} + \lambda\varepsilon_{xx}, \quad \tau_{xy} = 2\mu\varepsilon_{xy}, \tag{28a,b,c}$$

$$m_{xz} = 4\eta\kappa_{xz}, \quad m_{yz} = 4\eta\kappa_{yz}, \tag{29a,b}$$

whereas, the remaining components are given by

$$\tau_{zz} = -\frac{\lambda}{2(\lambda + \mu)}\left(\tau_{xx} + \tau_{yy}\right), \quad m_{zx} = \frac{\eta'}{\eta} m_{xz}, \quad m_{zy} = \frac{\eta'}{\eta} m_{yz}. \tag{30a,b,c}$$

Next, the non-vanishing components of the anti-symmetric part of the force-stress tensor are obtained from (8) as

$$\alpha_{xy} = -\alpha_{yx} = -\frac{1}{2}\left(\frac{\partial m_{xz}}{\partial x} + \frac{\partial m_{yz}}{\partial y}\right) = -2\eta\nabla^2\omega_z. \tag{31}$$

It should be noticed that the independence upon the coordinate z of *all* components of the force-stress and couple-stress tensors, under the assumption (24c), was proved by Muki and Sternberg (1965). Indeed, it is noteworthy that, contrary to the respective plane-strain case in the conventional theory, this independence is not obvious within the couple-stress theory.

Mindlin's stress functions

As Mindlin (1963) indicated, the equations of equilibrium in (7) and (9), in a plane-strain state, are identically satisfied when the stresses are derived from two stress functions $\Phi(x, y)$ and $\Psi(x, y)$ in the following manner

$$\sigma_{xx} = \frac{\partial^2 \Phi}{\partial y^2} - \frac{\partial^2 \Psi}{\partial y \partial x}, \quad \sigma_{yy} = \frac{\partial^2 \Phi}{\partial x^2} + \frac{\partial^2 \Psi}{\partial x \partial y}, \tag{32a,b}$$

$$\sigma_{xy} = -\frac{\partial^2 \Phi}{\partial y \partial x} - \frac{\partial^2 \Psi}{\partial y^2}, \quad \sigma_{yx} = -\frac{\partial^2 \Phi}{\partial y \partial x} + \frac{\partial^2 \Psi}{\partial x^2}, \tag{33a,b}$$

$$m_{xz} = \frac{\partial \Psi}{\partial x}, \quad m_{yz} = \frac{\partial \Psi}{\partial y}. \tag{34a,b}$$

where the functions Φ and Ψ satisfy the following PDEs

$$\nabla^4\Phi = 0, \quad \nabla^2\left(\ell^2\nabla^2 - 1\right)\Psi = 0. \tag{35a,b}$$

According to the compatibility equations between curvature and strain in (17), the stress functions are related through the following equations

$$\frac{\partial}{\partial x}\left(\Psi - \ell^2\nabla^2\Psi\right) = -2(1 - \nu)\,\ell^2\frac{\partial}{\partial y}\left(\nabla^2\Phi\right), \tag{36}$$

$$\frac{\partial}{\partial y}\left(\Psi - \ell^2\nabla^2\Psi\right) = 2(1 - \nu)\,\ell^2\frac{\partial}{\partial x}\left(\nabla^2\Phi\right), \tag{37}$$

where ν is the Poisson's ratio and $\ell \equiv (\eta/\mu)^{1/2}$ is a *characteristic material length.*

3.2 Anti-plane strain

For a body occupying a domain in the (x, y)-plane under conditions of anti-plane strain, the displacement field takes the general form

$$u_x \equiv 0, \quad u_y \equiv 0, \quad u_z = w(x, y) \neq 0. \tag{38a,b,c}$$

Again, by virtue of (13)–(16), the non-vanishing components of strain, rotation and curvature are given as

$$\varepsilon_{xz} = \varepsilon_{zx} = \frac{1}{2}\frac{\partial w}{\partial x}, \quad \varepsilon_{yz} = \varepsilon_{zy} = \frac{1}{2}\frac{\partial w}{\partial y}, \tag{39a,b}$$

$$\omega_x = \omega_{yz} = \frac{1}{2}\frac{\partial w}{\partial y}, \quad \omega_y = \omega_{xz} = -\frac{1}{2}\frac{\partial w}{\partial x}, \tag{40a,b}$$

$$\kappa_{xx} = -\kappa_{yy} = \frac{1}{2}\frac{\partial^2 w}{\partial x \partial y}, \quad \kappa_{xy} = -\frac{1}{2}\frac{\partial^2 w}{\partial x^2},$$
$$\kappa_{yx} = \frac{1}{2}\frac{\partial^2 w}{\partial y^2}. \tag{41a,b,c}$$

Then, the constitutive equations in 21 and 22 provide

$$\tau_{xz} = 2\mu\varepsilon_{xz} = \mu\frac{\partial w}{\partial x}, \quad \tau_{yz} = 2\mu\varepsilon_{yz} = \mu\frac{\partial w}{\partial y}, \tag{42a,b}$$

$$m_{xx} = 4\left(\eta+\eta'\right)\kappa_{xx} = 2\left(\eta+\eta'\right)\frac{\partial^2 w}{\partial x \partial y}, \tag{43a}$$

$$m_{yy} = 4\left(\eta+\eta'\right)\kappa_{yy} = -2\left(\eta+\eta'\right)\frac{\partial^2 w}{\partial x \partial y} = -m_{xx}, \tag{43b}$$

$$m_{xy} = 4\eta\kappa_{xy} + 4\eta'\kappa_{yx} = -2\eta\frac{\partial^2 w}{\partial x^2} + 2\eta'\frac{\partial^2 w}{\partial y^2}, \tag{43c}$$

$$m_{yx} = 4\eta\kappa_{yx} + 4\eta'\kappa_{xy} = 2\eta\frac{\partial^2 w}{\partial y^2} - 2\eta'\frac{\partial^2 w}{\partial x^2}. \tag{43d}$$

Further, the non-vanishing components of the anti-symmetric part of the force-stress tensor are obtained from (8)

$$\alpha_{zx} = -\alpha_{xz} = \frac{1}{2}\left(\frac{\partial m_{xy}}{\partial x} + \frac{\partial m_{yy}}{\partial y}\right) = \eta\frac{\partial}{\partial x}\left(\nabla^2 w\right), \tag{44a}$$

$$\alpha_{zy} = -\alpha_{yz} = -\frac{1}{2}\left(\frac{\partial m_{xx}}{\partial x} + \frac{\partial m_{yx}}{\partial y}\right) = \eta\frac{\partial}{\partial y}\left(\nabla^2 w\right). \tag{44b}$$

Finally, by taking into account (5) and (44), the components of the force-stress tensor can be written as

$$\sigma_{xz} = \mu\frac{\partial}{\partial x}\left(w - \ell^2\nabla^2 w\right),$$
$$\sigma_{zx} = \mu\frac{\partial}{\partial x}\left(w + \ell^2\nabla^2 w\right), \tag{45a,b}$$

$$\sigma_{yz} = \mu\frac{\partial}{\partial y}\left(w - \ell^2\nabla^2 w\right),$$
$$\sigma_{zy} = \mu\frac{\partial}{\partial y}\left(w + \ell^2\nabla^2 w\right). \tag{46a,b}$$

In view of the above and by enforcing equilibrium, a single PDE of the fourth order for the displacement component is obtained

$$\nabla^2 w - \ell^2\nabla^4 w = 0. \tag{47}$$

4 Discrete dislocations in couple-stress elasticity

4.1 Glide dislocation

Consider a glide dislocation with Burgers vector $\mathbf{b} = (b, 0, 0)$ imposed in an infinite medium along the plane $x > 0$, $y = 0$. The appropriate Mindlin's stress functions for this problem were given by Cohen (1966), Knesl and Semela (1972), and Nowacki (1974)

$$\Phi = -\frac{\mu b r}{4\pi(1-\nu)}(2\ln r + 1)\sin\theta, \tag{48}$$

$$\Psi = \frac{2\mu b\ell}{\pi}\left[K_1(r/\ell) - \ell/r\right]\cos\theta, \tag{49}$$

where $r = \left(x^2+y^2\right)^{1/2}$, $\theta = \tan^{-1}(y/x)$ and $K_i(r/\ell)$ is the ith-order modified Bessel function of the second kind. Further, the stresses induced at a point (x, y) may be found from the above stress functions by using Eqs. 32–34

$$\sigma_{xx} = -\frac{\mu b}{4\pi(1-\nu)r}(3\sin\theta + \sin 3\theta) + \frac{2\mu b}{\pi r}\left[\frac{2\ell^2}{r^2} - K_2(r/\ell)\right]\sin 3\theta - \frac{\mu b}{4\pi\ell^2}r\left[K_2(r/\ell) - K_0(r/\ell)\right] \times (\sin\theta + \sin 3\theta), \tag{50}$$

$$\sigma_{yy} = \frac{\mu b}{4\pi(1-\nu)r}(\sin 3\theta - \sin\theta) - \frac{2\mu b}{\pi r}\left[\frac{2\ell^2}{r^2} - K_2(r/\ell)\right]\sin 3\theta + \frac{\mu b}{4\pi\ell^2}r\left[K_2(r/\ell) - K_0(r/\ell)\right] \times (\sin\theta + \sin 3\theta), \tag{51}$$

$$\sigma_{xy} = \frac{\mu b}{4\pi(1-\nu)r}(\cos\theta + \cos 3\theta) - \frac{2\mu b}{\pi r}\left[\frac{2\ell^2}{r^2} - K_2(r/\ell)\right]\cos 3\theta - \frac{\mu b}{4\pi\ell^2}r\left[K_2(r/\ell) - K_0(r/\ell)\right] \times (\cos\theta - \cos 3\theta), \tag{52}$$

$$\sigma_{yx} = \frac{\mu b}{4\pi(1-\nu)r}(\cos\theta + \cos 3\theta) - \frac{2\mu b}{\pi r}\left[\frac{2\ell^2}{r^2} - K_2(r/\ell)\right]\cos 3\theta + \frac{\mu b}{4\pi\ell^2}r\left[K_2(r/\ell) - K_0(r/\ell)\right] \times (3\cos\theta + \cos 3\theta), \tag{53}$$

$$m_{xz} = \frac{\mu b}{\pi}\left[\frac{2\ell^2}{r^2} - K_2(r/\ell)\right]\cos 2\theta - \frac{\mu b}{\pi}K_0(r/\ell), \tag{54}$$

$$m_{yz} = \frac{\mu b}{\pi}\left[\frac{2\ell^2}{r^2} - K_2(r/\ell)\right]\sin 2\theta. \tag{55}$$

Examining now the asymptotic behavior of the above stress field (to determine the possibility of singularities), we note that as $r \to 0$ the following asymptotic relations hold (see e.g. Abramowitz and Stegun 1964)

$$\frac{1}{r}\left(\frac{2\ell^2}{r^2} - K_2(r/\ell)\right) = O\left(r^{-1}\right),$$
$$r\left[K_2(r/\ell) - K_0(r/\ell)\right] = O\left(r^{-1}\right),$$
$$K_0(r/\ell) = O(\ln r). \tag{56a,b,c}$$

In view of (56), as the dislocation core ($r \to 0$) is approached, the components of the force-stress tensor $\left(\sigma_{xx}, \sigma_{yy}, \sigma_{xy}, \sigma_{yx}\right)$ exhibit a Cauchy singularity (just as in classical elasticity), the couple-stress m_{xz} becomes logarithmically unbounded, while m_{yz} remains bounded. Finally, when $\ell \to 0$ the stress field of classical elasticity for a discrete glide dislocation is recovered.

4.2 Screw dislocation

For a screw dislocation with strength b the displacement field in couple-stress elasticity is given as (see our derivation in Appendix A)

$$w = \frac{b}{2\pi}\theta - \frac{b}{4\pi}(1+\beta)\left[\frac{2\ell^2}{r^2} - K_2(r/\ell)\right]\sin 2\theta, \tag{57}$$

where the ratio $\beta \equiv \eta'/\eta$ should satisfy the following inequality $-1 < \beta < 1$. The stress and couple-stress fields corresponding to (57) are obtained from Eqs. 42–46 as

$$\tau_{xz} = -\frac{\mu b}{2\pi r}\sin\theta, \quad \tau_{yz} = \frac{\mu b}{2\pi r}\cos\theta, \tag{58a,b}$$

$$\sigma_{xz} = -\frac{\mu b}{2\pi r}\sin\theta + \frac{\mu b\ell^2(1+\beta)}{\pi r^3}\sin 3\theta,$$
$$\sigma_{yz} = \frac{\mu b}{2\pi r}\cos\theta - \frac{\mu b\ell^2(1+\beta)}{\pi r^3}\cos 3\theta, \tag{59a,b}$$

$$m_{yy} = -m_{xx} = \frac{\mu\ell^2(1+\beta)b}{\pi r^2}\cos 4\theta - \frac{3\mu b\ell^2(1+\beta)^2}{\pi r^2}\left(\frac{2\ell^2}{r^2} - K_2(r/\ell)\right)\cos 4\theta + \frac{\mu b(1+\beta)^2}{2\pi}K_2(r/\ell)\cos 4\theta - \frac{\mu b(1+\beta)^2}{8\pi}K_0(r/\ell)(3\cos 4\theta + 1), \tag{60a,b}$$

$$m_{yx} = \frac{3\mu b\ell^2(1+\beta)^2}{\pi r^2}\left(\frac{2\ell^2}{r^2} - K_2(r/\ell)\right)\sin 4\theta - \frac{\mu b(1+\beta)^2}{4\pi}K_2(r/\ell)(2\sin 4\theta + \sin 2\theta) + \frac{3\mu b(1+\beta)^2}{8\pi}K_0(r/\ell)\sin 4\theta - \frac{\mu b(1+\beta)}{\pi}\left(\frac{2\ell^2}{r^2} - K_2(r/\ell)\right)\sin 2\theta, \tag{61}$$

$$m_{xy} = m_{yx} - 2\mu\ell^2(1-\beta)\nabla^2 w, \tag{62}$$

The following points are of notice now: (i) Using the well known asymptotic properties of the modified Bessel functions, we conclude that as $r \to 0$ the asymmetric and the symmetric shear stresses behave as $\sim r^{-3}$ and $\sim r^{-1}$, respectively, whereas the couple-stresses behave as $\sim r^{-2}$. (ii) When $\beta = -1$ (i.e. when $\eta = -\eta'$), the above stress field degenerates into the respective one in classical elasticity for a screw dislocation.

5 Formulation of crack problems by a distribution of dislocations

5.1 Mode II crack

Consider a straight crack of length $2a$ embedded in the xy-plane of infinite extend in a field of pure shear (Fig. 1). The crack faces are traction free and the body is considered to be in plane-strain conditions. The crack faces are defined by $\mathbf{n} = (0, \pm 1)$. Then, according to (18) and (19), the boundary conditions along the crack faces are written as

$$\sigma_{yx} = 0, \quad \sigma_{yy} = 0, \quad m_{yz} = 0 \quad \text{for } |x| < a, \tag{63a,b,c}$$

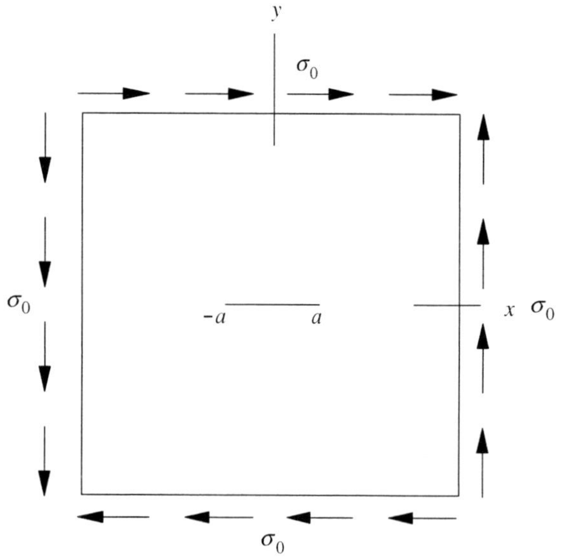

Fig. 1 Cracked body under remote shear in plane strain

whereas the regularity conditions at infinity are

$$\sigma_{yx}^{\infty} = \sigma_{xy}^{\infty} = \tau_{xy}^{\infty} \to \sigma_0, \quad \sigma_{yy}^{\infty}, \sigma_{xx}^{\infty} \to 0,$$
$$m_{xz}^{\infty}, m_{yz}^{\infty} \to 0 \quad as \quad r \to \infty, \tag{64a,b,c}$$

where $r = \left(x^2 + y^2\right)^{1/2}$ now is the distance from the origin and the constant σ_0 denotes the remotely applied shear loading.

Then, the crack problem is decomposed into the following two auxiliary problems.

The un-cracked body

It can readily be verified that the appropriate Mindlin's stress functions for the un-cracked body of infinite extent subjected to boundary conditions (64a,b,c) are as follows

$$\Phi = -\sigma_0 xy, \quad \Psi = 0. \tag{65a,b}$$

The stress field that corresponds to the above stress functions can be found from (32)–(34) as

$$\sigma_{yx}(x, y) = \sigma_{xy}(x, y) = \sigma_0,$$
$$\sigma_{xx} = \sigma_{yy} = 0, \quad m_{xz} = m_{yz} = 0. \tag{66a,b,c}$$

Notice, that there are no couple-stresses induced in the un-cracked body, the body being in a state of pure shear.

The corrective solution

Consider a body geometrically identical to the initial cracked body (Fig. 1) but with no remote loading now. The only loading applied is along the crack faces. This consists of equal and opposite tractions to those generated in the un-cracked body. The boundary conditions along the faces of the crack are written as

$$\sigma_{yx} = -\sigma_0, \quad \sigma_{yy} = 0, \quad m_{yz} = 0$$
$$\text{for } |x| < a. \tag{67a,b,c}$$

The corrective stresses (67a,b,c) may be generated by a continuous distribution of discrete glide dislocations along the crack faces. The stresses and couple-stresses induced by the continuous distribution of dislocations can be derived by integrating the effect of a discrete glide dislocation (i.e. by the use of Eqs. 50–55). We note that (67b,c) are automatically satisfied since a discrete glide dislocation does not produce normal stresses σ_{yy} or couple-stresses m_{yz} along the crack-line. Then, satisfaction of the boundary condition (67a) leads to a *single* IE. Separating the singular part from the regular part of the kernels, we obtain the governing SIE of the mode II problem in couple-stress elasticity as

$$-\sigma_0 = \frac{\mu(3-2\nu)}{2\pi(1-\nu)} \fint_{-a}^{a} \frac{B(\xi)}{x-\xi} d\xi + \frac{\mu}{\pi} \int_{-a}^{a} B(\xi) k(x,\xi) d\xi, \quad |x| < a, \tag{68}$$

where $\fint$ signifies Cauchy principal value integration and $B(\xi) = db/d\xi$ is the dislocation density at a point ξ ($|\xi| < a$), this density being defined in the same way as in classical elasticity (see e.g. Hills et al. 1996).

The kernel $k(x, \xi)$ is defined as

$$k(x,\xi) = -\frac{2}{x-\xi}\left[\frac{2\ell^2}{(x-\xi)^2} - K_2(|x-\xi|/\ell) - \frac{1}{2}\right] - \frac{(x-\xi)}{\ell^2}\left[\frac{2\ell^2}{(x-\xi)^2} - K_2(|x-\xi|/\ell) + K_0(|x-\xi|/\ell)\right]. \tag{69}$$

To show that $k(x, \xi)$ is regular, we expand the latter in series as $x \to \xi$ (see e.g. Abramowitz and Stegun 1964) and obtain

$$k(x,\xi) = (a_1 + a_2 \ln|x-\xi|)(x-\xi) + O\left((x-\xi)^3 \ln|x-\xi|\right), \tag{70}$$

where a_i are constants depending on the characteristic material length ℓ. Since $\lim_{x\to\xi}(x-\xi)^n \ln|x-\xi| = 0$ for $n > 0$, we conclude that $k(x, \xi)$ is regular in the closed domain $-a \le (x, \xi) \le a$.

The solution $B(\xi)$ in (68) is determined in the class of Hoelder continuous functions and may be written as a product of a regular bounded function and a fundamental solution. Asymptotic analysis, within the framework of the couple-stress elasticity, showed that the displacement u_x behaves as $\sim r^{1/2}$ in the crack tip region, where r denotes now the polar distance from the crack tip (Huang et al. 1997). Consequently, the dislocation density is expressed in the form

$$B(\xi) = f(\xi)\left(\alpha^2 - \xi^2\right)^{-1/2}, \tag{71}$$

where $f(\xi)$ is bounded and continuous in the interval $|\xi| \le \alpha$. Further, in order to render the problem determinate, the dislocation density should also satisfy an *auxiliary* condition expressing the requirement that there be no net relative tangential displacement between one end of the crack and the other, i.e.

$$\int_{-a}^{a} B(\xi)d\xi = 0. \tag{72}$$

Before proceeding with the solution of the governing integral equation, it is interesting to consider two limit cases concerning the behavior of (68) w.r.t. ℓ. By letting $\ell \to 0$ and noting that $\lim_{\ell\to 0} k(x,\xi) = -1/(x-\xi)$, Eq. 68 degenerates into the counterpart equation governing the mode II problem in classical elasticity, i.e.

$$-\sigma_0 = \frac{\mu}{2\pi(1-\nu)} \fint_{-a}^{a} \frac{B(\xi)}{x-\xi} d\xi, \qquad |x| < a. \tag{73}$$

On the other hand, by letting $\ell \to \infty$ and noting that $\lim_{\ell\to\infty} k(x,\xi) = 0$, (68) takes the form

$$-\sigma_0 = \frac{\mu(3-2\nu)}{2\pi(1-\nu)} \fint_{-a}^{a} \frac{B(\xi)}{x-\xi} d\xi, \qquad |x| < a. \tag{74}$$

It can readily be shown, that the ratio of the crack-face displacements obtained by the solution of (74) and (73), respectively, is $1/(3-2\nu)$. Equation 74 shows mathematically that there is a lower bound for the crack-face displacement u_x when $\ell \to \infty$. The same ratio of displacements was also obtained by Sternberg and Muki (1967) for a mode I crack in couple-stress elasticity.

For the numerical solution of the SIE in (68), the Gauss-Chebyshev quadrature developed by Erdogan and Gupta (1972) is used. After the appropriate normalization over the interval $[-1, 1]$, the integral equation takes the discretized form

$$-\sigma_0 = \frac{\mu}{n}\sum_{i=1}^{n}\left[\frac{(3-2\nu)}{2(1-\nu)(t_k - s_i)} + k(t_k, s_i)\right] f(s_i), \tag{75}$$

where

$$k(t_k, s_i) = -\frac{2}{t_k - s_i} \left[\frac{2}{p^2(t_k-s_i)^2} - K_2(p|t_k - s_i|) - \frac{1}{2}\right] - p^2(t_k - s_i)\left(\frac{2}{p^2(t_k-s_i)^2} - K_2(p|t_k - s_i|) + K_0(p|t_k - s_i|)\right), \tag{76}$$

with $p = a/\ell$, $t = x/a$, and $s = \xi/a$. The integration and collocation points are given, respectively, as

$$T_n(s_i) = 0,\ \ s_i = \cos[(2i-1)\pi/2n],\ \ i = 1, \ldots, n, \tag{77a}$$

$$U_{n-1}(t_k) = 0,\ \ t_k = \cos[k\pi/n],\ \ k = 1, \ldots, n-1, \tag{77b}$$

where $T_n(x)$ and $U_n(x)$ are the Chebyshev polynomials of the first and second kind, respectively. Formula (75) is a standard Gauss–Chebyshev quadrature with the requirement that the collocation points t_k must satisfy (77b), i.e. that t_k be the roots of U_{n-1}. The auxiliary condition in (72) can be written in discretized form as

$$\frac{\pi}{n}\sum_{i=1}^{n} f(s_i) = 0. \tag{78}$$

Equations 75 and 78 provide an algebraic system of n equations in the n unknown functions $f(s_i)$. A computer program was written that solves the above system of equations.

Some numerical results are presented now. In Fig. 2 the dependence of the tangential crack-face displacement on the ratio a/ℓ in couple-stress elasticity is depicted. It is noteworthy that as the crack length becomes comparable to the characteristic length ℓ, the material exhibits a more *stiff* behavior, i.e. the tangential crack-face displacements become smaller and smaller in magnitude. Finally, we note that the displacements obtained within the classical theory of elasticity serve as an upper *bound* of couple-stress elasticity.

Next, the near-tip behavior of the shear stress σ_{yx} given as the expression in the RHS of (68) plus σ_0, is determined. Due to the symmetry of the problem (in geometry and loading) with respect to y-axis we confine attention only to the right crack tip. Now, as

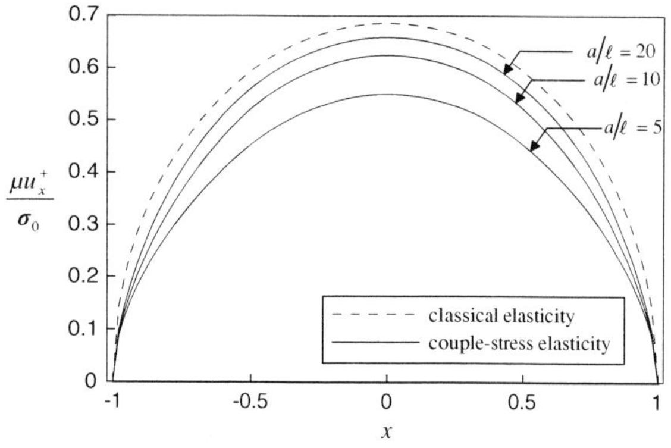

Fig. 2 Normalized upper-half tangential crack displacement profile ($\nu = 0.3$)

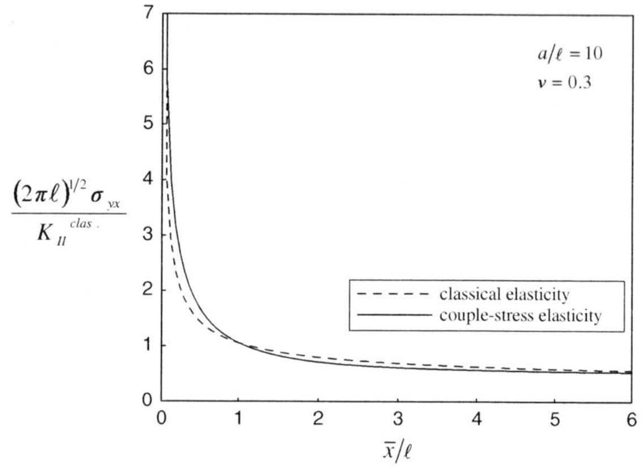

Fig. 4 Distribution of the shear stress ahead of the cracktip

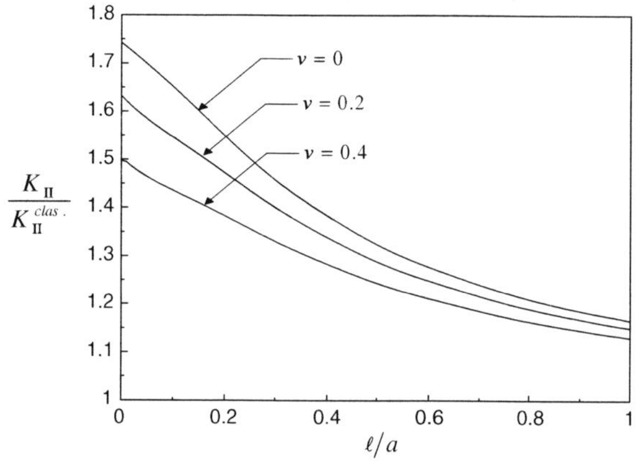

Fig. 3 Variation of the ratio of stress intensity factors in couple-stress elasticity and in classical elasticity

$x \to a^+$ the following asymptotic relations hold

$$\int_{-a}^{a} \frac{B(\xi)}{x-\xi} d\xi = \mathrm{O}\left((x-a)^{-1/2}\right),$$

$$\int_{-\alpha}^{\alpha} B(\xi)\, k(x,\xi)\, d\xi = \mathrm{O}(1), \qquad x > a, \qquad (79\text{a,b})$$

where the dislocation density is defined in (71). Thus, we conclude that σ_{yx} exhibits a square root singularity at the crack tip. In light of the above, we define the stress intensity factor in couple-stress elasticity as $K_{II} = \lim_{x \to a^+} [2\pi (x-a)]^{1/2} \sigma_{yx}(x, y=0)$ for the right crack tip ($x > a$) The dependence of the ratio of the stress intensity factor in couple-stress elasticity K_{II} to the one in classical elasticity upon ℓ/a is given in Fig. 3.

It is observed that for $\ell/a \to 0$ and Poisson's ratio $\nu = 0.4$, there is a 50% increase in K_{II} when couple-stress effects are taken into account, while for $\nu = 0.2$ and $\nu = 0$ the increase is 62% and 73%, respectively. It should be noted that when $\ell/a = 0$ (no couple-stress effects) the above ratio becomes evidently $K_{II}/K_{II}^{clas.} = 1$. Therefore, the ratio plotted in Fig. 3 exhibits a finite jump discontinuity at $\ell/a = 0$; the ratio at the tip of the crack rises abruptly as ℓ/a departures from zero. The same discontinuity was observed by Sternberg and Muki (1967), who attributed that kind of behavior to the severe *boundary-layer* effects predicted by the couple-stress elasticity in stress-concentration problems. Finally, it can be shown that the ratio decreases monotonically with increasing values of ℓ/a and tends to unity as $\ell/a \to \infty$. The case $\ell/a \to \infty$ is rather impractical since generally the relation between lengths in a usual crack problem will be $\ell << a$, i.e. the crack length will be much greater than the material length. However, in an attempt to explain the latter finding, we note that the case $\ell \to \infty$, with $a \neq 0$, resembles a situation where, in a sense, there is *no* microstructure in the body, since the 'building blocks' of the material are of infinite size. Of course, this case has an obscure physical meaning, but, as far as stresses are concerned, the solution shows that the material exhibits a behavior similar to the one for a material governed by the classical theory.

Further, the distribution of the shear stress σ_{yx} ahead of the crack tip (see Fig. 4) shows that the couple-stress effects are dominant for $x < \ell$, whereas outside this zone σ_{yx} gradually approaches the distribution of the classical solution. For convenience, a new variable $\bar{x} = x - a$ is introduced measuring now distance from the crack tip in the RHS of Fig. 1.

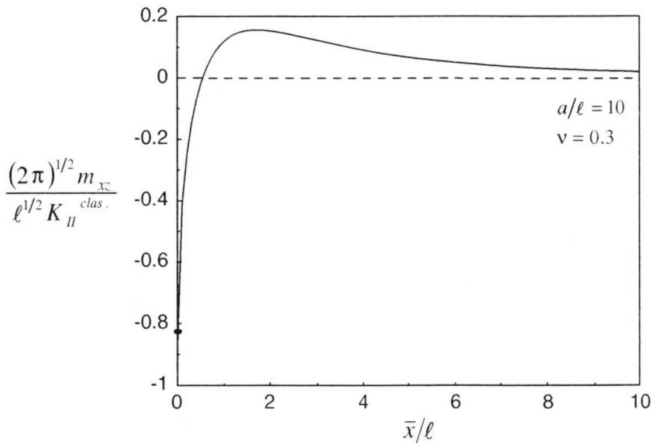

Fig. 5 Distribution of the couple-stress ahead of the crack tip

Finally, taking into account that m_{xz} exhibits a logarithmic singularity in the case of a glide dislocation and that

$$\int_{-\alpha}^{\alpha} \ln|x-\xi|\, B(\xi)\, d\xi = O(1) \quad \text{as} \quad x \to a^{+}, \tag{80}$$

we conclude that m_{xz} given as the integral of (54) is bounded at the crack tip. This observation is in agreement with the asymptotic results of Huang et al. (1997) for a mode II crack. Figure 5 depicts the distribution of the couple-stress m_{xz} ahead of the crack tip. In particular, we observe that m_{xz} takes finite *negative* values immediately ahead of the crack tip in the RHS. Then, as the position (observation point) moves away from the crack tip, m_{xz} changes sign and gradually reaches zero for $x > 10\ell$. It should be noted, though, that m_{xz} exhibits the property of anti-symmetry w.r.t. the y- axis (see Fig. 1): Therefore, m_{xz} is *positive* immediately ahead of the LHS crack tip. An anti-symmetric distribution of the couple-stress is required for the moment equation in (7) to be satisfied.

Finally, as we show in Appendix C, the orders of singularities of the above stress and couple-stress fields lead to an integrable strain-energy density in the vicinity of crack tips and also lead to a bounded value of the J-integral.

5.2 Mode III crack

Consider a straight crack of length $2a$ embedded in the (x, y)-plane of infinite extent under a remotely applied anti-plane shear loading (see Fig. 6). The crack faces are assumed to be traction free. The boundary conditions along the crack faces are written as (cf. (18) and (19))

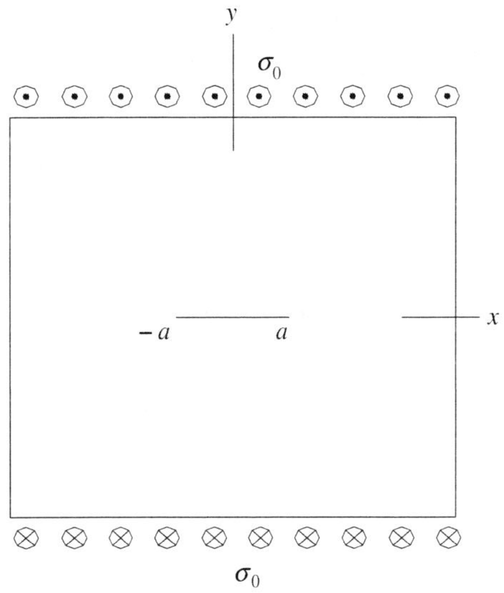

Fig. 6 Cracked body under remote shear in anti-plane strain

$$\sigma_{yz} + \frac{1}{2}\partial_x m_{yy} = 0, \qquad m_{yx} = 0 \qquad \text{for} \quad |x| < a, \tag{81a,b}$$

whereas the regularity conditions at infinity are given as

$$\sigma_{yz}^{\infty} = \tau_{yz}^{\infty} \to \sigma_0, \quad \sigma_{xz}^{\infty} \to 0,$$
$$m_{xx}^{\infty}, m_{yy}^{\infty}, m_{yx}^{\infty}, m_{xy}^{\infty} \to 0 \qquad \text{as} \quad r \to \infty, \tag{82a,b,c}$$

The 'reduced' boundary condition in (81a) is also justified physically from the fact that the displacement w and the rotation $\omega_y = -(1/2)(\partial w/\partial x)$ cannot be prescribed independently on the crack faces. This situation is analogous to the one in Kirchhoff's plate theory regarding the effective shear force.

Again, the crack problem is decomposed into the following two auxiliary problems.

The *un-cracked* body

It can be readily shown that the un-cracked body subjected to the boundary conditions (82a,b,c) is in a state of pure anti-plane shear. The only non-zero stresses are

$$\sigma_{yz}(x, y) = \tau_{yz}(x, y) = \sigma_0. \tag{83}$$

Note that there are no couple-stresses induced in the un-cracked body.

The corrective solution

Consider a body geometrically identical to the initial cracked body in Fig. 6 but with no remote loading now. The applied loading along the crack faces consists of equal and opposite tractions to those generated in the un-cracked body, i.e.

$$\sigma_{yz} + \frac{1}{2}\partial_x m_{yy} = -\sigma_0, \quad m_{yx} = 0 \qquad \text{for } |x| < a \text{ and } y = 0. \tag{84a,b}$$

The corrective stresses in (84a,b) may be generated by a continuous distribution of discrete screw dislocations along the crack faces. The stresses induced by the continuous distribution of dislocations are obtained as integrals of Eqs. 59–62. Note that (84b) is automatically satisfied since a discrete screw dislocation does not give rise to couple-stresses m_{yx} along the crack line. Then, satisfaction of the boundary condition (84a) leads, after lengthy calculations, to the governing hypersingular IE of the mode III problem in couple-stress elasticity ($|x| < a$)

$$-\sigma_0 = \fint_{-a}^{a} \left[\frac{c_1}{x-\xi} + \frac{c_2 \ell^2}{(x-\xi)^3} + c_3 k(x,\xi) \right] B(\xi)\, d\xi \tag{85}$$

where $\fint$ signifies Hadamard's finite-part integration (see e.g. Kutt 1975; Paget 1981), $B(\xi)$ is the dislocation density function at the point ξ ($|\xi| < a$), and

$$c_1 = \frac{\mu\left(\beta^2 + 2\beta + 9\right)}{16\pi}, \quad c_2 = \frac{\mu(1+\beta)(\beta-3)}{2\pi}, \quad c_3 = \frac{\mu(1+\beta)^2}{\pi}. \tag{86}$$

Further, the kernel $k(x,\xi)$ is defined as

$$k(x,\xi) = -\frac{\ell^2}{(x-\xi)^3}\left[6\left(K_2(|x-\xi|/\ell) - \frac{2\ell^2}{(x-\xi)^2}\right) + \frac{1}{2}\right] + \frac{1}{4(x-\xi)}\left[3K_0(|x-\xi|/\ell) - 5K_2(|x-\xi|/\ell) - \frac{1}{4}\right]. \tag{87}$$

Expanding $k(x,\xi)$ in series as $x \to \xi$ and using the asymptotic properties of the modified Bessel functions, it can be readily shown that $k(x,\xi)$ is regular in the closed domain $-a \leq (x,\xi) \leq a$. We also note that when $\beta = -1$ (i.e. when $\eta = -\eta'$), Eq. 85 degenerates into the SIE that governs the counterpart problem in classical elasticity.

In addition, Zhang et al. (1998) showed, by using the Williams eigenfunction asymptotic analysis, that the crack face displacement behaves as $\sim r^{3/2}$ in the crack tip region, where r denotes the polar distance from the crack tip. Thus, the dislocation density $B(\xi)$ can be expressed as

$$B(\xi) = f(\xi)\left(a^2 - \xi^2\right)^{1/2}, \tag{88}$$

where $f(\xi)$ is a continuous bounded function in $\xi \leq |a|$. Finally, to ensure uniqueness the dislocation density must satisfy the following auxiliary condition stemming from the requirement of single-valuedness of the displacement along a closed loop around the crack

$$\int_{-a}^{a} B(x)dx = 0. \tag{89}$$

Now, the *near-tip* behavior of the stress and couple-stress field for the mode III problem can be determined from the singular nature of the respective stress and couple-stress field of a discrete screw dislocation. Again, confining our attention to the RHS crack tip and taking into account the following result (Chan et al. 2003)

$$\frac{\partial^n}{\partial x^n} \fint_{-a}^{a} \frac{B(\xi)}{x-\xi} d\xi = O\left((x-a)^{1/2-n}\right), \text{ for } n \geq 0 \text{ as } x \to a^+, x > a \tag{90}$$

with the dislocation density being given by (88), we conclude that (τ_{yz}, σ_{yz}) given as the integrals of (58b) and (59b) behave as $\sim \bar{x}^{-3/2}$ and $\sim \bar{x}^{1/2}$, respectively, whereas the couple-stresses (m_{xx}, m_{yy}) given by the integration of (60a,b) exhibit a square root singularity at the crack tip. Again, $\bar{x} = x - a$ is the distance from the RHS crack tip along the crack line. Finally, in light of the above, the *total* shear stress defined as $t_{yz} = \sigma_{yz} + (1/2)\,\partial_x m_{yy}$ has the following asymptotic behavior $t_{yz} \sim \bar{x}^{-3/2}$ near the crack tip. Such a behavior was detected before in the mode III crack problem of gradient elasticity (Georgiadis 2003). The two problems present similarities in their mathematical analysis. Finally, as we show in Appendix C, despite the hypersingular nature of the above stress field, the strain-energy density is integrable in the vicinity of crack tips and, also, the J-integral takes a bounded value.

For the numerical solution of the hypersingular integral equation in 85, the appropriate quadrature is constructed here by taking into account the cubic singularity of the integral equation and the endpoint behavior of the dislocation density (details are given in Appendix B). Equation 85 after the appropriate normalization over the interval $[-1, 1]$ takes the discretized form

$$-\sigma_0 = -\frac{\pi c_2}{p^2} \sum_{i=1}^{n} (-1)^{i+k} \left[\frac{t_k}{2\left(1-t_k^2\right)} - \frac{1}{t_k - s_i} \right] \times \frac{f(s_i)\left(1-s_i^2\right)}{(t_k - s_i)\left(1-t_k^2\right)^{1/2}} + \sum_{i=1}^{n} \frac{\pi}{(1+n)} \times \left[\frac{c_1}{(t_k - s_i)} + \frac{c_2}{p^2 (t_k - s_i)^3} + c_3 k(t_k, s_i) \right] \times f(s_i)\left(1 - s_i^2\right), \tag{91}$$

where

$$k(t_k, s_i) = -\frac{1}{p^2 (t_k - s_i)^3} \left[6\left(K_2(p|t_k - s_i|) - \frac{2}{p^2 (t_k - s_i)^2} + \frac{1}{2} \right) \right] + \frac{1}{4(t_k - s_i)} \left[3K_0(p|t_k - s_i|) - 5K_2(p|t_k - s_i|) - \frac{1}{4} \right], \tag{92}$$

with $p = a/\ell$, and the set of the n discrete integration points are given by

$$s_i : U_n(s_i) = 0, \quad s_i = \cos(i\pi/(n+1)), \; i = 1, \ldots, n, \tag{93a}$$

while the $n+1$ collocation points are given by

$$t_k : T_{n+1}(t_k) = 0, \quad t_k = \cos\left((2k-1)\pi/2(n+1)\right), \; k = 1, \ldots, n+1. \tag{93b}$$

The auxiliary condition in (89) can be written in discretized form as

$$\sum_{i=1}^{n} \frac{\left(1 - s_i^2\right) f(s_i)}{1+n} = 0, \tag{94}$$

Then, Eqs. 91 and 94 provide a system of $n+2$ algebraic equations. The system is solved in the least-squares sense.

In Fig. 7, the crack-face displacements are shown for the special case $\beta = 0$ (i.e. $\eta' = 0$). It is observed that in the crack-tip vicinity, the crack closes more smoothly as compared to the classical result. Further, it is also noted that when the characteristic material length ℓ becomes comparable to the crack length the material behaves in a more rigid way (having increased stiffness).

Both couple-stress and classical elasticity ($K_{III}^{clas.}$ field) distributions ahead of the right crack tip are shown in Fig. 8. The total shear stress t_{yz} is employed to depict the couple-stress elasticity solution. As in the analogous gradient elasticity solution (Georgiadis 2003), we observe that for a very small zone in the crack-tip region ($x < 0.5\ell$) the total stress t_{yz} takes on negative values exhibiting therefore a *cohesive-traction* character along the prospective fracture zone. Also, t_{yz} exhibits a bounded maximum. As $\beta \to -1$, the *cohesive* zone becomes significantly smaller whereas the maximum value of the total shear stress increases. The behavior of t_{yz} reminds typical *boundary-layer* behavior as, e.g., that found for the surface pressure near the leading edge of Joukowski airfoil (Van Dyke 1964). Finally, we note that at points lying outside the domain where the effects of

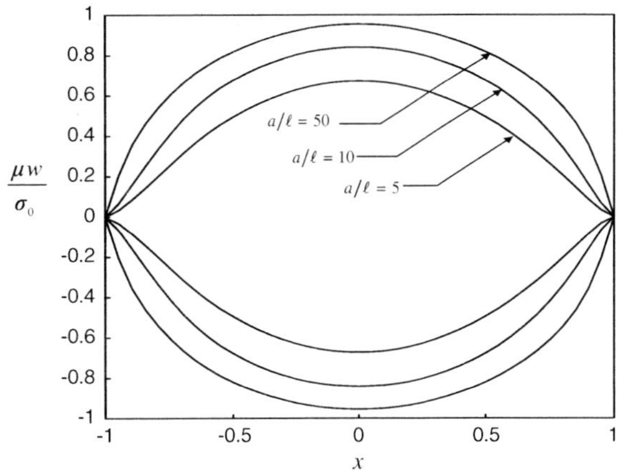

Fig. 7 Normalized upper and lower crack displacement profiles under remote mode III loading ($\beta = 0$)

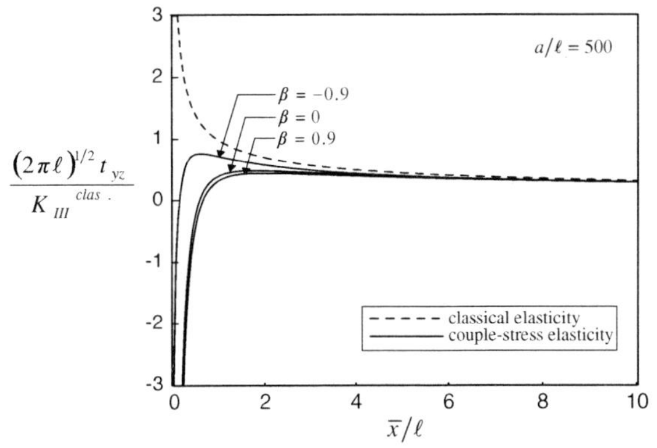

Fig. 8 Distribution of the total shear stress ahead of the crack tip

microstructure are pronounced (i.e. for $x > \ell$) the total shear stress tends to the classical $K_{III}^{clas.}$ shear stress.

6 Concluding remarks

In this paper, the technique of the distributed dislocations was used in order to solve finite-length shear crack problems in couple-stress elasticity. The technique provides an alternative approach to the elaborate analytical method of dual integral equations used before to attack asymptotically the mode I crack problem. Moreover, the present approach is capable to provide a full-field solution. In fact, we have obtained here the stress distribution ahead of the crack tips and the crack-face displacements (i.e. our results are not restricted to the crack-tip region). Also, our solution to the finite-length crack in mode III is quite novel in the literature.

The governing integral equations are derived using the discrete-dislocation stress fields in couple-stress elasticity, as the Green's functions of crack problems. In particular, it is shown that the mode II problem is governed by a single singular integral equation. In the mode III case, the governing integral equation is found to be hypersingular with a cubic singularity. For the solution of the latter equation, a new efficient quadrature is constructed.

The results of our analysis indicate that when the microstructure of the material is taken into account the material behaves in a more rigid way. In particular, in the mode II problem, the crack face displacements become significantly smaller than their counterparts in classical elasticity, when the length of the crack is comparable to the characteristic length ℓ of the material. Further, stresses retain the same order of singularity as in the classical theory, while the couple-stress field is found to be bounded in the crack-tip region. In the mode III problem, the results for the near-tip field show significant departure from the predictions of classical fracture mechanics. It is shown that cohesive stresses develop in the immediate vicinity of the crack-tip and that, ahead of the small cohesive zone, the stress distribution exhibits a local maximum that is bounded. This maximum value may serve, therefore, as a measure of the critical stress level at which further advancement of the crack may occur. In addition, in the vicinity of the crack-tip, the crack-face displacement closes more smoothly as compared to the classical result.

Acknowledgements This paper is a partial result of the Project PYTHAGORAS II / EPEAEK II [Title of the individual program: "Application of gradient theory for the solution of boundary value problems by the use of analytical methods and mixed adaptive finite elements" (# 68/8213)]. This Project is co-funded by the European Social Fund (75%) of the European Union and by National Resources (25%) of the Greek Ministry of Education.

Appendix A: The screw dislocation in couple-stress elasticity

Let the direct Fourier transform and its inverse be defined as

$$w^*(\xi, y) = \frac{1}{(2\pi)^{1/2}} \int_{-\infty}^{\infty} w(x, y)\, e^{ix\xi} dx, \qquad \text{(A1a)}$$

$$w(x, y) = \frac{1}{(2\pi)^{1/2}} \int_{-\infty}^{\infty} w^*(\xi, y)\, e^{-ix\xi} d\xi, \qquad \text{(A1b)}$$

where $i \equiv (-1)^{1/2}$. Transforming the field equation (47) with (A1a) gives the following ODE

$$\ell^2 \frac{d^4 w^*}{dy^4} - \left(2\ell^2\xi^2 + 1\right) \frac{d^2 w^*}{dy^2} + \left(\ell^2\xi^4 + \xi^2\right) w^* = 0, \qquad \text{(A2)}$$

and, further, the general transformed solution for $y \geq 0$

$$w^*(\xi, y) = A_1(\xi)\, e^{-|\xi| y} + A_2(\xi)\, e^{-y\frac{(1+\ell^2\xi^2)^{1/2}}{\ell}} \qquad \text{(A3)}$$

Now, we impose at the origin of the infinite (x, y)-plane a single screw dislocation with Burger's vector $b = (0, 0, b)$. In the upper half-plane, the screw dislocation gives rise to the following boundary value problem

$$w\left(x, 0^+\right) = -\frac{b}{2} H(x), \qquad \text{(A4a)}$$

$$m_{yx}\left(x, 0^+\right) = 0, \qquad \text{(A4b)}$$

where $H(x)$ is the Heaviside step function and the minus sign in (A4a) is justified from the sign convention that is adopted in dislocation theory. In view now of the constitutive equation (43d) and the properties of the Fourier transform, the boundary conditions (A4a,b) furnish in the transform domain

$$w^*\left(\xi, 0^+\right) = -b\,(\pi/2)^{1/2}\, \delta_+(\xi), \qquad \text{(A5a)}$$

$$m^*_{yx}\left(\xi, 0^+\right) = 2\eta \frac{d^2 w^*}{dy^2} + 2\eta' \xi^2 w^* = 0, \qquad \text{(A5b)}$$

where $\delta_+(\xi) = 1/2\,(\delta(\xi) + i/\pi\xi)$ is the Heisenberg delta function (Roos 1969) and $\delta(\xi)$ is the Dirac distribution. The constants $A_1(\xi)$ and $A_2(\xi)$ are now computed using the transformed boundary conditions i.e.

$$A_1(\xi) = -(\pi/2)^{1/2}\, b\left(1 + \ell^2(1+\beta)\xi^2\right)\delta_+(\xi),$$
$$A_2(\xi) = (\pi/2)^{1/2}\, b\ell^2(1+\beta)\xi^2\delta_+(\xi), \qquad \text{(A6a,b)}$$

where $\beta = \eta'/\eta$. With the aid of the inversion formula in (A1b), we obtain the integral representation for the displacement field due to a screw dislocation

$$w(x,y) = -\frac{b}{2}\int_{-\infty}^{\infty}\left[e^{-y|\xi|} - (\xi\ell)^2(1+\beta)\left(e^{-y|\xi|} - e^{-\frac{y\left(1+\ell^2\xi^2\right)^{1/2}}{\ell}}\right)\right]\delta_+(\xi)\,e^{-ix\xi}d\xi. \qquad \text{(A7)}$$

Using the properties of the Heisenberg delta function and the Dirac distribution, we finally obtain

$$w(x,y) = -\frac{b}{4} - \frac{b}{2\pi}\int_0^{\infty}\frac{e^{-y\xi}}{\xi}\sin(\xi x)d\xi - \frac{b\ell^2(1+\beta)}{2\pi}\int_0^{\infty} \xi\left(e^{-y\xi} - e^{-\frac{y\left(1+\ell^2\xi^2\right)^{1/2}}{\ell}}\right)\sin(\xi x)\,d\xi. \qquad \text{(A8)}$$

The above integrals can be determined in closed form. In particular, we have

$$\int_0^{\infty}\frac{e^{-y\xi}}{\xi}\sin(\xi x)\,d\xi = \tan^{-1}\frac{x}{y},$$
$$\int_0^{\infty}\xi e^{-y\xi}\sin(\xi x)d\xi = 2\frac{xy}{r^4},$$
$$\int_0^{\infty}\xi e^{-\frac{y\left(1+\ell^2\xi^2\right)^{1/2}}{\ell}}\sin(\xi x)\,d\xi = \frac{xy}{(r\ell)^2}K_2\left(\frac{r}{\ell}\right). \qquad \text{(A9a,b,c)}$$

In light of the above results, the displacement can be written as

$$w = \frac{b}{2\pi}\theta - \frac{b}{4\pi}(1+\beta)\left[\frac{2\ell^2}{r^2} - K_2(r/\ell)\right]\sin 2\theta. \qquad \text{(A10)}$$

Appendix B: Construction of numerical quadrature

The problem of finding a numerical quadrature for integrals with order of singularity greater than two ($a > 2$) arises naturally in generalized continuum theories where the field equations and the boundary conditions are of higher order than the respective ones in classical elasticity. Although a lot of work has been done in the literature for Hadamard type integrals ($a = 2$) (see e.g. Kutt 1975; Paget 1981; Ioakimidis 1983, 1995; Kaya and Erdogan 1987; Monegato 1987, 1994; Tsamasphyros and Dimou 1990; Korsunsky 1998; Kabir et al. 1998; Hui and Shia 1999), only a few papers have been published concerning integrals with $a > 2$. In a recent work by Chan et al. (2003), a systematic treatment of hypersingular integrals was presented based on the Kaya/Erdogan approach. This approach leads to very good results, with the only caveat that when the kernel cannot be explicitly given in terms of a sum of the hypersingular part and a remainder, the extraction of a strong singularity may lead to a loss of accuracy. Our intention here is to derive a numerical quadrature for the hypersingular integral

$$S(t) = \neq\!\!\!\!\!\!\int_{-1}^{1} f(s)\,w(s)\frac{ds}{(s-t)^3} \qquad \text{for } |t| < 1, \quad \text{(B.1)}$$

where $f(s)$ is a bounded and continuous function in the interval $[-1, 1]$, and $w(s) = \left(1 - s^2\right)^{1/2}$ is the weight function corresponding to the second-kind Chebyshev polynomials U_j. The integral in (B.1) is to be understood in the Hadamard finite-part sense (Kutt 1975; Paget 1981). The basic steps in the development of the quadrature follow the strategy introduced by Korsunsky (1998).

The unknown function can be approximated with a sufficient degree of accuracy by a truncated series of second-kind Chebyshev polynomials

$$f(s) \cong \sum_{j=0}^{p} B_j U_j(s). \qquad \text{(B.2)}$$

Making use of the relation for the Cauchy principal-value integral (Abramowitz and Stegun 1964)

$$\fint_{-1}^{1} U_j(s)\left(1-s^2\right)^{1/2}\frac{ds}{(s-t)} = -\pi T_{j+1}(t), \quad \text{(B.3)}$$

(B.1) can be rewritten as

$$S(t) = \oint_{-1}^{1} f(s)\, w(s) \frac{ds}{(s-t)^3} = \frac{1}{2}\frac{d^2}{dt^2} \times \oint_{-1}^{1} f(s) \left(1-s^2\right)^{1/2} \frac{ds}{(s-t)}$$
$$= \frac{1}{2}\frac{d^2}{dt^2} \sum_{j=0}^{p} B_j \oint_{-1}^{1} U_j(s) \left(1-s^2\right)^{1/2} \times \frac{ds}{(s-t)} = -\frac{\pi}{2} \sum_{j=0}^{p} B_j T''_{j+1}(t), \qquad \text{(B.4)}$$

where the prime denotes differentiation with respect to t. We note that the interchange of the order of differentiation and integration in (B.4) is valid in view of results by Monegato (1994).

Next, we establish the following identity

$$-\frac{T_{j+1}(t)\, U_n(t) - T_{n+1}(t)\, U_j(t)}{U_n(t)} = \sum_{i=1}^{n} \frac{a_i}{s_i - t} \quad \text{for } j < n, s_i : U_n(s_i) = 0, \qquad \text{(B.5)}$$

where the partial-fraction expansion above is possible because the degree of the numerator in the left hand side of (B.5) is less than that of the denominator. It can easily be found (Korsunsky 1998) that the coefficients a_i in (B.5) are given by the relation

$$a_i = \frac{\left(1-s_i^2\right)}{1+n} U_j(s_i). \qquad \text{(B.6)}$$

Equation (B.5) takes now the form

$$-T_{j+1}(t) + \frac{T_{n+1}(t)\, U_j(t)}{U_n(t)} = \sum_{i=1}^{n} \frac{1-s_i^2}{(1+n)(s_i-t)} U_j(s_i). \qquad \text{(B.7)}$$

Differentiating (B.7) twice with respect to t and selecting a discrete set of points t_k, $k = 1, \ldots, n+1$ such that $T_{n+1}(t) = 0$, we obtain

$$-T''_{j+1}(t_k) + T''_{n+1}(t_k) \frac{U_j(t_k)}{U_n(t_k)} + 2T'_{n+1}(t_k) \frac{U'_j(t_k)\, U_n(t_k) - U_j(t_k)\, U'_n(t_k)}{U_n^2(t_k)} = 2\sum_{i=1}^{n} \frac{1-s_i^2}{(1+n)(s_i-t)^3} U_j(s_i). \qquad \text{(B.8)}$$

Further, employing the well known identities about the derivatives of Chebyshev polynomials

$$T'_{n+1}(t) = (n+1)\, U_n(t), \quad n \geq 0, \qquad \text{(B.9)}$$

$$U'_n(t) = \left(1-t^2\right)^{-1} \left[-(n+1)\, T_{n+1}(t) + t U_n(t)\right], \quad n \geq 0, \qquad \text{(B.10)}$$

we write (B.8), after some lengthy algebra, under the form

$$-T''_{j+1}(t_k) + 2(1+n)\, U'_j(t_k) - (1+n) \frac{t_k}{1-t_k^2} \times U_j(t_k) = 2\sum_{i=1}^{n} \frac{1-s_i^2}{(1+n)(s_i-t_k)^3} U_j(s_i). \qquad \text{(B.11)}$$

Using (B.2), multiplying (B.9) by $\frac{\pi}{2} B_j$ and summing over j from 0 to p, we then get

$$S(t_k) \cong -\pi(1+n)\, f'(t_k) + \frac{\pi}{2}(1+n) \frac{t_k}{1-t_k^2} f(t_k) + \pi \sum_{i=1}^{n} \frac{1-s_i^2}{(1+n)(s_i-t_k)^3} f(s_i). \qquad \text{(B.12)}$$

One further step is needed now that would lead to the evaluation of the right hand side of (B.12) only at n points $s_i : U_n(s_i) = 0$. This can be done with the aid of the Lagrange interpolation formula, which will be exact within the class of polynomials chosen to represent $f(t)$

$$f(t) = \sum_{i=1}^{n} \frac{U_n(t)}{U'_n(s_i)(t-s_i)} f(s_i). \qquad \text{(B.13)}$$

Differentiating (B.13) with respect to t and then substituting t with $t = t_k : T_{n+1}(t) = 0$, we get

$$f'(t_k) = \sum_{i=1}^{n} U_n(t_k) \left[\frac{t_k}{1-t_k^2} - \frac{1}{(t_k-s_i)}\right] \times \frac{f(s_i)}{U'_n(s_i)(t_k-s_i)}. \qquad \text{(B.14)}$$

In light of the above analysis, (B.12) can be written as

$$S(t_k) \cong \pi \sum_{i=1}^{n} \left\{ -U_n(t_k) \left[\frac{1}{(s_i-t_k)} + \frac{t_k}{2\left(1-t_k^2\right)}\right] \times \frac{1}{(s_i-t_k)\, T_{n+1}(s_i)} + \frac{1}{(1+n)(s_i-t_k)^3} \right\} f(s_i) \left(1-s_i^2\right), \qquad \text{(B.15)}$$

where

$$t_k : T_{n+1}(t_k) = 0,\ t_k = \cos\left(\frac{(2k-1)\pi}{2(1+n)}\right), \quad k = 1, \ldots, n+1,$$

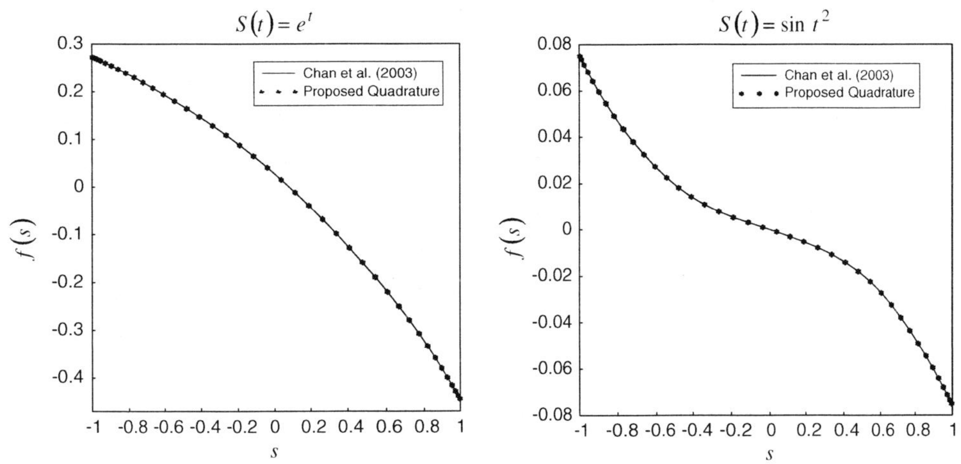

Fig. B1 Solution of the hypersingular integral equation (B.1) using the proposed quadrature and comparison with the semi-analytical method of Chan et al. (2003)

$$s_i : U_n(s_i) = 0, \quad s_i = \cos\left(\frac{i\pi}{1+n}\right), i = 1, \ldots, n.$$

Finally, taking into account that

$$U_n(t_k) = (-1)^{k+1} \Big/ \left(1 - t_k^2\right)^{1/2}, \quad T_{n+1}(s_i) = (-1)^i, \tag{B.16}$$

we write the resulting formula under the form

$$\begin{aligned} &\fint_{-1}^{1} f(s)\left(1 - s^2\right)^{1/2} \frac{ds}{(s-t)^3} \cong \\ &\pi \sum_{i=1}^{n} \left\{ (-1)^{i+k} \left[\frac{1}{(s_i - t_k)} + \frac{t_k}{2\left(1 - t_k^2\right)} \right] \right. \\ &\times \left. \frac{1}{(s_i - t_k)\left(1 - t_k^2\right)^{1/2}} + \frac{1}{(1+n)(s_i - t_k)^3} \right\} \\ &\times f(s_i)\left(1 - s_i^2\right). \end{aligned} \tag{B.17}$$

It is noteworthy, that formula (B.17) also holds in precisely the same form for the more general case when the integral kernel is split up into a hypersingular part of order $a = 3$ and a remainder

$$K(s, t) = \frac{1}{(s-t)^3} + k(s, t), \tag{B.18}$$

where the remainder may consist of Cauchy type and regular kernels. In that case, (B.17) takes the form

$$\begin{aligned} &\fint_{-1}^{1} K(s, t) f(s)\left(1 - s^2\right)^{1/2} ds \\ &\cong \pi \sum_{i=1}^{n} \left\{ (-1)^{i+k} \left[\frac{1}{(s_i - t_k)} + \frac{t_k}{2\left(1 - t_k^2\right)} \right] \right. \\ &\times \left. \frac{1}{(s_i - t_k)\left(1 - t_k^2\right)^{1/2}} + \frac{K(s_i, t_k)}{(1+n)} \right\} f(s_i)\left(1 - s_i^2\right). \end{aligned} \tag{B.19}$$

To check the validity of the proposed quadrature, we solve the hypersingular integral equation in (B.1) for two cases, i.e. for the loading function $S(t)$ being defined as: (i) $S(t) = e^t$, and (ii) $S(t) = \sin t^2$. For single-valuedness, the following auxiliary condition should also be taken into account

$$\int_{-1}^{1} f(s)\left(1 - s^2\right)^{1/2} ds = 0. \tag{B.20}$$

Then, (B.17) and (B.20) form a system of $n + 2$ equations in n unknowns which is solved in the least-squares sense. It is shown (see Fig. B1) that our results are in excellent agreement with the ones obtained by using the semi-analytical method of Chan et al. (2003).

Appendix C: Evaluation of the strain-energy density at crack tips and the *J*-integral

Our aim here is to show the orders of singularities of the stress and couple-stress fields obtained in the main body of the paper lead to an integrable strain-energy density in the vicinity of crack tips and also lead to a bounded value of the J-integral. The procedure followed is analogous in many respects with the one adopted in the work by Georgiadis (2003).

The strain-energy density function in (20) reads, in terms of stresses

$$W = \frac{1}{2\mu}\left[\tau_{ij}\tau_{ij} - \frac{\nu}{1+\nu}\tau_{ii}\tau_{jj} + \frac{1}{4\ell^2\left(1-\beta^2\right)}\left(m_{ij}m_{ij} - \beta m_{ij}m_{ji}\right)\right], \tag{C1}$$

where β is the ratio of the couple-stress moduli defined as $\beta = \eta'/\eta$.

Further, the path-independent J-integral within the couple-stress theory is given by (Atkinson and Leppington 1974; Lubarda and Markenskoff 2000)

$$J = \int_\Gamma \left[W n_x - P_q \frac{\partial u_q}{\partial x} - R_q \frac{\partial \omega_q}{\partial x}\right] d\Gamma = \int_\Gamma \left(W dy - \left[P_q \frac{\partial u_q}{\partial x} + R_q \frac{\partial \omega_q}{\partial x}\right] d\Gamma\right), \tag{C2}$$

where a Cartesian rectangular coordinate system is attached to the RHS crack tip with the distance x measured now from the tip, Γ is a piece-wise smooth simple two-dimensional contour surrounding the crack-tip, W is the strain-energy density, u_q is the displacement, ω_q is the rotation, P_q is the force-traction defined in (18), and R_q is the couple-traction defined in (19).

For the evaluation of the J-integral, we consider the rectangular-shaped contour Γ in Fig. C1 with vanishing "height" along the y- direction and with $\varepsilon \to +0$. This type of contour permits using solely the asymptotic near-tip stress and displacement fields. It is noted that upon this choice of contour, the integral $\int_\Gamma W dy$ in (C2) becomes zero if we allow the 'height' of the rectangle to vanish. In this way, the expression for the J-integral becomes

$$J = \lim_{\varepsilon \to +0}\left\{2\int_{-\varepsilon}^{\varepsilon}\left(P_q \frac{\partial u_q}{\partial x} + R_q \frac{\partial \omega_q}{\partial x}\right) dx\right\}. \tag{C3}$$

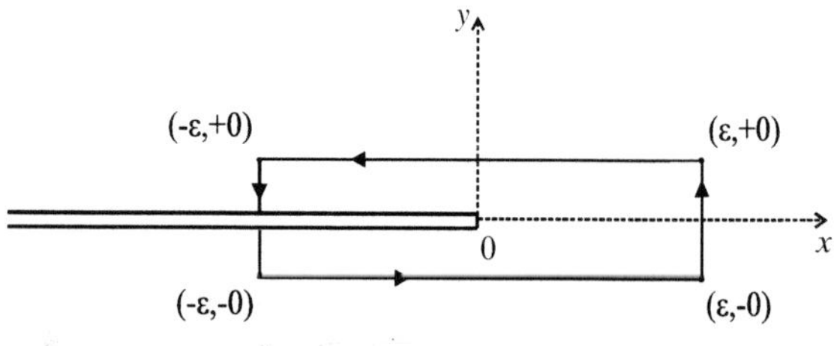

Fig. C1 Rectangular-shaped contour surrounding the cracktip

The cases of mode II and mode III cracks are examined in what follows.

Mode II

In the case of plane-strain, the strain-energy density reads

$$W = \frac{1}{4\mu}\left[(1-\nu)\left(\tau_{xx}^2 + \tau_{yy}^2\right) + 2\tau_{xy}^2 - 2\nu\tau_{xx}\tau_{yy}\right] + \frac{1}{8\mu\ell^2}\left(m_{xz}^2 + m_{yz}^2\right). \tag{C4}$$

As shown before, the couple-stresses (m_{xz}, m_{yz}) are bounded (non-singular) in the crack-tip vicinity in the mode II case, whereas both the asymmetric and symmetric stresses exhibit a square root singularity (see also Huang et al. 1997). Now, the term in square brackets in (C4) is the same as in classical elasticity and behaves in exactly the same way, while the second term (the one involving couple-stresses) is bounded in the crack-tip vicinity. Therefore, by following the standard procedure to check upon the integrability of the strain-energy density around a singularity (see e.g. Barber 1992), we conclude that the strain-energy density is *integrable* indeed in the crack-tip vicinity.

Further, taking into account that in the mode II case both the normal stress σ_{yy} and the couple-stress m_{yz} are zero along the crack line $\left(y = 0^{\pm}\right)$ and that the crack-faces are defined by $\mathbf{n} = (0, \pm 1)$, the J-integral in (C3) finally takes the form

$$J = \lim_{\varepsilon \to +0}\left\{2\int_{-\varepsilon}^{\varepsilon}\left(\sigma_{yx}(x, y=0) \cdot \frac{\partial u_x(x, y=0)}{\partial x}\right) \times dx\right\}. \tag{C5}$$

Now, in view of the asymptotic behavior of the fields entering (C5), we obtain

$$J = \lim_{\varepsilon \to +0}\left\{\frac{A_{II}^2}{\mu}\int_{-\varepsilon}^{\varepsilon}(x_+)^{-1/2}(x_-)^{-1/2} dx\right\} = \frac{\pi A_{II}^2}{2\mu}, \tag{C6}$$

where the product of distributions inside the integral was obtained by the use of Fisher's theorem (see e.g. Georgiadis 2003), i.e. the operational relation $(x_-)^{\lambda}(x_+)^{-1-\lambda} = -\pi\delta(x)\left[2\sin(\pi\lambda)\right]^{-1}$ with $\lambda \neq -1, -2, -3, \ldots$ and $\delta(x)$ being the Dirac delta distribution. Finally, we note that the amplitude factor A_{II} is connected with the asymptotic results of Huang et al. (1997), in the mode II case, through the relation $A_{II} = 2^{1/2}\left[(3-2\nu)(1-\nu)\right]^{1/2} B_{II}$.

Mode III

In this case, the strain-energy density is given by

$$W = \frac{1}{2\mu}\left(\tau_{xz}^2 + \tau_{yz}^2\right) + \frac{1}{8\mu\ell^2\left(1-\beta^2\right)}\left[(1-\beta) \times \left(m_{xx}^2+m_{yy}^2\right) + \left(m_{xy}^2+m_{yx}^2\right) - 2\beta m_{xy}m_{yx}\right]. \quad \text{(C7)}$$

Based on the results of our analysis for the mode III case, we notice that the couple-stresses behave as $\sim r^{-1/2}$ around the crack tip, while the symmetric stresses behave as $\sim r^{1/2}$. Thus, by invoking again the standard procedure involving the evaluation of a volume integral around the singularity (see e.g. Barber 1992), we conclude that the strain-energy density in (C7) is integrable in the crack-tip vicinity and the strain energy itself is bounded.

Next, taking into account that the couple-stress m_{yx} is identically zero along the crack line in the mode III problem, the J-integral takes the following form

$$J = \lim_{\varepsilon\to+0}\left\{2\int_{-\varepsilon}^{\varepsilon}\left(t_{yz}(x, y=0) \cdot \frac{\partial w(x, y=0)}{\partial x}\right)dx\right\}, \quad \text{(C8)}$$

where $t_{yz} = \sigma_{yz} + (1/2)\,\partial_x m_{yy}$ is the total shear stress which, as shown before, exhibits a near-tip behavior as $-(r^{-3/2})$. In light of the above, we obtain

$$J = \lim_{\varepsilon\to+0}\left\{3\mu A_{III}^2\int_{-\varepsilon}^{\varepsilon} -(x_+)^{-3/2}(x_-)^{1/2}dx\right\} = \frac{3\pi\mu A_{III}^2}{2}, \quad \text{(C9)}$$

where A_{III} is an amplitude factor (constant) dependent upon both couple-stress moduli and the remote loading. The above result shows that the J-integral is also bounded in the mode III case (despite the hypersingular nature of the near-tip total shear stress). Finally, we note that in the special case where the second couple-stress modulus is set equal to zero (i.e. $\beta = 0$), A_{III} above is connected with the amplitude factor B in the work by Zhang et al. (1998) through the relation $A_{III} = 2B\ell$.

References

Abramowitz M, Stegun IA (1964) Handbook of mathematical functions. National Bureau of Standards, Appl Math Series 55

Aero EL, Kuvshinskii EV (1960) Fundamental equations of the theory of elastic media with rotationally interacting particles. Fiz Tverd Tela 2:1399–1409, Translated in Soviet Physics—Solid State 2:1272–1281 (1961)

Anthoine A (2000) Effect of couple-stresses on the elastic bending of beams. Int J Solids Struct 37:1003–1018

Anthony KH (1970) Die theorie der dislokationen. Arch Ration Mech Anal 39:43–88

Atkinson C, Leppington FG (1974) Some calculations of the energy-release rate G for cracks in micropolar and couple-stress elastic media. Int J Frac 10:599–602

Atkinson C, Leppington FG (1977) The effect of couple stresses on the tip of a crack. Int J Solids Struct 13:1103–1122

Barber JR (1992) Elasticity. Kluwer Academic Publishers

Bardet JP, Vardoulakis I (2001) The asymmetry of stress in granular media. Int J Solids Struct 38:353–367

Bilby BA, Cottrell AH, Swinden KH (1963) The spread of plastic yield from a notch. Proc Roy Soc Lond A272:304–314

Bilby BA, Eshelby JD (1968) Dislocations and the theory of fracture. In: Liebowitz H (ed) Fracture, vol I. Academic Press, New York

Bogy DB, Sternberg E (1967a) The effect of couple-stresses on singularities due to discontinuous loadings. Int J Solids Struct 3:757–770

Bogy DB, Sternberg E (1967b) The effect of couple-stresses on the corner singularity due to an asymmetric shear loading. Int J Solids Struct 4:159–174

Chan Y, Fannjiang A, Paulino G (2003) Integral equations with hypersingular kernels–theory and applications to fracture mechanics. Int J Eng Sci 41:683–720

Chen JY, Huang Y, Ortiz M (1998) Fracture analysis of cellular materials: A strain gradient model. J Mech Phys Solids 46:789–828

Cohen H (1966) Dislocations in couple-stress elasticity. J Math Phys 45:35–44

Cosserat E, Cosserat F (1909) Theorie des Corps Deformables. Hermann et Fils, Paris

Ejike UBCO (1969) The plane circular crack problem in the linearized couple-stress theory. Int J Eng Sci 7:947–961

Erdogan F, Gupta GD (1972) On the numerical solution of singular integral equations. Q Appl Math 30:525–534

Georgiadis HG (2003) The mode-III crack problem in microstructured solids governed by dipolar gradient elasticity: Static and dynamic analysis. ASME J Appl Mech 70:517–530

Georgiadis HG, Velgaki EG (2003) High-frequency Rayleigh waves in materials with microstructure and couple-stress effects. Int J Solids Struct 40:2501–2520

Green AE, Rivlin RS (1964) Multipolar continuum mechanics. Arch Ration Mech Anal 17:113–147

Grentzelou CG, Georgiadis HG (2005) Uniqueness for plane crack problems in dipolar gradient elasticity and in couple-stress elasticity. Int J Solids Struct 42:6226–6244

Gourgiotis PA, Georgiadis HG (2007) An approach based on distributed dislocations and disclinations for crack problems in couple-stress elasticity. submitted

Hills DA, Kelly PA, Dai DN, Korsunsky AM (1996) Solution of crack problems, the distributed dislocation technique. Kluwer Academic Publishers

Huang Y, Zhang L, Guo TF, Hwang KC (1997) Mixed mode near tip fields for cracks in materials with strain-gradient effects. J Mech Phys Solids 45:439–465
Huang Y, Chen JY, Guo TF, Zhang L, Hwang KC (1999) Analytic and numerical studies on mode I and mode II fracture in elastic-plastic materials with strain gradient effects. Int J Frac 100:1–27
Hui CY, Shia D (1999) Evaluations of hypesingular integrals using Gaussian quadrature. Int J Numer Methods Eng 44:205–214
Ioakimidis NI (1983) A direct method for the construction of Gaussian quadrature rules for Cauchy type and finite-part integrals. Anal Numer Theor Approx 12:131–141
Ioakimidis NI (1995) Remarks on the Gauss quadrature rule for a particular class of finite-part integrals. Int J Numer Methods Eng 38:2433–2448
Kabir H, Madenci E, Ortega A (1998) Numerical solution of integral equations with logarithmic-, Cauchy- and Hadamard-type singularities. Int J Numer Methods Eng 41:617–638
Kaya AC, Erdogan F (1987) On the solution of integral equations with strongly singular kernels. Q Appl Math XLV 1:105–122
Knesl Z, Semela F (1972) The influence of couple-stresses on the elastic properties of an edge dislocation. Int J Eng Sci 10:83–91
Koiter WT (1964) Couple-stresses in the theory of elasticity, Part I. Proc Ned Akad Wet B67:17–29, II:30–44
Korsunsky AM (1998) Gauss-Chebyshev quadrature formulae for strongly singular integrals. Q Appl Math LVI 3:461–472
Kroener E (1963) On the physical reality of torque stresses in continuum mechanics. Int J Eng Sci 1:261–278
Kutt HR (1975) The numerical evaluation of principal value integrals by finite-part integration. Numer Math 24:205–210
Lubarda VA, Markenskoff X (2000) Conservation integrals in couple stress elasticity. J Mech Phys Solids 48:553–564
Mindlin RD (1964) Micro-structure in linear elasticity. Arch Ration Mech Anal 16:51–78
Mindlin RD (1963) Influence of couple-stresses on stress concentrations. Exp Mech 3:1–7
Mindlin RD, Tiersten HF (1962) Effects of couple-stresses in linear elasticity. Arch Ration Mech Anal 11:415–448
Misicu M (1965) On a general solution of the theory of singular dislocations of media with couple-stresses. Rev Roum Sci Techn, Ser Mec Appl 10:35–46
Monegato G (1987) On the weights of certain quadratures for the numerical evaluation of Cauchy principal value integrals and their derivatives. Numer Math 50:273–281
Monegato G (1994) Numerical evaluation of hypersingular integrals. J Comp Appl Math 50:9–31
Muki R, Sternberg E (1965) The influence of couple-stresses on singular stress concentrations in elastic solids. ZAMP 16:611–618
Nowacki W (1972) Theory of micropolar elasticity. CISM International Centre for Mechanical Sciences No. 25. Springer-Verlag
Nowacki W (1974) On discrete dislocations in micropolar elasticity. Arch Mech 26:3–11
Paget DF (1981) The numerical evaluation of Hadamard finite-part integrals. Numer Math 36:447–453
Paul HS, Sridharan K (1980) The penny-shaped crack problem in micropolar elasticity. Int J Eng Sci 18:651–664
Paul HS, Sridharan K (1981) The problem of a Griffith crack on micropolar elasticity. Int J Eng Sci 19:563–579
Palmov VA (1964) The plane problem of non-symmetrical theory of elasticity. Appl Math Mech (PMM) 28:1117–1120
Roos BW (1969) Analytic functions and distributions in physics and engineering. Wiley
Sternberg E, Muki R (1967) The effect of couple-stresses on the stress concentration around a crack. Int J Solids Struct 3:69–95
Teodosiu C (1965) The determination of stresses and couple-stresses generated by dislocations in isotropic media. Rev Roum Sci Techn, Ser Mec Appl 10:1462–1480
Toupin RA (1962) Perfectly elastic materials with couple stresses. Arch Ration Mech Anal 11:385–414
Tsamasphyros G, Dimou G (1990) Gauss quadrature rules for finite part integrals. Int J Numer Methods Eng 30:13–26
Van Dyke M (1964) Perturbation Methods in Fluid Mechanics. Academic Press, New York
Vardoulakis I, Sulem J (1995) Bifurcation analysis in geomechanics. Blackie Academic and Professional, Chapman and Hall, London
Voigt W (1887) Theoritiscke Studien uber die Elastizitatsverhaltnisse der Krystalle. Abhandl Ges Wiss Gottingen 34:3–51
Weitsman Y (1965) Couple-stress effects on stress concentration around a cylindrical inclusion in a field of uniaxial tension. ASME J Appl Mech 32:424–428
Zhang L, Huang Y, Chen JY, Hwang KC (1998) The mode-III full-field solution in elastic materials with strain gradient effects. Int J Frac 92:325–348

Bifurcation of equilibrium solutions and defects nucleation

Claude Stolz

Abstract The purpose of this article is to revise some concepts on defects nucleation based on bifurcation of equilibrium solutions. Equilibrium solutions are obtained on a homogeneous body and on a body with an infinitesimal defect such as cavity under the same prescribed dead load. First void formation and growth in non linear mechanics are examined. A branch of radial transformation bifurcates from the undeformed configuration in presence of a small cavity. Two cases of behaviour are examined. One case is the growth of the cavity by only the deformation of the shell. In another modelling the cavity evolves like a damaged zone, the transition between the sound part and the damaged one is governed by a local criterium. Each configuration leads to the definition of a nucleation criterion based on a presence of a bifurcation state, common state of the homogeneous body and a body with an infinitesimal defect.

Keywords Bifurcation · Nucleation of defects · Hyperelasticity · Local damage · Composite sphere

1 Introduction

We consider a composite sphere with external radius R_e composed by a cavity of radius R_i surrounded by a material. Under a uniform radial tensile loading p, the solution is assumed to be radial. We investigate the possible branches of radially symmetric configurations involving a traction free internal cavity bifurcation from the homogeneous sphere. At the bifurcation point p_c, the equilibrium solution for the porous sphere under the load p_c tends to the equilibrium solution of the homogeneous sphere when the volume of the cavity tends to zero.

Ball (1982) and Horgan and Pence (1989) have studied different classes of bifurcation problems for non linear incompressible elasticity, which simulate the appearance of a cavity inside a homogeneous sphere. Example in finite incompressible elastoplasticity have been also investigated by Chung et al. (1987). In both cases for all values of p one possible solution is that the sphere remains solid, this is due to incompressibility. On the other hand, for a certain range of loading p one has another configuration with an internal infinitesimal cavity. But as pointed out by Ball, only specific classes of hyperelastic incompressible materials have the possibility of such cavitation under radial dead load. However if the core of the sphere is made by a damaged material which does not suffer any tensile stress after a critical stretch, we obtain a critical pressure for all classes of hyperelastic incompressible materials. This point generalizes the point of view of Ball and gives new definition of nucleation of defects.

In Sect. 3, the possibility of cavitation is generalized to other classes of materials in small strains.

C. Stolz (✉)
Laboratoire de Mécanique des Solides, Ecole Polytechnique, CNRS, UMR7649, Palaiseau 91128, France
e-mail: stolz@lms.polytechnique.fr

2 Extension of the Ball's approach

Consider a sphere with internal cavity of radius R_i and external radius R_e. Under the radial loading, the actual state is defined by the transformation $r = f(R)$. The deformation gradient tensor, denoted $\mathbb{F}$, is then given by

$$\mathbb{F} = f' e_r \otimes E_R + \frac{f}{R}\left(e_\theta \otimes E_\Theta + e_\phi \otimes E_\Phi\right). \quad (1)$$

Taking the incompressibility condition into account, the transformation f is reduced to

$$r^3 = R^3 + A \quad (2)$$

where A is a constant whose value depends on the loading.

For an isotropic hyperelastic incompressible material the strain energy W is a symmetric function of the principal stretches λ_k, k $= 1, 2, 3$: $W(\lambda_1, \lambda_2, \lambda_3)$.

2.1 The homogeneous sphere

When the sphere is homogeneous, for any dead load p the stress $\sigma = -p\ \mathbb{I}$ is a solution of the equilibrium state equation with $\lambda_k = 1,\ k = 1, 2, 3$. There is no strain, that is a consequence of incompressibility of the material.

2.2 The porous sphere

When a spherical cavity of radius R_i exists, the solution has the form (2), the principal stretches are now

$$\lambda_1 = \frac{1}{\lambda^2},\ \lambda_2 = \lambda_3 = \frac{r}{R} = \lambda. \quad (3)$$

For this transformation, the strain energy has a value governed only by the stretch λ,

$$\psi(\lambda) = W\left(\frac{1}{\lambda^2}, \lambda, \lambda\right) \quad (4)$$

and the relation between the stretches and the external load P_e is given by

$$P_e = \int_{\lambda_i}^{\lambda^e} \frac{d\psi}{d\lambda}\frac{d\lambda}{\lambda^3 - 1} = I(\lambda_i, \lambda_e). \quad (5)$$

This equation is derived from the equilibrium elastic problem for the radial deformation of a sphere when the inner surface is stress free. The inflation of a thick walled sphere is an universal deformation in the sense of Rivlin-Eriksen and is a classical and well-known solution (Ogden 1997; Ball 1982). The values λ_i, λ_e are linked by the relation (2)

$$(\lambda^3(R) - 1)R^3 = (\lambda_i^3 - 1)R_i^3 = (\lambda_e^3 - 1)R_e^3 = A \quad (6)$$

2.3 Discussion

We must compare these results with those of the homogeneous sphere. For a given P_e, $A \neq 0$ and the possibility to λ_e to tends towards 1 is that R_e tends to infinity. Simultaneously when the cavity is reduced to a point, λ_i tends to infinity. In fact the cavity is embedded in an infinite matrix and for some value P_e, the cavity is not detectable from an external point of view. From this constatation, Ball studies the convergence of the integral I with the bounds $\lambda_i = \infty, \lambda_e = 1$.

For example, let us consider a density of energy of the form $W = \mu_\alpha \sum_k \lambda_k^\alpha$, where μ_α is a constant modulus associated to the power α of the principal stretches. In this case, the pressure $P_e = P_c$ is finite under the condition $\lambda_i = \infty, \lambda_e = 1$ if $-1 < \alpha < 2$.

The point (λ_e, P_e) describes a loading path for some range of P_e which intersects the loading path of the homogeneous sphere at point $(1, P_c)$. This is the critical value for cavitation. But locally the strain energy is not finite and the local solution with fixed R does not tend to the homogeneous one.

The Neohookean material has a critical pressure equal to $5\mu_o$, the Mooney-Rivlin material can not have cavitation under finite pressure.

2.4 Additional assumptions and remarks

However, the hyperelastic media cannot suffer infinite stretch without rupture. Assume now that the core of the sphere is made by a damaged material which cannot suffer any tensile stress after some critical value given in terms of stretches. The boundary between the sound material and the damaged one is now a spherical interface of radius R_i along which the stress vector is null and the stretch r_i/R_i has the critical value λ_c. The inner surface is stress free.

With these additional assumptions, the external load P_e is always finite for a finite porous sphere and is given by the preceding results

$$P_e = \int_{\lambda_c}^{\lambda_e} \frac{d\psi}{d\lambda}\frac{d\lambda}{\lambda^3 - 1} = I(\lambda_c, \lambda_e), \quad (7)$$

and the link between the stretches is preserved:

$$A = (\lambda_c^3 - 1)R_i^3 = (\lambda_e^3 - 1)R_e^3 = (\lambda^3(R) - 1)R^3. \tag{8}$$

Now when the radius of the damaged zone R_i tends to zero, A tends to zero, λ_e tends to 1 and the pressure P_e tends to the finite value given by

$$P_c = \int_{\lambda_c}^{\lambda_e} \frac{\mathrm{d}\psi}{\mathrm{d}\lambda} \frac{\mathrm{d}\lambda}{\lambda^3 - 1} = I(\lambda_c, 1). \tag{9}$$

We have now a finite value for all hyperelastic incompressible materials which does not support stretching greater than λ_c. Moreover, when A tends to zero, it is easy to show that $\lambda(R)$ tends to 1 at fixed material point R, the local solution tends to the solution of the homogeneous sphere.

Introducing the concept of local damage extends the point of view of Ball of bifurcated equilibrium solution and gives rise to an upper bound for the critical value for nucleation of a spherical defect in a homogeneous sphere.

We examine now the case of elastic brittle material in small strain.

3 Case of elastic brittle material

Consider a composite sphere, whose shell (material 1) and core (material 2) are linear elastic materials with different elastic modulus. The bulk moduli are denoted by κ_i and the shear moduli are μ_i, i $= 1, 2$. κ_1 is supposed greater than κ_2.

The sphere is submitted to an isotropic loading, the radial displacement is prescribed on the external boundary ($R = R_e$) (Fig. 1). The solution of the elasticity problem is given considering a radial displacement

$$u = u_i(R)e_r, u_i(R) = A_i R + \frac{B_i}{R^2}, \quad i = 1, 2. \tag{10}$$

The boundary conditions imply:

$$u_1(R_e) = E\ R_e, u_2(0) = 0, B_2 = 0. \tag{11}$$

For a given history of E, the external surface is submitted to a radial force:

$$\sigma_1(R_e) \cdot e_r = \Sigma e_r. \tag{12}$$

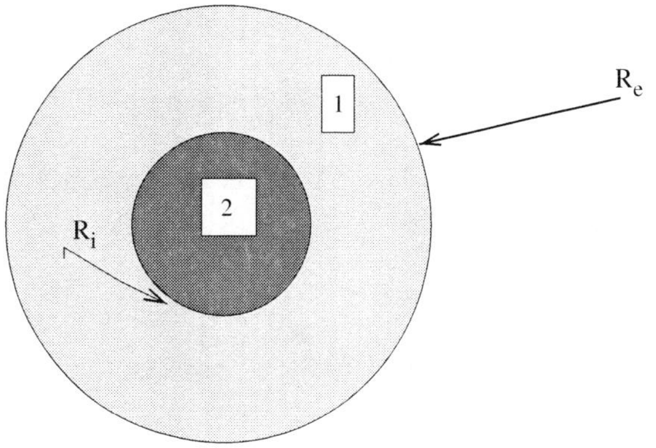

Fig. 1 The composite sphere

For given volume fraction of material 2, $c = \frac{R_i^3}{R_e^3}$, the solution of heterogeneous elastic sphere is

$$b = \frac{B_1}{R_i^3} = \frac{3(\kappa_1 - \kappa_2)A_1}{3\kappa_2 + 4\mu_1}$$

$$E = \frac{D(c)}{3\kappa_2 + 4\mu_1} A_1$$

$$\Sigma = \sigma_{rr}(R_e) = \frac{3E}{D(c)}((3\kappa_2 + 4\mu_1)\kappa_1 - 4\mu_1 c(\kappa_1 - \kappa_2))$$

where

$$D(c) = 3\kappa_2 + 4\mu_1 + 3c(\kappa_1 - \kappa_2), \quad c = \frac{R_i^3}{R_e^3}.$$

The last equation defines the global behaviour of the composite sphere as having an effective bulk modulus

$$\kappa^{eff} = \frac{(3\kappa_2 + 4\mu_1)\kappa_1 - 4\mu_1 c(\kappa_1 - \kappa_2)}{3\kappa_2 + 4\mu_1 + 3c(\kappa_1 - \kappa_2)} \tag{13}$$

Then it is obvious that when c tends to zero, κ^{eff} tends to κ_1.

When the radius R_i increases, the rigidity of the composite sphere decreases and some dissipation occurs. The dissipation is given by the rate

$$4\pi R_i^2 \mathcal{G}(R_i, E)\dot{R}_i = -\frac{\partial W}{\partial R_i}\dot{R}_i \tag{14}$$

This defines the analogous way the energy release rate associated with the dissipation along a moving surface (Stolz and Pradeilles-Duval 1997, 2004).

Along the interface Γ the energy release rate has the value

$$\mathcal{G}(R_i, E) = \frac{9E^2}{D^2(c)}(\kappa_1 - \kappa_2)(3\kappa_2 + 4\mu_1)(3\kappa_1 + 4\mu_1) \tag{15}$$

3.1 The response under monotonic loading

The loading parameter E is increasing. Initially, the core does not evolve, the critical value G_c is not reached. At one time the critical value G_c is reached, the strain is $E_c(c_o)$. After that the actual value of R_i is determined by the implicit equation

$$\mathcal{G}(R_i(t), E(t)) = G_c \tag{16}$$

this is the consistency condition. During this phase the internal radius R_i increases monotonical with E and attains the value R_e at the value E_T of the loading.

To any choosen critical value G_c corresponds a Griffith type local criterion for fracture, and this induces that the local stretch $u(R_i)/R_i$ is a constant proportional of the square root of G_c. From Eqs. 15 and 16 we deduce that when the damage occurs

$$\frac{E}{D(c)} = \alpha_c \tag{17}$$

where α_c is a constant. We remark that D is an increasing function of c. During the damage evolution $A_1 = E_c$

$$\alpha_c = \frac{E_c}{3\kappa_2 + 4\mu_1} = \frac{1}{3\kappa_1 + 4\mu_1}\frac{u(R_i)}{R_i}. \tag{18}$$

At intermediate time $t < T$ the sphere is not completely transformed, $\mathcal{G}(R_i(t), E) < G_c$ for any $E < E(t)$, then the composite sphere has the answer of an elastic heterogeneous medium with new concentration $c = R_i^3/R_e^3$. The global bulk modulus decreases with the transformation.

With the given propagation law of the interface, from it's initial position determined by $c_o = R_i^3(O)/R_e^3$, we have successively for an increasing function $E(t)$:

$$\begin{cases} E(t)<E_c(c_o), & \mathcal{G}(R_i, E(t))<G_c, & R_i(t)=R_i(0) \\ E(t)\geq E_c(c_o), & \mathcal{G}(R_i(t), E(t))=G_c, & R_i(t)=f(E(t)), \\ E(t) = E_T, & \mathcal{G}(R_e, E_T) = G_c & R_i(T)=R_e \\ E(t) \geq E_T, & & R_i(t)=R_e \end{cases} \tag{19}$$

and the answer can be plotted as in Fig. 2.

Conserving the value $\mathcal{G} = G_c$ when the concentration of damaged zone tends to zero ($c_o \to O^+$), the global bulk modulus tends to κ_1, $E_c(c_o) \to E_c^o$ and Σ tends to $P_c = 3\kappa_1 E_c^o$. The local energy remains finite and the local stresses at fixed R tend to the response of the homogeneous sphere.

This defines a new manner to consider a critical value for nucleation of defects. Let us assume the existence of a small defect of volume ω with a given evolution law. Conserving the evolution law, when the volume ω tends to zero, we consider the applied load as the critical load for nucleation of defect if this value coincides with the value of the original volume without defect for the same value of global strain.

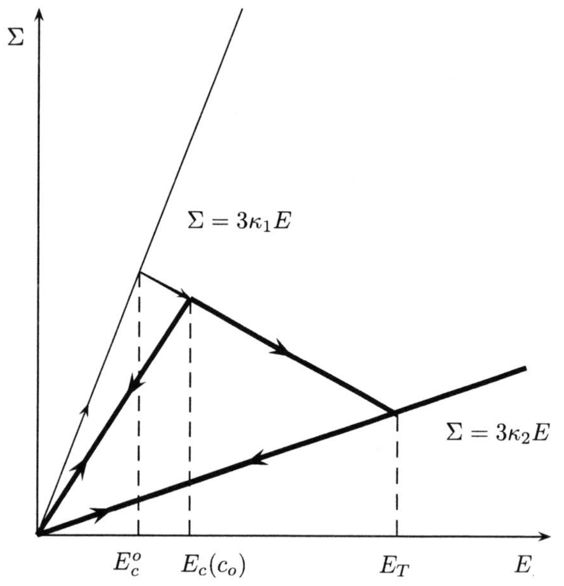

Fig. 2 The response of the composite sphere

In this sense, we cannot distinguish at the global level for this state of equilibrium if there exists or not an infinitesimal defect.

4 Conclusion

We have investigated the approach of a critical value for nucleation of defects inside an homogeneous body based on a bifurcation analysis of equilibrium solutions.

The nucleation of defects is considered with two differents points of view. One is the point of view of Ball and other authors, with locally infinite stretches and energy. This point of view defines different classes of materials, on one side we can have for same range cavitation, on the other side the cavitation is prohibited.

By introducing the concept of local damage, the nucleation of defects is possible for any materials, whose strains are limited in amplitude. Using this characteristic for governing the propagation of the damaged zone, a new point of view has been presented for critical value of defects nucleation.

We point out that this point of view gives for each configuration an upper bound for the critical value of nucleation because the approach is based on an analysis of bifurcation, then the value depends on the geometry and of the path of the loading.

References

Ball JM (1982) Discontinuous equilibrium solutions and cavitation in non linear elastiticty. Phil Trans R Soc Lond A 306:557–610

Chung DT, Horgan CO, Abeyaratne R (1987) A note on a bifurcation problem in finite plasticity related to void nucleation. Int J Solids Struct 23:983–988

Horgan CO, Pence TJ (1989) Void nuleation in tensile dead loading of a composite incompressible non linearly elastic sphere. J Elast 21:61–82

Ogden RW (1997) Non linear elastic deformations. Dover Publications

Stolz C, Pradeilles-Duval RM (1997) Thermomechanical approach of running discontinuities. In: Variations of domains and free-boundary problems in solids mechanics. IUTAM, Springer, pp 245–251

Stolz C, Pradeilles-Duval RM (2004) Stability and bifurcation with moving discontinuities. In: Mechanics of material forces, AMMA11, Springer, Chap 26, 261–268

Theoretical and numerical aspects of the material and spatial settings in nonlinear electro-elastostatics

Duc Khoi Vu · Paul Steinmann

Abstract The formulation of the spatial and material motion problem in nonlinear electro-elastostatics is revisited in this work. A finite element discretization is realized and a numerical example is presented to demonstrate possible application of the formulation in studying the closing process of cracks.

Keywords Nonlinear electro-elastostatics · Electromechanical coupling · Material forces

1 Introduction

The nonlinear electro-elastic behavior of electro-sensitive materials that exhibit large displacements and change their mechanical properties in response to the application of electric field was and still is the subject of many researches due to the interesting application of these materials in developing artificial muscles (Bar-Cohen 2002). Because of their great potential, the analysis of defects of electro-sensitive materials is of high interest. Maugin (1993) considered the material defect problem (in form of material inhomogeneity) related to nonlinear electro-elasticity with the help of the material forces, where the balance equation of linear momentum of the spatial motion problem is transformed into an appropriate form (material motion form) and in an appropriate continuum mechanical setting (the material setting). This material form of the balance equation of linear momentum has proved to be very useful in fracture mechanics. Concerning with material defects, the problem of material forces in nonlinear electro-elasticity was also studied by many other authors (Epstein and Maugin 1990; Epstein and Maugin 1991; Huang and Batra 1996; Kalpakides and Agiasofitou 2002; Maugin and Epstein 1991; Pak and Herrmann 1986a, b; Trimarco 2007). In a recent study (Vu and Steinmann 2007), the material force problem in nonlinear electro-elastostatics was revisited. By using a variational approach, the striking similarity between the spatial and material motion stresses as well as the similarity between the spatial and material governing equations of the two problems of elastostatics and electro-elastostatics were revealed. With the introduction of some material tractions and body forces to capture the energetic changes that are associated with changes in material configuration and material motions of defects relative to the ambient material, it was shown that the two problems of material and spatial motions are only equivalent for defect free bodies. In this work, the formulations presented in the work of Vu and Steinmann (2007) are discretized by the finite element method for the case of a crack. A numerical example is presented to highlight the possible application of the material force method in studying the closing process of cracks in cracked structures made of electro-sensitive materials undergoing large deformation. For details about the material force method and its

D. K. Vu (✉) · P. Steinmann
Chair of Applied Mechanics, University of Kaiserslautern,
P.O. Box 3049, 67653 Kaiserslautern, Germany
e-mail: vuduc@rhrk.uni-kl.de

Defect and Material Mechanics. C. Dascalu, G.A. Maugin & C. Stolz (eds.),
doi: 10.1007/978-1-4020-6929-1_11, © Springer Science+Business Media B.V. 2007

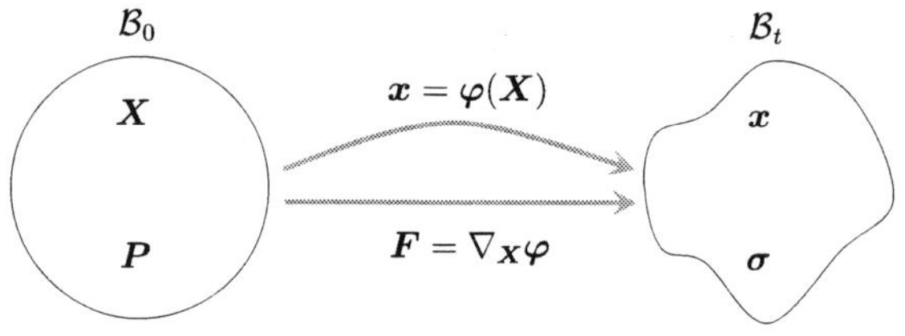

Fig. 1 Spatial motion problem in nonlinear elastostatics

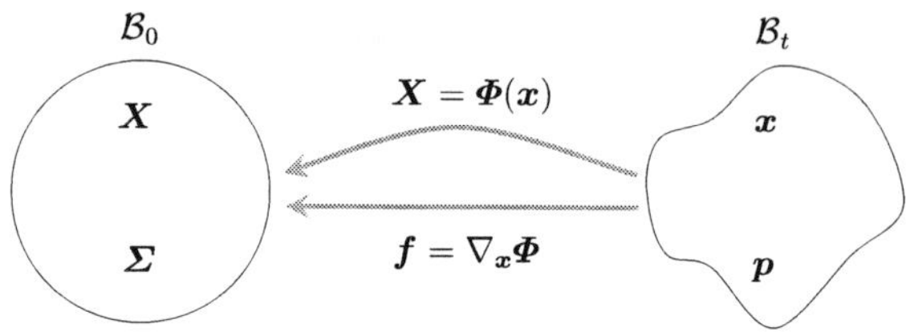

Fig. 2 Material motion problem in nonlinear elastostatics

applications, the readers are referred to, for example, the works (Denzer 2006; Denzer et al. 2003; Kienzler and Herrmann 2000; Kuhl and Steinmann 2004; Liebe et al. 2003; Steinmann 2000, 2002a, b, c, 2005; Steinmann et al. 2001).

2 Material and spatial settings in nonlinear electro-elastostatics

2.1 Spatial and material motion problem in nonlinear elastostatics

The two problems that are concerned with the material force method are the spatial and the material motion problems. For nonlinear elastostatics, in the spatial motion problem (Fig. 1), the position vector $\boldsymbol{x}$ of a point in the spatial (or deformed) configuration is described by a nonlinear spatial motion map $\boldsymbol{\varphi}$ that maps a point $\boldsymbol{X}$ from the material (or undeformed) configuration $\mathcal{B}_0$ into the spatial configuration $\mathcal{B}_t$. The deformation is characterized by the spatial motion deformation gradient $\boldsymbol{F}$ defined as the partial derivative of the spatial position vector with respect to the material one. Spatial motion stresses are represented by the two spatial motion stress tensors Cauchy $\boldsymbol{\sigma}$ and Piola $\boldsymbol{P}$. Corresponding to these two spatial motion stress tensors, two governing systems, each consists of one balance equation and one boundary condition, can be established. The first one is constructed in reference to the spatial configuration and uses the spatial motion Cauchy stress tensor as variable. The second one is constructed in reference to the material configuration and uses the spatial motion Piola stress tensor as variable.

In the material motion problem (Fig. 2), the position vector $\boldsymbol{X}$ of a point in the material configuration is described by a nonlinear material motion map $\boldsymbol{\Phi}$ that maps a point $\boldsymbol{x}$ from the spatial into the material configuration. The deformation is characterized by the material motion deformation gradient $\boldsymbol{f}$ defined as the partial derivative of the material position vector with respect to the spatial one. In the material motion problem, material motion stresses are defined as material motion Cauchy (Eshelby) stress $\boldsymbol{\Sigma}$ (in reference to the material configuration) and material motion Piola stress $\boldsymbol{p}$ (in reference to the spatial configuration). Again, corresponding to these two material motion stress tensors, two governing systems can be established. The first one is constructed in reference to the material configuration and uses the material motion Cauchy stress tensor as variable. The second one is constructed in reference to the spatial configuration and uses the material motion Piola stress tensor as variable.

In nonlinear elasticity, it is well-known that by the direct approach, the balance equations of linear momentum of the material motion problem (in reference to the spatial or material configuration) can be derived directly from their counterparts in the spatial motion problem. This transformation can be realized by multiplying the spatial motion balance equations of linear momentum by the deformation gradient $\boldsymbol{F}$. As an alternative to the direct approach, the variational approach can be used to derive the governing equations of the spatial and material motion problem. As a first step toward a more complete discussion on the formulations of spatial and material motion problem, in this work the system under consideration is assumed to be conservative in the sense that the elastic material response can be characterized by some internal potential energy densities and the conservative loading can be characterized by some external potential energy densities per unit volume. Furthermore, only zero spatial motion tractions are considered. The whole system is then characterized by some total potential energy densities $U_{0F}\,(\boldsymbol{F}, \boldsymbol{\varphi}, \boldsymbol{X})$, $U_{0f}\,(\boldsymbol{f}, \boldsymbol{x}, \boldsymbol{\Phi})$ in reference to the material configuration, or $U_{tF}(\boldsymbol{F}, \boldsymbol{\varphi}, \boldsymbol{X})$, $U_{tf}(\boldsymbol{f}, \boldsymbol{x}, \boldsymbol{\Phi})$ in reference to the spatial configuration such that $U_{0F}|_{\boldsymbol{F},\boldsymbol{\varphi},\boldsymbol{X}} = U_{0f}|_{\boldsymbol{f},\boldsymbol{x},\boldsymbol{\Phi}}$ and that $U_{tF}|_{\boldsymbol{F},\boldsymbol{\varphi},\boldsymbol{X}} = U_{tf}|_{\boldsymbol{f},\boldsymbol{x},\boldsymbol{\Phi}}$. These total potential energy densities are defined as the sum of the corresponding internal and external potential energy densities. The assumption about the conservative property of the system under consideration is made in order to make it possible to derive all stresses and all body forces from internal and external potential energy densities. For the derivation of the material motion formulations, the core argument here comes from the work of Steinmann (2005), who noted that taking a variation of a total potential energy functional at fixed spatial placement $\boldsymbol{x}$ only leads to a stationary point for the case of configurational equilibrium. In more general cases, configurational or rather material tractions acting on the material boundary and material forces, acting on defects such as vacancies, interfaces, dislocations, cracks and the like must be considered. These material forces capture the energetic changes that go along with material motions of the defects relative to the ambient material.

2.2 Spatial and material motion problem in nonlinear electro-elastostatics

For the nonlinear electro-elastostatic problem, a variational formulation for the spatial motion problem is built from the basic equations of electrostatics. The obtained variational formulation states that the governing equations of the nonlinear electro-elastostatic problem is equivalent to the stationary condition of some augmented total potential energy functionals, wherein the augmented total potential energy densities $\hat{U}_{0\boldsymbol{F}}$, $\hat{U}_{0f}$ (in reference to the material configuration) or $\hat{U}_{t\boldsymbol{F}}$, $\hat{U}_{tf}$ (in reference to the spatial configuration) are computed by taking into account the elastic energy stored in the bulk of the material and the electric energy contributed from the free space. Based on these augmented total potential energy functionals, the balance equations of linear momentum and boundary conditions of the material motion problem can be derived.

In electro-elastostatics, the electric field acting on material is governed by Faraday's law, Gauss' law for electricity, and a relationship describing the link between the electric polarization, the electric displacement and the electric field. Because of the electric polarization, the electric field exerts on matter a body force that can be considered as a function of the electric field vector and the electric polarization. With this electric body force, the balance equation of linear momentum is the same as that of a normal nonlinear elastic system except the fact that the Cauchy stress tensor is not symmetric. Besides the non-symmetric property of the Cauchy stress, it is noted that difficulties also appear in dealing with the jump conditions for Cauchy stress at the boundary of the considered body or across a surface of discontinuity within the body. This is due to the fact that on the one hand the Cauchy stress difference across a surface must balance both electrical and mechanical surface tractions. On the other hand, any traction measured by mechanical means is related to the contribution of both mechanical and electrical effects, since no available experiment can separate the effects of the Cauchy and Maxwell stresses unambiguously (McMeeking and Landis 2005). This leads to the definition of the so-called spatial motion total stress tensor $\hat{\boldsymbol{\sigma}}$, which is the combination of the Cauchy and Maxwell stresses.

For the sake of simplicity, let us assume that we only have homogeneous Neumann boundary conditions such that: $\mathbb{d} \cdot \boldsymbol{n} = 0$, $\hat{\boldsymbol{\sigma}} \cdot \boldsymbol{n} = \boldsymbol{0}$, where $\mathbb{d}$ is the electric displacement and $\boldsymbol{n}$ is the outward pointing unit normal at the boundary $\partial\mathcal{B}_t$. In the spatial motion problem (Fig. 3), the balance equations of linear momentum and the (homogeneous) boundary conditions can be derived from the stationary condition of the augmented total potential energy functionals, which requires the first variations of these functionals at fixed material placement $\boldsymbol{X}$ to vanish:

$$\delta_x \int_{\mathcal{B}_0} \hat{U}_{0\boldsymbol{F}} dV = 0 \quad \text{and} \quad \delta_x \int_{\mathcal{B}_t} \hat{U}_{t\boldsymbol{F}} dv = 0 \quad (1.1\text{–}2)$$

where $\hat{U}_{0\boldsymbol{F}} = \hat{U}_{0\boldsymbol{F}}\ (\boldsymbol{F}, \mathbb{E}, \boldsymbol{\varphi}, \boldsymbol{X})$, $\hat{U}_{t\boldsymbol{F}} = \hat{U}_{t\boldsymbol{F}}(\boldsymbol{F}, \mathbb{E}, \boldsymbol{\varphi}, \boldsymbol{X})$, $\mathbb{e}$ and $\mathbb{E}$ denote, respectively, the electric field vectors in reference to the spatial and material configurations. In reference to the material configuration, $\hat{U}_{0\boldsymbol{F}}$ can be computed as:

$$\hat{U}_{0\boldsymbol{F}} = W_{0\mathbb{E}}\,(\boldsymbol{F}, \mathbb{E}; \boldsymbol{X}) - \frac{1}{2}\varepsilon_0 J \boldsymbol{C}^{-1} : [\mathbb{E} \otimes \mathbb{E}] + V_{0\boldsymbol{F}}\,(\boldsymbol{\varphi}; \boldsymbol{X}) \quad (2)$$

where the first term $W_{0\mathbb{E}}\,(\boldsymbol{F}, \mathbb{E}; \boldsymbol{X})$ is the elastic energy density stored in the bulk of the material, the second term represents the contribution of the free space, the third term $V_{0\boldsymbol{F}}\,(\boldsymbol{\varphi}; \boldsymbol{X})$ is the external potential energy density characterizing external loads, ε_0 is the vacuum permittivity, $J = \det(\boldsymbol{F})$ and $\boldsymbol{C} = \boldsymbol{F}^t \cdot \boldsymbol{F}$. The conditions (1.1–2) lead to the governing equations:

$$\nabla_x \cdot \hat{\boldsymbol{\sigma}} + \boldsymbol{b}_t = \boldsymbol{0} \quad \text{and} \quad \hat{\boldsymbol{\sigma}} \cdot \boldsymbol{n} = \boldsymbol{0} \quad (3)$$

in reference to the spatial configuration, or:

$$\nabla_X \cdot \hat{\boldsymbol{P}} + \boldsymbol{b}_0 = \boldsymbol{0} \quad \text{and} \quad \hat{\boldsymbol{P}} \cdot \boldsymbol{N} = \boldsymbol{0} \quad (4)$$

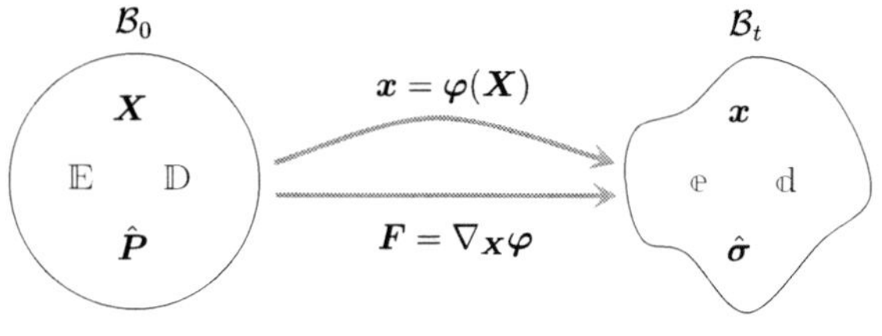

Fig. 3 Spatial motion problem in nonlinear electro-elastostatics

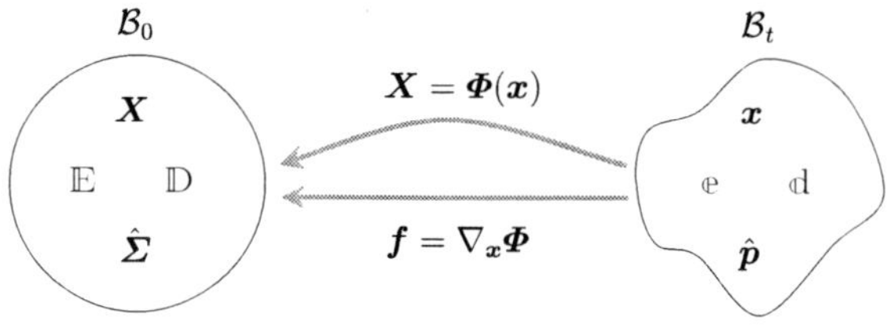

Fig. 4 Material motion problem in nonlinear electro-elastostatics

in reference to the material configuration, where $\hat{\boldsymbol{P}}$ denotes the counterpart of the spatial motion Piola stress tensor in elastostatics, $\boldsymbol{b}_t$ and $\boldsymbol{b}_0$ denote the spatial body forces, and $\boldsymbol{N}$ is the outward pointing unit normal at the boundary $\partial\mathcal{B}_0$.

In the material motion problem (Fig. 4), if only distributed defects are considered, the energetic changes can be captured by some distributed configurational forces $\hat{\boldsymbol{B}}_t^d$ and material tractions $\hat{\boldsymbol{T}}_t^d$ in reference to the spatial configuration or $\hat{\boldsymbol{B}}_0^d$ and $\hat{\boldsymbol{T}}_0^d$ in reference to the material configuration. Correspondingly, the first variations of the augmented total potential energy functionals at fixed spatial placement $\boldsymbol{x}$ become equal to the energy changes associated with these material forces:

$$\delta_X \int_{\mathcal{B}_t} \hat{U}_{tf} dv =: \int_{\mathcal{B}_t} \hat{\boldsymbol{B}}_t^d \cdot \delta\boldsymbol{\Phi} dv + \int_{\partial\mathcal{B}_t} \hat{\boldsymbol{T}}_t^d \cdot \delta\boldsymbol{\Phi} ds \leq 0 \quad (5)$$

in reference to the spatial configuration, or:

$$\delta_X \int_{\mathcal{B}_0} \hat{U}_{0f} dV =: \int_{\mathcal{B}_0} \hat{\boldsymbol{B}}_0^d \cdot \delta\boldsymbol{\Phi} dV + \int_{\partial\mathcal{B}_0} \hat{\boldsymbol{T}}_0^d \cdot \delta\boldsymbol{\Phi} dS \leq 0 \quad (6)$$

in reference to the material configuration, where $\hat{U}_{0f} = \hat{U}_{0f}(\boldsymbol{f}, \mathbb{e}, \boldsymbol{x}, \boldsymbol{\Phi})$, $\hat{U}_{tf} = \hat{U}_{tf}(\boldsymbol{f}, \mathbb{e}, \boldsymbol{x}, \boldsymbol{\Phi})$, $\hat{U}_{0f}|_{\boldsymbol{f},\mathbb{e},\boldsymbol{x},\boldsymbol{\Phi}} = \hat{U}_{0F}|_{\boldsymbol{F},\mathbb{E},\boldsymbol{\varphi},\boldsymbol{X}}$, $\hat{U}_{tf}|_{\boldsymbol{f},\mathbb{e},\boldsymbol{x},\boldsymbol{\Phi}} = \hat{U}_{tF}|_{\boldsymbol{F},\mathbb{E},\boldsymbol{\varphi},\boldsymbol{X}}$. The conditions (5) and (6) lead to the balance equations of linear momentum and boundary conditions of the material motion problem written as:

$$\nabla_x \cdot \hat{\boldsymbol{p}} + \hat{\boldsymbol{B}}_t =: -\hat{\boldsymbol{B}}_t^d \quad \text{and} \quad \hat{\boldsymbol{p}} \cdot \boldsymbol{n} =: \hat{\boldsymbol{T}}_t^d \quad (7.1\text{–}2)$$

in reference to the spatial configuration, or:

$$\nabla_X \cdot \hat{\boldsymbol{\Sigma}} + \hat{\boldsymbol{B}}_0 =: -\hat{\boldsymbol{B}}_0^d \quad \text{and} \quad \hat{\boldsymbol{\Sigma}} \cdot \boldsymbol{N} =: \hat{\boldsymbol{T}}_0^d \quad (8.1\text{–}2)$$

in reference to the material configuration, where $\hat{\boldsymbol{p}}$ and $\hat{\boldsymbol{\Sigma}}$ denote the material motion total stress tensors, $\hat{\boldsymbol{B}}_t$ and $\hat{\boldsymbol{B}}_0$ denote the material body forces. Based on these formulations, it can be shown that the two problems of material and spatial motions are only equivalent for defect free bodies, wherein: $\hat{\boldsymbol{B}}_t^d = \hat{\boldsymbol{B}}_0^d = \boldsymbol{0}$. Note that the stresses in the above systems can be computed through the corresponding potential energy densities as:

$$\hat{\boldsymbol{P}} = J\hat{\boldsymbol{\sigma}} \cdot \boldsymbol{F}^{-t} = \partial_{\boldsymbol{F}} \hat{U}_{0\boldsymbol{F}} \quad \text{and}$$
$$\hat{\boldsymbol{p}} = J^{-1}\hat{\boldsymbol{\Sigma}} \cdot \boldsymbol{f}^{-t} = \partial_{\boldsymbol{f}} \hat{U}_{t\boldsymbol{f}} \quad (9.1\text{–}2)$$

Furthermore, the spatial motion stress $\hat{\boldsymbol{\sigma}}$ and the material motion stress $\hat{\boldsymbol{\Sigma}}$ (Eshelby stress) have a similar energy-momentum type:

$$\hat{\boldsymbol{\sigma}} = \hat{U}_{t\boldsymbol{f}} \boldsymbol{I} - \boldsymbol{f}^t \cdot \hat{\boldsymbol{p}} + \mathbb{e} \otimes \mathbb{d} \quad \text{and}$$
$$\hat{\boldsymbol{\Sigma}} = \hat{U}_{0\boldsymbol{F}} \boldsymbol{I} - \boldsymbol{F}^t \cdot \hat{\boldsymbol{P}} + \mathbb{E} \otimes \mathbb{D} \quad (10.1\text{–}2)$$

where $\mathbb{D}$ denotes the electric displacement vector in reference to the material configuration. For more details about the material and spatial settings in nonlinear electro-elastostatics, the readers are referred to the work of Vu and Steinmann (2007).

3 Finite element discretization

3.1 Material force method

Application of the material motion formulations above in fracture mechanics is directly related to the computation of the material surface force $\mathfrak{f}^{\text{sur}}$ defined as:

$$\mathfrak{f}^{\text{sur}} = \int_{\partial\mathcal{B}_0} \hat{\boldsymbol{\Sigma}} \cdot \boldsymbol{N} \, dS \quad (11)$$

In order to compute this material force by the finite element method, let us consider the balance equation of linear momentum (8.1) in the case $\hat{\boldsymbol{B}}_0 = \boldsymbol{0}$ and $\hat{\boldsymbol{B}}_0^d = \boldsymbol{0}$. By multiplying this equation with a test function (material virtual displacement) $\boldsymbol{W}$, under the necessary smoothness and appropriate boundary assumptions, we arrive at the virtual work equation:

$$\mathcal{W}^{\text{sur}} = \mathcal{W}^{\text{int}} \quad (12)$$

where $\mathcal{W}^{\text{sur}}$ denotes the material variation of the potential energy due to its complete dependence on the material position and $\mathcal{W}^{\text{int}}$ denotes the material variation of the potential energy due to its implicit dependence on the material position:

$$\mathcal{W}^{\text{sur}} = \int_{\partial \mathcal{B}_0} \boldsymbol{W} \cdot \hat{\boldsymbol{\Sigma}} \cdot \boldsymbol{N} \, dS \quad \text{and}$$

$$\mathcal{W}^{\text{int}} = \int_{\mathcal{B}_0} \nabla_X \boldsymbol{W} : \hat{\boldsymbol{\Sigma}} \, dV \qquad (13.1\text{–}2)$$

By using the finite element method, the virtual displacement field $\boldsymbol{W}$ can be discretized over the element e occupying the element domain $\mathcal{B}_0^e$ as:

$$\boldsymbol{W}^h \Big|_{\mathcal{B}_0^e} = \sum_{n=1}^{n_{en}} N^n \boldsymbol{W}_n \qquad (14)$$

where $\boldsymbol{W}_n$ is the nodal value of $\boldsymbol{W}$, N^n is the shape function at node n and n_{en} is the number of nodes of element e. To evaluate the gradient of $\boldsymbol{W}$ with respect to $\boldsymbol{X}$, the following formula is used:

$$\nabla_X \boldsymbol{W}^h \Big|_{\mathcal{B}_0^e} = \sum_{n=1}^{n_{en}} \boldsymbol{W}_n \otimes \nabla_X N^n \qquad (15)$$

Over the element e, the left- and right-hand sides of (12) can be discretized as:

$$\mathcal{W}^{\text{sur},h} \Big|_{\mathcal{B}_0^e} = \sum_{n=1}^{n_{en}} \boldsymbol{W}_n \cdot \int_{\partial \mathcal{B}_0^e} N^n [\hat{\boldsymbol{\Sigma}} \cdot \boldsymbol{N}] dS \qquad (16)$$

and:

$$\mathcal{W}^{\text{int},h} \Big|_{\mathcal{B}_0^e} = \sum_{n=1}^{n_{en}} \boldsymbol{W}_n \cdot \int_{\mathcal{B}_0^e} \hat{\boldsymbol{\Sigma}} \cdot \nabla_X N^n dV \qquad (17)$$

Because of the arbitrariness of the material virtual nodal displacement $\boldsymbol{W}_n$, the global discrete material nodal force characterizing the material surface loads at node n can be eventually computed as:

$$\mathfrak{f}_n^{\text{sur},h} = \mathbf{A}_{e=1}^{n_{el}} \int_{\mathcal{B}_0^e} \hat{\boldsymbol{\Sigma}} \cdot \nabla_X N^n dV \qquad (18)$$

where n_{el} is the number of elements used to discretize the domain $\mathcal{B}_0$. Note that in (18) the material motion stress tensor can be computed by (10) when the spatial motion problem is solved (Vu et al. 2007).

3.2 Relation to J-integral

Now consider the simple case of a structure with a single crack lying in a subdomain $\mathcal{V}_0$ of $\mathcal{B}_0$. A direct application of the material force method is based on the link between the vectorial generalization $\mathfrak{J}$ of the J-integral in fracture mechanics and the integration of the normal projection of the material motion Cauchy stress (Eshelby stress) over a surface (in 3D) or a line (in 2D) enclosing a crack tip:

$$\mathfrak{J} = \lim_{\partial \mathcal{V}_0^r \to 0} \int_{\partial \mathcal{V}_0^r} \hat{\boldsymbol{\Sigma}} \cdot \boldsymbol{N} dS \qquad (19)$$

where $\partial \mathcal{V}_0^r$ is the regular part of the boundary $\partial \mathcal{V}_0$. The singular part of the boundary, $\partial \mathcal{V}_0^s$, denotes the crack tip. The regular and singular parts of the boundary under consideration $\partial \mathcal{V}_0$ are defined such that: $\partial \mathcal{V}_0 = \partial \mathcal{V}_0^r \bigcup \partial \mathcal{V}_0^s$ and $\emptyset = \partial \mathcal{V}_0^r \bigcap \partial \mathcal{V}_0^s$.

It should be noted that the concept of J-integral (19) is an extension of the J-integral in elasticity to electroelasticity. This extension using the theory of material forces was given by, for example, Pak and Herrmann (1986a, b), Epstein and Maugin (1990, 1991), Maugin and Epstein (1991) and Dascalu and Maugin (1994). In order to maintain the similitude with the formulation of the J-integral in pure mechanics, or in other words, for $\mathfrak{J}$ to be path-independent, some special conditions must be specified, namely on the crack faces traction and charge must be zero: $\hat{\boldsymbol{\sigma}} \cdot \boldsymbol{n} = \boldsymbol{0}$, $\mathbb{d} \cdot \boldsymbol{n} = 0$, which are assumed above. For more details, see, for example, the work of Dascalu and Maugin (1994) or Abendroth et al. (2002). The electric boundary condition on the crack faces increases the complication in analyzing crack problems in electroelasticity. With respect to the electric boundary conditions, two types of crack are usually considered: impermeable (where electric displacements are zero on the crack surfaces) and permeable (where electric displacements are continuous through the crack surfaces) crack. It is well-known that electrically impermeable conditions lead to overestimating, while electrically permeable conditions lead to underestimating the influence of the electric field on crack propagation. In order to properly model cracks in electroelasticity, special attention should be paid to the effect of the medium filling the crack and a transition model bewteen permeable and impermeable crack models with increasing crack opening may be used, see for example (Qi et al. 2001; Wang and Jiang 2003). In this work, for the sake of simplicity we consider only the electrically impermeable condition $\mathbb{d} \cdot \boldsymbol{n} = 0$.

By taking the decomposition of the boundary $\partial \mathcal{V}_0$ into singular $\partial \mathcal{V}_0^s$ and regular parts $\partial \mathcal{V}_0^r$ into account, the resulting material force acting on the whole boundary $\partial \mathcal{V}_0$ is given by:

$$\mathfrak{f}^{\mathrm{sur}} = \mathfrak{f}^{\mathrm{sur},s} + \mathfrak{f}^{\mathrm{sur},r} = \mathbf{0} \tag{20}$$

where:

$$\mathfrak{f}^{\mathrm{sur},s} = \int_{\partial\mathcal{V}_0^s} \hat{\boldsymbol{\Sigma}} \cdot \boldsymbol{N} dS \quad \text{and}$$

$$\mathfrak{f}^{\mathrm{sur},r} = \int_{\partial\mathcal{V}_0^r} \hat{\boldsymbol{\Sigma}} \cdot \boldsymbol{N} dS \tag{21.1-2}$$

From (20), the vector $\mathfrak{J}$ can be computed as:

$$\mathfrak{J} = -\mathfrak{f}^{\mathrm{sur},s} = -\int_{\partial\mathcal{V}_0^s} \hat{\boldsymbol{\Sigma}} \cdot \boldsymbol{N} dS \tag{22}$$

The discretized version $\mathfrak{J}^h$ of $\mathfrak{J}$ can be therefore computed by the global discrete material nodal force at the node characterizing the crack tip. For details, see for example Steinmann et al. (2001).

4 Numerical example

As numerical example, a 2-D plain strain rectangular plate (length $a = 60\,\mathrm{mm}$, width $b = 40\,\mathrm{mm}$) with a crack at the center of the plate (crack length $c = 20\,\mathrm{mm}$) is considered. The plate is assumed to be made of a compressible neo-Hookean-like material, of which the elastic energy density $W_{0\mathbb{E}}$ is defined as:

$$W_{0\mathbb{E}} = \frac{\mu}{2}[\boldsymbol{C} : \boldsymbol{I} - 3] - \mu \ln J + \frac{\lambda}{2}[\ln J]^2$$
$$+ c_1 \boldsymbol{I} : [\mathbb{E} \otimes \mathbb{E}] + c_2 \boldsymbol{C} : [\mathbb{E} \otimes \mathbb{E}] \tag{23}$$

where the last term could be thought of as having the nature of electrostriction coupling. For the purpose of investigating quantitatively the closing of the crack, let us, for simplicity, ignore the contribution of the free space term ($\frac{1}{2}\varepsilon_0 J \boldsymbol{C}^{-1} : [\mathbb{E} \otimes \mathbb{E}]$) in the augmented total potential energy density $\hat{U}_{0\boldsymbol{F}}$ (Eq. 2) and use in the calculation:

$$\hat{U}_{0\boldsymbol{F}} \approx W_{0\mathbb{E}} \tag{24}$$

This energy density means that in the absence of an electric stimulation the material behaves exactly as a material of the compressible neo-Hookean type in non-linear elasticity. The elastic behavior of this material is controlled by two parameters μ and λ, which are actually two Lamé coefficients. Under the application of an electric field, the energy density (24) means that our material will exhibit a nonlinear coupling behavior through the term $\boldsymbol{C} : [\mathbb{E} \otimes \mathbb{E}]$. The material constants c_1 and c_2 would have to be determined experimentally. However, due to the lack of experimental results, material properties are inadequate at this point. Therefore for the testing purpose, in this example we assume that the plate's material has the following properties: bulk modulus $\kappa = 10\,\mathrm{MPa}$, shear modulus $\mu = 5\,\mathrm{MPa}$, $c_1 = 10^{-3}\,\mathrm{Pa} \cdot \mathrm{m}^2/V^2$ and $c_2 = 6 \times 10^{-5}\,\mathrm{Pa} \cdot \mathrm{m}^2/V^2$. The parameter λ is computed by using the relationship: $\lambda = \kappa - 2\mu/3$. By using these values, numerical results confirmed that (24) is actually a good approximation. Nevertheless, it should be noted that for other material properties, the contribution of the free electric field may become important to the crack closing and should be counted for. Because the contribution of the free electric field to the energy release rate (or J-integral) may be negative, as noted by Pak (1990) and Dascalu and Maugin (1994), the neglect of the free electric field contribution may lead to slowing down of the crack closing. However, with insufficient numerical evidence, we refrain from discussing further the effect of the free electric field.

Due to symmetry only half of the plate is modeled by 672 four-node quadrangular and 24 three-node triangular elements, in which the triangular elements are used to model the crack tip. For simplicity, it is assumed that when close, the two crack surfaces have no electrical contact. The loading process is divided into 36 steps. In the first 10 steps, increasing extension forces are applied near the two ends of the plate. In the next 26 steps, an increasing electric potential difference $\Delta\varphi$ is applied between the two ends of the plate while keeping the extension forces constant. Under the application of the extension forces, the crack is opened. With the electric potential loading, the electrostatic forces cause the plate to shrink and with a large enough $\Delta\varphi$, the electrostatic forces lead to closing of the crack. Figure 5 presents the closing process of the crack from maximum opening (step 10) to complete closing (step 36). Zoom-in pictures of the crack opening are shown in Fig. 6. Material forces at the crack tip at different steps are computed and normalized by the absolute value obtained at step 10 (where maximum crack tip opening displacement or CTOD occurs). The normalized material forces at the crack tip presented in Figs. 7 and 8 show clear changes in magnitude and direction of these forces. In the opening process of the crack, it is observed that the material force vector at the crack tip lies on the crack axis and points toward the ambient space (steps 1 to 10). The magnitude of this vector increases with increasing CTOD. Under the application

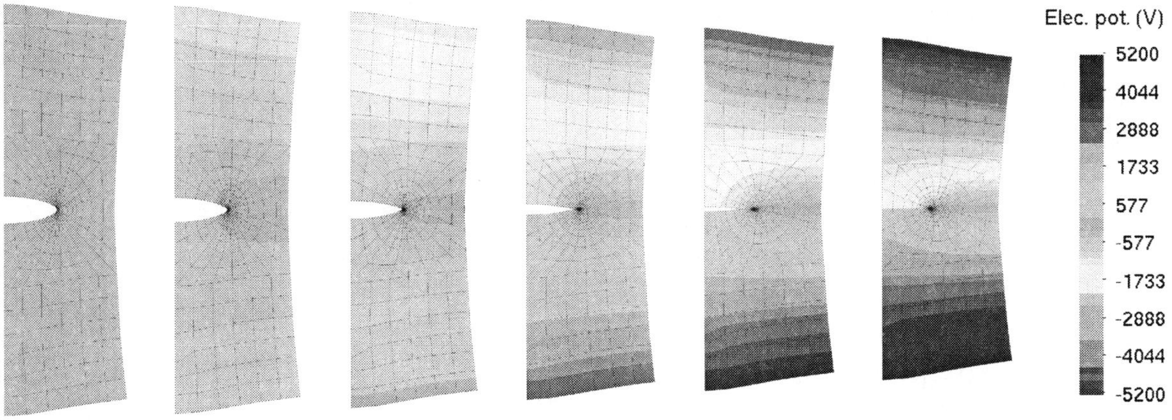

Fig. 5 Crack closing under electric loading (from left to right: $\Delta\varphi = 0$, 2000, 4000, 6000, 8000 and 10400 V): deformed configuration

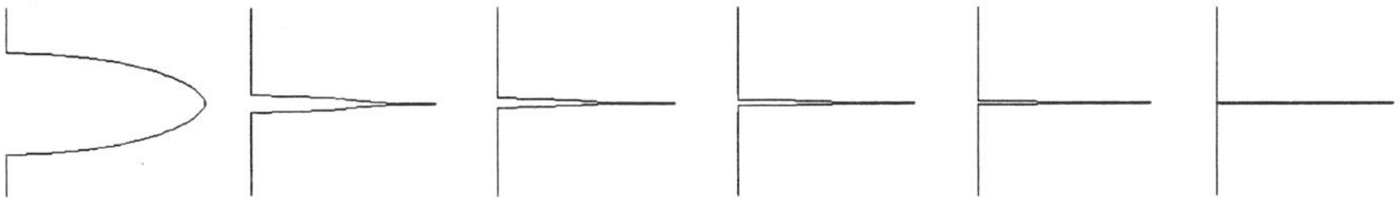

Fig. 6 Crack closing under electric loading (from left to right: $\Delta\varphi = 0$, 8000, 8800, 9200, 9600 and 10400 V): contour of the crack surfaces

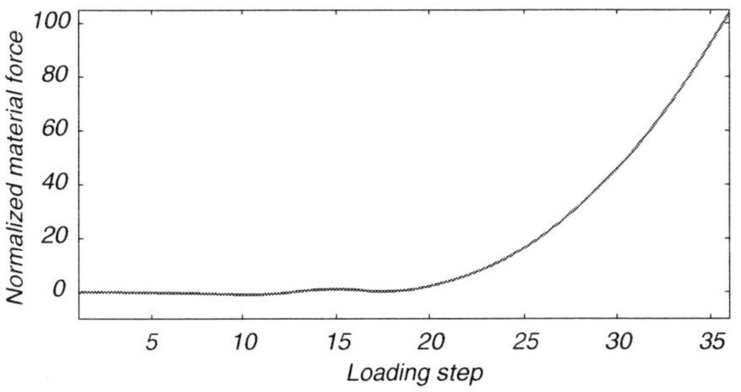

Fig. 7 Normalized material force at crack tip: loading steps 1–36

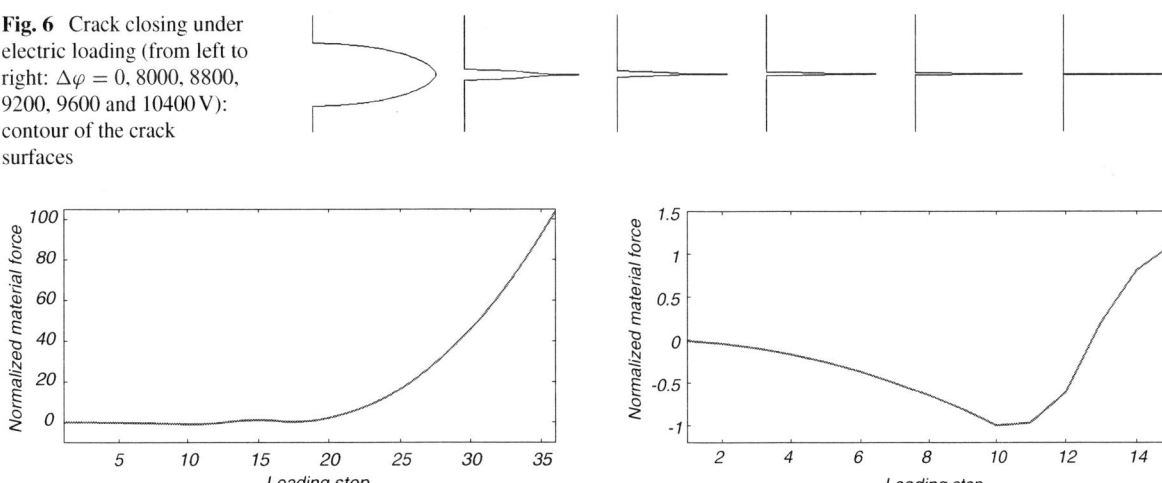

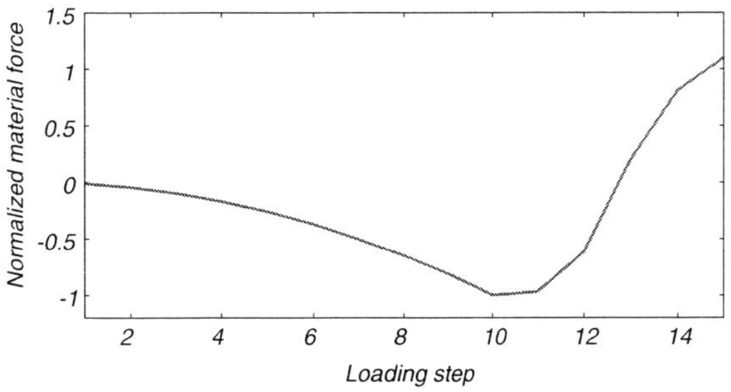

Fig. 8 Normalized material force at crack tip: zoom of loading steps 1–15

of the electric potential loading, this vector becomes smaller and eventually changes direction (in Figs. 7 and 8, positive values represent material forces pointing toward the ambient material). This observation suggests a useful application of the material force method in the study of crack closing.

5 Conclusion

The material force problem in nonlinear electro-elastostatics is considered in this work. By using the finite element method, material surface forces are discretized and computed for the case of a single crack structure made of electro-sensitive materials. The computed material forces at the tip of the crack offer useful information and can be used in prediction and analysis of crack closing.

Acknowledgements The work is funded by German Research Foundation (DFG) within the Research Training Groups 814, Engineering Materials at Multiple Scales: Experiments, Modeling and Simulation; Grant Number: GRK 814.

References

Abendroth M, Groh U, Kuna M, Ricoeur A (2002) Finite element-computation of the electromechanical J-integral for 2-D and 3-D crack analysis. Int J Fract 114:359–378

Bar-Cohen Y (2002) Electro-active polymers: current capabilities and challenges. In: Proceedings of SPIE – electroactive polymer actuators and devices

Dascalu C, Maugin GA (1994) Energy-release rates and path-independent integrals in electroelastic crack propagation. Int J Eng Sci 32(5):755–765

Denzer R (2006) Computational configurational mechanics. Dissertation, Technische Universität Kaiserslautern, Germany

Denzer R, Barth FJ, Steinmann P (2003) Studies in elastic fracture mechanics based on the material force method. Int J Numer Methods Eng 58:1817–1835

Epstein M, Maugin GA (1990) Inhomogeneities, Eshelby's tensor and J-integral in electroelasticity. In: Hsieh RKT (ed) Mechanical modelling of new electromagnetic materials. Elsevier, Amsterdam pp 253–258

Epstein M, Maugin GA (1991) Energy-momentum tensor and J-integral in electrodeformable bodies. Int J Appl Electromagn Mater 2:141–145

Huang Y-N, Batra RC (1996) Energy-momentum tensors in nonsimple elastic dielectrics. J Elasticity 42:275–281

Kalpakides VK, Agiasofitou EK (2002) On material equations in second gradient electroelasticity. J Elasticity 67:205–227

Kienzler R, Herrmann G (2000) Mechanics in material space with applications to defect and fracture mechanics. Springer, Berlin Heidelberg New York

Kuhl E, Steinmann P (2004) Computational modeling of healing: an application of the material force method. Biomech Model 2:187–203

Liebe T, Denzer R, Steinmann P (2003) Application of the material force method to isotropic continuum damage. Comput Mech 30:171–184

Maugin GA (1993) Material inhomogeneities in elasticity. Chapman and Hall

Maugin GA, Epstein M (1991) The electroelastic energy-momentum tensor. Proc Roy Soc London A 433:299–312

McMeeking RM, Landis CM (2005) Electrostatic forces and stored energy for deformable dielectric materials. ASME J Appl Mech 72:581–590

Pak YE (1990) Crack extension force in a piezoelectric material. ASME J Appl Mech 57:647–653

Pak YE, Herrmann G (1986a) Conservation laws and the material momentum tensor for the elastic dielectric. Int J Eng Sci 24:1365–1374

Pak YE, Herrmann G (1986b) Crack extension force in a dielectric medium. Int J Eng Sci 24:1375–1388

Qi H, Fang D, Yao Z (2001) Analysis of electric boundary condition effects on crack propagation in piezoelectric ceramics. Acta Mech Sinica 17(1):59–70

Steinmann P (2000) Application of material forces to hyperelastostatic fracture mechanics. I. Continuum mechanical setting. Int J Solids Struct 37:7371–7391

Steinmann P (2002a) On spatial and material setting of hyperelastodynamics. Acta Mech 156:193–218

Steinmann P (2002b) On spatial and material setting of thermo-hyperelastodynamics. J Elasticity 66:109–157

Steinmann P (2002c) On spatial and material settings of hyperelastostatic crystal defects. J Mech Phys Solids 50:1743–1766

Steinmann P (2005) On potential energy shifts in hyperelastic energy-momentum tensors. Technische Mechanik 25(3–4):174–181

Steinmann P, Ackermann D, Barth FJ (2001) Application of material forces to hyperelastostatic fracture mechanics. II. Computational setting. Int J Solids Struct 38:5509–5526

Trimarco C (2007) Material electromagnetic fields and material forces. Arch Appl Mech 77(2–3):177–184

Vu DK, Steinmann P (2007) Nonlinear electro- and magneto-elastostatics: material and spatial settings. Int J Solids Struct (accepted)

Vu DK, Steinmann P, Possart G (2007) Numerical modelling of non-linear electroelasticity. Int J Numer Methods Eng 70:685–704

Wang XD, Jiang LY (2003) The effective electroelastic property of piezoelectric media with parallel dielectric cracks. Int J Solids Struct 40:5287–5303

Energy-based r-adaptivity: a solution strategy and applications to fracture mechanics

Michael Scherer · Ralf Denzer · Paul Steinmann

Abstract This paper deals with energy based r-adaptivity in finite hyperelastostatics. The focus lies on the development of a numerical solution strategy. Although the concept of improving the accuracy of a finite element solution by minimizing the discrete potential energy with respect to the material node point positions is well-known, the numerical implementation of the underlying minimization problem is difficult. In this paper, energy based r-adaptivity is defined as a minimization problem with inequality constraints. The constraints are introduced to restrict the maximum distortion of the finite element mesh. As a solution strategy for the constrained problem, we use a classical barrier method. Beside the theoretical aspects and the implementation, a numerical experiment is presented. We illustrate the performance of the proposed r-adaptivity in the case of a cracked specimen.

M. Scherer (✉) · R. Denzer
Department of Mechanical Engineering,
Applied Mechanics, University of Kaiserslautern,
P.O. Box 3049, 67653 Kaiserslautern, Germany
e-mail: mscherer@rhrk.uni-kl.de

R. Denzer
e-mail: denzer@rhrk.uni-kl.de

P. Steinmann
Department of Mechanical Engineering,
Applied Mechanics, University of Erlangen–Nürnberg,
Egerlandstr. 5, 91058 Erlangen, Germany
e-mail: steinmann@ltm.uni-erlangen.de

Keywords Mesh optimization · Hyperelasticity · Constrained energyminimization · Material forces · Barrier method

1 Introduction

The finite element solution of a boundary-value problem in solid mechanics is based on a variational representation of the problem. In this paper, the focus lies on problems that are associated with the principle of minimum potential energy. In a standard finite element method the deformation is interpolated by a linear combination of fixed shape functions that are predetermined by the material node point positions and the connectivity of the mesh. Following a Ritz method, the unknown nodal deformations, the spatial node point positions, are determined by minimizing the potential energy of the discrete mechanical system. Since the necessary condition for a minimum of the discrete potential energy with respect to the spatial node point positions requires the equilibrium of internal and external spatial node point forces, it is denoted as spatial equilibrium condition.

The discrete potential energy in the state of equilibrium solely depends on the finite element approximation, i.e. the element connectivity and the material node point positions that determine the shape functions of the finite element approximation. In linear elasticity it is a proven fact, that the reduction of the potential energy corresponds to a reduction of the discretization error measured by the energy norm. The concept of

Defect and Material Mechanics. C. Dascalu, G.A. Maugin & C. Stolz (eds.),
doi: 10.1007/978-1-4020-6929-1_12, © Springer Science+Business Media B.V. 2007

energy based mesh optimization has been transferred to finite elasticity by several authors, see Askes et al. (2004), Kuhl et al. (2004), Mosler and Ortiz (2006), Thoutireddy (2003) and Thoutireddy and Ortiz (2004). A discrete deformation $\boldsymbol{\varphi}^{h1}$ is regarded as more accurate than $\boldsymbol{\varphi}^{h2}$, if $\mathcal{I}(\boldsymbol{\varphi}^{h1}) < \mathcal{I}(\boldsymbol{\varphi}^{h2})$, where $\mathcal{I}$ denote the potential energy. Numerical experiments justify this reasonable approach, but a strict mathematical proof is not given in the cited literature. Throughout this paper, we deal with a pure r-adaptive scheme, i.e. the discrete potential energy is minimized with respect to the material node point positions, whereas the element connectivity is fixed. An advantage of the considered r-adaptivity is that the computation of the gradient of the discrete potential energy with respect to the material node point positions is straightforward.

As mentioned before, the concept of energy-based r-adaptivity is well-known. However, the numerical implementation represents an ongoing research area in computational mechanics, see Askes et al. (2004), Kuhl et al. (2004), Mosler and Ortiz (2006) and Thoutireddy and Ortiz (2004). The solution strategy presented in this paper follows the concept of Felippa (1976b). We define energy-based r-adaptivity as a minimization problem with inequality constraints which are introduced to restrict the maximum distortion of the finite element mesh. As a solution strategy for the constrained problem, we use a classical barrier method. Similar to a penalty method, the barrier method transforms the constrained problem into a sequence of unconstrained problems. Each problem of the resulting sequence is solved by a staggered Newton method that alternately shifts the material node points and recovers the state of equilibrium.

Earlier publications dealing with linear elasticity already present several solution strategies. Carrol and Barker (1973), McNeice and Marcal (1973), Felippa (1976a, b), Bathe and Sussman (1983) and Sussman and Bathe (1985) used staggered schemes, steepest descent, conjugate gradient (cg) or derivative free methods. Carpenter and Zendegui (1982) proposed a cg method that determines the displacements and nodal positions simultaneously.

Braun (1997) emphasized the relation between configurational mechanics and energy-based r-adaptivity. He showed that the gradient of the discrete potential energy with respect to the material node point positions can be expressed in terms of the Eshelby stress tensor and denoted the components of the gradient associated with the nodes as discrete configurational forces. Mueller et al. (2002, 2004) and Mueller and Maugin (2002) investigated the application of configurational node point forces in the context of r- and h-adaptivity, fracture mechanics and inhomogeneities. As a solution strategy for the r-adaptivity, they used a staggered steepest descent method.

In the context of finite elasticity, energy-based mesh optimization was investigated by Thoutireddy (2003) and Thoutireddy and Ortiz (2004). They denoted their scheme as a variational arbitrary Lagrangian Eulerian method (VALE). To solve the underlying minimization problem, they used a staggered cg method and incorporated connectivity changes of the mesh to improve the performance of their scheme. Mosler and Ortiz (2006) focused on the numerical implementation of the VALE method, they proposed a simultaneous solution strategy based on the regularization of the configurational forces. Moreover, they developed a strategy to minimize the potential energy by changing the mesh connectivity. In the two-part publication of Askes et al. (2004) and Kuhl et al. (2004), the considered r-adaptivity is embedded into an ALE formulation for nonlinear elasticity. Therein, the variational formulation is derived from the principle of stationary potential energy. The finite element discretization of the variational formulation renders two coupled systems of equations which correspond to the derivatives of the discrete potential energy with respect to the spatial and material node point positions. To solve the resulting equations, the authors proposed staggered and simultaneous Newton schemes.

This paper is structured as follows. Section 2 outlines the considered class of boundary value problems and the variational formulation that is used for the numerical analysis. Section 3 deals with the finite element approach, i.e. the design of a finite element approximation for the spatial deformation and its usage in the framework of a Ritz method. Section 4 is concerned with the main topic: Energy-based r-adaptivity. In Sect. 5, we present a numerical example, a cracked specimen, that illustrates the performance of the solution strategy proposed in Sect. 4. A discussion in Sect. 6 completes the paper.

2 Boundary value problem and variational formulation

Mainly to introduce terminology and the notation, we start with a brief reiteration of the considered class

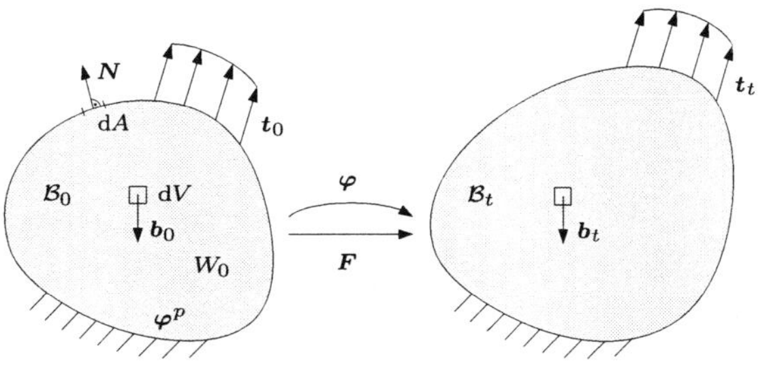

Fig. 1 Boundary value problem of elastostatics

of boundary value problems and the variational formulation that serves as a basis for the numerical analysis. For detailed information about the named topics we refer to Holzapfel (2000), Marsden and Hughes (1983) and Wriggers (2001). Figure 1 depicts the considered problem. A body B occupying the material configuration $\mathcal{B}_0$ in the undeformed state is subjected to conservative body forces and surface tractions acting on the Neumann boundary $\partial\mathcal{B}_0^t$.

The time independent spatial deformation map of the body $\boldsymbol{\varphi} : \mathcal{B}_0 \to \mathcal{B}_t$ with prescribed values $\boldsymbol{\varphi} = \boldsymbol{\varphi}^p$ on the Dirichlet boundary $\partial\mathcal{B}_0^{\varphi}$ is the primal unknown. We assume that the body consists of a homogeneous hyperelastic material with a strain energy density $W_0(\boldsymbol{F})$ per unit volume of the material configuration, where $\boldsymbol{F} = \nabla_X \boldsymbol{\varphi}$ denotes the spatial deformation gradient. The theory of nonlinear continuum mechanics yields the following boundary value problem of elastostatics

$$\begin{array}{ll} \mathrm{Div}\boldsymbol{P} + \boldsymbol{b}_0 = 0 & \text{on } \mathcal{B}_0 \\ \boldsymbol{\varphi} = \boldsymbol{\varphi}^p & \text{on } \partial\mathcal{B}_0^{\varphi} \\ \boldsymbol{P} \cdot \boldsymbol{N} = \boldsymbol{t}_0 & \text{on } \partial\mathcal{B}_0^{t}, \end{array} \tag{1}$$

where $\boldsymbol{N}$ denotes the outward normal to the boundary $\partial\mathcal{B}_0^t$ and $\boldsymbol{P}$ the Piola stress tensor that is given by

$$\boldsymbol{P} = \frac{\partial W_0}{\partial \boldsymbol{F}}. \tag{2}$$

The total potential energy of the system reads as

$$\mathcal{I}(\boldsymbol{\varphi}) = \int_{\mathcal{B}_0} W_0(\boldsymbol{F}(\boldsymbol{\varphi}))\,\mathrm{d}V - \int_{\mathcal{B}_0} \boldsymbol{\varphi} \cdot \boldsymbol{b}_0\,\mathrm{d}V - \int_{\partial B_0^t} \boldsymbol{\varphi} \cdot \boldsymbol{t}_0\,\mathrm{d}A. \tag{3}$$

The principle of minimum potential energy

$$\mathcal{I}(\boldsymbol{\varphi}) \to \mathrm{Min}, \quad \text{where } \boldsymbol{\varphi} = \boldsymbol{\varphi}^p \text{ on } \partial\mathcal{B}_0^{\varphi}. \tag{4}$$

serves as a basis for

1. the finite element analysis of the considered boundary value problem and
2. the r-adaptivity that improves the accuracy of the numerical solution.

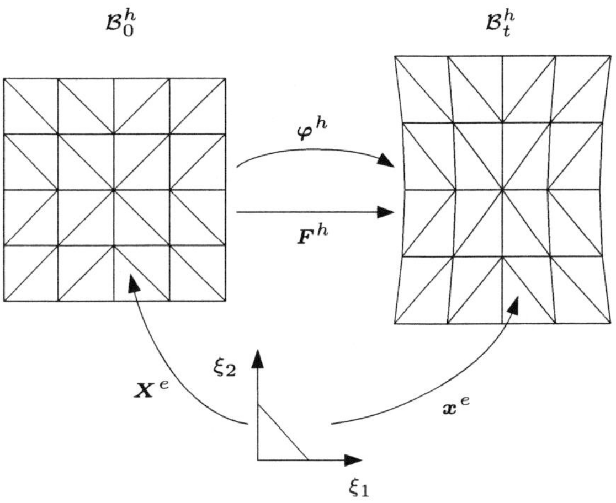

Fig. 2 Discretization and isoparametric concept

3 Finite element approach

In order to solve the variational problem (4) numerically, the spatial deformation of the body is approximated by a finite element approach that is subsequently used in a Ritz method. The design of finite element approximation $\boldsymbol{\varphi}^h : \mathcal{B}_0^h \to \mathcal{B}_t^h$ of the spatial deformation is based on the isoparametric concept, see Hughes (1987) and Wriggers (2001).

The material domain $\mathcal{B}_0$ and the spatial domain $\mathcal{B}_t$ are approximated by finite elements $\mathcal{B}_0^h = \cup_{e=1}^{n_{\mathrm{el}}} \mathcal{B}_0^e$ and $\mathcal{B}_t^h = \cup_{e=1}^{n_{\mathrm{el}}} \mathcal{B}_t^e$, see Fig. 2. The spatial deformation map of each material element to its spatial counterpart

$$\boldsymbol{\varphi}^e : \mathcal{B}_0^e \to \mathcal{B}_t^e, \qquad \boldsymbol{\varphi}^e = \boldsymbol{x}^e \circ \boldsymbol{X}^{e,-1} \tag{5}$$

is based on two bijective mappings

$$\boldsymbol{X}^e : \mathcal{B}_\xi^e \to \mathcal{B}_0^e, \qquad \boldsymbol{X}^e(\boldsymbol{\xi}) = \sum_{i=1}^{n_{\mathrm{en}}} N_i(\boldsymbol{\xi})\,\boldsymbol{X}_i, \tag{6}$$

$$\boldsymbol{x}^e : \mathcal{B}_\xi^e \to \mathcal{B}_t^e, \qquad \boldsymbol{x}^e(\boldsymbol{\xi}) = \sum_{i=1}^{n_{\mathrm{en}}} N_i(\boldsymbol{\xi})\,\boldsymbol{x}_i, \tag{7}$$

where $\mathcal{B}_\xi^e$ denotes the so called parent domain in $\boldsymbol{\xi}$-space, $i = 1, \ldots, n_{\mathrm{en}}$ is the local node numbering, N_i are isoparametric shape functions associated with the element nodes, $\boldsymbol{X}_i$ denotes known material node point positions and $\boldsymbol{x}_i$ are the unknown spatial node point positions. The assembly of all deformations $\boldsymbol{\varphi}^e(\boldsymbol{X})$ yields the global approximation of the spatial deformation map $\boldsymbol{\varphi}^h$ which can be written in the form

$$\boldsymbol{\varphi}^h(\boldsymbol{X}) = \sum_{i=1}^{n_n} N_i^g(\boldsymbol{X})\,\boldsymbol{x}_i, \tag{8}$$

where $i = 1, \dots n_n$ denotes the global node numbering, N_i^g are global C^0-continuous shape functions associated with the nodes of the mesh and $\boldsymbol{X}$ is the position vector of an arbitrary material point within the discrete material domain. For a compact notation, we introduce the column vector

$$\mathbf{x} = \left[[\boldsymbol{x}_1], \dots, [\boldsymbol{x}_{n_{nx}}]\right]^T \tag{9}$$

that contains all variable spatial node point positions. Note that the number n_{nx} of unknown spatial node point positions is smaller than the total number of nodes n_n, since $\boldsymbol{x}_i = \boldsymbol{\varphi}^p(\boldsymbol{X}_i)$ if $\boldsymbol{X}_i \in \partial\mathcal{B}_0^{\varphi}$. If r-adaptivity is excluded, the approximation of the spatial deformation solely depends on the choice of the spatial node points.

$$\boldsymbol{\varphi}^h = \boldsymbol{\varphi}^h(\boldsymbol{X}, \mathbf{x}) \tag{10}$$

Substituting the approximation $\boldsymbol{\varphi}^h$ into the potential energy $\mathcal{I}$, we obtain the discrete potential energy

$$\mathcal{I}^h(\mathbf{x}) = \mathcal{I}(\boldsymbol{\varphi}^h(\boldsymbol{X}, \mathbf{x})), \tag{11}$$

which is a nonlinear scalar function of the unknown spatial node point positions. The Ritz method transforms the variational problem into a minimization problem of real variables

$$\mathcal{I}^h(\mathbf{x}) \quad \rightarrow \quad \text{Min.} \tag{12}$$

If $\mathbf{x}^*$ is a solution of the minimization problem (12), then $\boldsymbol{\varphi}^h(\boldsymbol{X}, \mathbf{x}^*)$ is a numerical solution of the variational problem (4). The solution $\mathbf{x}^*$ fulfills the first order necessary condition

$$\mathbf{r} = \left[\frac{\partial \mathcal{I}^h}{\partial \mathbf{x}}\right]^T = \mathbf{0}. \tag{13}$$

which is is denoted as discrete spatial equilibrium condition. The components

$$\boldsymbol{r}_i = \mathop{\mathbf{A}}_{e=1}^{n_{\text{el}}} \int_{\mathcal{B}_0^e} \boldsymbol{P} \cdot \nabla_X N_i^e \, \mathrm{d}V - \int_{\mathcal{B}_0^e} N_i^e \, \boldsymbol{b}_0 \, \mathrm{d}V - \int_{\partial\mathcal{B}_0^{et}} N_i^e \, \boldsymbol{t}_0 \, \mathrm{d}A \tag{14}$$

of the spatial residual

$$\mathbf{r} = \left[[\boldsymbol{r}_1], \dots, [\boldsymbol{r}_{n_{nx}}]\right]^T \tag{15}$$

can be interpreted as discrete spatial node point forces. Thereby $\mathbf{A}_{e=1}^{n_{\text{el}}}$ denotes the assembly of all element contributions to $\boldsymbol{r}_i$.

4 Energy based r-adaptivity

So far we have been concerned with a standard finite element method. In the following sections, we deal with energy based r-adaptivity. We give a motivation, illustrate the concept, define the underlying problem of nonlinear programming, specify a set of constraints for linear triangular elements and describe the barrier method that is used to solve the problem.

4.1 Motivation

The motivation for energy-based mesh optimization, originates from the linear theory, see Felippa (1976b). In linear problems the internal energy of a system

$$E(\boldsymbol{u}, \boldsymbol{u}) = \frac{1}{2} \int_{\mathcal{B}} \boldsymbol{\epsilon}(\boldsymbol{u}) : \mathbb{E} : \boldsymbol{\epsilon}(\boldsymbol{u}) \tag{16}$$

is a bilinear form that induces the so-called energy norm

$$\|\boldsymbol{u}\|_E = E(\boldsymbol{u}, \boldsymbol{u})^{1/2}, \tag{17}$$

where $\boldsymbol{u}$ is the displacement vector, $\boldsymbol{\epsilon}$ denotes the linear strain tensor and $\mathbb{E}$ is the elastic tangent of the linear theory. Due to the Galerkin orthogonality, the discretization error of a numerical solution $\boldsymbol{u}^h$ measured by this energy norm can be expressed in terms of the potential energy

$$\|\boldsymbol{u}^h - \boldsymbol{u}^e\|_E^2 = \mathcal{I}(\boldsymbol{u}^h) - \mathcal{I}(\boldsymbol{u}^e) \geq 0, \tag{18}$$

where $\boldsymbol{u}^e$ is the exact solution. Since the energy of the exact solution is constant and independent of the discretization, a mesh with a smaller discrete energy yields a more accurate numerical solution in terms of the energy norm. R-adaptivity based on energy minimization is a problem of nonlinear programming. The numerical scheme does not require an estimate of the exact solution. This advantage becomes a drawback if one wants to evaluate the level of optimization. Without the knowledge of an estimate of $\mathcal{I}^h(\boldsymbol{u}^e)$ the potential energy is just an error indicator. It is necessary to accomplish an evaluation of the optimization process in a postprocessing step using an additional error estimator.

4.2 Concept

In contrast to Mosler and Ortiz (2006) and Thoutireddy and Ortiz (2004) we restrict ourselves to a pure r-adaptive scheme. The material node point positions of the finite element mesh are introduced as additional variables in order to further minimize the discrete potential energy, but the connectivity of the mesh is fixed. For notational convenience, we store the coordinates of variable material node point positions in a column vector

$$\mathbf{X} = \left[[\boldsymbol{X}_1], \ldots, [\boldsymbol{X}_{n_{nX}}]\right]^T. \tag{19}$$

Note that the number of variable material node point positions n_{nX} is smaller than the total number of nodes. To ensure that the discrete material domain $\mathcal{B}_0^h$ is an approximation of the continuous counterpart $\mathcal{B}_0$ the motion of the nodes is restricted to certain constraints, namely: Nodes on the boundaries are allowed to change their material positions along the boundary and nodes situated on vertices are fixed. For simplicity, positions of nodes on Dirichlet boundaries and nodes on Neumann boundaries with nonvanishing tractions are also fixed.

Since each global shape function N_i^g depends on all material node points that determine the shape of the support of N_i^g, the approximation of the spatial deformation depends on the material and spatial node point position

$$\boldsymbol{\varphi}^h = \boldsymbol{\varphi}^h(\boldsymbol{X}, \mathbf{x}, \mathbf{X}). \tag{20}$$

Thus, the discrete potential energy can be regarded as a function of both sets of variables

$$\mathcal{I}^h(\mathbf{x}, \mathbf{X}) = \mathcal{I}(\boldsymbol{\varphi}^h(\boldsymbol{X}, \mathbf{x}, \mathbf{X})). \tag{21}$$

This interpretation was proposed by several authors in the case of linear (Carpenter and Zendegui 1982) and nonlinear elasticity (Braun 1997; Mosler and Ortiz 2006; Thoutireddy and Ortiz 2004) and suggests a simultaneous minimization with respect to the material and spatial node point positions. In general, $\mathcal{I}^h(\mathbf{x}, \mathbf{X})$ is nonconvex since the Hessian matrix has negative eigenvalues, see Mosler and Ortiz (2006). This complicates the numerical treatment of the simultaneous minimization. For instance, if a Newton method it used, it is not guaranteed that the Newton direction is a descent direction. Another problem arising with the simultaneous approach using a Newton method is that the Hessian matrix of $\mathcal{I}^h$ can be singular. Moreover, if the simultaneous minimization of $\mathcal{I}^h$ fails, then a numerical solution of the variational problem is not at hand, not to mention a mesh improvement.

The staggered solution strategy that is used in this paper is based on an interpretation that was proposed by Felippa (1976b). We assume that for all admissible material node point positions $\mathbf{X}$ a unique solution $\mathbf{x}^*$ of the minimization problem (12) exists. The spatial node point positions $\mathbf{x}^*$ are considered to be an implicit differentiable function of the material node point positions

$$\mathbf{x}^* = \mathbf{x}^*(\mathbf{X}). \tag{22}$$

Substituting Eq. 22 into Eq. 21 we obtain the discrete potential energy in the state of spatial equilibrium

$$\mathcal{I}^{h*}(\mathbf{X}) = \mathcal{I}^h(\mathbf{x}^*(\mathbf{X}), \mathbf{X}). \tag{23}$$

This new function $\mathcal{I}^{h*}$ depending solely on $\mathbf{X}$ is considered to be the objective function of energy based mesh optimization. According to our experience, the Hessian of $\mathcal{I}^{h*}$ is nonsingular, but suffers from negative eigenvalues, i.e. $\mathcal{I}^{h*}$ is also nonconvex. The derivative of $\mathcal{I}^{h*}$ with respect to the material node point positions equals the explicit derivative of $\mathcal{I}^h$ with respect to $\mathbf{X}$ evaluated in the state of equilibrium[1]

$$\begin{aligned}\left[\mathbf{R}^*\right]^T &= \frac{\partial \mathcal{I}^{h*}}{\partial \mathbf{X}} = \left.\frac{\partial \mathcal{I}^h}{\partial \mathbf{x}}\right|_{\mathbf{x}^*,\mathbf{X}} \cdot \frac{\partial \mathbf{x}^*}{\partial \mathbf{X}} + \left.\frac{\partial \mathcal{I}^h}{\partial \mathbf{X}}\right|_{\mathbf{x}^*,\mathbf{X}} \\ &= \left.\frac{\partial \mathcal{I}^h}{\partial \mathbf{X}}\right|_{\mathbf{x}^*,\mathbf{X}} = \left[\mathbf{R}\big|_{\mathbf{x}^*,\mathbf{X}}\right]^T.\end{aligned} \tag{24}$$

The gradient $\mathbf{R}^*$ consists of components $\boldsymbol{R}_i^*$ that are associated with the nodes, i.e.

$$\mathbf{R}^* = \left[[\boldsymbol{R}_1^*], \ldots, [\boldsymbol{R}_{n_{nX}}^*]\right]^T. \tag{25}$$

Since the components $\boldsymbol{R}_i^*$ can be expressed in terms of the Eshelby stress tensor $\boldsymbol{\Sigma}$, they are denoted as discrete configurational forces or material node point forces, see Braun (1997), Mueller et al. (2002) and Steinmann et al. (2001). If body forces and surface tractions are absent, an explicit expression for the material node point forces reads as

$$\boldsymbol{R}_i^* = \mathop{\mathsf{A}}_{e=1}^{n_{\mathrm{el}}} \int_{\mathcal{B}_0^e} \boldsymbol{\Sigma} \cdot \nabla_X N_i^e \, \mathrm{d}V. \tag{26}$$

where

$$\boldsymbol{\Sigma} = W_0 \boldsymbol{I} - \boldsymbol{F}^t \cdot \boldsymbol{P}. \tag{27}$$

The material residual $\mathbf{R}^*$ can be used to formulate a steepest descent method for minimizing $\mathcal{I}^{h*}$. By an iteration step of the form

$$\mathbf{X}_{n+1} = \mathbf{X}_n - \alpha_n \mathbf{R}^* \tag{28}$$

[1] In Eq. 24, we exploit that $\left.\frac{\partial \mathcal{I}^h}{\partial \mathbf{x}}\right|_{\mathbf{x}^*,\mathbf{X}} = \mathbf{0}$

the material nodes are shifted in the opposite direction of the material node point forces. Thereby, the stepsize $\alpha_n > 0$ is chosen such that

$$\mathcal{I}^{h*}(\mathbf{X}_{n+1}) < \mathcal{I}^{h*}(\mathbf{X}_n). \tag{29}$$

Note that this steepest descent method is a staggered scheme: After shifting the node points the equilibrium has to be recovered. The outlined scheme and variations thereof, e.g. cg methods, were proposed in several references as solution strategies for energy-based mesh optimization in linear (Bathe and Sussman 1983; Carroll and Barker 1973; Felippa 1976a; McNeice and Marcal 1973; Muller et al. 2002; Sussman and Bathe 1985) and nonlinear elasticity (Mueller and Maugin 2002; Rajagopal et al. 2006; Thoutireddy and Ortiz 2004).

Assuming that the minimization of $\mathcal{I}^{h*}$ with respect to the material node point positions is an unconstrained problem, the first order necessary condition reads as

$$\mathbf{R}^* = \mathbf{0}. \tag{30}$$

Results of numerical experiments presented in Sect. 5 suggest that the existence of a material mesh that fulfills Eq. 30 is questionable for some problems/meshes, since the mobility of the material nodes is restricted by the mesh. Elements with negative Jacobians at the Gauss points are inadmissible and lead to an abortion of a finite element computation. Therefore, we explicitly take into account the naturally given restrictions of the finite element mesh by introducing inequality constraints which restrict the distortion of each element.

4.3 Minimization problem with constraints

A general formulation of the constrained minimization problem reads as

$$\mathcal{I}^{h*}(\mathbf{X}) \rightarrow \text{Min} \tag{31}$$

subjected to

$$g^l(\mathbf{X}) \leq 0, \quad l = 1, \dots, n_g. \tag{32}$$

The first order necessary condition of the constrained problem, the so called Kuhn-Tucker condition, reads as

$$\mathbf{R}^{\mathrm{KT}} = \begin{bmatrix} \mathbf{R}^* + \sum_{l=1}^{n_g} \lambda^l \left[\frac{\partial g^l}{\partial \mathbf{X}}\right]^T \\ \lambda^1 \, g^1 \\ \vdots \\ \lambda^{n_g} \, g^{n_g} \end{bmatrix} = \mathbf{0} \tag{33}$$

with Lagrange multipliers

$$\lambda^l \geq 0, \quad l = 1, \dots, n_g. \tag{34}$$

The constraints which are not yet specified define a feasible domain restricting the mobility of the material node points. The constraints

- have to be defined properly such that an adequate optimization of the mesh is possible, but an overly distortion of the elements is avoided.
- have to be adapted to the different element types, since different element geometries require individual strategies to avoid an inadmissible shape.

Concerning the applicability of various algorithms of nonlinear programming, it is preferable that the functions g^l characterizing the constraints are twice continuously differentiable.

4.3.1 Inequality constraints

In the following, we define a set of constraints that is especially designed for linear triangular elements. In contrast to Felippa (1976b), our constraints are not directly based on the Jacobian. We restrict the volume-preserving (distortional) deformation of each material element. Following a standard finite element procedure the deformation of a triangular element is described via linear isoparametric shape functions and node point coordinates. As a measure for the distortional deformation we use a scalar invariant of the deformation gradient. Since the deformation gradient and the derived strains are constant, one constraint $g^e \leq 0$ per element is sufficient to avoid an inadmissible change of the shape and, consequently, a negative volume.

The deformation of each linear triangular element e is described by the map

$$\boldsymbol{\Phi}_r^e : \mathcal{B}_{0r}^h \rightarrow \mathcal{B}_0^h, \tag{35}$$

where $\mathcal{B}_{0r}^e$ and $\mathcal{B}_0^e$ denote the material domains that the element e occupies in the original material mesh and the current adapted material mesh, see Fig. 3. The deformation map

$$\boldsymbol{\Phi}_r^e = \boldsymbol{X}^e \circ \boldsymbol{X}_r^{e,-1}, \tag{36}$$

is based on two bijective mappings

$$\boldsymbol{X}_r^e : \mathcal{B}_\xi^e \rightarrow \mathcal{B}_{0r}^e, \qquad \boldsymbol{X}_r^e(\boldsymbol{\xi}) = \sum_{i=1}^{n_{\mathrm{en}}} N_i(\boldsymbol{\xi}) \boldsymbol{X}_{r,i}, \tag{37}$$

$$\boldsymbol{X}^e : \mathcal{B}_\xi^e \rightarrow \mathcal{B}_0^e, \qquad \boldsymbol{X}^e(\boldsymbol{\xi}) = \sum_{i=1}^{n_{\mathrm{en}}} N_i(\boldsymbol{\xi}) \boldsymbol{X}_i, \tag{38}$$

where $\boldsymbol{X}_{r,i}$ and $\boldsymbol{X}_i$ denote the original and adapted material node point positions. The gradient of the deformation with respect to the material points $\boldsymbol{X}_r$ of the original element domain and the corresponding Jacobian are denoted by

$$\boldsymbol{f}_r^e = \nabla_{X_r} \boldsymbol{\Phi}_r^e \quad \text{and} \quad j_r^e = \det \boldsymbol{f}_r^e. \tag{39}$$

A matrix representation of the deformation gradient $\boldsymbol{f}_r^e$ using cartesian coordinates (X, Y) of the original and adapted material node points reads as

$$[\boldsymbol{f}_r^e] = \frac{\partial \boldsymbol{X}^e}{\partial \boldsymbol{\xi}} \cdot \left[\frac{\partial \boldsymbol{X}_r^e}{\partial \boldsymbol{\xi}}\right]^{-1} \tag{40}$$

where

$$\frac{\partial \boldsymbol{X}^e}{\partial \boldsymbol{\xi}} = \begin{bmatrix} X_2^e - X_1^e & X_3^e - X_1^e \\ Y_2^e - Y_1^e & Y_3^e - Y_1^e \end{bmatrix} \tag{41}$$

and

$$\frac{\partial \boldsymbol{X}_r^e}{\partial \boldsymbol{\xi}} = \begin{bmatrix} X_{r,2}^e - X_{r,1}^e & X_{r,3}^e - X_{r,1}^e \\ Y_{r,2}^e - Y_{r,1}^e & Y_{r,3}^e - Y_{r,1}^e \end{bmatrix}. \tag{42}$$

The constraints that restrict the deformation of the elements read as

$$g^e = \delta - \gamma^{-1} \leq 0, \quad e = 1, \dots n_{\mathrm{el}} \tag{43}$$

where γ is a measure of the deformation based on scalar invariants of the gradient $\boldsymbol{f}_r^e$, and the constant δ allows to adjust the maximum admissible distortion. Note that the restrictions introduced by the constraints can be chosen arbitrarily small by reducing δ.

Following the works of Knupp (2000a, b) and Armero and Love (2003) we use the invariant[2]

$$\gamma = \frac{\boldsymbol{f}_r^e : \boldsymbol{f}_r^e}{j_r}. \tag{44}$$

as a measure for the distortional deformation of the two-dimensional elements. The invariant γ is not influenced by volume changes of an element and is comparable with the norm of the isochoric part of the right Cauchy Green tensor (in 3 dimensions).[3] Provided that $j_r > 0$, γ is bounded below

$$\gamma \geq 2. \tag{46}$$

[2] The operator : represents the double contraction of two second order tensors, i.e. $\boldsymbol{A} : \boldsymbol{B} = A_{ij} B_{ij}$.

[3] We assume that a set of constaints based on

$$\gamma^{3D} = \frac{\boldsymbol{f}_r^e : \boldsymbol{f}_r^e}{j_r^{2/3}} \tag{45}$$

is suitable for linear tetrahedral elements.

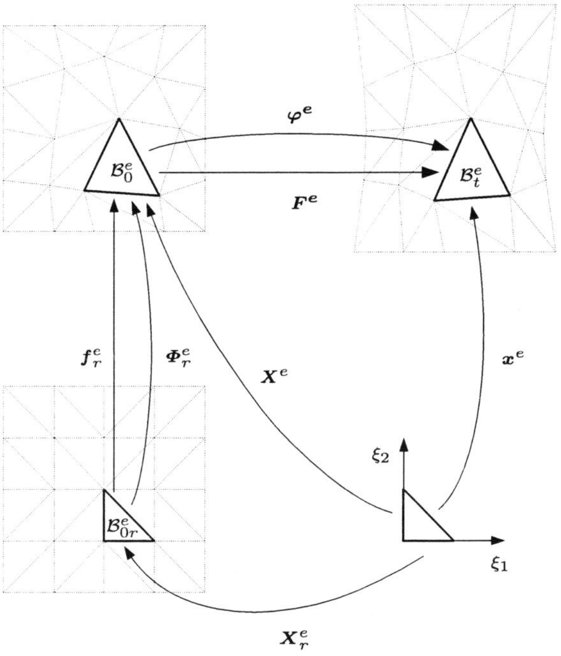

Fig. 3 Kinematics of constraints

Therefore, the constant δ has to be chosen

$$0 < \delta < 0.5. \tag{47}$$

If the adapted triangular element has the same shape as the original element, i.e. if $\boldsymbol{f}_r^e = \beta \mathbf{1}$, $\beta > 0$, then $\gamma = 2$. Note that γ is objective in the sense that γ is invariant under superposed (material) rigid body motions of a triangular element. Moreover, γ is twice continuously differentiable. Explicit expressions for the first and second derivative of g^e with respect to the material node point positions $\mathbf{X}$ are given in the appendix. The implementation of these derivatives is straightforward in the framework of a finite element method.

Remark 1 The presented constraints restrict the deformation of each element with respect to the original element. To design constraints that are independent of the original mesh, one can replace each original element by a so called reference element, for instance an equilateral triangle or an isosceles triangle like the reference element of the isoparametric concept, see Fig. 3. Since γ is independent of volume changes the scaling of the side length plays no role and can be normalized, i.e. set to one.

4.4 Solution strategy

4.4.1 Barrier method

As a solution strategy for the constrained problem (31), we propose a barrier method. A barrier method works by establishing a barrier on the boundary of the feasible domain that prevents an iterative solution strategy from leaving the region, see Bertsekas (1995), Gill et al. (1981), Jarre (2004) and Luenberger (1973). This barrier is realized with an additional function, a barrier function $\mathcal{I}^+(\mathbf{X})$ that is defined in the interior of the feasible domain and goes to infinity, if $\mathbf{X}$ reaches the boundary of the feasible domain via its interior. Similar to a penalty method, the barrier method transforms the constrained problem into a sequence of unconstrained problems

$$\mathcal{I}^k(\mathbf{X}) = \mathcal{I}^{h*}(\mathbf{X}) + \alpha^k\, \mathcal{I}^+(\mathbf{X}), \tag{48}$$

whereas the sequence of the scaling parameters tends to zero

$$0 < \alpha^{k+1} < \alpha^k, \qquad k = 1, 2, \ldots \qquad \alpha^k \to 0. \tag{49}$$

We use a classical inverse barrier function that reads as

$$\mathcal{I}^+(\mathbf{X}) = \sum_{k=1}^{n} -\frac{1}{g^e(\mathbf{X})}. \tag{50}$$

Remark 2 A feature of the considered barrier method is that

$$\frac{\alpha^k}{g^e(\mathbf{X}^k)^2} \to \lambda^e, \quad e = 1, \ldots, n_{\text{el}}, \tag{51}$$

if $\mathbf{X}^k \to \mathbf{X}^*$, where $\mathbf{X}^k$ denotes the solution of the k-th minimization problem of the barrier method, $\mathbf{X}^*$ is a solution of the constrained problem and λ^e denotes the Lagrange multipliers. Since the barrier method provides approximations for the Lagrange multipliers of the problem, a stop criterion can be formulated: Terminate the barrier method, if $\|\mathbf{R}^{\text{KT}}\|$ is smaller than a given tolerance, i.e. if $\mathbf{X}^k$ is close to a Kuhn-Tucker point of the constrained problem.

Since we solve each minimization problem of the sequence (48) with a Newton method, we need explicit expressions for the gradient and the Hessian of $\mathcal{I}^+$ with respect to $\mathbf{X}$. The gradient of the barrier function with respect to $\mathbf{X}$ is given by

$$\mathbf{R}^+ = \left[\frac{\partial \mathcal{I}^+}{\partial \mathbf{X}}\right]^T = \sum_{i=1}^{n_{\text{el}}} \frac{1}{[g^e]^2}\left[\frac{\partial g^e}{\partial \mathbf{X}}\right]^T, \tag{52}$$

By differentiating the gradient, we obtain the Hessian

$$\frac{\partial \mathbf{R}^+}{\partial \mathbf{X}} = \sum_{e=1}^{n_{el}} -\frac{2}{[g^e]^3}\left[\frac{\partial g^e}{\partial \mathbf{X}}\right]^T \cdot \frac{\partial g^e}{\partial \mathbf{X}} + \frac{1}{[g^e]^2}\frac{\partial^2 g^e}{\partial \mathbf{X}^2}. \tag{53}$$

The following three sections deal with the development of a solution strategy for the sequence of problems (48). Thereby we proceed as follows. First, we derive a staggered Newton scheme that solves each problem of the sequence. Then, we specify an algorithm that determines the sequence of the scaling parameters. And finally, we propose a predictor that provides an estimator for the solution of the $k+1$-th problem based on the solution of the k-th problem.

4.4.2 Staggered Newton scheme

A special characteristic of each minimization problem of the sequence (48) is that an explicit expression for the discrete potential energy in the state of equilibrium $\mathcal{I}^{h*}$ is not known. Nevertheless it is possible to implement a Newton scheme to solve the problem, a staggered scheme that alternately shifts the node points and recovers the equilibrium. The method is based on an iteration step of the form

$$\mathbf{X}^k_{n+1} = \mathbf{X}^k_n + \mathbf{d}^k_n, \tag{54}$$

Starting from a state of equilibrium, we perform the n-th iteration step according to (54) and obtain $\mathbf{X}^k_{n+1}$. As a result, the old spatial node point positions $\mathbf{x}^{*,k}_n$ does not fit to the new material node point positions $\mathbf{X}^k_{n+1}$ in the sense that the spatial equilibrium is fulfilled. To recover the equilibrium at least within a certain error tolerance, we compute $\mathbf{x}^{*,k}_{n+1}$ by iteratively solving the spatial equilibrium equation with a standard Newton method. If the iteration that recovers the equilibrium terminates successfully, the new Newton direction $\mathbf{d}^k_{n+1}$ can be calculated, and the $(n+1)$-th step according to Eq. 54 can be performed. Step (54) and the subsequent computations are iterated until the norm of the gradient of $\mathcal{I}^k$ is smaller than a given tolerance.

So far, we have assumed that for a given α^k the presented algorithm works properly. Possible errors that can occur are:

1. An iteration step according to Eq. 54 leads to a violation of the constraints, i.e. an infeasible material mesh.

2. The Newton method that is used to recover the spatial equilibrium fails to converge.
3. The norm of the gradient of $\mathcal{I}^k$ in the $(n+1)$-th step is higher than in the n-th step.

If one of the errors listed above occurs, the k-th iteration is stopped, α^k is increased, the predictor $\mathbf{X}^k_{\text{pr}}$ is recalculated and the k-th iteration is restarted.

After specifying the fundamental flow of the algorithm, we focus on the computation of the Newton direction $\mathbf{d}$. First, we derive the gradient and the Hessian of $\mathcal{I}^k$ with respect to $\mathbf{X}$. Considering the Eqs. 24 and 48, the gradient reads as

$$\mathbf{R}^k = \left[\frac{\partial \mathcal{I}^k}{\partial \mathbf{X}}\right]^T = \left[\frac{\partial \mathcal{I}^{h*}}{\partial \mathbf{X}} + \alpha^k \frac{\partial I^+}{\partial \mathbf{X}}\right]^T$$
$$= \mathbf{R}^* + \alpha^k \mathbf{R}^+ \tag{55}$$

By differentiating the gradient, we obtain the Hessian

$$\frac{\partial^2 \mathcal{I}^k}{\partial \mathbf{X}^2} = \frac{\partial \mathbf{R}^k}{\partial \mathbf{X}}$$
$$= \left.\frac{\partial \mathbf{R}}{\partial \mathbf{x}}\right|_{\mathbf{x}^*,\mathbf{X}} \cdot \frac{\partial \mathbf{x}^*}{\partial \mathbf{X}} + \left.\frac{\partial \mathbf{R}}{\partial \mathbf{X}}\right|_{\mathbf{x}^*,\mathbf{X}} + \alpha^k \frac{\partial \mathbf{R}^+}{\partial \mathbf{X}}. \tag{56}$$

To calculate $\partial \mathbf{x}^*/\partial \mathbf{X}$, we differentiate the spatial equilibrium equation (13) with respect to $\mathbf{X}$

$$\left.\frac{\partial \mathbf{r}}{\partial \mathbf{x}}\right|_{\mathbf{x}^*,\mathbf{X}} \cdot \frac{\partial \mathbf{x}^*}{\partial \mathbf{X}} + \left.\frac{\partial \mathbf{r}}{\partial \mathbf{X}}\right|_{\mathbf{x}^*,\mathbf{X}} = \mathbf{0}, \tag{57}$$

and obtain

$$\frac{\partial \mathbf{x}^*}{\partial \mathbf{X}} = -\left.\frac{\partial \mathbf{r}}{\partial \mathbf{x}}\right|^{-1}_{\mathbf{x}^*,\mathbf{X}} \cdot \left.\frac{\partial \mathbf{r}}{\partial \mathbf{X}}\right|_{\mathbf{x}^*,\mathbf{X}}. \tag{58}$$

Explicit expressions for the derivatives of the spatial and material residual with respect to the material and spatial node point positions in Eqs. 58 and 56 are given in the references (Kuhl et al. 2004; Thoutireddy 2003; Thoutireddy and Ortiz 2004). Finally, we obtain the Newton direction $\mathbf{d}^k_n$ by solving the linear system of equations

$$\left.\frac{\partial \mathbf{R}^k}{\partial \mathbf{X}}\right|_{\mathbf{X}^k_n} \cdot \mathbf{d}^k_n = -\mathbf{R}^k\big|_{\mathbf{X}^k_n}. \tag{59}$$

The presented solution strategy is computationally expensive, since each step of the proposed Newton scheme involves an additional Newton iteration to recover the spatial equilibrium. But this additional Newton iteration can be accelerated by means of the matrix $\partial \mathbf{x}^*/\partial \mathbf{X}$. In the n-th step of the Newton scheme of the mesh optimization, $\partial \mathbf{x}^*/\partial \mathbf{X}$ is necessary in order to calculate $\mathbf{d}^k_n$. Moreover, it can be used to accomplish the predictor step

$$\mathbf{x}^{*,k}_{n+1} \approx \mathbf{x}^k_{n,pr} = \mathbf{x}^k_n + \left.\frac{\partial \mathbf{x}^*}{\partial \mathbf{X}}\right|_{\mathbf{X}^k_n} \cdot \left[\mathbf{X}^k_{n+1} - \mathbf{X}^k_n\right] \tag{60}$$

which provides a good estimator for $\mathbf{x}^{*,k}_{n+1}$.

4.4.3 Sequence α^k

The calculation of the scaling parameter α^k is based on

$$\alpha^k = \eta^\kappa \alpha^{k-1}, \tag{61}$$

where $0 < \eta < 1$ and $\kappa > 0$. The parameter η is constant during the overall numerical procedure, whereas κ is adapted according to the following rules. If the k-th iteration has to be stopped (since one of the three listed errors occurs), κ is reduced by multiplication with a constant factor $0 < c_{\text{red}} < 1$

$$\kappa \rightarrow c_{\text{red}}\kappa, \tag{62}$$

and α^k is recalculated until the k-th iteration terminates successfully. If the k-th iteration succeeds at the first try, κ is increased according to

$$\kappa \rightarrow c_{\text{inc}}\kappa, \tag{63}$$

where $c_{\text{inc}} > 1$ is a constant factor. Then, the new scaling parameter α^{k+1} of the next minimization problem is determined. Note that $0 < \eta^\kappa < 1$ holds.

Remark 3 A serious problem often arising with the barrier method is that the first subproblem is difficult to solve numerically. A feature of the constraints presented in Sect. 4.3.1 is that the inverse barrier function has a minimum, if the original and the adapted mesh coincide This is an advantage, since the material node point positions of the original are a good estimator for the solution of the first subproblem $\mathbf{X}^0$, if the value of α^0 is chosen high enough. The barrier function regularizes the problem. If the constraints are based on a reference triangle as described in Remark 1 of Sect. 4.3.1, the barrier function seem to have the character of an energy for mesh smoothing. (For more detailled informations about energies designed for mesh smoothing we refer to Knupp (2000a, b).) Consequently, the mesh that minimizes the barrier function does not coincide with the original mesh. Therefore, the set constraints based on the the original mesh is preferred here.

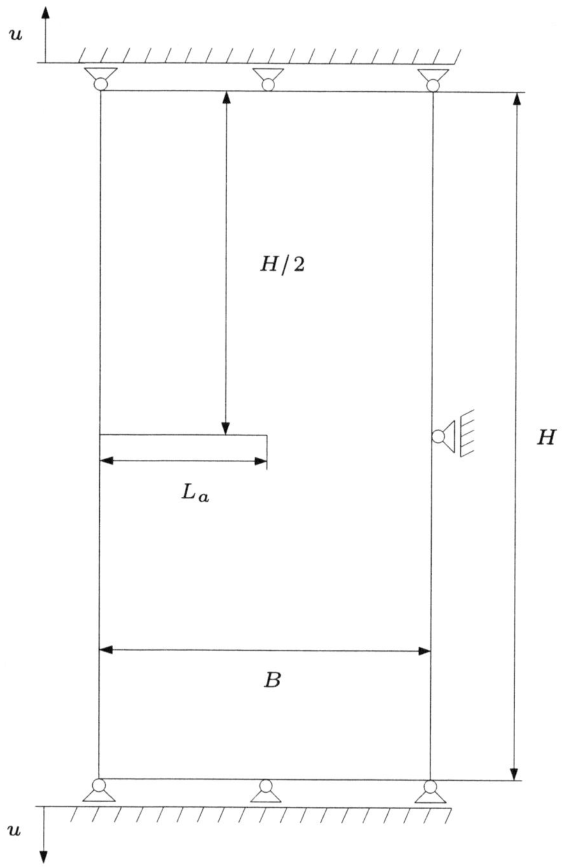

Fig. 4 Cracked specimen

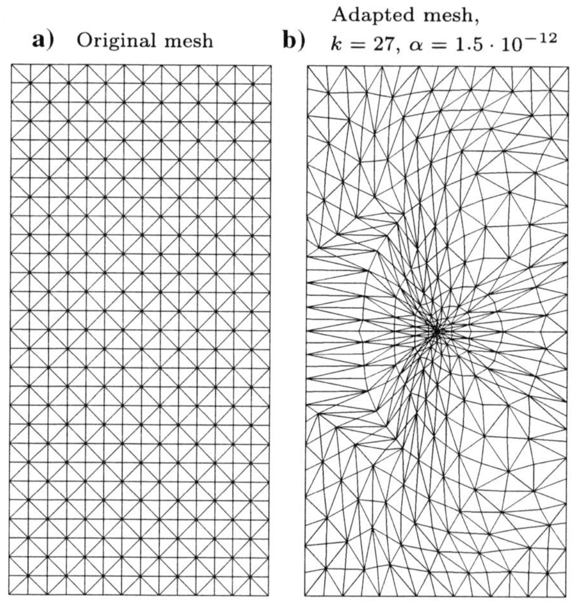

Fig. 5 Cracked specimen—entire original and adapted material mesh. (**a**) Orignal mesh; (**b**) adapted mesh, $k = 27$, $\alpha = 1.5 \times 10^{-12}$

4.4.4 Predictor for $\mathbf{X}^{k+1}$

The predictor $\mathbf{X}_{\text{pr}}^{k+1}$ for the solution of the $k+1$-th minimization problem is based on the assumption that the solutions $\mathbf{X}^k$ are discrete points of a differentiable curve $\mathbf{X}^*(\alpha)$. Given the tangent $\mathbf{t}^k = \mathrm{d}\mathbf{X}^*/\mathrm{d}\alpha$ of this curve at the point $\mathbf{X}^k = \mathbf{X}^*(\alpha^k)$, the predictor reads as

$$\mathbf{X}_{\text{pr}}^{k+1} = [\alpha^{k+1} - \alpha^k]\,\mathbf{t}^k + \mathbf{R}^k. \tag{64}$$

To compute $\mathbf{t}^k$, we solve the linear system of equations

$$\left.\frac{\partial \mathbf{R}^k}{\partial \mathbf{X}}\right|_{\mathbf{X}^k} \cdot \mathbf{t}^k = -\mathbf{R}^+|_{\mathbf{X}^k}. \tag{65}$$

5 Numerical experiment

Figure 4 illustrates the example, the stretching of a cracked specimen of dimensions $B = 10 \times H = 20$, $L_a = 5$. The finite element mesh used for the numerical analysis consists of 784 elements and 442 nodes, see Fig. 5a. Note that the computation does not take any advantage of the symmetry of the problem. On the upper and lower boundary of the specimen constant vertical displacements $u = 0.5$ are prescribed which corresponds to a 5% elongation. Only the middle node on the right boundary is fixed in horizontal direction. The specimen consists of a compressible hyperelastic material with a strain energy density

$$W_0 = \frac{\lambda}{2}\ln^2(J) + \frac{\mu}{2}[\boldsymbol{I} : \boldsymbol{C} - 3] - \mu\ln(J) \tag{66}$$

of neo–Hookean type, where $\boldsymbol{C} = \boldsymbol{F}^t \cdot \boldsymbol{F}$ is the right Cauchy-Green strain tensor and λ, μ denote the Lamé parameters. The material properties are $\lambda = 1.15 \times 10^4$ and $\mu = 7.69 \times 10^3$ which corresponds to a Young's modulus $E = 2.0 \times 10^4$ and a Poisson's ratio $\nu = 0.3$. The parameters of the algorithm are set to: $\text{tol}_1 = 1 \times 10^{-6}$, $\text{tol}_2 = 1 \times 10^{-8}$, $\eta = 0.5$, $c_{\text{red}} = 0.6$, $c_{\text{inc}} = 1/c_{\text{red}}$. With the exception of the meshes depicted in Fig. 6, the results presented in this section refer to a set of constraints that is characterized by $\delta = 0.2$. Concerning the r-adaptive procedure, the motion of interior nodes is unconstrained, nodes on the upper and lower boundary and nodes on vertices are fixed and nodes on the left and right boundary as well as nodes on the edges of the crack are allowed to change their positions along the boundary. Therefore, only the

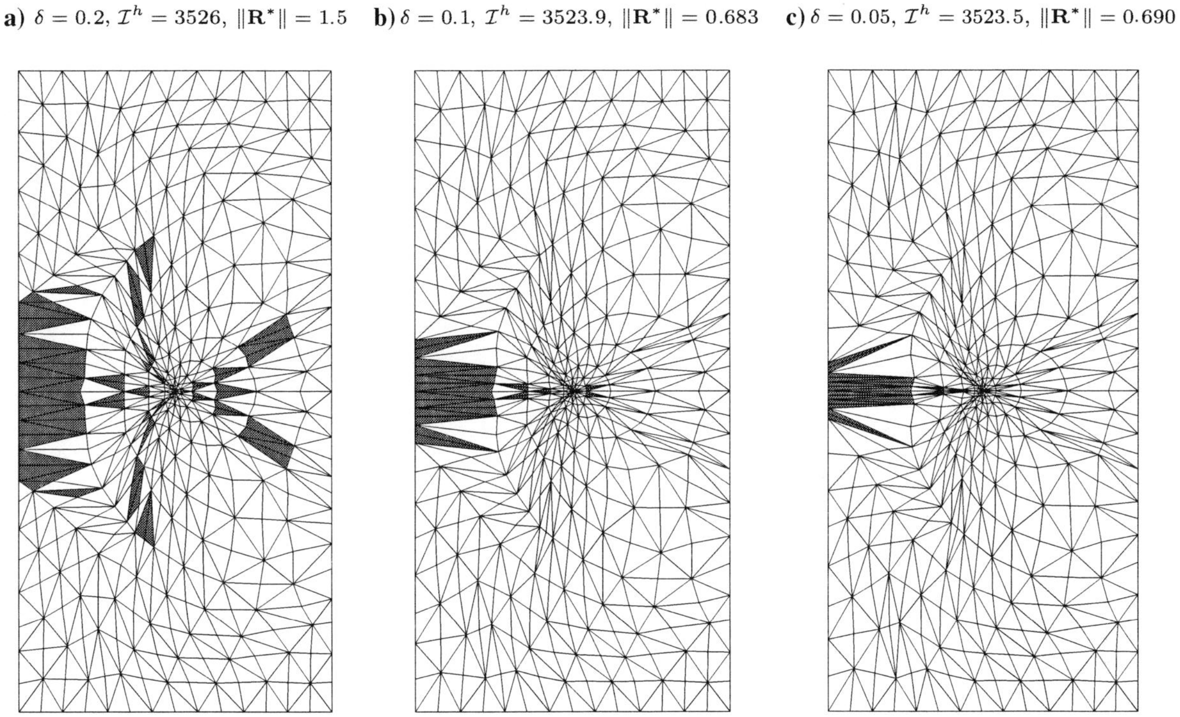

Fig. 6 Cracked specimen—adapted material meshes for varying δ, termination criterion of the barrier method: $\|\mathbf{R}^{\mathrm{KT}}\| < 1\times 10^{-5}$, elements with active constraints (numerically characterized by $g^e > -5 \times 10^{-5}$) are gray colored. (**a**) $\delta = 0.2$, $\mathcal{I}^h = 3526$, $\|\mathbf{R}^*\| = 1.5$; (**b**) $\delta = 0.1$, $\mathcal{I}^h = 3523.9$, $\|\mathbf{R}^*\| = 0.683$; (**c**) $\delta = 0.05$, $\mathcal{I}^h = 3523.5$, $\|\mathbf{R}^*\| = 0.690$

tangential components of the material node point forces at the boundray influence the r-adaptivity. These components resulting from the numerical error of the finite element method have no physical interpretation. In the continuous case, the material surface forces acting on traction free boundaries are normal to the boundary.

Within $k = 27$ iterations the parameter α decreases from $\alpha^0{=}5$ to $\alpha^{27}{=}1.5 \times 10^{-12}$. In Fig. 7, $\|\mathbf{R}^{\mathrm{KT}}\|$ is plotted versus α. The termination criterion was $\|\mathbf{R}^{\mathrm{KT}}\| < 10^{-5}$, whereas $\|\mathbf{R}^{\mathrm{KT}}\| = 467$ for $k = 1$ and $\|\mathbf{R}^{\mathrm{KT}}\| = 5.7 \times 10^{-6}$ for $k = 27$. This shows that $\mathbf{X}^{27}$ is close to a Kuhn-Tucker point. Moreover, the value

$$\max_{e=1,\dots,784} (g^e) = -1.2 \times 10^{-6}, \quad k = 27 \tag{67}$$

and other values (of g^e) of the same order of magnitude indicate that several constraints are active, i.e. the solution of the constrained problem is situated on the boundary of the feasible domain. Figure 6a shows the positions of elements with an active constraint in the adapted material mesh ($k = 27$). These elements (numerically characterized by $g^e > -5 \times 10^{-5}$) are gray colored.

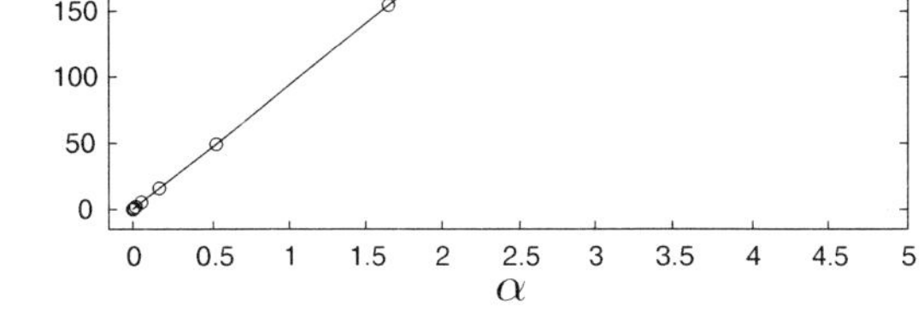

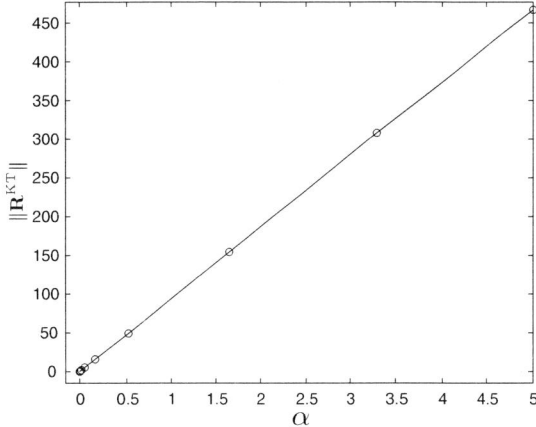

Fig. 7 Cracked specimen—evolution of $\|\mathbf{R}^{\mathrm{KT}}\|$

Figure 5 shows the original and the adapted material mesh after $k = 27$ iterations. One can observe that the node points concentrate in the vicinity of the crack tip, where high stress gradients appear. Due to the symmetry of the problem and the symmetry of the original

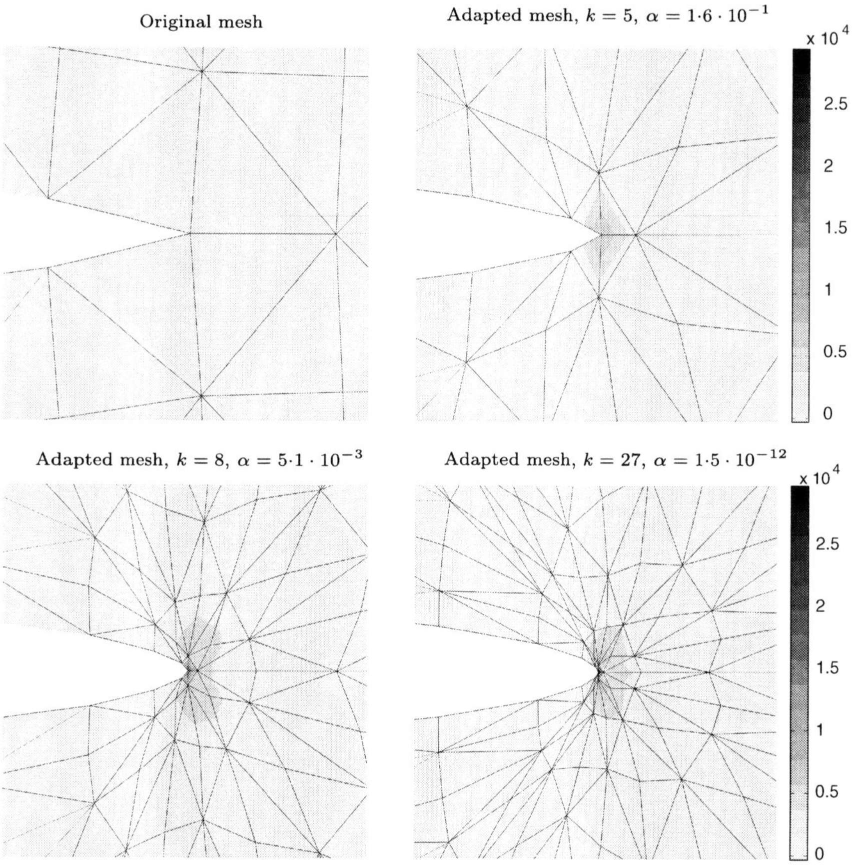

Fig. 8 Cracked specimen—evolution of the spatial mesh and the Cauchy stresses σ_{22} in the vicinity of the crack tip

mesh, the adapted mesh is also symmetric. The evolution of the tensile stress σ_{22} and the spatial mesh in the vicinity of the crack tip is shown in Fig. 8. The maximal tensile stress occurring at the crack tip of the adapted mesh ($k = 27$) is $\sigma_{22} = 2.96 \times 10^4$ which corresponds to an 800% increase with respect to the original mesh. Note that the crack tip is a singularity where the stresses tend to infinity. Despite the singularity, the presented r-adaptivity stops, since the barrier method converges to a Kuhn-Tucker point. Figure 6 shows two additional meshes obtained for $\delta = 0.1$, $\delta = 0.05$. Note that both meshes represent the final state of the iteration, whereas $\|\mathbf{R}^{\mathrm{KT}} < 10^{-5}\|$. The evolution of the active constraints (represented by gray colored elements) indicate that for $\delta = 0.05$ the r-adaptive process in the vicinity of the crack tip is terminated but not on the flanks of the crack. The level of the discrete potential energy and the norm of material residual stagnates. In the present example, the barrier method does not converge to a Kuhn-Tucker point if δ is further reduced. But if smaller displacement $u = 0.25$ are prescribed on the upper and lower boundary, a further reduction up to $\delta = 0.0031$ is possible (a further reduction has not been investigated). Once more, the r-adaptivity in the vicinity of the crack tip seems to be terminated but the number and position of elements with active constraints approximately coincides with mesh depicted in Fig. 6c. Badly shaped, acicular elements occur having a length-to-height ratio of approximately 300. Again, the level of the material residual and the potential energy stagnates. Similar results have been obtained for an example without a singularity, a plate with a hole subjected to tension which is not further discussed in this paper. As mentioned in Sect. 4.2, this suggests that the existence of a material mesh with completely vanishing material node point forces is questionable at least for some problems/meshes. Known exceptions are meshes with a very small number of nodes or problems that are exactly solved by the finite element method. In the latter case the node point positions can usually be chosen arbitrarily.

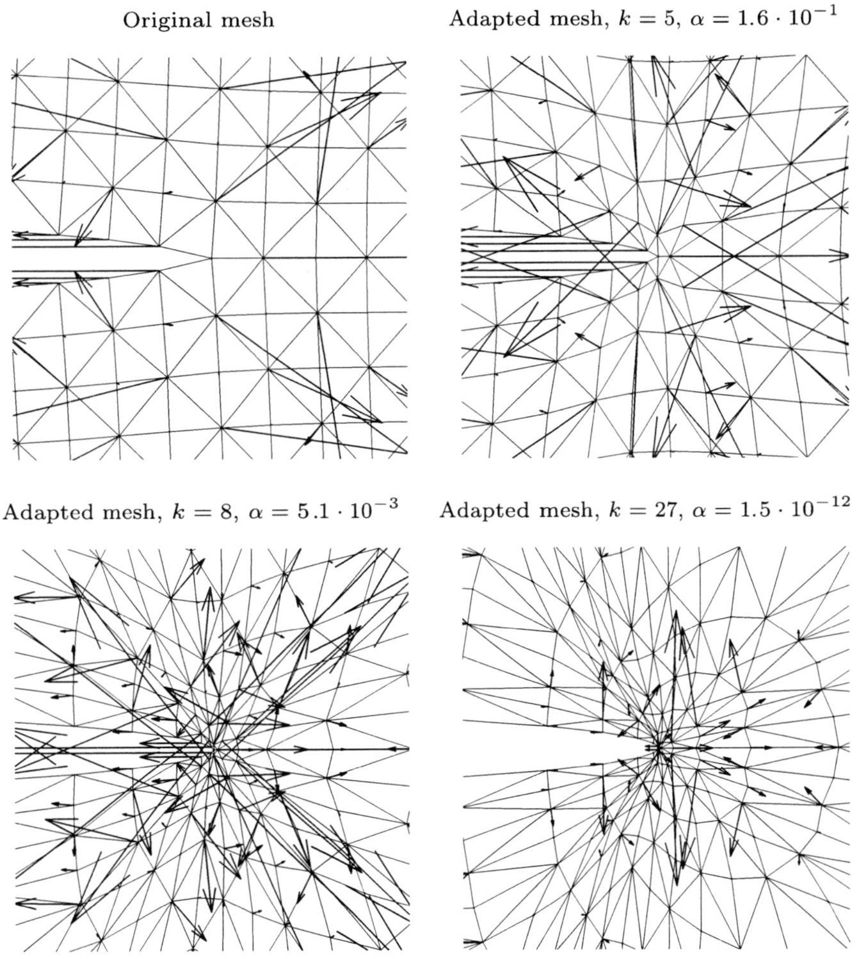

Fig. 9 Cracked specimen—evolution of the material node point forces near the crack

Figure 9 illustrates the evolution of the configurational node point forces in the vicinity of the crack tip. Note that the arrows representing these forces are equally scaled in all figures and that material node point forces assigned to fixed nodes, e.g. the crack tip node, are not depicted. The norm of the material residual vector of the adapted mesh ($k = 27$) is $\|\mathbf{R}^*\| = 1.5$ which corresponds to a 98% reduction with respect to the original mesh. Since Lagrange multipliers $\lambda^e > 0$ exist, the material residual does not vanish completely. The active constraints provide "material reaction forces" such that the sum of the material node point forces and the reaction forces vanishes, see Eq. 33. Figure 10 shows the entire original and adapted material mesh ($k = 27$), the material node point forces that are assigned to variable material node points and the values of the functions g^e characterizing the inequality constraints. Note that nonvanishing material node point forces of the adapted mesh always occur at the node points of the darkest elements which are associated with active constraints. Only elements with active constraints provide "reaction forces" that are in equilibrium with the material node point forces. Material node point forces of nodes that are not associated with active elements tend to vanish completely.

In Fig. 11 the discrete potential energy $\mathcal{I}^h$ and the additional energy $\mathcal{I}^+$ is plotted versus α. The graphs show that the discrete potential energy $\mathcal{I}^h$ decreases and the additional energy $\mathcal{I}^+$ increases monotonically as α decreases. The energy obtained for $k = 27$ is $\mathcal{I}^h = 3526$. This corresponds to a relative decrease

$$\frac{\mathcal{I}^{h,27} - \mathcal{I}^{h,\text{ref}}}{\mathcal{I}^{h,\text{orig}} - \mathcal{I}^{h,\text{ref}}} = 79\% \tag{68}$$

where $\mathcal{I}^{h,\text{ref}}$ is the discrete potential energy of a much finer so called reference mesh. The reference mesh is

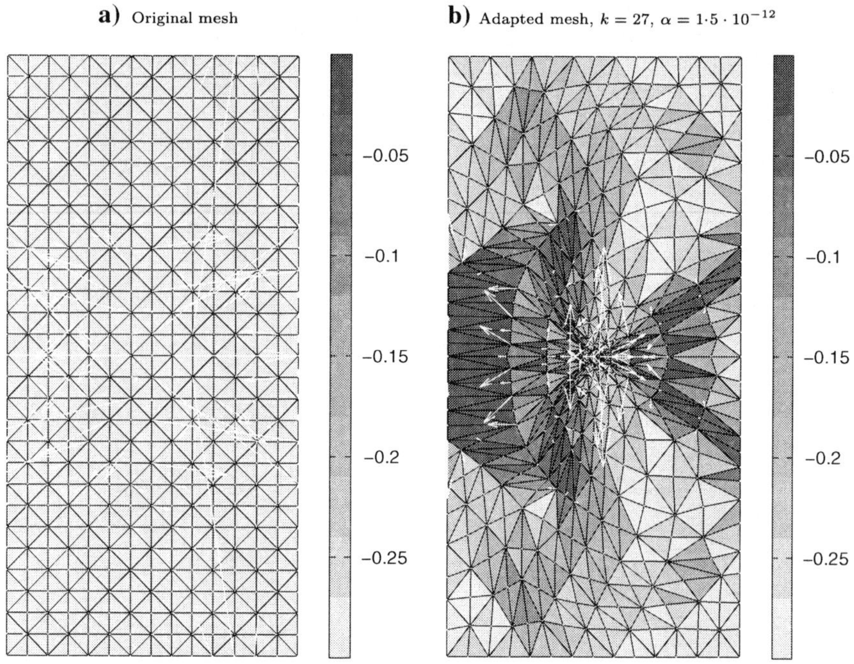

Fig. 10 Cracked specimen—material forces and values of the functions g^e. (**a**) Original mesh; (**b**) adapted mesh, $k = 27$, $\alpha = 1.5 \times 10^{-12}$

refined near the crack tip of the specimen and consists of 9,648 triangular elements and 4,915 nodes. In the left diagram of Fig. 12 the distance

$$\|\boldsymbol{\varphi}_k^h - \boldsymbol{\varphi}_{\text{ref}}^h\|_2 = \sqrt{\int_{\mathcal{B}_0} (\boldsymbol{\varphi}_k^h - \boldsymbol{\varphi}_{\text{ref}}^h)^2 \,\mathrm{d}V} \approx L^2\text{-error} \qquad (69)$$

between the spatial deformation of the adapted mesh $\boldsymbol{\varphi}_k^h$ and the spatial deformation of the reference mesh $\boldsymbol{\varphi}_{\text{ref}}^h$ is plotted versus α. The distance between the two functions measured by the L^2-norm is an approximation for the L^2-error, i.e the distance between the numerical and the exact solution. For $k = 27$, the approximated L^2-error is $\|\boldsymbol{\varphi}_{27}^h - \boldsymbol{\varphi}_{\text{ref}}^h\|_2 = 5.5 \times 10^{-2}$ which corresponds to a 71% reduction with respect to the original mesh. Note that within $k = 7$ steps a 68% reduction is achieved.

6 Conclusions

This paper describes a solution strategy for energy-based r-adaptivity in finite elasticity. Energy-based r-adaptivity is defined as a problem of nonlinear programming, a minimization problem with inequality constraints. The objective function of the problem is the discrete potential energy in the state of equilibrium which is considered to be a function of the material node point positions. The inequality constraints are introduced to restrict the distortion of the material elements during the r-adaptive process. We have presented a set of constraints designed for linear triangular elements that restrict the distortional deformation of the elements. As a solution strategy for the constrained problem, we use a classical barrier method with an inverse barrier function.

Due to the complexity of the underlying minimization problem, the scheme is computationally expensive, but the results of the presented numerical example are promising. The proposed barrier method is suitable to optimize meshes with a high number of nodes. Moreover, an appreciable improvement of the finite element solution has been achieved. Unfortunately, the barrier method does not always succeed to get as close to a Kuhn-Tucker point as in the presented example. Below a given point the scaling parameter of the barrier method cannot be further reduced. Therefore, the performance of the Newton method that is used to solve the unconstrained subproblems of the barrier method should be improved. For instance, by using special line search schemes that take advantage of the known form of the singularity in a barrier function. Due to the nonconvexity of the discrete potential energy ($\mathcal{I}^{h*}$), it is reasonable to apply a modified Newton method that works with a modified, positive definite Hessian

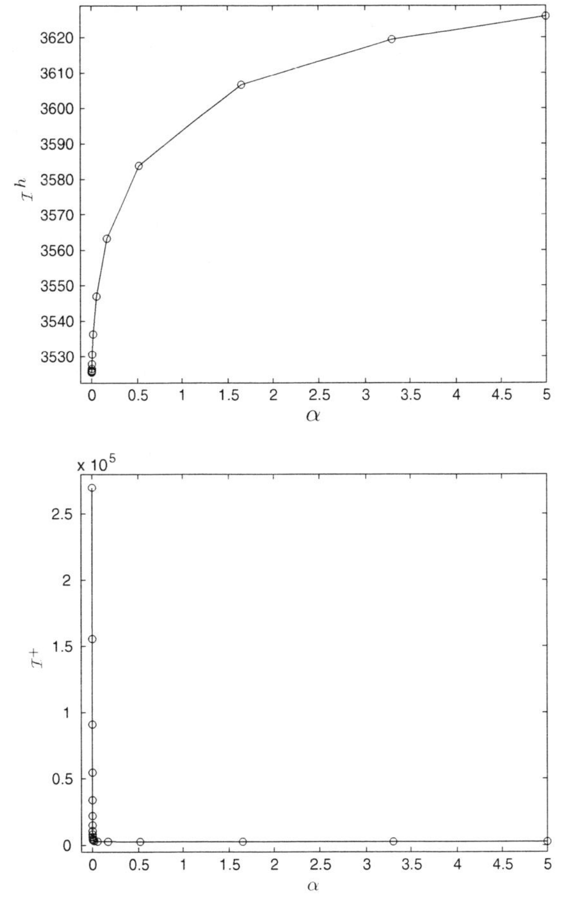

Fig. 11 Cracked specimen—discrete potential energy $\mathcal{I}^h$ and additional energy $\mathcal{I}^+$ versus α

matrix. Moreover, we want to investigate, if alternative algorithms of nonlinear programming, e.g. primal-dual interior point method, penalty barrier methods or penalty methods, are more suitable to solve constrained minimization problem that is associated with the presented r-adaptivity. To enhance the applicability of the method, we want to develop constraints for other types of elements, i.e. two-dimensional elements of higher order and three-dimensional elements.

Acknowledgements The authors would like to thank the "Deutsche Forschungsgemeinschaft" (DFG) for their support of this work under the Grant STE 544/24-1.

Appendix A: Derivatives of the constraints

For the implementation of the presented barrier method, it is necessary to know the first and second derivatives

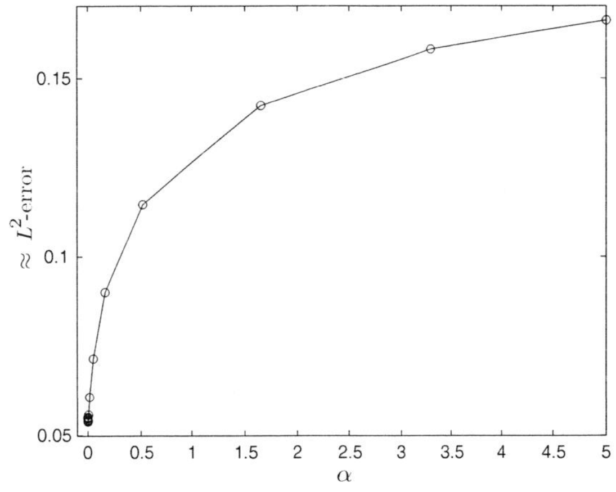

Fig. 12 Cracked specimen—approximated L^2-error versus α

of the functions g^e with respect to $\mathbf{X}$. The first derivative reads as

$$\frac{\partial g^e}{\partial \mathbf{X}} = \left[\left[\frac{\partial g^e}{\partial \boldsymbol{X}_1}\right], \ldots, \left[\frac{\partial g^e}{\partial \boldsymbol{X}_{nX}}\right]\right], \tag{70}$$

where

$$\frac{\partial g^e}{\partial \boldsymbol{X}_i} = \begin{cases} \gamma^{-2} \dfrac{\partial \gamma}{\partial \boldsymbol{f}_r^e} \cdot \nabla_{X_r} N_{r,i}^e & \text{if } \boldsymbol{X}_i \in \partial \mathcal{B}_0^e \\ \mathbf{0} & \text{else} \end{cases}, \tag{71}$$

with

$$\frac{\partial \gamma}{\partial \boldsymbol{f}_r^e} = -\gamma \left[\boldsymbol{f}_r^e\right]^{-t} + \frac{2}{j_r^e} \boldsymbol{f}_r^e. \tag{72}$$

And the second derivative is given by

$$\frac{\partial^2 g^e}{\partial \mathbf{X}^2} = \begin{bmatrix} \left[\dfrac{\partial^2 g^e}{\partial \boldsymbol{X}_1 \partial \boldsymbol{X}_1}\right] & \cdots & \left[\dfrac{\partial^2 g^e}{\partial \boldsymbol{X}_{n_{nX}} \partial \boldsymbol{X}_1}\right] \\ \vdots & \ddots & \vdots \\ \left[\dfrac{\partial^2 g^e}{\partial \boldsymbol{X}_1 \partial \boldsymbol{X}_{n_{nX}}}\right] & \cdots & \left[\dfrac{\partial^2 g^e}{\partial \boldsymbol{X}_{n_{nX}} \partial \boldsymbol{X}_{n_{nX}}}\right] \end{bmatrix} \tag{73}$$

In addition, if $\boldsymbol{X}_i, \boldsymbol{X}_j \in \partial \mathcal{B}_h^e$, then[4]

$$\frac{\partial^2 g^e}{\partial \boldsymbol{X}_j \partial \boldsymbol{X}_i} = \nabla_{X_r} N_{r,i}^e * \frac{\partial}{\partial \boldsymbol{f}_r^e}\left[\gamma^{-2} \frac{\partial \gamma}{\partial \boldsymbol{f}_r^e}\right] \cdot \nabla_{X_r} N_{r,j}^e \tag{74}$$

else

$$\frac{\partial^2 g^e}{\partial \boldsymbol{X}_j \partial \boldsymbol{X}_i} = 0, \tag{75}$$

[4] The nonstandard contraction $\boldsymbol{a} * \mathbb{C} \cdot \boldsymbol{b}$ of the cartesian vectors $\boldsymbol{a} = a_i \boldsymbol{e}_i$, $\boldsymbol{b} = b_j \boldsymbol{e}_j$ and the forth order tensor $\mathbb{C} = C_{klmn} \boldsymbol{e}_k \otimes \boldsymbol{e}_l \otimes \boldsymbol{e}_m \otimes \boldsymbol{e}_n$ is defined as $\boldsymbol{a} * \mathbb{C} \cdot \boldsymbol{b} = a_i C_{kimj} b_j \boldsymbol{e}_k \otimes \boldsymbol{e}_m$.

where[5]

$$
\begin{aligned}
\frac{\partial}{\partial \boldsymbol{f}_r^e}\left[\gamma^{-2}\frac{\partial \gamma}{\partial \boldsymbol{f}_r^e}\right] &= -2\gamma^{-3}\frac{\partial \gamma}{\partial \boldsymbol{f}_r^e} \otimes \frac{\partial \gamma}{\partial \boldsymbol{f}_r^e} \\
&\quad + \gamma^{-2}\Bigg[-\left[\boldsymbol{f}_r^e\right]^{-t} \otimes \frac{\partial \gamma}{\partial \boldsymbol{f}_r^e} \\
&\quad + \gamma \left[\boldsymbol{f}_r^e\right]^{-t} \underline{\otimes} \left[\boldsymbol{f}_r^e\right]^{-1} \qquad (76) \\
&\quad - \frac{2}{j_r^e}\boldsymbol{f}_r^e \otimes \left[\boldsymbol{f}_r^e\right]^{-t} + \frac{2}{j_r^e}\mathbf{1}\overline{\otimes}\mathbf{1}\Bigg].
\end{aligned}
$$

References

Armero F, Love E (2003) An arbitrary Lagrangian–Eulerian finite element method for finite strain plasticity. Int J Numer Meth Eng 57:471–508

Askes H, Kuhl E, Steinmann P (2004) An ALE formulation based on spatial and material settings of continuum mechanics. Part 2: classification and applications. Comput Methods Appl Mech Eng 193:4223–4245

Bathe KJ, Sussman TD (1983) An algorithm for the construction of optimal finite element meshes in linear elasticity. Comput Methods Nonlinear Solids Struct Mech 54:15–36, ASME, AMD

Bertsekas DP (1995) Nonlinear programming. Athena Scientific, Belmont, MA

Braun M (1997) Configurational forces induced by finite-element discretization. Proc Estonian Acad Sci Phys Math 46:24–31

Carpenter WC, Zendegui S (1982) Optimum nodal locations for a finite element idealization. Eng Optim 5:215–221

Carroll WE, Barker RM (1973) A theorem for optimum finite-element idealizations. Int J Solids Struct 9:883–895

Felippa CA (1976a) Numerical experiments in finite element grid optimization by direct energy search. Appl Math Model 1:239–244

Felippa CA (1976b) Optimization of finite element grids by direct energy search. Appl Math Model 1:93–96

Gill PE, Murray W, Wright MH (1981) Practical optimization. Academic Press

Holzapfel GA (2000) Nonlinear solid mechanics. Wiley, Chichester

Hughes TJR (1987) The finite element method. Dover Publications, New York

Jarre S (2004) Optimierung. Springer-Verlag, Berlin

Knupp PM (2000a) Achieving finite element mesh quality via optimization of the Jacobian matrix norm and associated quantities. Part I – a framework for surface mesh optimization. Int J Numer Meth Eng 48:401–420

Knupp PM (2000b) Achieving finite element mesh quality via optimization of the Jacobian matrix norm and associated quantities. Part II – a framework for volume mesh optimization and the condition number of the Jacobian matrix. Int J Numer Meth Eng 48:1165–1185

Kuhl E, Askes H, Steinmann P (2004) An ALE formulation based on spatial and material settings of continuum mechanics. Part 1: generic hyperelastic formulation. Comput Methods Appl Mech Eng 193:4207–4222

Luenberger DG (1973) Introduction to linear and nonlinear programming. Addison-Wesley Publishing Company, Reading

Marsden JE, Hughes TJR (1983) Mathematical foundations of elasticity. Dover Publications, New York

McNeice GM, Marcal PV (1973) Optimization of finite element grids based on minimum potential energy. Trans ASME J Eng Ind 95(1):186–190

Mosler J, Ortiz M (2006) On the numerical implementation of varational arbitrary Lagrangian–Eulerian (VALE) formulations. Int J Numer Meth Eng 67:1272–1289

Mueller R, Maugin GA (2002) On material forces and finite element discretizations. Comput Mech 29:52–60

Mueller R, Kolling S, Gross D (2002) On configurational forces in the context of the finite element method. Int J Numer Meth Eng 53:1557–1574

Mueller R, Gross D, Maugin GA (2004) Use of material forces in adaptive finite element methods. Comput Mech 33: 421–434

Rajagopal A, Gangadharan R, Sivakumar SM (2006) On material forces and finite element discretizations. Int J Comput Meth Eng Sci Mech 7:241–262

Steinmann P, Ackermann D, Barth FJ (2001) Application of material forces to hyperelastostatic fracture mechanics. Part II: computational setting. Int J Solids Struct 38:5509–5526

Sussman T, Bathe KJ (1985) The gradient of the finite element variational indicator with respect to nodal point co-ordinates: an explicit calculation and application in fracture mechanics and mesh optimization. Int J Numer Meth Eng 21:763–774

Thoutireddy P (2003) Variational arbitrary Lagrangian–Eulerian method. Ph.D. Thesis, California Institute of Technology, Passadena, California

Thoutireddy P, Ortiz M (2004) A variational r-adaption and shape-optimization method for finite-deformation elasticity. Int J Numer Meth Eng 61:1–21

Wriggers P (2001) Nichtlineare finite-element-methoden. Springer-Verlag, Berlin

[5] The special dyadic products $\boldsymbol{A}\overline{\otimes}\boldsymbol{B}$, $\boldsymbol{A}\underline{\otimes}\boldsymbol{B}$ of two second order tensors $\boldsymbol{A} = A_{ij}\boldsymbol{e}_i \otimes \boldsymbol{e}_j$, $\boldsymbol{B} = B_{kl}\boldsymbol{e}_k \otimes \boldsymbol{e}_l$ read as $\boldsymbol{A}\overline{\otimes}\boldsymbol{B} = A_{ik}B_{jl}\boldsymbol{e}_i \otimes \boldsymbol{e}_j \otimes \boldsymbol{e}_k \otimes \boldsymbol{e}_l$, $\boldsymbol{A}\underline{\otimes}\boldsymbol{B} = A_{il}B_{jk}\boldsymbol{e}_i \otimes \boldsymbol{e}_j \otimes \boldsymbol{e}_k \otimes \boldsymbol{e}_l$.

Variational design sensitivity analysis in the context of structural optimization and configurational mechanics

Daniel Materna · Franz-Joseph Barthold

Abstract Variational design sensitivity analysis is a branch of structural optimization. We consider variations of the material configuration and we are interested in the change of the state variables and the objective functional due to these variations. In the same manner in configurational mechanics we are interested in changes of the material body. In this paper, we derive the physical and material residual problem by using standard optimization procedures and we investigate sensitivity relations for the physical and material problem. These sensitivity relations are used in order to solve the coupled physical and material problem. Both problems are coupled by the pseudo load operator, which play an important role for the solution of structural optimization problems. Furthermore, we derive explicit formulations for the variations of the physical and material problem and propose different solution algorithms for the coupled problem.

Keywords Variational sensitivity analysis · Structural optimization · Configurational mechanics · Mesh optimization · Shape optimization

D. Materna (✉) · F.-J. Barthold
Numerical Methods and Information Processing,
University of Dortmund, August-Schmidt-Straße 8,
44227 Dortmund, Germany
e-mail: daniel.materna@uni-dortmund.de

F.-J. Barthold
e-mail: franz-joseph.barthold@uni-dortmund.de

1 Introduction

Configurational mechanics is a branch of continuum mechanics and the so-called configurational or material forces are used in the context of material inhomogeneities as well as any kind of material defects, see e.g. Maugin (1993), Gurtin (2000), Kienzler and Herrmann (2000) and Steinmann (2000). The notion of configurational forces was introduced by the fundamental work of Eshelby (1951, 1975). In a series of recently published papers other applications with configurational forces are treated. Procedures for the optimization of finite element meshes based on the use of configurational forces were proposed for instance by Braun (1997), Mueller and Maugin (2002), Kuhl et al. (2004), Askes et al. (2004), Thoutireddy (2003) and Kalpakides and Balassas (2005). The mesh optimization problem has a long tradition. First steps for the optimization of finite element meshes based on a discrete formulation of energy minimization were outlined for instance in Carroll and Barker (1973), McNeice and Marcal (1973) and Carpenter and Zendegui (1982). In fact, the authors obtained the same discrete indicators for mesh optimization like the above mentioned approach from configurational mechanics, but they have not called them material or configurational forces.

On the other hand, variational design sensitivity analysis is a branch of structural optimization, e.g. shape or topology optimization, see e.g. Kamat (1993) and Choi and Kim (2005a, b) and the reference therein for

Defect and Material Mechanics. C. Dascalu, G.A. Maugin & C. Stolz (eds.),
doi: 10.1007/978-1-4020-6929-1_13,

an overview. In these disciplines we consider variations of the material configuration and we are interested in the change of the state variables and the objective functional due to these variations. The variations are required in order to solve the corresponding Lagrangian equation using nonlinear programming algorithms.

In many engineering applications, the energy functional of the problem is used as the objective functional of the optimization problem. The relations of configurational mechanics and its application to mesh optimization can be obtained by using techniques from variational design sensitivity analysis applied to the energy functional (Barthold 2003, 2005; Materna and Barthold 2006a, b, 2007a). The energy depends on the state function and the design function. The design function specifies the shape and/or the topology of the material body. The first variation of the energy functional with respect to changes in the design leads to the material residual or weak form of the material or configurational force equilibrium. In the context of sensitivity analysis, we can interpret the configurational forces as the sensitivity of the energy with respect to variations in the design. The classical physical problem and the material problem are coupled by the pseudo load operator, which is used for the solution of structural optimization problems. A simple application is the above mentioned mesh optimization problem. For mesh optimization, we choose simply the nodal coordinates as design variables. The optimal nodal positions (the design) of a given mesh (the reference configuration) can be computed using standard optimization algorithms. Furthermore, the material residual is directly related to the J-integral or energy release rate and we can perform a sensitivity analysis for these quantities.

The minimization of the energy is directly related to the minimization of the compliance of the system or equivalently to the maximization of the stiffness and is used as objective in shape and topology optimization, see e.g. Dems and Mróz (1978) and Bendsøe and Sigmund (2003). In this context, the configurational forces on the design boundary are indicators in which direction the boundary has to move in order to minimize the compliance or to maximize the structural stiffness.

The paper is organized as follows. After some preliminaries about the notation, we begin with an abstract setting of a classical structural optimization problem and formulate an energy minimization problem. We obtain the classic physical residual problem and the material residual problem in one step. In Sect. 4, sensitivity relations for the physical and material problem are investigated and we derive explicit formulations for the variations of the physical and material residuals. The finite element approximation of the sensitivity relations is discussed in Sect. 5. Finally, we propose different solution algorithms for the optimization problem and summarize the theoretical as well as computational treatment by means of selected examples from mesh optimization and shape optimization.

2 Preliminaries

This section is concerned with some settings and definitions. We consider an open bounded material body with an undeformed reference placement $\Omega_R \subset \mathbb{E}^3$ with a piecewise smooth, polyhedral and Lipschitz-continuous boundary $\Gamma = \partial\Omega_R$ such that $\Gamma = \Gamma_\mathrm{D} \cup \Gamma_\mathrm{N}$ and $\Gamma_\mathrm{D} \cap \Gamma_\mathrm{N} = \emptyset$, where Γ_D denotes the Dirichlet boundary and Γ_N the Neumann boundary, respectively. The corresponding deformed current placement is denoted by $\Omega_t \subset \mathbb{E}^3$. The deformation of the material body from Ω_R into Ω_t is given by the nonlinear mapping $\boldsymbol{\varphi} : \Omega_R \to \Omega_t$, $\boldsymbol{x} = \boldsymbol{\varphi}(\boldsymbol{X})$. Here, $\boldsymbol{\varphi}$ maps the material particle $\boldsymbol{X}$ from the reference placement Ω_R to the spatial coordinates $\boldsymbol{x}$ in the deformed placement Ω_t for any fixed time $t \in I_t$. The corresponding deformation gradient, i.e. the tangent map of $\boldsymbol{\varphi}$ from the material tangent space $T_X\Omega_R$ to the spatial tangent space $T_x\Omega_t$, as well as its Jacobian J are given by $\boldsymbol{F} = \nabla_X\boldsymbol{\varphi} : T_X\Omega_R \to T_x\Omega_t$ and $J = \det \boldsymbol{F}$. Furthermore, we assume that the deformation is injective, sufficiently smooth and that $J > 0$, such that there exist the inverse deformation mapping $\boldsymbol{\phi} : \Omega_t \to \Omega_R$, $\boldsymbol{X} = \boldsymbol{\phi}(\boldsymbol{x})$. The corresponding deformation gradient $\boldsymbol{f}$ and its Jacobian j are given by $\boldsymbol{f} = \nabla_x\boldsymbol{\phi}$ and $j = \det\boldsymbol{f}$.

The above introduced kinematical settings in the referential placement Ω_R and the current placement Ω_t could be enhanced by using the *intrinsic formulation* by Noll (1972). This is based on the concept of a *differentiable manifold*, see for instance Truesdell and Noll (2004), Bertram (1989) and Marsden and Hughes (1994) for details. An enhancement of the above mentioned approach by Noll in the context of variational design sensitivity analysis was proposed by Barthold and Stein (1996) and Barthold (2002, 2007). Following the intrinsic concept, a given manifold can be described locally using an intrinsic coordinate system defined on an independent continuous parameter space P_Θ with local coordinates $\mathbf{\Theta}$. Without going into detail, this leads to two fundamental mappings, a design dependent *local reference placement mapping* $\boldsymbol{\kappa} : P_\Theta \times I_s \rightarrow \Omega_R$, $\boldsymbol{X} = \boldsymbol{\kappa}(\mathbf{\Theta}, s)$ and a time dependent *local current placement mapping* $\boldsymbol{\mu} : P_\Theta \times I_t \rightarrow \Omega_t$, $\boldsymbol{x} = \boldsymbol{\mu}(\mathbf{\Theta}, t)$ for any fixed time $t \in I_t$ and any design $s \in I_s$[1]. The corresponding tangent maps and its Jacobians are given by $\boldsymbol{K} = \nabla_\Theta \boldsymbol{\kappa} : T_\Theta P_\Theta \rightarrow T_X \Omega_R$ and $J_K = \det \boldsymbol{K}$ as well as $\boldsymbol{M} = \nabla_\Theta \boldsymbol{\mu} : T_\Theta P_\Theta \rightarrow T_x \Omega_t$ and $J_M = \det \boldsymbol{M}$, respectively. With these mappings, the deformation map $\boldsymbol{\varphi}$ and its tangent map can be written in the form $\boldsymbol{\varphi} = \boldsymbol{\mu} \circ \boldsymbol{\kappa}^{-1}$ and $\boldsymbol{F} = \nabla_X \boldsymbol{\varphi} = \boldsymbol{M}\boldsymbol{K}^{-1}$ and for the inverse mapping $\boldsymbol{\phi}$ follow $\boldsymbol{\phi} = \boldsymbol{\kappa} \circ \boldsymbol{\mu}^{-1}$ and $\boldsymbol{f} = \nabla_x \boldsymbol{\phi} = \boldsymbol{K}\boldsymbol{M}^{-1}$, respectively. The difference vector between the reference and current placements is the displacement $\boldsymbol{u} = \boldsymbol{x} - \boldsymbol{X}$. The corresponding *local displacement mapping* $\boldsymbol{v}$ written in terms of the local mappings $\boldsymbol{\kappa}$ and $\boldsymbol{\mu}$ is given with $\boldsymbol{u} = \boldsymbol{v}(\mathbf{\Theta}, s, t) = \boldsymbol{\mu}(\mathbf{\Theta}, t) - \boldsymbol{\kappa}(\mathbf{\Theta}, s)$.

Different gradient operators can be defined, i.e. grad $:= \nabla_x$, Grad $:= \nabla_X$ and GRAD $:= \nabla_\Theta$ corresponding to the variables $\boldsymbol{x}$, $\boldsymbol{X}$ and $\mathbf{\Theta}$ of the considered domains Ω_R, Ω_t and P_Θ, respectively. Overall, $\boldsymbol{K}$, $\boldsymbol{M}$ and $\boldsymbol{F}$ are used to perform pull back and push forward transformations between current placement, reference placement and parameter space.

For problems with changes in the material configuration, e.g. shape optimization or problems from configurational mechanics, the intrinsic formulation in local coordinates has advantages because we can deal with two independent placement mappings, see Barthold and Stein(1996) and Barthold (2002, 2007) for more details.

Remark 1 The *local reference placement mapping* is parameterized by a design parameter s, i.e. $\boldsymbol{X} = \boldsymbol{\kappa}(\mathbf{\Theta}, s)$. With this in mind, we introduce in an abstract sense a *generalized design* or *control function* $\boldsymbol{s} \in \mathcal{D}$, which specifies the current reference configuration Ω_R, i.e. $\Omega_R = \Omega_R(\boldsymbol{s})$ or $\boldsymbol{X} = f(\boldsymbol{s})$. Here, $\mathcal{D}$ denotes the space with all admissible design or control functions. Therefore, the material particle $\boldsymbol{X}$ as well as the spatial particle $\boldsymbol{x}$ depend on the design variable, i.e. $\boldsymbol{X} = \boldsymbol{X}(\boldsymbol{s})$ and $\boldsymbol{x} = \boldsymbol{x}(\boldsymbol{X}(\boldsymbol{s}))$. The explicit coupling between $\boldsymbol{X}$ and $\boldsymbol{s}$, i.e. the explicit structure of $f(\boldsymbol{s})$ depend on the particular problem. The *local current placement mapping* $\boldsymbol{x} = \boldsymbol{\mu}(\mathbf{\Theta}, t)$ can be expressed by the *local displacement mapping* $\boldsymbol{u} = \boldsymbol{v}(\mathbf{\Theta}, s, t)$. With these, a quantity $(\cdot)$, which depends on the state variable $\boldsymbol{u}$ and the design variable $\boldsymbol{s}$ is denoted by $(\cdot)(\boldsymbol{u}, \boldsymbol{s})$.

[1] Here, s is used as a general scalar (time-like) design variable, which parameterizes in an abstract sense the material body in the reference configuration Ω_R, i.e. $\Omega_R = \Omega_R(s)$. This could be a parametrization for the material points $\boldsymbol{X} = \boldsymbol{X}(s)$, density $\rho = \rho(s)$, mass $m = m(s)$, material properties etc. In this paper, we consider only an abstract parametrization for the material points, see Remark 1.

The variation of a quantity $(\cdot)(\boldsymbol{\varphi}; \boldsymbol{X}, t)$ with respect to $\boldsymbol{x}$ at fixed $\boldsymbol{X}$ and t is denoted by $\delta_x(\cdot) =: (\cdot)'_x$ and the variation of $(\cdot)(\boldsymbol{\phi}; \boldsymbol{x}, t)$ with respect to $\boldsymbol{X}$ at fixed $\boldsymbol{x}$ and t by $\delta_X(\cdot) =: (\cdot)'_X$. In order to avoid confusion between the small $\boldsymbol{x}$ and the capital $\boldsymbol{X}$ and due to the fact that the reference configuration of the material body is prescribed by a design function $\boldsymbol{s}$, we prefer to use $\delta_u(\cdot)$ instead of $\delta_x(\cdot)$ and $\delta_s(\cdot)$ instead of $\delta_X(\cdot)$, respectively. Furthermore, the short notation $(\cdot)'$ denote the total variation of a quantity $(\cdot)$ and $(\cdot)'_u$ as well as $(\cdot)'_s$ the partial variation with respect to $\boldsymbol{u}$ and $\boldsymbol{s}$, i.e. $(\cdot)'_u := (\cdot)'_x = \delta_x(\cdot)$ and $(\cdot)'_s := (\cdot)'_X = \delta_X(\cdot)$. With this notation, the total variation of a quantity $(\cdot)(\boldsymbol{u}, \boldsymbol{s})$, which depends on the deformation and the design, is given by the partial variation with respect to $\boldsymbol{u}$ and a fixed design $\hat{\boldsymbol{s}}$ as well as the partial variation with respect to $\boldsymbol{s}$ and a fixed deformation $\hat{\boldsymbol{u}}$, i.e.

$$(\cdot)'(\boldsymbol{u}, \boldsymbol{s}) = (\cdot)'_u(\boldsymbol{u}, \hat{\boldsymbol{s}}) + (\cdot)'_s(\hat{\boldsymbol{u}}, \boldsymbol{s}) \tag{1}$$

In the same way we define the second variations $(\cdot)''_{uu} := \delta^2_{xx}(\cdot)$ and $(\cdot)''_{ss} := \delta^2_{XX}(\cdot)$ as well as the mixed variations $(\cdot)''_{us} := \delta^2_{xX}(\cdot)$ and $(\cdot)''_{su} := \delta^2_{Xx}(\cdot)$, respectively.

3 Abstract setting of a structural optimization problem

3.1 General setting for optimal design

We consider a classical structural optimization problem in an abstract setting. In general, we are interested in the solution of the following optimization problem:

Problem 1 Find $\{\boldsymbol{u}, \boldsymbol{s}\} \in \mathcal{V} \times \mathcal{D}$ of the objective functional $\mathcal{J} : \mathcal{V} \times \mathcal{D} \rightarrow \mathbb{R}$ such that

$$\mathcal{J}(\boldsymbol{u}, \boldsymbol{s}) \rightarrow \min_{\boldsymbol{u}, \boldsymbol{s} \in \mathcal{V} \times \mathcal{D}} \tag{2}$$

subject to the constraints

$$\left.\begin{array}{lll} \boldsymbol{A}(\boldsymbol{u},s)=\boldsymbol{0} & \text{in} & \Omega \\ \boldsymbol{u}=\bar{\boldsymbol{u}} & \text{on} & \Gamma_{\mathrm{D}_u} \\ s=\bar{s} & \text{on} & \Gamma_{\mathrm{D}_s} \end{array}\right\}. \tag{3}$$

Here, $\boldsymbol{A}$ is an elliptic differential operator for the state function $\boldsymbol{u} \in \mathcal{V}$ and $s \in \mathcal{D}$ is a generalized design function. The space $\mathcal{V}$ denotes the usual Sobolev space and $\mathcal{D}$ the space with all admissible designs. Additionally we have to fulfil boundary conditions on the Dirichlet boundary Γ_{D_u} for the state function and on the corresponding boundary Γ_{D_s} for the design function.

To solve this problem we introduce the corresponding Lagrangian functional

$$\begin{aligned} \mathcal{L}(\boldsymbol{u},s,\boldsymbol{\lambda}) &:= \mathcal{J}(\boldsymbol{u},s) + < \boldsymbol{A}(\boldsymbol{u},s), \boldsymbol{\lambda} > \\ &= \mathcal{J}(\boldsymbol{u},s) + \mathcal{R}(\boldsymbol{u},s;\boldsymbol{\lambda}), \end{aligned} \tag{4}$$

where

$$\mathcal{R}(\boldsymbol{u},s;\boldsymbol{\lambda}) := a(\boldsymbol{u},s;\boldsymbol{\lambda}) - F(\boldsymbol{\lambda}) = 0 \qquad \forall\, \boldsymbol{\lambda} \in \mathcal{V} \tag{5}$$

is the weak form of the differential operator $\boldsymbol{A}$ written in terms of the physical residual $\mathcal{R}(\boldsymbol{u},s;\cdot)$. We seek for stationary points of $\mathcal{L}$ which are candidates for optimal solutions of the system

$$\mathcal{L}'(\boldsymbol{u},s,\boldsymbol{\lambda})(\boldsymbol{\eta},\boldsymbol{\psi},\boldsymbol{\nu}) := \nabla\mathcal{L} = \frac{\partial \mathcal{L}(\boldsymbol{u},s,\boldsymbol{\lambda})}{\partial(\boldsymbol{u},s,\boldsymbol{\lambda})} = 0.$$

This is a boundary value problem for the triple $\{\boldsymbol{u},s,\boldsymbol{\lambda}\} \in \mathcal{V}\times\mathcal{D}\times\mathcal{V}$,

$$\mathcal{L}' = \left\{\begin{array}{l} \mathcal{J}'_u(\boldsymbol{u},s;\boldsymbol{\eta}) + \mathcal{R}'_u(\boldsymbol{u},s;\boldsymbol{\lambda},\boldsymbol{\eta}) \\ \mathcal{J}'_s(\boldsymbol{u},s;\boldsymbol{\psi}) + \mathcal{R}'_s(\boldsymbol{u},s;\boldsymbol{\lambda},\boldsymbol{\psi}) \\ \mathcal{R}(\boldsymbol{u},s;\boldsymbol{\nu}) \end{array}\right\} = \boldsymbol{0} \quad \forall\,\{\boldsymbol{\eta},\boldsymbol{\psi},\boldsymbol{\nu}\} \in \mathcal{V}\times\mathcal{D}\times\mathcal{V}, \tag{6}$$

where $\mathcal{J}'_u, \mathcal{J}'_s, \mathcal{R}'_u, \mathcal{R}'_s$ are the partial variations of the objective functional $\mathcal{J}$ and the physical residual $\mathcal{R}$ with respect to $\boldsymbol{u}$ and s. The last equation of (6) is simply the physical residual (5) for the state $\boldsymbol{u}$.

Remark 2 For the solution $\boldsymbol{u} \in \mathcal{V}$ of (5) with a fixed $\hat{s}$, the corresponding adjoint or dual variable $\boldsymbol{\lambda} \in \mathcal{V}$ is the solution of the so-called adjoint or dual problem Eq. 6_1 at the current linearization point $\hat{\boldsymbol{u}}$, i.e. $\boldsymbol{\lambda}$ is the solution of

$$k(\hat{\boldsymbol{u}},\hat{s};\boldsymbol{\lambda},\boldsymbol{\eta}) = -\mathcal{J}'_u(\hat{\boldsymbol{u}},\hat{s};\boldsymbol{\eta}) \qquad \forall\,\boldsymbol{\eta} \in \mathcal{V}. \tag{7}$$

The bilinear form $k(\hat{\boldsymbol{u}},\hat{s};\boldsymbol{\lambda},\boldsymbol{\eta}) := \mathcal{R}'_u(\hat{\boldsymbol{u}},\hat{s};\boldsymbol{\lambda},\boldsymbol{\eta})$ is the usual tangent stiffness operator at the current linearization point for the solution of the primal problem.

3.2 A special objective function

The objective function of the optimization problem is arbitrary and depends on the particular application. In many engineering applications, the energy functional of the problem is used as the objective functional, because there is a relation between the overall minimization of the energy and the maximization of the stiffness. Therefore, the energy is an interesting objective in mechanics and engineering and is used as objective functional in shape and topology optimization, see e.g. Dems and Mróz (1978) and Bendsøe and Sigmund (2003). Let

$$\begin{aligned} E(\boldsymbol{u},s) &:= \int_{\Omega_R} U_R(\boldsymbol{X};\boldsymbol{\varphi},\boldsymbol{F})\,d\Omega \\ &= \int_{\Omega_t} U_t(\boldsymbol{x};\boldsymbol{\phi},\boldsymbol{f})\,d\Omega \\ &= \int_{P_\Theta} U_\Theta(\boldsymbol{\Theta};\boldsymbol{\mu},\boldsymbol{\kappa},\boldsymbol{M},\boldsymbol{K})\,d\Omega \end{aligned} \tag{8}$$

be the total potential energy of a hyperelastic body, where U_R is the energy density in Ω_R, U_t the energy written in terms of the inverse deformation in Ω_t and U_Θ is the energy density defined on a continuous parameter space P_Θ. Here, $U = W + V$ consists of two parts, the stored strain energy W and the external potential energy V, which can be defined on the different configurations. We assume that the body forces are conservative forces.

With these assumptions, the Problem 1 can be reformulated in the following form:

Problem 2 Find $\{\boldsymbol{u},s\} \in \mathcal{V}\times\mathcal{D}$ of the objective function $\mathcal{J}:\mathcal{V}\times\mathcal{D}\to\mathbb{R}$ such that

$$\mathcal{J}(\boldsymbol{u},s) = E(\boldsymbol{u},s) \to \min_{\{\boldsymbol{u},s\}\in\mathcal{V}\times\mathcal{D}} \tag{9}$$

subject to the constraints (3).

For this problem, we have $\mathcal{J}'_u = E'_u = \mathcal{R} = 0$ and the adjoint equation (7) can be rewritten in the form

$$k(\boldsymbol{u},\hat{s};\boldsymbol{\lambda},\boldsymbol{\eta}) = -\mathcal{J}'_u(\hat{\boldsymbol{u}},\hat{s};\boldsymbol{\eta}) = -\mathcal{R} = 0 \tag{10}$$

and therefore the Lagrangian variable $\boldsymbol{\lambda}$ becomes zero as well as the Lagrangian (4) remains $\mathcal{L}(\boldsymbol{u},s) = \mathcal{J}(\boldsymbol{u},s)$. Furthermore, the system (6) is reduced to the following variational problem.

Problem 3 Find $\{\boldsymbol{u},s\} \in \mathcal{V}\times\mathcal{D}$ such that

$$\mathcal{L}'(\boldsymbol{u},s)(\boldsymbol{\eta},\boldsymbol{\psi}) = \left\{\begin{array}{l} \mathcal{J}'_u(\boldsymbol{u},s;\boldsymbol{\eta}) \\ \mathcal{J}'_s(\boldsymbol{u},s;\boldsymbol{\psi}) \end{array}\right\} = \left\{\begin{array}{l} \mathcal{R}(\boldsymbol{u},s;\boldsymbol{\eta}) \\ \mathcal{G}(\boldsymbol{u},s;\boldsymbol{\psi}) \end{array}\right\} = \boldsymbol{0} \quad \forall\,\{\boldsymbol{\eta},\boldsymbol{\psi}\} \in \mathcal{V}\times\mathcal{D}. \tag{11}$$

The partial variation of $\mathcal{J}$ with respect to $\boldsymbol{u}$ leads to the physical residual (5) $\mathcal{R} : \mathcal{V} \to \mathbb{R}$

$$\mathcal{R}(\boldsymbol{u}, s; \boldsymbol{\eta}) := a(\boldsymbol{u}, s; \boldsymbol{\eta}) - F(s; \boldsymbol{\eta}). \tag{12}$$

In the same manner, variation with respect to changes in the design s leads to the material residual $\mathcal{G} : \mathcal{D} \to \mathbb{R}$ in the form

$$\mathcal{G}(\boldsymbol{u}, s; \boldsymbol{\psi}) := b(\boldsymbol{u}, s; \boldsymbol{\psi}) - L(s; \boldsymbol{\psi}). \tag{13}$$

The semilinear forms $a : \mathcal{V} \times \mathcal{V} \to \mathbb{R}$ and $b : \mathcal{D} \times \mathcal{D} \to \mathbb{R}$ contain the parts of the partial variations with respect to deformation gradients, i.e.

$$\begin{aligned} a(\boldsymbol{u}, s; \boldsymbol{\eta}) &:= \frac{d}{d\varepsilon} \int_{P_\Theta} U_\Theta(\boldsymbol{\Theta}; \boldsymbol{\mu}, \boldsymbol{\kappa}, \boldsymbol{M} \\ &\quad + \varepsilon \nabla \tilde{\boldsymbol{\eta}}, \boldsymbol{K})\, d\Omega \Big|_{\varepsilon=0} \\ &= \frac{d}{d\varepsilon} \int_{\Omega_R} U_R(\boldsymbol{X}; \boldsymbol{\varphi}, \boldsymbol{F} + \varepsilon \nabla \boldsymbol{\eta})\, d\Omega \Big|_{\varepsilon=0} \end{aligned} \tag{14}$$

$$\begin{aligned} b(\boldsymbol{u}, s; \boldsymbol{\psi}) &:= \frac{d}{d\varepsilon} \int_{P_\Theta} U_\Theta(\boldsymbol{\Theta}; \boldsymbol{\mu}, \boldsymbol{\kappa}, \boldsymbol{M}, \boldsymbol{K} \\ &\quad + \varepsilon \nabla \tilde{\boldsymbol{\psi}})\, d\Omega \Big|_{\varepsilon=0} \\ &= \frac{d}{d\varepsilon} \int_{\Omega_t} U_t(\boldsymbol{x}; \boldsymbol{\phi}, \boldsymbol{f} + \varepsilon \nabla \boldsymbol{\psi})\, d\Omega \Big|_{\varepsilon=0} \end{aligned} \tag{15}$$

The linear functionals $F : \mathcal{V} \to \mathbb{R}$ and $L : \mathcal{D} \to \mathbb{R}$ contain the parts of the partial variations with respect to deformations

$$\begin{aligned} F(s; \boldsymbol{\eta}) &:= \frac{d}{d\varepsilon} \int_{P_\Theta} U_\Theta(\boldsymbol{\Theta}; \boldsymbol{\mu} + \varepsilon\, \tilde{\boldsymbol{\eta}}, \boldsymbol{\kappa}, \boldsymbol{M}, \boldsymbol{K})\, d\Omega \Big|_{\varepsilon=0} \\ &= \frac{d}{d\varepsilon} \int_{\Omega_R} U_R(\boldsymbol{X}; \boldsymbol{\varphi} + \varepsilon\, \boldsymbol{\eta}, \boldsymbol{F})\, d\Omega \Big|_{\varepsilon=0} \end{aligned} \tag{16}$$

$$\begin{aligned} L(s; \boldsymbol{\psi}) &:= \frac{d}{d\varepsilon} \int_{P_\Theta} U_\Theta(\boldsymbol{\Theta}; \boldsymbol{\mu}, \boldsymbol{\kappa} + \varepsilon\, \tilde{\boldsymbol{\psi}}, \boldsymbol{M}, \boldsymbol{K})\, d\Omega \Big|_{\varepsilon=0} \\ &= \frac{d}{d\varepsilon} \int_{\Omega_t} U_t(\boldsymbol{x}; \boldsymbol{\phi} + \varepsilon\, \boldsymbol{\psi}, \boldsymbol{f})\, d\Omega \Big|_{\varepsilon=0}. \end{aligned} \tag{17}$$

Using standard pull back and push forward operations we can transform all quantities into the different domains. For details about the kinematical settings and the variational techniques for shape sensitivity analysis based on the intrinsic formulation in local coordinates see Barthold and Stein (1996) and Barthold (2002, 2007).

The contributions of the physical residual $\mathcal{R}(\boldsymbol{u}, s; \boldsymbol{\eta})$ in terms of the reference configuration are given by

$$a(\boldsymbol{u}, s; \boldsymbol{\eta}) = \int_{\Omega_R} \boldsymbol{P} : \operatorname{Grad} \boldsymbol{\eta}\, d\Omega \tag{18}$$

$$F(s; \boldsymbol{\eta}) = \int_{\Omega_R} \boldsymbol{b}_R \cdot \boldsymbol{\eta}\, d\Omega \tag{19}$$

where $\boldsymbol{P}$ is the first Piola-Kirchhoff stress tensor and $\boldsymbol{b}_R = -\partial_x V_R$ are the body forces per unit volume in the reference configuration derived from the potential V_R.

For the material residual $\mathcal{G}(\boldsymbol{u}, s; \boldsymbol{\psi})$ follows after a pull back to the reference configuration

$$b(\boldsymbol{u}, s; \boldsymbol{\psi}) = \int_{\Omega_R} (\boldsymbol{\Sigma} + V_R \boldsymbol{1}) : \operatorname{Grad} \boldsymbol{\psi}\, d\Omega \tag{20}$$

$$L(s; \boldsymbol{\psi}) = \int_{\Omega_R} \boldsymbol{b}_R^{inh} \cdot \boldsymbol{\psi}\, d\Omega \tag{21}$$

where $\boldsymbol{\Sigma} = W_R \boldsymbol{1} - \boldsymbol{F}^T \boldsymbol{P}$ is the well-known energy momentum or Eshelby tensor (Eshelby 1951, 1975). Furthermore, $\boldsymbol{b}_R^{inh} = -\partial_X W_R|_{expl.}$ denotes the inhomogeneity force given by the explicit derivative of W_R with respect to $\boldsymbol{X}$.

Remark 3 The material residual is also referred to as the weak form of the material or configurational force equilibrium as well as the weak form of the pseudo-momentum equation (Maugin 1993; Gurtin 2000; Kienzler and Herrmann 2000). In the case of a homogeneous elastic body the material residual is the weak form of the inverse deformation problem. It should be noted, that the material residual can be the weak form of the direct deformation problem when the role of the spatial and material coordinates are interchanged. On the other hand, the inverse deformation problem and its weak form can be obtained by a re-parametrization of the direct deformation problem in terms of the inverse deformation $\boldsymbol{\phi}$ (Govindjee and Mihalic 1996). For a detailed discussion about the direct and inverse deformation problem and its duality see for instance (Shield 1967; Chadwick 1975; Steinmann 2000; Kuhl et al. 2004; Kalpakides and Balassas 2005).

Remark 4 In the context of design sensitivity analysis, we can interpret $\boldsymbol{\Sigma}$ as the configurational forces associated to the variation in the design (the configuration) or rather they are the sensitivities of the energy functional with respect to variations in the design. In the case of a homogeneous elastic body and pure design variations, the inhomogeneity force vanishes, i.e. $L(\cdot) = 0$.

Remark 5 The semilinear forms $a(\boldsymbol{u}, s; \boldsymbol{\eta})$ and $b(\boldsymbol{u}, s; \boldsymbol{\psi})$ could be expressed in terms of the symmetric second Piola-Kirchhoff stress tensor $\boldsymbol{S}$ in the form

$$a(\boldsymbol{u}, s; \boldsymbol{\eta}) = \int_{\Omega_R} \boldsymbol{S} : \boldsymbol{E}'_u(\boldsymbol{u}, \boldsymbol{\eta})\, d\Omega \tag{22}$$

$$b(\boldsymbol{u}, s; \boldsymbol{\psi}) = \int_{\Omega_R} \boldsymbol{S} : \boldsymbol{E}'_s(\boldsymbol{u}, \boldsymbol{\psi})\, d\Omega + \int_{\Omega_R} (W_R + V_R)\mathbf{1} : \text{Grad}\, \boldsymbol{\psi}\, d\Omega \tag{23}$$

where the relations

$$\begin{aligned} \boldsymbol{P} : \text{Grad}\, \boldsymbol{\eta} &= \boldsymbol{F}\boldsymbol{S} : \text{Grad}\, \boldsymbol{\eta} \\ &= \boldsymbol{S} : \text{sym}\{\boldsymbol{F}^T\, \text{Grad}\, \boldsymbol{\eta}\} = \boldsymbol{S} : \boldsymbol{E}'_u(\boldsymbol{u}, \boldsymbol{\eta}) \end{aligned} \tag{24}$$

and

$$\begin{aligned} -\boldsymbol{F}^T \boldsymbol{P} : \text{Grad}\, \boldsymbol{\psi} &= -\boldsymbol{F}^T \boldsymbol{F}\boldsymbol{S} : \text{Grad}\, \boldsymbol{\psi} \\ &= -\boldsymbol{S} : \text{sym}\{\boldsymbol{F}^T \boldsymbol{F}\, \text{Grad}\, \boldsymbol{\psi}\} \\ &= \boldsymbol{S} : \boldsymbol{E}'_s(\boldsymbol{u}, \boldsymbol{\psi}) \end{aligned} \tag{25}$$

have been used. The terms $\boldsymbol{E}'_u(\boldsymbol{u}, \cdot)$ and $\boldsymbol{E}'_s(\boldsymbol{u}, \cdot)$ denote the partial variations of the Green-Lagrange strain tensor $\boldsymbol{E}$ with respect to $\boldsymbol{u}$ and s, respectively, see Appendix A for the explicit form of these quantities. The terms $a(\cdot, \cdot)$ and $b(\cdot, \cdot)$ have the same structure with different variations of the Green-Lagrange strain tensor $\boldsymbol{E}$, but $b(\cdot, \cdot)$ has an additional term as a result of the variation of the domain Ω_R. We will see in the following that it is very useful to use this expressions because we utilize some symmetry properties and get similar structures for the variations of the physical and material problem. Furthermore, for the computational treatment it is helpful to deal with symmetric quantities.

Remark 6 In this paper we consider a pure variational setting for elasticity. The theory of configurational forces is not restricted to elastic problems and variational approaches. Inelastic problems play an important role in many fields, e.g. elastic-plastic fracture mechanics. For an overview of different applications and approaches see e.g. Steinmann and Maugin (2005) and the references therein.

4 Variational design sensitivity analysis

4.1 Sensitivity of the objective functional

The solution of structural optimization problems require the variations of the objective functional and the constraints due to variations in the design. In the context of structural optimization this is termed as *design sensitivity analysis*.

The total partial derivative $D_s f$ of a function $f(s, \boldsymbol{u}) = f(s, \boldsymbol{u}(s))$ with respect to s and the explicit partial derivative $\partial_s f$ are connected by the relation

$$D_s f(s, \boldsymbol{u}) = \partial_s f + \frac{\partial f}{\partial \boldsymbol{u}} \frac{\partial \boldsymbol{u}}{\partial s}. \tag{26}$$

With this, we introduce the total partial variation of a functional $\mathcal{J}(\boldsymbol{u}, s) = \mathcal{J}(\boldsymbol{u}(s), s)$ with respect to s in the form

$$D_s \mathcal{J}(\boldsymbol{u}, s) \cdot \delta s = \partial_s \mathcal{J} \cdot \delta s + \frac{\partial \mathcal{J}}{\partial \boldsymbol{u}} \frac{\partial \boldsymbol{u}}{\partial s} \cdot \delta s. \tag{27}$$

The above sensitivity relation depends on the partial variation $\partial_s \mathcal{J} \cdot \delta s = \mathcal{J}'_s(\boldsymbol{u}, s, \delta s)$ and a second part with the derivatives $\partial_u \mathcal{J}$ and the changes in the state $\partial_s \boldsymbol{u}$, which is a result of the functional dependencies. The derivatives $\partial_s \mathcal{J}$ and $\partial_u \mathcal{J}$ are given explicitly, but the derivative of the deformation $\partial_s \boldsymbol{u}$ with respect to s is given implicitly, because they are the results of an analysis for the deformation, see Sect. 4.2. This is a classical well-known expression in structural optimization for the design sensitivity of the objective functional.

In the case of $\mathcal{J}(\boldsymbol{u}, s) = E(\boldsymbol{u}, s)$, i.e. we choose the energy as objective functional, we have

$$\mathcal{J}'_u = E'_u = \partial_u \mathcal{J} \cdot \delta \boldsymbol{u} = \mathcal{R}(\boldsymbol{u}, s; \delta \boldsymbol{u}) = 0 \tag{28}$$

and $\mathcal{J}'_s = \partial_s \mathcal{J} \cdot \delta s = \mathcal{G}(\boldsymbol{u}, s, \delta s)$. With this, the second part of the above sensitivity relation vanishes and it remains only the material residual, i.e.

$$D_s \mathcal{J}(\boldsymbol{u}(s); s) \cdot \delta s = \mathcal{J}'_s(\boldsymbol{u}, s; \delta s) = \mathcal{G}(\boldsymbol{u}, s; \delta s). \tag{29}$$

Therefore, in the context of structural optimization, we can interpret the material residual or configurational forces as the sensitivity of the energy with respect to variations in the design, see Remark 4.

4.2 Sensitivity of the physical residual

The solution of the Lagrangian requires the partial variations of the physical residual with respect to $\boldsymbol{u}$ and s, respectively. The total variation of the physical residual $\mathcal{R} = a(\boldsymbol{u}, s; \boldsymbol{\eta}) - F(s; \boldsymbol{\eta}) = 0$ reads

$$\mathcal{R}' = \mathcal{R}'_u(\boldsymbol{u}, s; \boldsymbol{\eta}, \delta \boldsymbol{u}) + \mathcal{R}'_s(\boldsymbol{u}, s; \boldsymbol{\eta}, \delta s) = 0 \tag{30}$$

where the partial variations are given by

$$\mathcal{R}'_u(\boldsymbol{u}, s; \boldsymbol{\eta}, \delta \boldsymbol{u}) := a'_u(\boldsymbol{u}, s; \boldsymbol{\eta}, \delta \boldsymbol{u}) - F'_u(s; \boldsymbol{\eta}, \delta \boldsymbol{u}) \tag{31}$$

$$\mathcal{R}'_s(\boldsymbol{u}, s; \boldsymbol{\eta}, \delta s) := a'_s(\boldsymbol{u}, s; \boldsymbol{\eta}, \delta s) - F'_s(s; \boldsymbol{\eta}, \delta s). \tag{32}$$

Note, that $\mathcal{R}'_u = \mathcal{J}''_{uu} = E''_{uu}$ and $\mathcal{R}'_s = \mathcal{J}''_{us} = E''_{us}$, respectively.

We introduce for the variations of the physical residual $\mathcal{R}$ with respect to $\boldsymbol{u}$ and $\boldsymbol{s}$ the operators

$$k(\boldsymbol{u}, \boldsymbol{s}; \boldsymbol{\eta}, \delta\boldsymbol{u}) := \mathcal{R}'_u(\boldsymbol{u}, \boldsymbol{s}; \boldsymbol{\eta}, \delta\boldsymbol{u}) \tag{33}$$

and

$$p(\boldsymbol{u}, \boldsymbol{s}; \boldsymbol{\eta}, \delta\boldsymbol{s}) := \mathcal{R}'_s(\boldsymbol{u}, \boldsymbol{s}; \boldsymbol{\eta}, \delta\boldsymbol{s}) \tag{34}$$

where $k(\boldsymbol{u}, \boldsymbol{s}; \cdot, \cdot)$ is the well-known *tangent physical stiffness operator* and we call $p(\boldsymbol{u}, \boldsymbol{s}; \cdot, \cdot)$ the *tangent pseudo or fictitious load operator* for the physical problem. Both operators are bilinear forms $k : \mathcal{V} \times \mathcal{V} \to \mathbb{R}$ and $p : \mathcal{V} \times \mathcal{D} \to \mathbb{R}$.

With these notations the total variation yields the form

$$\mathcal{R}' = k(\boldsymbol{u}, \boldsymbol{s}; \boldsymbol{\eta}, \delta\boldsymbol{u}) + p(\boldsymbol{u}, \boldsymbol{s}; \boldsymbol{\eta}, \delta\boldsymbol{s}) = 0. \tag{35}$$

After rearranging the above terms we can formulate the following sensitivity equation for the physical problem.

Problem 4 Let $\delta\hat{\boldsymbol{s}} \in \mathcal{D}$ be a given fixed design variation. Find $\delta\boldsymbol{u} \in \mathcal{V}$ such that

$$k(\boldsymbol{u}, \boldsymbol{s}; \boldsymbol{\eta}, \delta\boldsymbol{u}) = -Q(\boldsymbol{u}, \boldsymbol{s}; \boldsymbol{\eta}, \delta\hat{\boldsymbol{s}}) \qquad \forall\, \boldsymbol{\eta} \in \mathcal{V}, \tag{36}$$

where

$$Q(\boldsymbol{u}, \boldsymbol{s}; \boldsymbol{\eta}, \delta\hat{\boldsymbol{s}}) := p(\boldsymbol{u}, \boldsymbol{s}; \boldsymbol{\eta}, \delta\hat{\boldsymbol{s}}) = \mathcal{R}'_s(\boldsymbol{u}, \boldsymbol{s}; \boldsymbol{\eta}, \delta\hat{\boldsymbol{s}}) \tag{37}$$

is the *pseudo load* of the physical problem for the variation $\delta\hat{\boldsymbol{s}}$.

This is a variational equation for the sensitivity of the deformation due to changes in the design. For a given variation in the design $\delta\hat{\boldsymbol{s}}$, we can calculate the variation in the state $\delta\boldsymbol{u}$.

Remark 7 In general the pseudo load operator $p(\boldsymbol{u}, \boldsymbol{s}; \boldsymbol{\eta}, \delta\boldsymbol{s})$ is a bilinear form $p : \mathcal{V} \times \mathcal{D} \to \mathbb{R}$. For a chosen fixed $\delta\hat{\boldsymbol{s}}$ it becomes a linear functional $Q : \mathcal{V} \to \mathbb{R}$ and is called *pseudo load* because it plays the role of a *load* in the sensitivity equation (36) and is denoted by $Q(\boldsymbol{u}, \boldsymbol{s}; \cdot, \delta\hat{\boldsymbol{s}})$, i.e. $Q(\boldsymbol{u}, \boldsymbol{s}; \cdot, \delta\hat{\boldsymbol{s}}) = p(\boldsymbol{u}, \boldsymbol{s}; \cdot, \delta\hat{\boldsymbol{s}})$.

A straightforward calculation by using Eq. 22 gives the tangent physical stiffness (33) and the pseudo load operator (34) in the form

$$\begin{aligned} k(\boldsymbol{u}, \boldsymbol{s}; \boldsymbol{\eta}, \delta\boldsymbol{u}) = & \int_{\Omega_R} \boldsymbol{S} : \boldsymbol{E}''_{uu}(\boldsymbol{\eta}, \delta\boldsymbol{u})\, d\Omega \\ & + \int_{\Omega_R} \boldsymbol{E}'_u(\boldsymbol{u}, \boldsymbol{\eta}) : \mathbb{C} : \boldsymbol{E}'_u(\boldsymbol{u}, \delta\boldsymbol{u})\, d\Omega. \end{aligned} \tag{38}$$

and

$$\begin{aligned} p(\boldsymbol{u}, \boldsymbol{s}; \boldsymbol{\eta}, \delta\boldsymbol{s}) = & \int_{\Omega_R} \boldsymbol{S} : \boldsymbol{E}''_{us}(\boldsymbol{u}, \boldsymbol{\eta}, \delta\boldsymbol{s})\, d\Omega \\ & + \int_{\Omega_R} \boldsymbol{E}'_u(\boldsymbol{u}, \boldsymbol{\eta}) : \mathbb{C} : \boldsymbol{E}'_s(\boldsymbol{u}, \delta\boldsymbol{s})\, d\Omega \\ & + \int_{\Omega_R} \boldsymbol{S} : \boldsymbol{E}'_u(\boldsymbol{u}, \boldsymbol{\eta})\ \mathrm{Div}\, \delta\boldsymbol{s}\, d\Omega \\ & - \int_{\Omega_R} \boldsymbol{b}_R \cdot \boldsymbol{\eta}\ \mathrm{Div}\, \delta\boldsymbol{s}\, d\Omega \end{aligned} \tag{39}$$

where $\boldsymbol{S}$ is the second Piola-Kirchhoff stress tensor, $\mathbb{C}$ the second elasticity tensor and $\boldsymbol{E}'_u(\boldsymbol{u}, \cdot)$ as well as $\boldsymbol{E}'_s(\boldsymbol{u}, \cdot)$ are the partial variations of the Green-Lagrange strain tensor $\boldsymbol{E}$ with respect to $\boldsymbol{u}$ and $\boldsymbol{s}$, respectively. Furthermore, $\boldsymbol{E}''_{uu}(\boldsymbol{u}, \cdot, \cdot)$ denote the second and $\boldsymbol{E}''_{us}(\boldsymbol{u}, \cdot, \cdot)$ the mixed variation of $\boldsymbol{E}$, see Appendices A and B.1 for the explicit formulations of the above terms.

The derived tangent operators $k(\cdot, \cdot)$ and $p(\cdot, \cdot)$ have the same structure with different variations of the Green-Lagrange strain tensor $\boldsymbol{E}$. The pseudo load operator $p(\cdot, \cdot)$ has an additional term as a result of the variation of the domain Ω_R.

4.3 Sensitivity of the material residual

In the same way we can perform the total variation of the material residual $\mathcal{G} = b(\boldsymbol{u}, \boldsymbol{s}; \boldsymbol{\psi}) - L(\boldsymbol{s}; \boldsymbol{\psi}) = 0$. This reads

$$\mathcal{G}' = \mathcal{G}'_u(\boldsymbol{u}, \boldsymbol{s}; \boldsymbol{\psi}, \delta\boldsymbol{u}) + \mathcal{G}'_s(\boldsymbol{u}, \boldsymbol{s}; \boldsymbol{\psi}, \delta\boldsymbol{s}) = 0 \tag{40}$$

where the partial variations are given by

$$\mathcal{G}'_u(\boldsymbol{u}, \boldsymbol{s}; \boldsymbol{\psi}, \delta\boldsymbol{u}) := b'_u(\boldsymbol{u}, \boldsymbol{s}; \boldsymbol{\psi}, \delta\boldsymbol{u}) - L'_u(\boldsymbol{s}; \boldsymbol{\psi}, \delta\boldsymbol{u}) \tag{41}$$

$$\mathcal{G}'_s(\boldsymbol{u}, \boldsymbol{s}; \boldsymbol{\psi}, \delta\boldsymbol{s}) := b'_s(\boldsymbol{u}, \boldsymbol{s}; \boldsymbol{\psi}, \delta\boldsymbol{s}) - L'_s(\boldsymbol{s}; \boldsymbol{\psi}, \delta\boldsymbol{s}). \tag{42}$$

Note, that $\mathcal{G}'_u = \mathcal{J}''_{su} = E''_{su}$ and $\mathcal{G}'_s = \mathcal{J}''_{ss} = E''_{ss}$, respectively.

We introduce for the variations of the material residual $\mathcal{G}$ with respect to $\boldsymbol{u}$ and $\boldsymbol{s}$ the operators

$$d(\boldsymbol{u}, \boldsymbol{s}; \boldsymbol{\psi}, \delta\boldsymbol{s}) := \mathcal{G}'_s(\boldsymbol{u}, \boldsymbol{s}; \boldsymbol{\psi}, \delta\boldsymbol{s}) \tag{43}$$

and

$$t(\boldsymbol{u}, \boldsymbol{s}; \boldsymbol{\psi}, \delta\boldsymbol{u}) := \mathcal{G}'_u(\boldsymbol{u}, \boldsymbol{s}; \boldsymbol{\psi}, \delta\boldsymbol{u}) \tag{44}$$

where $d(\cdot; \cdot)$ is the so-called *tangent material stiffness* or *tangent material operator* in order to highlight the duality to the tangent physical stiffness (33) and we call $t(\cdot;\cdot)$ the *tangent pseudo or fictitious load operator* for the material problem, compare with Eq. 34. Both operators are bilinear forms $d : \mathcal{D} \times \mathcal{D} \to \mathbb{R}$ and $t : \mathcal{D} \times \mathcal{V} \to \mathbb{R}$.

As a result of the permutableness of variations, i.e.

$$\mathcal{G}'_u = \mathcal{J}''_{su} = \mathcal{J}''_{us} = \mathcal{R}'_s \tag{45}$$

and the symmetry of the bilinear forms, i.e. $a'_s(\boldsymbol{u}, \boldsymbol{s}; \cdot, \boldsymbol{\eta}) = a'_s(\boldsymbol{u}, \boldsymbol{s}; \boldsymbol{\eta}, \cdot)$, we obtain for the variation of $\mathcal{G}$ with respect to $\boldsymbol{u}$

$$\begin{aligned}\mathcal{G}'_u(\boldsymbol{u}, \boldsymbol{s}; \cdot, \delta\boldsymbol{u}) &= \mathcal{R}'_s(\boldsymbol{u}, \boldsymbol{s}; \delta\boldsymbol{u}, \cdot) = p(\boldsymbol{u}, \boldsymbol{s}; \delta\boldsymbol{u}, \cdot) \\ &= a'_s(\boldsymbol{u}, \boldsymbol{s}; \delta\boldsymbol{u}, \cdot) - F'_s(\boldsymbol{s}; \delta\boldsymbol{u}, \cdot).\end{aligned} \tag{46}$$

Thus, due to symmetry, the partial variation $\mathcal{G}'_u$ leads to the tangent pseudo load operator of the physical problem (34), i.e.

$$t(\boldsymbol{u}, \boldsymbol{s}; \boldsymbol{\psi}, \delta\boldsymbol{u}) = p(\boldsymbol{u}, \boldsymbol{s}; \delta\boldsymbol{u}, \boldsymbol{\psi}). \tag{47}$$

Therefore, we have additional to specify only the material tangent operator $d(\boldsymbol{u}, \boldsymbol{s}; \boldsymbol{\psi}, \delta\boldsymbol{s}) = \mathcal{G}'_s$.

With these notations the total variation yields the form

$$\mathcal{G}' = p(\boldsymbol{u}, \boldsymbol{s}; \delta\boldsymbol{u}, \boldsymbol{\psi}) + d(\boldsymbol{u}, \boldsymbol{s}; \boldsymbol{\psi}, \delta\boldsymbol{s}) = 0. \tag{48}$$

After rearranging the above terms we can formulate the following sensitivity equation for the material problem.

Problem 5 Let $\delta\hat{\boldsymbol{u}} \in \mathcal{V}$ be a given fixed variation in the state. Find $\delta\boldsymbol{s} \in \mathcal{D}$ such that

$$d(\boldsymbol{u}, \boldsymbol{s}; \boldsymbol{\psi}, \delta\boldsymbol{s}) = -Q(\boldsymbol{u}, \boldsymbol{s}; \delta\hat{\boldsymbol{u}}, \boldsymbol{\psi}) \qquad \forall\, \boldsymbol{\psi} \in \mathcal{D}, \tag{49}$$

where

$$Q(\boldsymbol{u}, \boldsymbol{s}; \delta\hat{\boldsymbol{u}}, \boldsymbol{\psi}) := p(\boldsymbol{u}, \boldsymbol{s}; \delta\hat{\boldsymbol{u}}, \boldsymbol{\psi}) = \mathcal{G}'_u(\boldsymbol{u}, \boldsymbol{s}; \boldsymbol{\psi}, \delta\hat{\boldsymbol{u}}) \tag{50}$$

is the *pseudo load* of the material problem for the variation $\delta\hat{\boldsymbol{u}}$.

This is a variational equation for the sensitivity of the design due to changes in the deformation. For a given variation in the state $\delta\hat{\boldsymbol{u}}$, we can calculate the variation in the design $\delta\boldsymbol{s}$.

Remark 8 It is interesting to note, that due to symmetry both the sensitivity of the deformation and the sensitivity of the design depend on the pseudo load operator $p(\cdot, \cdot)$. Therefore, this operator plays an important role for the solution of the minimization problem for $\boldsymbol{u}$ and $\boldsymbol{s}$.

Finally, using Eq. 23, we obtain the material pseudo load operator (44) and the tangent material stiffness (43) in the form

$$\begin{aligned}t(\boldsymbol{u}, \boldsymbol{s}; \boldsymbol{\psi}, \delta\boldsymbol{u}) = &\int_{\Omega_R} \boldsymbol{S} : \boldsymbol{E}''_{su}(\boldsymbol{u}, \boldsymbol{\psi}, \delta\boldsymbol{u})\, d\Omega \\ &+ \int_{\Omega_R} \boldsymbol{E}'_s(\boldsymbol{u}, \boldsymbol{\psi}) : \mathbb{C} : \boldsymbol{E}'_u(\boldsymbol{u}, \delta\boldsymbol{u})\, d\Omega \\ &+ \int_{\Omega_R} \boldsymbol{S} : \boldsymbol{E}'_u(\boldsymbol{u}, \delta\boldsymbol{u}) \operatorname{Div} \boldsymbol{\psi}\, d\Omega \\ &- \int_{\Omega_R} \boldsymbol{b}_R \cdot \delta\boldsymbol{u} \operatorname{Div} \boldsymbol{\psi}\, d\Omega.\end{aligned} \tag{51}$$

and

$$\begin{aligned}d(\boldsymbol{u}, \boldsymbol{s}; \boldsymbol{\psi}, \delta\boldsymbol{s}) = &\int_{\Omega_R} \boldsymbol{S} : \boldsymbol{E}''_{ss}(\boldsymbol{u}, \boldsymbol{\psi}, \delta\boldsymbol{s})\, d\Omega \\ &+ \int_{\Omega_R} \boldsymbol{E}'_s(\boldsymbol{u}, \boldsymbol{\psi}) : \mathbb{C} : \boldsymbol{E}'_s(\boldsymbol{u}, \delta\boldsymbol{s})\, d\Omega \\ &+ \int_{\Omega_R} \boldsymbol{S} : \boldsymbol{E}'_s(\boldsymbol{u}, \boldsymbol{\psi}) \operatorname{Div} \delta\boldsymbol{s}\, d\Omega \\ &+ \int_{\Omega_R} \boldsymbol{S} : \boldsymbol{E}'_s(\boldsymbol{u}, \delta\boldsymbol{s}) \operatorname{Div} \boldsymbol{\psi}\, d\Omega \\ &+ \int_{\Omega_R} W_R \operatorname{Div} \boldsymbol{\psi} \operatorname{Div} \delta\boldsymbol{s}\, d\Omega \\ &- \int_{\Omega_R} W_R\, \mathbf{1} : \operatorname{Grad} \boldsymbol{\psi} \operatorname{Grad} \delta\boldsymbol{s}\, d\Omega.\end{aligned} \tag{52}$$

Here, $\boldsymbol{E}''_{ss}(\boldsymbol{u}, \cdot, \cdot)$ denote the second variation of $\boldsymbol{E}$ with respect to $\boldsymbol{s}$. See Appendices A and B.2 for the explicit formulations of the above terms. As a result of symmetry follow $t(\boldsymbol{u}, \boldsymbol{s}; \boldsymbol{\psi}, \delta\boldsymbol{u}) = p(\boldsymbol{u}, \boldsymbol{s}; \delta\boldsymbol{u}, \boldsymbol{\psi})$, see (39). Furthermore, $d(\cdot, \cdot)$ has a similar structure like $k(\cdot, \cdot)$, but additional terms occur as a result of the second variation of Ω_R, see (38).

5 The finite element approximation

5.1 The discrete optimization problem

The finite element formulation is based on a conforming Galerkin method defined on meshes $\mathcal{T}_h = \{K\}$ with a mesh parameter h consisting of closed cells K which are either triangles or quadrilaterals. The boundary ∂K of each element K is assumed to be Lipschitz-continuous. On the mesh $\mathcal{T}_h$ we define finite dimensional element spaces $\mathcal{V}_h \subset \mathcal{V}$ and $\mathcal{D}_h \subset \mathcal{D}$ consisting of cellwise polynomial functions.

The corresponding discrete state $\boldsymbol{u}_h \in \mathcal{V}_h \subset \mathcal{V}$ and discrete design $\boldsymbol{s}_h \in \mathcal{D}_h \subset \mathcal{D}$ are determined by the following discrete version of Problem 2.

Problem 6 Find $\{\boldsymbol{u}_h, \boldsymbol{s}_h\} \in \mathcal{V}_h \times \mathcal{D}_h$ of the objective function $\mathcal{J} : \mathcal{V}_h \times \mathcal{D}_h \to \mathbb{R}$ such that

$$\mathcal{J}(\boldsymbol{u}_h, \boldsymbol{s}_h) = E(\boldsymbol{u}_h, \boldsymbol{s}_h) \to \min_{\{\boldsymbol{u}_h, \boldsymbol{s}_h\} \in \mathcal{V}_h \times \mathcal{D}_h} \tag{53}$$

subject to the constraint

$$< \boldsymbol{A}(\boldsymbol{u}_h, \boldsymbol{s}_h), \boldsymbol{\eta}_h > = \mathcal{R}(\boldsymbol{u}_h, \boldsymbol{s}_h; \boldsymbol{\eta}_h) = 0 \quad \forall\, \boldsymbol{\eta}_h \in \mathcal{V}_h. \tag{54}$$

For the optimal solutions $\boldsymbol{u}_h^* \in \mathcal{V}_h$ and $\boldsymbol{s}_h^* \in \mathcal{D}_h$ in the chosen approximation spaces $\mathcal{V}_h$ and $\mathcal{D}_h$ we require that

$$\mathcal{J}(\boldsymbol{u}_h^*, \boldsymbol{s}_h^*) \leq \mathcal{J}(\boldsymbol{u}_h, \boldsymbol{s}_h) \qquad \forall\, \boldsymbol{u}_h, \boldsymbol{s}_h \in \mathcal{V}_h \times \mathcal{D}_h. \tag{55}$$

The error of the state $\boldsymbol{e}_{u,h}$ and the design $\boldsymbol{e}_{s,h}$ in the chosen approximation spaces are introduced by

$$\boldsymbol{e}_{u,h} := \boldsymbol{u}_h^* - \boldsymbol{u}_h \quad \text{and} \quad \boldsymbol{e}_{s,h} := \boldsymbol{s}_h^* - \boldsymbol{s}_h, \tag{56}$$

respectively.

For a matrix description of the derived sensitivity operators we introduce the discrete approximations for displacements and design, i.e. the nodal displacement vector $\boldsymbol{u} \in \mathbb{R}^n$ and the vector of design variables $\boldsymbol{s} \in \mathbb{R}^m$. Here, n and m are the dimensions of the introduced approximation spaces. We introduce in the same manner the discrete approximations for the corresponding variations, i.e. $\delta\boldsymbol{u} \in \mathbb{R}^n$ and $\delta\boldsymbol{s} \in \mathbb{R}^m$.

A standard finite element discretization of the tangent operators lead to the usual tangent stiffness matrix $\boldsymbol{K} \in \mathbb{R}^{n\times n}$, the tangent pseudo load operator matrix $\boldsymbol{P} \in \mathbb{R}^{n\times m}$ as well as the tangent material stiffness matrix $\boldsymbol{D} \in \mathbb{R}^{m\times m}$ associated to the bilinear forms $k(\cdot,\cdot)$, $p(\cdot,\cdot)$ and $d(\cdot,\cdot)$. Furthermore, $\boldsymbol{R} \in \mathbb{R}^n$ denotes the physical residual vector and $\boldsymbol{G} \in \mathbb{R}^m$ denotes the material residual vector associated to the functionals $\mathcal{R}(\boldsymbol{u}_h, \boldsymbol{s}_h; \cdot)$ and $\mathcal{G}(\boldsymbol{u}_h, \boldsymbol{s}_h; \cdot)$, respectively.

5.2 The error in the material residual

The fulfillment of the physical residual for every admissible design is an important constraint of the optimization problem, i.e.

$$\mathcal{R}(\boldsymbol{u}_h, \boldsymbol{s}_h; \boldsymbol{\eta}_h) = 0 \qquad \forall\, \boldsymbol{s}_h \in \mathcal{D}_h. \tag{57}$$

The optimality condition (11) for the material problem holds only for stationary points $\boldsymbol{u}_h^*$ and $\boldsymbol{s}_h^*$ of the Lagrangian, i.e. $\mathcal{L}'_s = \mathcal{G}(\boldsymbol{u}_h^*, \boldsymbol{s}_h^*; \boldsymbol{\psi}_h) = 0$.

This means, that the material residual is not fulfilled for every $\boldsymbol{u}_h \neq \boldsymbol{u}_h^* = \boldsymbol{u}_h(\boldsymbol{s}_h^*)$, i.e.

$$\mathcal{G}(\boldsymbol{u}_h, \boldsymbol{s}_h; \boldsymbol{\psi}_h) \neq 0 \qquad \forall\, \boldsymbol{u}_h \neq \boldsymbol{u}_h^* = \boldsymbol{u}_h(\boldsymbol{s}_h^*) \in \mathcal{V}_h. \tag{58}$$

Hence, we obtain a material residual as a result of the non-optimal solution $\boldsymbol{u}_h$, which is a result of the non-optimal design $\boldsymbol{s}_h$ in the sense of the minimization of Problem 6. With the definition of the errors (56) we have $\boldsymbol{s}_h^* = \boldsymbol{s}_h + \boldsymbol{e}_{s,h}$ and therefore

$$\mathcal{G}(\boldsymbol{u}_h^*, \boldsymbol{s}_h^*; \boldsymbol{\psi}_h) = \mathcal{G}(\boldsymbol{u}_h(\boldsymbol{s}_h + \boldsymbol{e}_{s,h}), \boldsymbol{s}_h + \boldsymbol{e}_{s,h}; \boldsymbol{\psi}_h) = 0. \tag{59}$$

From this, a suitable approximation for the error could be obtained from the linearization

$$\mathcal{G}(\hat{\boldsymbol{u}}_h, \hat{\boldsymbol{s}}_h; \boldsymbol{\psi}_h) + D_s\mathcal{G}(\hat{\boldsymbol{u}}_h, \hat{\boldsymbol{s}}_h; \boldsymbol{\psi}_h) \cdot \boldsymbol{e}_{s,h} + \mathcal{O} = 0. \tag{60}$$

Finally, with the total partial derivative

$$D_s\mathcal{G}(\boldsymbol{u}(\boldsymbol{s}), \boldsymbol{s}) = \frac{\partial \mathcal{G}}{\partial \boldsymbol{s}} + \frac{\partial \mathcal{G}}{\partial \boldsymbol{u}}\frac{\partial \boldsymbol{u}}{\partial \boldsymbol{s}} \tag{61}$$

follow the tangent operator of the material problem

$$\begin{aligned} b_T(\hat{\boldsymbol{u}}_h, \hat{\boldsymbol{s}}_h; \boldsymbol{\psi}_h, \boldsymbol{e}_{s,h}) &= D_s\mathcal{G}(\hat{\boldsymbol{u}}_h, \hat{\boldsymbol{s}}_h; \boldsymbol{\psi}_h) \cdot \boldsymbol{e}_{s,h} \\ &= \frac{d}{d\varepsilon}\left[\mathcal{G}(\hat{\boldsymbol{u}}_h, \boldsymbol{s}_h + \varepsilon \boldsymbol{e}_{s,h}; \boldsymbol{\psi}_h)\right]\Big|_{\varepsilon=0} \\ &\quad + \frac{d}{d\varepsilon}\left[\mathcal{G}(\boldsymbol{u}_h(\boldsymbol{s}_h + \varepsilon \boldsymbol{e}_{s,h}), \hat{\boldsymbol{s}}_h; \boldsymbol{\psi}_h)\right]\Big|_{\varepsilon=0}. \end{aligned} \tag{62}$$

We obtain an equation for the error in the design or rather for the error in the material residual in the form

$$b_T(\hat{\boldsymbol{u}}_h, \hat{\boldsymbol{s}}_h; \boldsymbol{\psi}_h, \boldsymbol{e}_{s,h}) = -\mathcal{G}(\hat{\boldsymbol{u}}_h, \hat{\boldsymbol{s}}_h; \boldsymbol{\psi}_h) \ \ \forall\, \boldsymbol{\psi}_h \in \mathcal{D}_h. \tag{63}$$

This is used for a staggered solution algorithm, see Sect. 6.2. Furthermore, it could be used for a mesh optimization algorithm, see Sect. 7.1.

5.3 The discrete sensitivity equations

With the above definitions, the discrete versions of the sensitivity equations for the physical (35) and material (48) residuals become

$$\boldsymbol{K}\delta\boldsymbol{u} + \boldsymbol{P}\delta\boldsymbol{s} = \boldsymbol{0} \qquad \text{or} \qquad \delta\boldsymbol{u} = -\boldsymbol{K}^{-1}\boldsymbol{P}\delta\boldsymbol{s} \tag{64}$$

and

$$P^T \delta u + D\delta s = \mathbf{0} \quad \text{or} \quad \delta s = -D^{-1} P^T \delta u. \quad (65)$$

For chosen fixed variations $\delta\hat{s}$ and $\delta\hat{u}$ the discrete versions of the sensitivity equations for the physical (36) and material (49) problem follow in the form

$$K\delta u = -Q_p \quad \text{with} \quad Q_p := P\delta\hat{s} \quad (66)$$

and

$$D\delta s = -Q_m \quad \text{with} \quad Q_m := P^T \delta\hat{u}, \quad (67)$$

respectively. Here, $Q_p \in \mathbb{R}^n$ is the pseudo load vector of the physical residual problem associated to the functional $Q(u_h, s_h; \cdot, \delta\hat{s}_h)$ and $Q_m \in \mathbb{R}^m$ is the pseudo load vector of the material residual problem associated to the functional $Q(u_h, s_h; \delta\hat{u}_h, \cdot)$.

Remark 9 It is important to note, that we obtain with the relations from (64) directly a connection between the physical and the material space. Let $S_p := -K^{-1}P \in \mathbb{R}^{n\times m}$ be the discrete sensitivity operator, the physical and material space are connected by the transformation

$$\delta u = S_p \, \delta s. \quad (68)$$

We call the matrix S_p the *sensitivity matrix* of the physical problem. With the knowledge of the pseudo load operator matrix P, we can evaluate the sensitivity equation for arbitrary admissible variations $\delta\hat{s}$ in the material space. In the same manner, the discrete sensitivity operator of the material problem, i.e. the *sensitivity matrix* $S_m := -D^{-1}P^T \in \mathbb{R}^{m\times n}$ follows from (65) and we obtain the transformation

$$\delta s = S_m \, \delta u. \quad (69)$$

With this, we can perform the sensitivity analysis for arbitrary admissible variations $\delta\hat{u}$ in the physical space.

Finally, we have summarized the most important variational and discrete sensitivity relations in Table 1.

6 Solution algorithms

6.1 Full Newton method

We apply the Newton method on the continuous level in order to get the solution of (11). A Taylor expansion with $y := \{u, s\}^T$ reads

$$\mathcal{L}'(y + \Delta y) = \mathcal{L}'(y) + \nabla\mathcal{L}'(y)(\Delta y) + \mathcal{O} = \mathbf{0}. \quad (70)$$

The remainder $\mathcal{O}$ contains higher order terms which can usually be neglected. Each Newton step requires the solution of the linear system

$$H(y)(\Delta y) = \begin{Bmatrix} \mathcal{R}'_u & \mathcal{R}'_s \\ \mathcal{G}'_u & \mathcal{G}'_s \end{Bmatrix} = -\mathcal{L}'(y), \quad (71)$$

where $H(y)(\Delta y) := \nabla\mathcal{L}'(y)(\Delta y)$ is the Hessian matrix of the Lagrangian $\mathcal{L}(u, s)$. The Hessian contains the partial variations of $\mathcal{R}$ and $\mathcal{G}$, i.e.

$$\mathcal{R}'_u = a'_u(u, s; \eta, \Delta u) - F'_u(s; \eta, \Delta u) \quad (72)$$
$$\mathcal{R}'_s = a'_s(u, s; \eta, \Delta s) - F'_s(s; \eta, \Delta s) \quad (73)$$
$$\mathcal{G}'_u = b'_u(u, s; \Delta u, \psi) - L'_u(s; \Delta u, \psi) \quad (74)$$
$$\mathcal{G}'_s = b'_s(u, s; \psi, \Delta s) - L'_s(s; \psi, \Delta s). \quad (75)$$

With the definitions of the sensitivities of the physical (35) and material residual (48)

$$\mathcal{R}' = \mathcal{R}'_u + \mathcal{R}'_s = k(u, s; \eta, \delta u) + p(u, s; \eta, \delta s)$$
$$\mathcal{G}' = \mathcal{G}'_u + \mathcal{G}'_s = p(u, s; \delta u, \psi) + d(u, s; \psi, \delta s)$$

follows the Hessian

$$H(y)(\Delta y) := \begin{Bmatrix} k(u, s; \eta, \Delta u) & p(u, s; \eta, \Delta s) \\ p(u, s; \Delta u, \psi) & d(u, s; \psi, \Delta s) \end{Bmatrix}. \quad (76)$$

Finally, the solution of the above problem requires the solution of the linear system

$$\begin{Bmatrix} k(u, s; \eta, \Delta u) & p(u, s; \eta, \Delta s) \\ p(u, s; \Delta u, \psi) & d(u, s; \psi, \Delta s) \end{Bmatrix} = -\begin{Bmatrix} \mathcal{R}(u, s; \eta) \\ \mathcal{G}(u, s; \psi) \end{Bmatrix} \quad (77)$$

in each Newton step.

Remark 10 The diagonal elements of the above saddle point problem are the pseudo load operators of the physical (34) and material residual (44). Hence, the physical and the material problem are coupled by the pseudo load operator $p(\cdot, \cdot)$, which is used for the solution of structural optimization problems.

The finite element approximation of (77) is given by

$$\begin{Bmatrix} k(u_h, s_h; \eta_h, \Delta u_h) & p(u_h, s_h; \eta_h, \Delta s_h) \\ p(u_h, s_h; \Delta u_h, \psi_h) & d(u_h, s_h; \psi_h, \Delta s_h) \end{Bmatrix} = -\begin{Bmatrix} \mathcal{R}_h \\ \mathcal{G}_h \end{Bmatrix} \quad (78)$$

where $\mathcal{R}_h = \mathcal{R}(u_h, s_h; \eta_h)$ and $\mathcal{G}_h = \mathcal{G}(u_h, s_h; \psi_h)$. After a standard finite element discretization, the system (78) takes the form

$$\begin{bmatrix} K & P \\ P^T & D \end{bmatrix} \begin{bmatrix} \Delta u \\ \Delta s \end{bmatrix} = -\begin{bmatrix} R \\ G \end{bmatrix}. \quad (79)$$

Table 1 Summary of variational and discrete sensitivity relations

Relation	Variational formulation	Discrete formulation
Physical residual	$\mathcal{R}' = k(\boldsymbol{u}, \boldsymbol{s}; \boldsymbol{\eta}, \delta\boldsymbol{u}) + p(\boldsymbol{u}, \boldsymbol{s}; \boldsymbol{\eta}, \delta\boldsymbol{s})$	$\delta\boldsymbol{R} = \boldsymbol{K}\delta\boldsymbol{u} + \boldsymbol{P}\delta\boldsymbol{s}$
Material residual	$\mathcal{G}' = p(\boldsymbol{u}, \boldsymbol{s}; \delta\boldsymbol{u}, \boldsymbol{\psi}) + d(\boldsymbol{u}, \boldsymbol{s}; \boldsymbol{\psi}, \delta\boldsymbol{s})$	$\delta\boldsymbol{G} = \boldsymbol{P}^T\delta\boldsymbol{u} + \boldsymbol{D}\delta\boldsymbol{s}$
Pseudo load operator	$p(\boldsymbol{u}, \boldsymbol{s}; \boldsymbol{\eta}, \delta\boldsymbol{s})$	$\boldsymbol{\eta}^T\boldsymbol{P}\delta\boldsymbol{s}$
Pseudo load (physical)	$Q(\boldsymbol{u}, \boldsymbol{s}; \boldsymbol{\eta}, \delta\hat{\boldsymbol{s}}) = p(\boldsymbol{u}, \boldsymbol{s}; \boldsymbol{\eta}, \delta\hat{\boldsymbol{s}})$	$\boldsymbol{Q}_p = \boldsymbol{P}\delta\hat{\boldsymbol{s}}$
Pseudo load (material)	$Q(\boldsymbol{u}, \boldsymbol{s}; \delta\hat{\boldsymbol{u}}, \boldsymbol{\psi}) = p(\boldsymbol{u}, \boldsymbol{s}; \delta\hat{\boldsymbol{u}}, \boldsymbol{\psi})$	$\boldsymbol{Q}_m = \boldsymbol{P}^T\delta\hat{\boldsymbol{u}}$
Deformation sensitivity	$k(\boldsymbol{u}, \boldsymbol{s}; \boldsymbol{\eta}, \delta\boldsymbol{u}) = -Q(\boldsymbol{u}, \boldsymbol{s}; \boldsymbol{\eta}, \delta\hat{\boldsymbol{s}})$	$\boldsymbol{K}\delta\boldsymbol{u} = -\boldsymbol{Q}_p$
Design sensitivity	$d(\boldsymbol{u}, \boldsymbol{s}; \boldsymbol{\psi}, \delta\boldsymbol{s}) = -Q(\boldsymbol{u}, \boldsymbol{s}; \delta\hat{\boldsymbol{u}}, \boldsymbol{\psi})$	$\boldsymbol{D}\delta\boldsymbol{s} = -\boldsymbol{Q}_m$

If we substitute the first equation of (79) into the second and vice versa, we can eliminate $\Delta\boldsymbol{u}$ or $\Delta\boldsymbol{s}$ by using the Schur complement in order to solve the problem one after the other and to minimize the size of the system. We obtain a formulation for the state variables

$$[\,\boldsymbol{K} - \boldsymbol{P}\boldsymbol{D}^{-1}\boldsymbol{P}^T\,]\Delta\boldsymbol{u} = \boldsymbol{P}\boldsymbol{D}^{-1}\boldsymbol{G} - \boldsymbol{R} \tag{80}$$

or for the design variables

$$[\,\boldsymbol{D} - \boldsymbol{P}^T\boldsymbol{K}^{-1}\boldsymbol{P}\,]\Delta\boldsymbol{s} = \boldsymbol{P}^T\boldsymbol{K}^{-1}\boldsymbol{R} - \boldsymbol{G}. \tag{81}$$

This requires well-conditioned matrices $\boldsymbol{K}$ and $\boldsymbol{D}$ in order to compute the corresponding inverse matrices accurately and to obtain a stable solution algorithm.

Remark 11 It seems, that the naturally best way to find a solution of the optimization problem (9) is a full Newton method, i.e. the simultaneous solution of the physical and material problem. We expect the best convergence speed and hence, the lowest computational cost. This optimization problem is non-convex in general and for real problems with a large number of design variables such an algorithm is very sensitive and not stable, see also Sect. 7.1. Furthermore, the system (79) has a typical saddle point structure and we have to be careful by solving this system. Often a preconditioner is needed in order to obtain a well-conditioned system matrix. See for instance Benzi and Golub (2004) and the references therein for preconditioners for this type of equations. Therefore, other solution algorithm should be considered, see e.g. Nocedal and Wright (1999) for algorithms used in classical structural optimization.

6.2 Staggered solution method

Alternatively, we propose a staggered solution method, i.e. we solve the physical problem with a fixed design $\hat{\boldsymbol{s}}$ and thereafter the material problem with a fixed deformation $\hat{\boldsymbol{u}}$. For a given solution $\hat{\boldsymbol{u}}$ of the physical residual $\mathcal{R}$, we have to solve $\mathcal{G}(\hat{\boldsymbol{u}}, \boldsymbol{s}; \boldsymbol{\psi}) = 0$ within a Newton algorithm. This requires the total partial derivative (26) of the functional $\mathcal{G}(\boldsymbol{u}, \boldsymbol{s}) = \mathcal{G}(\boldsymbol{u}(\boldsymbol{s}), \boldsymbol{s})$ with respect to $\boldsymbol{s}$, which is given by Eq. 61. With this, the linearization of $\mathcal{G}$ in the direction $\Delta\boldsymbol{s}$ yields

$$\mathcal{G}(\hat{\boldsymbol{u}}, \hat{\boldsymbol{s}}, \boldsymbol{\psi}) + D_s\mathcal{G}(\hat{\boldsymbol{u}}, \hat{\boldsymbol{s}}, \boldsymbol{\psi}) \cdot \Delta\boldsymbol{s} + \mathcal{O} = 0 \tag{82}$$

with the tangent operator

$$\begin{aligned} D_s\mathcal{G}(\boldsymbol{u}, \boldsymbol{s}, \boldsymbol{\psi}) \cdot \Delta\boldsymbol{s} &= b_T(\boldsymbol{u}(\boldsymbol{s}), \boldsymbol{s}; \boldsymbol{\psi}, \Delta\boldsymbol{s}) \\ &= \frac{d}{d\varepsilon}\Big[\,\mathcal{G}(\boldsymbol{u}, \boldsymbol{s} + \varepsilon\Delta\boldsymbol{s}; \boldsymbol{\psi}) \\ &\quad + \mathcal{G}(\boldsymbol{u}(\boldsymbol{s} + \varepsilon\Delta\boldsymbol{s}), \boldsymbol{s}; \boldsymbol{\psi})\Big]\Big|_{\varepsilon=0}. \end{aligned} \tag{83}$$

The remainder $\mathcal{O}$ can usually be neglected and we obtain a linear equation for $\Delta\boldsymbol{s}$ in the form

$$b_T(\hat{\boldsymbol{u}}, \hat{\boldsymbol{s}}; \boldsymbol{\psi}, \Delta\boldsymbol{s}) = -\mathcal{G}(\hat{\boldsymbol{u}}, \hat{\boldsymbol{s}}, \boldsymbol{\psi}) \qquad \forall\, \boldsymbol{\psi} \in \mathcal{D}. \tag{84}$$

The finite element approximation reads

$$b_T(\hat{\boldsymbol{u}}_h, \hat{\boldsymbol{s}}_h; \boldsymbol{\psi}_h, \Delta\boldsymbol{s}_h) = -\mathcal{G}(\hat{\boldsymbol{u}}_h, \hat{\boldsymbol{s}}_h, \boldsymbol{\psi}_h) \quad \forall\, \boldsymbol{\psi}_h \in \mathcal{D}_h. \tag{85}$$

This is equivalent to Eq. 63, in which we have replaced the error in the design $\boldsymbol{e}_{s,h}$ by the design increment $\Delta\boldsymbol{s}_h$.

Remark 12 In Sect. 5.2 we have mentioned, that the material residual is not fulfilled for every $\boldsymbol{u}_h \neq \boldsymbol{u}_h^* = \boldsymbol{u}_h(\boldsymbol{s}_h^*)$, i.e. $\mathcal{G}(\boldsymbol{u}_h, \boldsymbol{s}_h; \boldsymbol{\psi}_h) \neq 0$, $\forall\, \boldsymbol{u}_h \neq \boldsymbol{u}_h^* \in V_h$ and hence, for a given solution $\boldsymbol{u}_h$ of the physical problem, we have to solve the material problem with a Newton algorithm in order to find a $\boldsymbol{s}_h^*$ such that $\boldsymbol{u}_h(\boldsymbol{s}_h^*) = \boldsymbol{u}_h^*$ and hence $\mathcal{G}(\boldsymbol{u}_h^*, \boldsymbol{s}_h^*; \boldsymbol{\psi}_h) = 0$.

In the discrete case follows from Eq. 44 $\partial_u\boldsymbol{G} = \boldsymbol{P}^T$ and with Eq. 64 $\partial_s\boldsymbol{u} = -\boldsymbol{K}^{-1}\boldsymbol{P}$ and hence a matrix description

$$b_T(\hat{u}_h, \hat{s}_h; \psi_h, \Delta s_h) = \psi^T \left[\frac{\partial G}{\partial s} + \frac{\partial G}{\partial u} \frac{\partial u}{\partial s} \right] \Delta s$$
$$= \psi^T \left[D - P^T K^{-1} P \right] \Delta s. \tag{86}$$

Finally, we have to solve the discrete version of (84), i.e.

$$\left[D - P^T K^{-1} P \right] \Delta s = -G \tag{87}$$

in every Newton step. With the definition of the sensitivity operator S_p of the physical problem (68) this relation could be expressed as

$$\left[D + P^T S_p \right] \Delta s = -G. \tag{88}$$

Remark 13 Additional we have to fulfil in every step the constraint $A(u_h, s_h; \eta_h) = 0$, which means that we have to solve the equilibrium equation for the primal problem. Moreover, the computation of the accurate material residual $\mathcal{G}(\hat{u}_h, \hat{s}_h; \psi_h)$ for the next iteration step requires the solution of the state variable u_h for the updated current design, i.e. we have to solve $\mathcal{R}(u_h, s_h; \eta_h) = 0$. For a simple linear physical problem, only the solution of one linear equation is required. In the general nonlinear case, we have to perform some Newton iterations for the physical problem in order to find the new state of equilibrium for the current design, i.e. we have to solve the tangent version of the primal problem

$$k(\hat{u}_h, \hat{s}_h; \eta_h, \Delta u_h) = -\mathcal{R}(\hat{u}_h, \hat{s}_h; \eta_h) \tag{89}$$

in every Newton step. The best performance of the algorithm could be achieved, if we transfer beforehand the data of the state u_h to the new design. To do this, we can use the sensitivity relation for the physical problem (68) and obtain with a sufficient small Δs an approximation for the change in the state from

$$\Delta u = -K^{-1} P \, \Delta s = S_p \, \Delta s. \tag{90}$$

After this, the system is usually still unbalanced, but only few Newton iterations are required in order to find the new state of equilibrium.

Remark 14 In a classical staggered solution algorithm, the coupling terms P in (79) are usually simply neglected. As a result of the functional dependencies $\mathcal{G}(u, s) = \mathcal{G}(u(s), s)$ of the material residual, we have nevertheless the coupling terms in the tangent operator (83) for the solution of the material motion problem.

6.3 Steepest descent method

The simplest way to solve the optimization problem is the well-known steepest descent method. We rewrite the objective function to $\mathcal{J}(s) = \mathcal{J}(u(s), s)$.

For a given starting point s_0 and $u(s_0)$ we calculate a search direction $v := -\mathcal{J}'_s$ and perform a line search with $\varepsilon > 0$ such that $\mathcal{J}(s + \varepsilon v) < \mathcal{J}(s)$. This leads to the solution algorithm

$$s_{i+1} = s_i + \varepsilon_i \, v_i = s_i - \varepsilon_i \, \mathcal{J}'_s(u, s_i). \tag{91}$$

If we choose the energy of the system (8) as the objective functional, i.e. $\mathcal{J} = E$, we have in the discrete case

$$s_{i+1} = s_i + \Delta s = s_i - \varepsilon_i \, G(u_h, s_{h,i}) \tag{92}$$

where $G(u_h, s_{h,i}) \in \mathbb{R}^m$ is the material residual vector associated to the functional $\mathcal{G}(u_h, s_h; \cdot)$ at the current state $u_h(s_{h,i})$. This is a pure gradient based method and it is not necessary to calculate the second derivatives of the objective functional, but a large number of iterations and functional evaluations are required.

In order to update the state variables u_h during the iteration, we can use the same procedure as mentioned in Remark 13.

Remark 15 The steepest descent directions at two consecutive steps are orthogonal to each other if the line search is based on the minimization problem

$$\mathcal{J}(s_i + \varepsilon \, v_i) = \mathcal{J}(s_i - \varepsilon \, \mathcal{J}'_s(s_i)) \rightarrow \min_{\varepsilon} \tag{93}$$

in order to find a step size parameter ε_i. Furthermore, often the steepest descent exhibit 'zigzag' behavior and the solution oscillates, see e.g. Nocedal and Wright (1999) for details. This tends to slow down steepest descent method although it is convergent and hence a large number of iterations are required in order to find the solution. A simple and efficient improvement is the *conjugate gradient method*. This method is also based only on gradient information of the objective functional. The new design is given by $s_{i+1} = s_i + \varepsilon_i \, v_i$ with the conjugate direction

$$v_i := -\left[G(s_i) - \beta_i \, v_{i-1} \right]. \tag{94}$$

The scale factor β_i is determined for instance by the Fletcher-Reeves or Polak-Ribiere method, see Table 2. The conjugate gradient method should be preferred over the steepest descent method, because the rate of convergence could be improved without appreciable cost.

Table 2 Summary of discrete solution algorithms

Algorithm	Discrete formulation
Full Newton	$\begin{bmatrix} \boldsymbol{K} & \boldsymbol{P} \\ \boldsymbol{P}^T & \boldsymbol{D} \end{bmatrix} \begin{bmatrix} \Delta \boldsymbol{u} \\ \Delta \boldsymbol{s} \end{bmatrix} = - \begin{bmatrix} \boldsymbol{R} \\ \boldsymbol{G} \end{bmatrix}$
Staggered Newton	$\left[\boldsymbol{D} - \boldsymbol{P}^T \boldsymbol{K}^{-1} \boldsymbol{P} \right] \Delta \boldsymbol{s} = -\boldsymbol{G}$
Steepest descent method	$\Delta \boldsymbol{s} = -\varepsilon_i \, \boldsymbol{G}(\boldsymbol{u}_h, \boldsymbol{s}_{h,i})$
Conjugate gradient method	$\Delta \boldsymbol{s} = \varepsilon_i \, \boldsymbol{v}_i$ with $\boldsymbol{v}_i = - \left[\boldsymbol{G}(\boldsymbol{s}_i) - \beta_i \, \boldsymbol{v}_{i-1} \right]$
	$\beta_i = \dfrac{\boldsymbol{G}(\boldsymbol{s}_i)^T \boldsymbol{G}(\boldsymbol{s}_i)}{\boldsymbol{G}(\boldsymbol{s}_{i-1})^T \boldsymbol{G}(\boldsymbol{s}_{i-1})}$ (Fletcher-Reeves method)
	$\beta_i = \dfrac{\boldsymbol{G}(\boldsymbol{s}_i)^T \left[\boldsymbol{G}(\boldsymbol{s}_i) - \boldsymbol{G}(\boldsymbol{s}_{i-1}) \right]}{\boldsymbol{G}(\boldsymbol{s}_{i-1})^T \boldsymbol{G}(\boldsymbol{s}_{i-1})}$ (Polak-Ribiere method)

In Table 2 we have summarized the discrete solution algorithms for the optimization problem (53). These algorithms could be used in this pure form or could be enhanced by different modifications, see for instance Nocedal and Wright (1999).

7 Applications

7.1 Mesh optimization

An application of the proposed framework is the mesh optimization problem. In the case of mesh optimization the nodal coordinates $\boldsymbol{X}$ are chosen as design variables, i.e. $\boldsymbol{s} = \boldsymbol{X}$. All inner nodes or a subset of nodes are chosen to build up the design space $\mathcal{D}_h$. Additional we can include nodal coordinates in tangential direction on the boundary. Finally, it follows a r-adaption algorithm for mesh optimization.

The optimal nodal positions (the design) of a given mesh (the reference configuration) can be computed by using the solution algorithms which are listed in Table 2. For the full Newton method, Eq. 79, follows

$$\begin{bmatrix} \boldsymbol{K} & \boldsymbol{P} \\ \boldsymbol{P}^T & \boldsymbol{D} \end{bmatrix} \begin{bmatrix} \Delta \boldsymbol{u} \\ \Delta \boldsymbol{X} \end{bmatrix} = - \begin{bmatrix} \boldsymbol{R} \\ \boldsymbol{G} \end{bmatrix}. \tag{95}$$

The proposed staggered solution algorithm (87) yields the form

$$\left[\boldsymbol{D} - \boldsymbol{P}^T \boldsymbol{K}^{-1} \boldsymbol{P} \right] \Delta \boldsymbol{X} = -\boldsymbol{G} \tag{96}$$

and the steepest descent method (92) for mesh optimization follows in the form

$$\boldsymbol{X}_{i+1} = \boldsymbol{X}_i - \varepsilon \, \boldsymbol{G}(\boldsymbol{u}_h, \boldsymbol{X}_i). \tag{97}$$

Remark 16 The mesh optimization problem can be interpreted as an inverse problem. It requires the solution of the material residual problem (96) or the coupled system (95). Different numerical difficulties arise due to the solution of this problem. In many cases, the Hessian matrix of the system is ill-conditioned and becomes singular or close to singular during the iterations and therefore the Newton algorithm is not stable and fails. The optimization problem is non-convex in general and hence, the solution needs not to be unique. Therefore, the problem could be termed ill-posed in the sense of Hadamard and reasonable regularization methods should be used in order to regularize the problem, see for instance Engl et al. (2000) and the references therein. The numerical difficulties are also mentioned by Askes et al. (2004) and Mosler and Ortiz (2006). The authors have proposed different strategies in order to overcome these problems.

Remark 17 There exist some nodal coordinates (design variables), which do not contribute to a change in the energy during the mesh optimization process, even though the corresponding material residual is not zero. Hence, these nodal coordinates should not be considered. The sensitivity of the material residual

$$\delta \boldsymbol{G} = \left[\boldsymbol{D} + \boldsymbol{P}^T \boldsymbol{S}_p \right] \delta \boldsymbol{X} \tag{98}$$

could be used in order to find out the most relevant nodal coordinates (design variables).

The regularization and the identification of the most relevant design variables are discussed in more detail in Materna and Barthold (2007a, b).

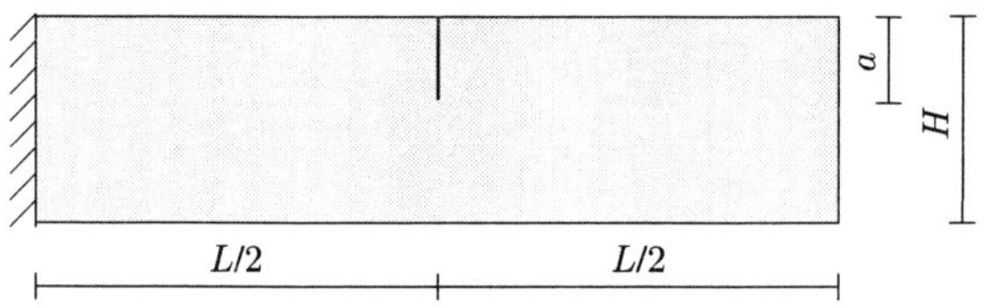

Fig. 1 Cantilever beam

Example: cantilever beam

We consider a cracked cantilever beam under self-weight loading $b_y = -10$, see Fig. 1. The dimensions of the rectangular domain are $L = 4$ and $H = 1$ as well as the crack length is $a = 0.4$. The considered model problem is given in Appendix B. We use a strain energy function of a classical compressible Neo-Hooke material given in Eq. 131. The Lamé parameters are chosen as $\lambda = 5.769 \cdot 10^3$ and $\mu = 3.846 \cdot 10^3$, which correspond to $E = 10^4$ and $\nu = 0.3$.

All inner nodes as well as all nodal coordinates in tangential direction on the free boundary are chosen as design variables. Only the nodal coordinates on the Dirichlet boundary are fixed. We use the staggered solution algorithm Eq. 96 in order to solve the problem.

Different numerical difficulties arise due to the solution of this problem, see Remark 16. We use a regularization based on a *singular value decomposition* (SVD) of the Hessian matrix. From this decomposition, we can compute a rank-r approximation to the original Hessian in order to overcome the ill-conditioning, see Materna and Barthold (2007a, b) for more details.

A second problem is the mesh distortion during the optimization progress. The quality of the mesh (the distortion of the mesh) has an important influence on the shape derivatives and hence on the results and the success of the optimization process. We use a simple geometrical distortion parameter ξ, which is used in shape optimization (Zhang and Belegundu 1993). The distortion parameter for a quadrilateral element with four nodes is given by

$$\xi = \frac{4}{A} \min(\det \boldsymbol{J}_i) \qquad i = 1, \ldots, 4, \tag{99}$$

where $\min(\det \boldsymbol{J}_i)$ denotes the minimum value of the Jacobian determinant and A the element area. The parameter ξ must be greater than zero to avoid degeneracy of the element. In our computations we use the condition $\xi \geq tol_\xi$ with a tolerance $tol_\xi = 0.1$.

The solution was attained within 16 iterations, see Fig. 2. The overall energy decreases from -3.23597 (initial mesh) to -3.33385 (optimized mesh). The norm of the material residual $||\boldsymbol{G}||_{L_2}$ decreases from $2.8793 \cdot 10^{-1}$ to $8.7777 \cdot 10^{-2}$, i.e. a reduction of 69.51 % with respect to the initial mesh. The material residual does not vanish completely as a result of the mesh distortion control. A remeshing strategy with a change of the nodal connectivities as well as h-adaptivity techniques for patches with distorted elements should be considered. Furthermore, other mesh distortion control criteria, e.g. energy based, could be used to improve the algorithm.

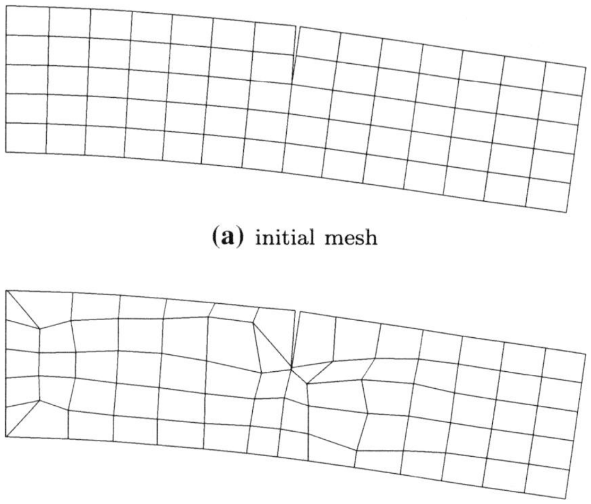

(a) initial mesh

(b) optimized mesh

Fig. 2 Cantilever beam: deformed initial and optimized mesh

In order to quantify the capability of the optimized mesh, we control as a quantity of interest the vertical displacement at the lower right corner. We use a reference solution $u_y^* = -0.440225$ obtained from a fine mesh with 4,258 nodes. The vertical displacement of the initial mesh is $u_y^0 = -0.406161$. The relative error with respect to u_y^* is given by 7.74 % For the optimized mesh follows the displacement $u_y = -0.420303$ and hence a relative error of 4.53 %. Finally, we obtain a reduction of 41.52 % in the relative error for the displacement at the lower right corner with respect to the reference solution.

7.2 Shape optimization

In order to highlight the relations between shape optimization and configurational mechanics we choose the strain energy or internal energy as objective functional. This is a measure for the mean compliance of the

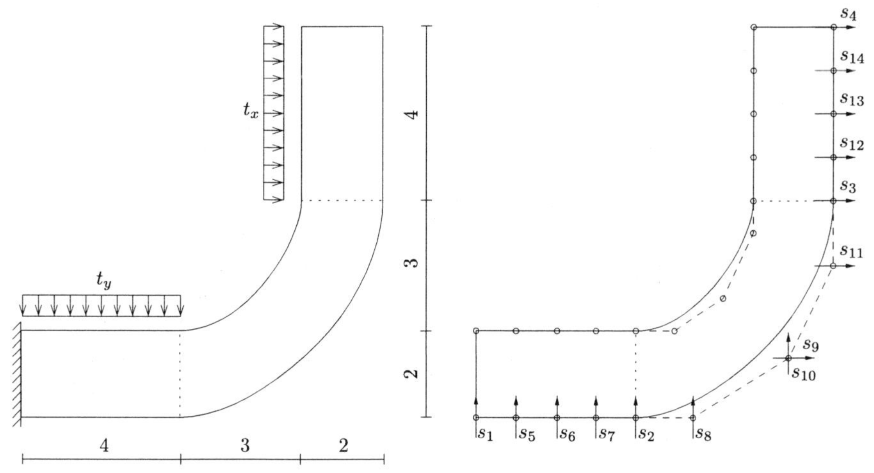

Fig. 3 L-shape: system (left) and geometry model with 14 design variables s_i (right). (**a**) initial mesh; (**b**) optimized mesh

structure. The minimization of the compliance is equivalent to the maximization of the stiffness. Let

$$C := \int_{\Omega_R} W_R \, d\Omega \tag{100}$$

be the strain energy of the structure. We are looking for the optimal design for which the structure attains a minimum of internal elastic energy among the structures of constant volume V_0 or material cost. We use a staggered solution algorithm and we formulate the optimization problem only in terms of the design. This ends in the following problem.

Problem 7 Find $s \in \mathcal{D}$ of the objective functional $\mathcal{J} : \mathcal{D} \to \mathbb{R}$ such that the strain energy

$$\mathcal{J}(s) = \mathcal{J}(\boldsymbol{u}(s), s) = C \to \min_{s \in \mathcal{D}} \tag{101}$$

subject to the constraints

$$\mathcal{R}(\boldsymbol{u}, s; \eta) = 0 \tag{102}$$

$$V - V_0 = 0. \tag{103}$$

Example: L-shape

We consider as an example from shape optimization a L-shaped cantilever problem, see Fig. 3. The optimization task is to generate the most efficient material distribution with respect to the overall stiffness of the structure. The nonlinear programming problem under consideration consists of the objective function (internal elastic energy), the constraint function (constant volume) and the geometrical design variables, i.e. we solve Problem 7.

The L-shape consists of three geometry patches as indicated in Fig. 3. The boundaries of the patches are modeled using Bézier curves each with three internal control points. We consider only the lower and right boundary of the L-shape as design boundary and keep the loaded boundary fixed. Including the corner points, overall 14 coordinates of the control points are design variables s_i. The L-shape is modeled with the plane strain condition with $E = 10^5$ and $\nu = 0.3$ and loaded by line loads $t_x = t_y = 5$.

The optimum was attained within 37 iterations. The initial and final solutions as well as a selected iteration are shown in Fig. 4. The optimal shape leads to a uniform distribution of the material residual $\boldsymbol{G}(\boldsymbol{X})$ on the design boundary. The norm of the material residual $\boldsymbol{G}(s)$ on the design variables decreases from 0.4009 to 0.1911, i.e. a reduction of 52 %. The material residual on the boundary does not vanish, because there is still an ambition to find a state with lower internal energy, i.e. a more stiffer structure. The side constraints for the design variables and the volume constraint avoid this movement.

The design variables s_2 and s_3 connect two geometry patches, respectively. Therefore, they have a large influence domain, i.e. small variations result in large changes in the geometry and hence in the internal energy. This is indicated by large material residuals $\boldsymbol{G}(s_i)$ on these design variables, see Fig. 4.

The internal energy decreases from 0.9015 to 0.5409, i.e. a reduction of 40 %. The distribution of the internal energy for the initial and optimized design is shown in Fig. 5. The energy distribution of the optimized shape is more smooth in comparison with the initial shape.

Remark 18 It is interesting to note, that the material residual $\mathcal{G}$ on the design boundary is the indicator in

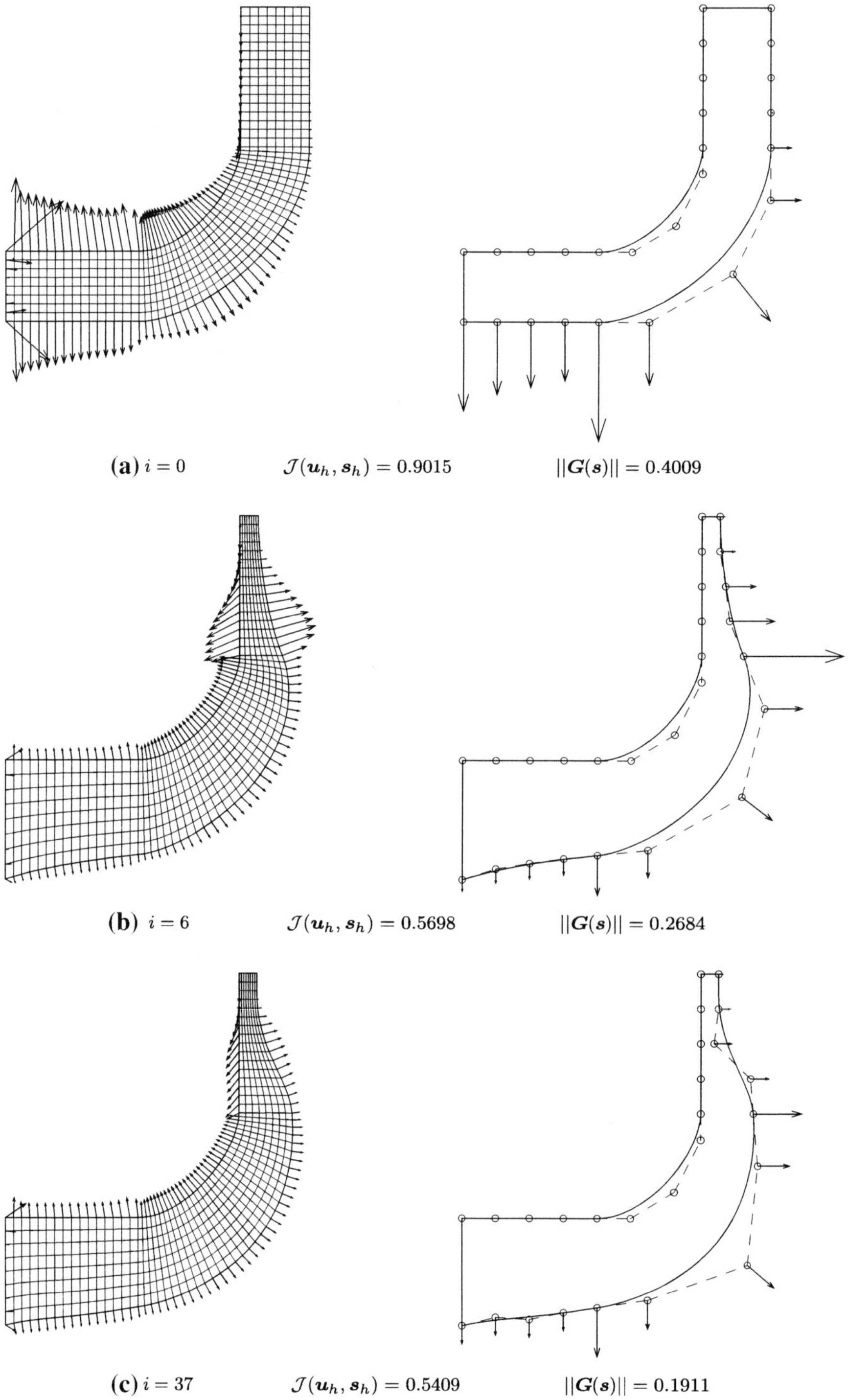

Fig. 4 Distribution of the material residuals on the mesh nodes $G(X)$ (left side) and design variables $G(s)$ (control points of the Bézier curves) (right side) during the optimization progress, (**a**) initial shape (**c**) final shape

(**a**) $i = 0$ $\mathcal{J}(\boldsymbol{u}_h, \boldsymbol{s}_h) = 0.9015$ $||\boldsymbol{G}(\boldsymbol{s})|| = 0.4009$

(**b**) $i = 6$ $\mathcal{J}(\boldsymbol{u}_h, \boldsymbol{s}_h) = 0.5698$ $||\boldsymbol{G}(\boldsymbol{s})|| = 0.2684$

(**c**) $i = 37$ $\mathcal{J}(\boldsymbol{u}_h, \boldsymbol{s}_h) = 0.5409$ $||\boldsymbol{G}(\boldsymbol{s})|| = 0.1911$

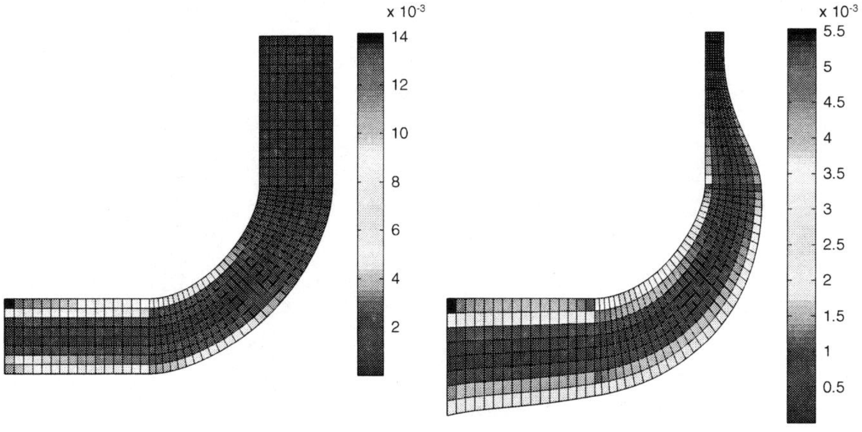

(a) initial design **(b)** optimized design

Fig. 5 Distribution of the internal energy. (**a**) initial design; (**b**) optimized design

which direction the boundary has to move in order to minimize the internal energy. The optimal shape is obtained, if the material residual is uniformly distributed over the design boundary, see Fig. 4c.

The mesh nodal coordinates $\boldsymbol{X}$ are connected with the geometrical design variables s_i by the so-called *design velocity fields* $\boldsymbol{V}_i$, see e.g. Choi and Kim (2005a, b). The design velocity field corresponding to the design variable s_i is given by

$$\boldsymbol{V}_i(\boldsymbol{X}) = \frac{\partial \boldsymbol{X}}{\partial s_i}. \tag{104}$$

The design velocity field $\boldsymbol{V}_i(\boldsymbol{X})$ characterize the changes of the finite element nodal point coordinates $\boldsymbol{X}$ with respect to the changes of arbitrary design parameter s_i.

For example, a variation of the design variables s_2 and s_{11} generate the fields shown in Fig. 6a and b, respectively. Hence, these fields reflect the influence domains of the design variations.

The velocity fields are also important and fundamental in the context of mesh updating and smoothing. Let $\boldsymbol{X}$ be the vector of nodal coordinates, than the new shape is obtained from

$$\boldsymbol{X}(\boldsymbol{s}) = \boldsymbol{X}(\boldsymbol{s}) + \varepsilon \Delta \boldsymbol{X}(\boldsymbol{s}) \tag{105}$$

with

$$\Delta \boldsymbol{X}(\boldsymbol{s}) = \frac{\partial \boldsymbol{X}}{\partial \boldsymbol{s}} \Delta \boldsymbol{s} = \sum_{i=1}^{m} \frac{\partial \boldsymbol{X}}{\partial s_i} \Delta s_i = \sum_{i=1}^{m} \boldsymbol{V}_i \, \Delta s_i. \tag{106}$$

Here, ε is a step size parameter, which controls the decrease in the objective and the mesh distortion as well as $\Delta \boldsymbol{s}$ is the increment of the design variables obtained from the solution at the current iteration.

Furthermore, the sensitivity of the energy with respect to a variation of the design variable s_i is given by

$$E'_{s_i} = \boldsymbol{G} \cdot \delta \boldsymbol{X}_i = \boldsymbol{G} \cdot \frac{\partial \boldsymbol{X}}{\partial s_i} \delta s_i = \boldsymbol{G} \cdot \boldsymbol{V}_i \, \delta s_i. \tag{107}$$

The design velocity field connects the variation in the geometrical design variable s_i with the variation in the energy.

7.3 Sensitivity of the energy release rate

For a homogeneous elastic body, the material residual or configurational forces are directly related to the well-known J-integral, which is in linear fracture mechanics equal to the energy release rate G, i.e.

$$J = G = -\mathcal{G}(\boldsymbol{u}, \boldsymbol{s}; \boldsymbol{\psi}). \tag{108}$$

Therefore, the sensitivity of the material residual (48) can be used to calculate the sensitivity of the J-integral or energy release rate G, which can be derived as

$$G' = -\mathcal{G}' = -(d(\boldsymbol{u}, \boldsymbol{s}; \boldsymbol{\psi}, \delta \boldsymbol{s}) + p(\boldsymbol{u}, \boldsymbol{s}; \delta \boldsymbol{u}, \boldsymbol{\psi})). \tag{109}$$

In order to obtain a dependency only from the variation in the design $\delta \boldsymbol{s}$, we can substitute the sensitivity of the deformation (36) to eliminate $\delta \boldsymbol{u}$.

With the notations from (64) and (65) the discrete approximation follows in the form

$$G'_h = -\mathcal{G}'(\boldsymbol{u}_h, \boldsymbol{s}_h; \boldsymbol{\psi}_h, \delta \boldsymbol{u}_h, \delta \boldsymbol{s}_h) = -\boldsymbol{\psi}^T \delta \boldsymbol{G} \tag{110}$$

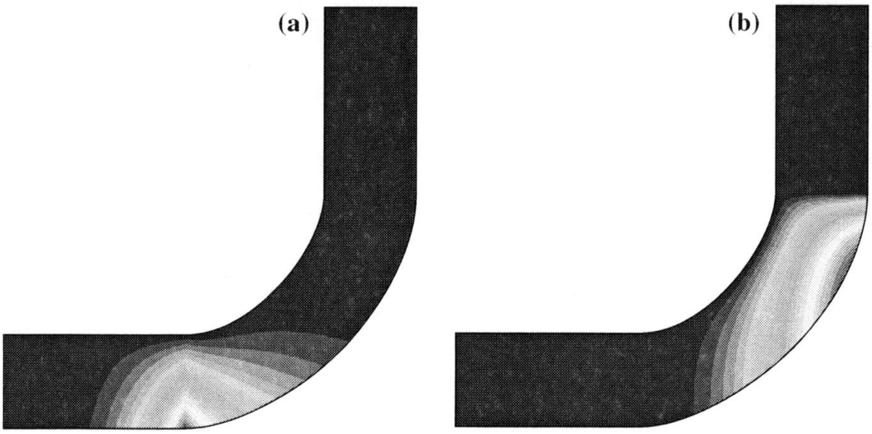

Fig. 6 Design velocity fields for s_2 (**a**) and s_{11} (**b**)

where

$$\begin{aligned}\delta \boldsymbol{G} &= \boldsymbol{D}\delta \boldsymbol{s} + \boldsymbol{P}^T \delta \boldsymbol{u} = \boldsymbol{D}\delta \boldsymbol{s} - \boldsymbol{P}^T \boldsymbol{K}^{-1} \boldsymbol{P}\, \delta \boldsymbol{s} \\ &= [\, \boldsymbol{D} - \boldsymbol{P}^T \boldsymbol{K}^{-1} \boldsymbol{P}\,]\, \delta \boldsymbol{s} \end{aligned} \tag{111}$$

is the variation of the material residual. With the sensitivity operator $\boldsymbol{S}_p$ of the physical problem (68) we obtain

$$\delta \boldsymbol{G} = \left[\, \boldsymbol{D} + \boldsymbol{P}^T \boldsymbol{S}_p \,\right] \delta \boldsymbol{s}. \tag{112}$$

Hence, for a given variation in the design $\delta \boldsymbol{s}$ we can calculate the variation in the energy release rate or rather the variation in the material or configurational residual $\boldsymbol{G}$.

An application for the sensitivity of the energy release rate of a problem from fracture mechanics is given in Barthold and Mesecke (1999).

8 Conclusions

In the present paper, we have investigated some aspects of variational design sensitivity analysis in the context of structural optimization and configurational mechanics. In both disciplines, we are interested in variations of the material body. We obtain the relations from configurational mechanics by using techniques from variational design sensitivity analysis applied to the energy functional of the problem. Furthermore, we have obtained sensitivity relations for the physical and material problem. Due to symmetry both sensitivity relations contain the pseudo load operator, which play an important role for the solution of classical structural optimization problems. The physical and material problem are coupled by this operator. The obtained sensitivity relations of the physical and material problem could be used for the optimization of finite element meshes, shape optimization or to perform a sensitivity analysis of the energy release rate.

Appendix A: Relations for variations in the physical and material space

The calculation of the sensitivities require the variations of the field quantities with respect to $\boldsymbol{u} \in \mathcal{V}$ and $\boldsymbol{s} \in \mathcal{D}$. Let $\boldsymbol{\eta},\ \boldsymbol{v} \in \mathcal{V}$ be admissible variations for the deformation and $\boldsymbol{\psi},\ \boldsymbol{\chi} \in \mathcal{D}$ be admissible design variations. Some important relations for the variations in the physical and material space are listed below.

A.1 Variations of strains

For the displacement gradient $\boldsymbol{H} = \operatorname{Grad} \boldsymbol{u}$ we have

$$\boldsymbol{H}'_u(\boldsymbol{\eta}) = \operatorname{Grad} \boldsymbol{\eta} \tag{113}$$

$$\boldsymbol{H}'_s(\boldsymbol{u}, \boldsymbol{\psi}) = -\operatorname{Grad} \boldsymbol{u}\ \operatorname{Grad} \boldsymbol{\psi} = -\boldsymbol{H}\ \operatorname{Grad} \boldsymbol{\psi}. \tag{114}$$

The variation of the deformation gradient $\boldsymbol{F} = \operatorname{Grad} \boldsymbol{\varphi}$ reads

$$\boldsymbol{F}'_u(\boldsymbol{\eta}) = \operatorname{Grad} \boldsymbol{\eta} \tag{115}$$

$$\boldsymbol{F}'_s(\boldsymbol{u}, \boldsymbol{\psi}) = -\operatorname{Grad} \boldsymbol{\varphi}\ \operatorname{Grad} \boldsymbol{\psi} = -\boldsymbol{F}\ \operatorname{Grad} \boldsymbol{\psi}. \tag{116}$$

Therefore, the variations of the Green-Lagrange strain tensor $\boldsymbol{E} = \frac{1}{2}\,(\boldsymbol{F}^T \boldsymbol{F} - \boldsymbol{1})$ follow from the above definitions in the form

$$\begin{aligned}\boldsymbol{E}'_u(\boldsymbol{u}, \boldsymbol{\eta}) &= \frac{1}{2}\left(\operatorname{Grad} \boldsymbol{\eta}^T \boldsymbol{F} + \boldsymbol{F}^T \operatorname{Grad} \boldsymbol{\eta}\right) \\ &= \operatorname{sym}\{\boldsymbol{F}^T \operatorname{Grad} \boldsymbol{\eta}\} \end{aligned} \tag{117}$$

$$
\begin{aligned}
\boldsymbol{E}'_s(\boldsymbol{u}, \boldsymbol{\psi}) &= -\frac{1}{2}\left((\boldsymbol{F} \text{ Grad } \boldsymbol{\psi})^T \boldsymbol{F} + \boldsymbol{F}^T \boldsymbol{F} \text{ Grad } \boldsymbol{\psi}\right) \\
&= -\frac{1}{2}\left(\text{Grad } \boldsymbol{\psi}^T \boldsymbol{F}^T \boldsymbol{F} + \boldsymbol{F}^T \boldsymbol{F} \text{ Grad } \boldsymbol{\psi}\right) \\
&= -\operatorname{sym}\{\boldsymbol{F}^T \boldsymbol{F} \text{ Grad } \boldsymbol{\psi}\}. \qquad (118)
\end{aligned}
$$

Furthermore, the second and mixed variations follow in a straightforward manner for variations $\boldsymbol{v} \in \mathcal{V}$ and $\boldsymbol{\chi} \in \mathcal{D}$

$$
\begin{aligned}
\boldsymbol{E}''_{uu}(\boldsymbol{\eta}, \boldsymbol{v}) &= \frac{1}{2}\left(\text{Grad } \boldsymbol{\eta}^T \text{ Grad } \boldsymbol{v} + \text{Grad } \boldsymbol{v}^T \text{ Grad } \boldsymbol{\eta}\right) \\
&= \operatorname{sym}\{\text{Grad } \boldsymbol{v}^T \text{ Grad } \boldsymbol{\eta}\} \qquad (119)
\end{aligned}
$$

$$
\begin{aligned}
\boldsymbol{E}''_{us}(\boldsymbol{u}, \boldsymbol{\eta}, \boldsymbol{\psi}) = -\frac{1}{2}\Big((&\text{Grad } \boldsymbol{\eta} \text{ Grad } \boldsymbol{\psi})^T \boldsymbol{F} \\
&+ \text{Grad } \boldsymbol{\eta}^T \boldsymbol{F} \text{ Grad } \boldsymbol{\psi} \\
&+ (\boldsymbol{F} \text{ Grad } \boldsymbol{\psi})^T \text{ Grad } \boldsymbol{\eta} \\
&+ \boldsymbol{F}^T \text{ Grad } \boldsymbol{\eta} \text{ Grad } \boldsymbol{\psi}\Big) \\
= -\operatorname{sym}\{&\text{Grad } \boldsymbol{\psi}^T \boldsymbol{F}^T \text{ Grad } \boldsymbol{\eta} \\
&+ \boldsymbol{F}^T \text{ Grad } \boldsymbol{\eta} \text{ Grad } \boldsymbol{\psi}\}. \qquad (120)
\end{aligned}
$$

$$
\begin{aligned}
\boldsymbol{E}''_{ss}(\boldsymbol{\eta}, \boldsymbol{\psi}, \boldsymbol{\chi}) = \frac{1}{2}\Big((&\text{Grad } \boldsymbol{\psi} \text{ Grad } \boldsymbol{\chi})^T \boldsymbol{F}^T \boldsymbol{F} \\
&+ \text{Grad } \boldsymbol{\psi}^T (\boldsymbol{F} \text{ Grad } \boldsymbol{\chi})^T \boldsymbol{F} \\
&+ \text{Grad } \boldsymbol{\psi}^T \boldsymbol{F}^T \boldsymbol{F} \text{ Grad } \boldsymbol{\chi} \\
&+ (\boldsymbol{F} \text{ Grad } \boldsymbol{\chi})^T \boldsymbol{F} \text{ Grad } \boldsymbol{\psi} \\
&+ \boldsymbol{F}^T \boldsymbol{F} \text{ Grad } \boldsymbol{\chi} \text{ Grad } \boldsymbol{\psi} \\
&+ \boldsymbol{F}^T \boldsymbol{F} \text{ Grad } \boldsymbol{\psi} \text{ Grad } \boldsymbol{\chi}\Big) \\
= \operatorname{sym}\{&\text{Grad } \boldsymbol{\chi}^T \boldsymbol{F}^T \boldsymbol{F} \text{ Grad } \boldsymbol{\psi} \\
&+ \boldsymbol{F}^T \boldsymbol{F} \text{ Grad } \boldsymbol{\chi} \text{ Grad } \boldsymbol{\psi} \\
&+ \boldsymbol{F}^T \boldsymbol{F} \text{ Grad } \boldsymbol{\psi} \text{ Grad } \boldsymbol{\chi}\} \qquad (121)
\end{aligned}
$$

$$
\begin{aligned}
\boldsymbol{E}''_{su}(\boldsymbol{u}, \boldsymbol{\psi}, \boldsymbol{\eta}) = -\frac{1}{2}\Big((&\text{Grad } \boldsymbol{\eta} \text{ Grad } \boldsymbol{\psi})^T \boldsymbol{F} \\
&+ \text{Grad } \boldsymbol{\psi}^T \boldsymbol{F}^T \text{ Grad } \boldsymbol{\eta} \\
&+ \text{Grad } \boldsymbol{\eta}^T \boldsymbol{F} \text{ Grad } \boldsymbol{\psi} \\
&+ \boldsymbol{F}^T \text{ Grad } \boldsymbol{\eta} \text{ Grad } \boldsymbol{\psi}\Big) \\
= -\operatorname{sym}\{&\text{Grad } \boldsymbol{\psi}^T \boldsymbol{F}^T \text{ Grad } \boldsymbol{\eta} \\
&+ \boldsymbol{F}^T \text{ Grad } \boldsymbol{\eta} \text{ Grad } \boldsymbol{\psi}\}. \qquad (122)
\end{aligned}
$$

Hence, as a result of symmetry follows

$$
\boldsymbol{E}''_{su}(\boldsymbol{u}, \boldsymbol{\psi}, \boldsymbol{\eta}) = \boldsymbol{E}''_{us}(\boldsymbol{u}, \boldsymbol{\eta}, \boldsymbol{\psi}). \qquad (123)
$$

A.2 Variations of stresses

We consider a hyperelastic material, i.e. there exist a free energy function Ψ such that

$$
\boldsymbol{P} = \rho_R \frac{\partial \Psi}{\partial \boldsymbol{F}} \quad \text{and} \quad \boldsymbol{S} = \rho_R \frac{\partial \Psi}{\partial \boldsymbol{E}}. \qquad (124)
$$

The variations of the first Piola-Kirchhoff stress tensor $\boldsymbol{P}$ read

$$
\begin{aligned}
\boldsymbol{P}'_u(\boldsymbol{u}, \boldsymbol{\eta}) &= \mathbb{A} : \boldsymbol{F}'_u = \mathbb{A} : \text{Grad } \boldsymbol{\eta} \qquad (125) \\
\boldsymbol{P}'_s(\boldsymbol{u}, \boldsymbol{\psi}) &= \mathbb{A} : \boldsymbol{F}'_s = -\mathbb{A} : \boldsymbol{F} \text{ Grad } \boldsymbol{\psi} \qquad (126)
\end{aligned}
$$

where

$$
\mathbb{A} = \frac{\partial \boldsymbol{P}}{\partial \boldsymbol{F}} = \rho_R \frac{\partial^2 \Psi}{\partial \boldsymbol{F}\, \partial \boldsymbol{F}} \qquad (127)
$$

is called the first elasticity tensor (Marsden and Hughes 1994). The variations of the second Piola-Kirchhoff stress tensor $\boldsymbol{S}$ read

$$
\begin{aligned}
\boldsymbol{S}'_u(\boldsymbol{u}, \boldsymbol{\eta}) &= \mathbb{C} : \boldsymbol{E}'_u = \mathbb{C} : \operatorname{sym}\{\boldsymbol{F}^T \text{ Grad } \boldsymbol{\eta}\} \qquad (128) \\
\boldsymbol{S}'_s(\boldsymbol{u}, \boldsymbol{\psi}) &= \mathbb{C} : \boldsymbol{E}'_s = -\mathbb{C} : \operatorname{sym}\{\boldsymbol{F}^T \boldsymbol{F} \text{ Grad } \boldsymbol{\psi}\} \qquad (129)
\end{aligned}
$$

where

$$
\mathbb{C} = \frac{\partial \boldsymbol{S}}{\partial \boldsymbol{E}} = \rho_R \frac{\partial^2 \Psi}{\partial \boldsymbol{E}\, \partial \boldsymbol{E}} \qquad (130)
$$

is the second elasticity tensor.

We consider as an example the case of isothermal hyperelasticity, i.e. thermal effects are ignored and the free energy Ψ coincides with the stored energy function W. The stored energy function $W_R = \rho_R \Psi$ of a classical compressible Neo-Hooke material in terms of the invariants I_C of $\boldsymbol{C} = \boldsymbol{F}^T \boldsymbol{F}$ and $J = \det \boldsymbol{F}$ can be expressed by

$$
W_R(I_C, J) = \frac{1}{2}\mu\,(I_C - 3 - 2 \ln J) + \frac{1}{2}\lambda\,(J-1)^2. \qquad (131)
$$

For this material follow the second Piola-Kirchhoff stress tensor

$$
\boldsymbol{S} = 2\frac{\partial W_R}{\partial \boldsymbol{C}} = \mu\,(\boldsymbol{1} - \boldsymbol{C}^{-1}) + \lambda\, J\,(J-1)\boldsymbol{C}^{-1} \qquad (132)
$$

and the second elasticity tensor

$$
\begin{aligned}
\mathbb{C} &= 4\frac{\partial^2 W_R}{\partial \boldsymbol{C}\, \partial \boldsymbol{C}} \\
&= \lambda\, J(2J-1)\boldsymbol{C}^{-1} \otimes \boldsymbol{C}^{-1} \\
&\quad + 2[\mu - \lambda\, J\,(J-1)]\mathbb{I}_{C^{-1}} \qquad (133)
\end{aligned}
$$

with the fourth order tensor $\mathbb{I}_{C^{-1}\,IJKL} = \frac{1}{2}\,(\boldsymbol{C}^{-1}_{IK}\boldsymbol{C}^{-1}_{JL} + \boldsymbol{C}^{-1}_{IL}\boldsymbol{C}^{-1}_{JK})$.

A.3 Variation of line, surface and volume

The variations of line, surface and volume elements with respect to s are given by

$$(dX)'_s = \text{Grad}\,\psi\, dX, \tag{134}$$
$$(dA)'_s = [\,\text{Div}\,\psi\,\mathbf{1} - \text{Grad}\,\psi^T\,]\,dA, \tag{135}$$
$$(dV)'_s = \text{Div}\,\psi\, dV. \tag{136}$$

Following the chain rule, the variation of a quantity $\mathcal{F} = \int_{\Omega_R}(\cdot)d\Omega$ is given by

$$\mathcal{F}'_u = \int_{\Omega_R} (\cdot)'_u\, d\Omega \tag{137}$$
$$\mathcal{F}'_s = \int_{\Omega_R} [\,(\cdot)'_s + (\cdot)\,\text{Div}\,\psi\,]\, d\Omega. \tag{138}$$

Appendix B: Variations of the physical and material residual

B.1 Variations of the physical residual

The physical residual (12) written in terms of the reference configuration is given by

$$\begin{aligned}\mathcal{R}(u, s; \eta) &= \int_{\Omega_R} P : \text{Grad}\,\eta\, d\Omega - \int_{\Omega_R} b_R \cdot \eta\, d\Omega \\ &= \int_{\Omega_R} S : E'_u(u, \eta)\, d\Omega - \int_{\Omega_R} b_R \cdot \eta\, d\Omega.\end{aligned} \tag{139}$$

Under the assumption that $(b_R)'_u = 0$ follow for (33) and (34)

$$\begin{aligned}k(u, s; \eta, \delta u) = &\int_{\Omega_R} S : E''_{uu}(\eta, \delta u)\, d\Omega \\ &+ \int_{\Omega_R} E'_u(u, \eta) : \mathbb{C} : E'_u(u, \delta u)\, d\Omega.\end{aligned} \tag{140}$$

$$\begin{aligned}p(u, s; \eta, \delta s) = &\int_{\Omega_R} S : E''_{us}(u, \eta, \delta s)\, d\Omega \\ &+ \int_{\Omega_R} E'_u(u, \eta) : \mathbb{C} : E'_s(u, \delta s)\, d\Omega \\ &+ \int_{\Omega_R} S : E'_u(u, \eta)\ \text{Div}\,\delta s\, d\Omega \\ &- \int_{\Omega_R} b_R \cdot \eta\ \text{Div}\,\delta s\, d\Omega\end{aligned} \tag{141}$$

The explicit forms of the tangent operators are given by

$$\begin{aligned}k(u, s; \eta, \delta u) = &\int_{\Omega_R} S : \text{Grad}\,\delta u^T\ \text{Grad}\,\eta\, d\Omega \\ &+ \int_{\Omega_R} F^T\ \text{Grad}\,\eta : \mathbb{C} : F^T \text{Grad}\,\delta u\, d\Omega\end{aligned} \tag{142}$$

$$\begin{aligned}p(u, s; \eta, \delta s) = &-\int_{\Omega_R} S : \text{Grad}\,\delta s^T\, F^T\ \text{Grad}\,\eta\, d\Omega \\ &- \int_{\Omega_R} S : F^T\ \text{Grad}\,\eta\ \text{Grad}\,\delta s\, d\Omega \\ &- \int_{\Omega_R} F^T\ \text{Grad}\,\eta : \mathbb{C} : F^T F\ \text{Grad}\,\delta s\, d\Omega \\ &+ \int_{\Omega_R} S : F^T\ \text{Grad}\,\eta\ \text{Div}\,\delta s\, d\Omega \\ &- \int_{\Omega_R} b_R \cdot \eta\ \text{Div}\,\delta s\, d\Omega.\end{aligned} \tag{143}$$

B.2 Variations of the material residual

The material residual (13) written in terms of the reference configuration by using of Eq. 25 is given by

$$\begin{aligned}\mathcal{G}(u, s; \psi) = &\int_{\Omega_R} (\Sigma + V_R \mathbf{1}) : \text{Grad}\,\psi\, d\Omega \\ &- \int_{\Omega_R} b_R^{inh} \cdot \psi\, d\Omega \\ = &\int_{\Omega_R} S : E'_s(u, \psi)\, d\Omega \\ &+ \int_{\Omega_R} (W_R + V_R)\mathbf{1} : \text{Grad}\,\psi\, d\Omega \\ &- \int_{\Omega_R} b_R^{inh} \cdot \psi\, d\Omega\end{aligned} \tag{144}$$

where $\Sigma = W_R\mathbf{1} - F^T P = W_R\mathbf{1} - F^T F S$. The pseudo load operator (44) for the material problem follow in the form

$$\begin{aligned}t(u, s; \psi, \delta u) = &\int_{\Omega_R} S : E''_{su}(u, \psi, \delta u)\, d\Omega \\ &+ \int_{\Omega_R} E'_s(u, \psi) : \mathbb{C} : E'_u(u, \delta u)\, d\Omega \\ &+ \int_{\Omega_R} S : E'_u(u, \delta u)\ \text{Div}\,\psi\, d\Omega \\ &- \int_{\Omega_R} b_R \cdot \delta u\ \text{Div}\,\psi\, d\Omega.\end{aligned} \tag{145}$$

As a result of the symmetry of the pseudo load operator for the physical (34) and material problem (44) follows $t(\boldsymbol{u}, s; \boldsymbol{\psi}, \delta\boldsymbol{u}) = p(\boldsymbol{u}, s; \delta\boldsymbol{u}, \boldsymbol{\psi})$.

If we consider a homogeneous elastic body without any material inhomogeneity, the variation of $L(\cdot)$ vanishes identically, i.e. $L'_s = 0$. Therefore, only the variation $b'_s(\boldsymbol{u}, s; \boldsymbol{\psi}, \delta s)$ contribute to the material tangent operator $d(\boldsymbol{u}, s; \boldsymbol{\psi}, \delta s)$. Following the chain rule, the variation is given by

$$d(\boldsymbol{u}, s; \boldsymbol{\psi}, \delta s) = \int_{\Omega_R} \left[\boldsymbol{\Sigma}'_s : \operatorname{Grad}\boldsymbol{\psi} + \boldsymbol{\Sigma} : (\operatorname{Grad}\boldsymbol{\psi})'_s\right] d\Omega + \int_{\Omega_R} \boldsymbol{\Sigma} : \operatorname{Grad}\boldsymbol{\psi} \operatorname{Div}\delta\boldsymbol{s}\, d\Omega. \quad (146)$$

A straightforward calculation by using the relations from Appendix A leads to

$$\begin{aligned} d(\boldsymbol{u}, s; \boldsymbol{\psi}, \delta s) = &\int_{\Omega_R} \boldsymbol{S} : \boldsymbol{E}''_{ss}(\boldsymbol{u}, \boldsymbol{\psi}, \delta\boldsymbol{s})\, d\Omega \\ &+ \int_{\Omega_R} \boldsymbol{E}'_s(\boldsymbol{u}, \boldsymbol{\psi}) : \mathbb{C} : \boldsymbol{E}'_s(\boldsymbol{u}, \delta\boldsymbol{s})\, d\Omega \\ &+ \int_{\Omega_R} \boldsymbol{S} : \boldsymbol{E}'_s(\boldsymbol{u}, \boldsymbol{\psi}) \operatorname{Div}\delta\boldsymbol{s}\, d\Omega \\ &+ \int_{\Omega_R} \boldsymbol{S} : \boldsymbol{E}'_s(\boldsymbol{u}, \delta\boldsymbol{s}) \operatorname{Div}\boldsymbol{\psi}\, d\Omega \\ &+ \int_{\Omega_R} W_R \operatorname{Div}\boldsymbol{\psi} \operatorname{Div}\delta\boldsymbol{s}\, d\Omega \\ &- \int_{\Omega_R} W_R \boldsymbol{1} : \operatorname{Grad}\boldsymbol{\psi} \operatorname{Grad}\delta\boldsymbol{s}\, d\Omega. \end{aligned} \quad (147)$$

Finally, the explicit forms of the tangent operators are given by

$$\begin{aligned} t(\boldsymbol{u}, s; \boldsymbol{\psi}, \delta\boldsymbol{u}) = &-\int_{\Omega_R} \boldsymbol{S} : \operatorname{Grad}\boldsymbol{\psi}^T \boldsymbol{F}^T \operatorname{Grad}\delta\boldsymbol{u}\, d\Omega \\ &- \int_{\Omega_R} \boldsymbol{S} : \boldsymbol{F}^T \operatorname{Grad}\delta\boldsymbol{u} \operatorname{Grad}\boldsymbol{\psi}\, d\Omega \\ &- \int_{\Omega_R} \boldsymbol{F}^T \boldsymbol{F} \operatorname{Grad}\boldsymbol{\psi} : \mathbb{C} : \boldsymbol{F}^T \operatorname{Grad}\delta\boldsymbol{u}\, d\Omega \\ &+ \int_{\Omega_R} \boldsymbol{S} : \boldsymbol{F}^T \operatorname{Grad}\delta\boldsymbol{u} \operatorname{Div}\boldsymbol{\psi}\, d\Omega \\ &- \int_{\Omega_R} \boldsymbol{b}_R \cdot \delta\boldsymbol{u} \operatorname{Div}\boldsymbol{\psi}\, d\Omega \end{aligned} \quad (148)$$

$$\begin{aligned} d(\boldsymbol{u}, s; \boldsymbol{\psi}, \delta s) = &\int_{\Omega_R} \boldsymbol{S} : \operatorname{Grad}\delta\boldsymbol{s}^T \boldsymbol{F}^T \boldsymbol{F} \operatorname{Grad}\boldsymbol{\psi}\, d\Omega \\ &+ \int_{\Omega_R} \boldsymbol{S} : \boldsymbol{F}^T \boldsymbol{F} \operatorname{Grad}\delta\boldsymbol{s} \operatorname{Grad}\boldsymbol{\psi}\, d\Omega \\ &+ \int_{\Omega_R} \boldsymbol{S} : \boldsymbol{F}^T \boldsymbol{F} \operatorname{Grad}\boldsymbol{\psi} \operatorname{Grad}\delta\boldsymbol{s}\, d\Omega \\ &+ \int_{\Omega_R} \boldsymbol{F}^T \boldsymbol{F} \operatorname{Grad}\boldsymbol{\psi} : \mathbb{C} : \boldsymbol{F}^T \boldsymbol{F} \operatorname{Grad}\delta\boldsymbol{s}\, d\Omega \\ &- \int_{\Omega_R} \boldsymbol{S} : \boldsymbol{F}^T \boldsymbol{F} \operatorname{Grad}\boldsymbol{\psi} \operatorname{Div}\delta\boldsymbol{s}\, d\Omega \\ &- \int_{\Omega_R} \boldsymbol{S} : \boldsymbol{F}^T \boldsymbol{F} \operatorname{Grad}\delta\boldsymbol{s} \operatorname{Div}\boldsymbol{\psi}\, d\Omega \\ &+ \int_{\Omega_R} W_R \operatorname{Div}\boldsymbol{\psi} \operatorname{Div}\delta\boldsymbol{s}\, d\Omega \\ &- \int_{\Omega_R} W_R \boldsymbol{1} : \operatorname{Grad}\boldsymbol{\psi} \operatorname{Grad}\delta\boldsymbol{s}\, d\Omega. \end{aligned} \quad (149)$$

Remark 19 All variations are obtained with respect to the deformation map $\boldsymbol{\varphi}$, i.e. all operators have to be evaluated at the current deformation $\boldsymbol{\varphi}$. It is also possible to express all the variations in terms of the actual displacement field $\boldsymbol{u}$. The variation of the deformation gradient $\boldsymbol{F} = \operatorname{Grad}\boldsymbol{\varphi} = \boldsymbol{1} + \boldsymbol{H}$ is then given by $\boldsymbol{F}'_s(\boldsymbol{u}, \boldsymbol{\psi}) = \boldsymbol{H}'_s(\boldsymbol{u}, \boldsymbol{\psi}) = -\boldsymbol{H} \operatorname{Grad}\boldsymbol{\psi}$. The variations of the Green-Lagrange strain tensor yield

$$\boldsymbol{E}'_s(\boldsymbol{u}, \boldsymbol{\psi}) = -\operatorname{sym}\{\boldsymbol{F}^T \boldsymbol{H} \operatorname{Grad}\boldsymbol{\psi}\} \quad (150)$$

$$\begin{aligned} \boldsymbol{E}''_{us}(\boldsymbol{u}, \boldsymbol{\eta}, \boldsymbol{\psi}) = -\operatorname{sym}\{&\operatorname{Grad}\boldsymbol{\psi}^T \boldsymbol{H}^T \operatorname{Grad}\boldsymbol{\eta} \\ &+ \boldsymbol{F}^T \operatorname{Grad}\boldsymbol{\eta} \operatorname{Grad}\boldsymbol{\psi}\} \end{aligned} \quad (151)$$

$$\begin{aligned} \boldsymbol{E}''_{ss}(\boldsymbol{\eta}, \boldsymbol{\psi}, \boldsymbol{\chi}) = \operatorname{sym}\{&\operatorname{Grad}\boldsymbol{\chi}^T \boldsymbol{H}^T \boldsymbol{H} \operatorname{Grad}\boldsymbol{\psi} \\ &+ \boldsymbol{F}^T \boldsymbol{H} \operatorname{Grad}\boldsymbol{\chi} \operatorname{Grad}\boldsymbol{\psi} \\ &+ \boldsymbol{F}^T \boldsymbol{H} \operatorname{Grad}\boldsymbol{\psi} \operatorname{Grad}\boldsymbol{\chi}\}. \end{aligned} \quad (152)$$

The tangent operators $p(\cdot;\cdot)$ and $d(\cdot;\cdot)$ follow in a straightforward manner by using these relations and have to be evaluated for the current displacement field $\boldsymbol{u}$.

This distinction is important for the evaluation of the sensitivity relation for the physical problem (68). Let $\boldsymbol{P}(\boldsymbol{\varphi})$ be the pseudo load operator matrix corresponding to the pseudo load operator $p(\cdot;\cdot)$ evaluated at the current deformation $\boldsymbol{\varphi}$, i.e. by using Eq. 118 and (120). Furthermore, let $\boldsymbol{P}(\boldsymbol{u})$ be the same operator in terms of the actual displacement field $\boldsymbol{u}$, i.e. by using Eq. 150 and 151. With these assumptions follow

$$\frac{d\boldsymbol{\varphi}}{ds} = \left(\mathbf{1} + \frac{d\boldsymbol{u}}{ds}\right) = -\boldsymbol{K}^{-1}\boldsymbol{P}(\boldsymbol{\varphi}) = \boldsymbol{S}(\boldsymbol{\varphi}) \tag{153}$$

and hence

$$\frac{d\boldsymbol{u}}{ds} = \boldsymbol{S}(\boldsymbol{\varphi}) - \mathbf{1} = \boldsymbol{S}(\boldsymbol{u}) = -\boldsymbol{K}^{-1}\boldsymbol{P}(\boldsymbol{u}). \tag{154}$$

References

Askes H, Kuhl E, Steinmann P (2004) An ALE formulation based on spatial and material settings of continuum mechanics. Part 2: classification and applications. Comput Methods Appl Mech Eng 193:4223–4245

Barthold FJ (2002) Zur Kontinuumsmechanik inverser Geometrieprobleme. Habilitationsschrift, Braunschweiger Schriften zur Mechanik 44-2002, TU Braunschweig, URL http://hdl.handle.net/2003/23095

Barthold FJ (2003) A structural optimization viewpoint on configurational mechanics. Proc Appl Math Mech 3:246–247

Barthold FJ (2005) On structural optimisation and configurational mechanics. In: Steinmann P, Maugin GA (eds) Mechanics of material forces, vol 11. Springer-Verlag, pp 219–228

Barthold FJ (2007) Remarks on variational shape sensitivity analysis based on local coordinates. Eng Anal Boundary Elem. Accepted for publication

Barthold FJ, Mesecke S (1999) Remarks on computing the energy release rate and its sensitivities. In: Mota Soares C, Mota Soares C, Freitas M (eds) Mechanics of composite materials and structures, NATO science series, Serie E: applied science, vol 361. Kluwer Academic Publishers, pp 341–350

Barthold FJ, Stein E (1996) A continuum mechanical based formulation of the variational sensitivity analysis in structural optimization. Part I: analysis. Struct Optim 11(1/2):29–42

Bendsøe MP, Sigmund O (2003) Topology optimization—theory, methods and applications. Springer-Verlag

Benzi M, Golub GH (2004) A preconditioner for generalized saddle point problems. Siam J Matrix Anal Appl 26(1):20–41

Bertram A (1989) Axiomatische Einführung in die Kontinuumsmechanik. BI-Wissenschaftsverlag, Mannheim

Braun M (1997) Configurational forces induced by finite-element discretisation. Proc Estonian Acad Sci Phys Math 46:24–31

Carpenter WC, Zendegui S (1982) Optimum nodal locations for a finite element idealization. Eng Optim 5:215–221

Carroll WE, Barker RM (1973) A theorem for optimum finite-element idealizations. Int J Solids Struct 9:883–895

Chadwick P (1975) Applications of an energy-momentum tensor in non-linear elastostatics. J Elast 5(3–4):249–258

Choi KK, Kim NH (2005a) Structural sensitivity analysis and optimization 1—linear systems. Mechanical engineering series. Springer-Verlag, Berlin

Choi KK, Kim NH (2005b) Structural sensitivity analysis and optimization 2—nonlinear systems and applications. Mechanical engineering series. Springer-Verlag, Berlin

Dems K, Mróz Z (1978) Multiparameter structural shape optimization by the finite element method. Int J Numer Methods Eng 13(2):247–263

Engl HW, Hanke M, Neubauer A (2000) Regularization of inverse problems. Kluwer Academic Publishers

Eshelby JD (1951) The force on an elastic singularity. Philos Trans R Soc Lond 244:87–112

Eshelby JD (1975) The elastic energy-momentum tensor. J Elast 5:321–335

Govindjee S, Mihalic PA (1996) Computational methods for inverse finite elastostatics. Comput Methods Appl Mech Eng 136(1–2):47–57

Gurtin ME (2000) Configurational forces as basic concepts of continuum physics. Springer-Verlag, New York

Kalpakides VK, Balassas KG (2005) The inverse deformation mapping in the finite element method. Philos Mag 85(33–35):4257–4275

Kamat MP (1993) Structural optimization: status and promise. American Institute of Aeronautics and Astronautics, Washington

Kienzler R, Herrmann G (2000) Mechanics in material space. Springer-Verlag, Berlin Heidelberg New York

Kuhl E, Askes H, Steinmann P (2004) An ALE formulation based on spatial and material settings of continuum mechanics. Part 1: Generic hyperelastic formulation. Comput Methods Appl Mech Eng 193:4207–4222

Marsden JE, Hughes TJR (1994) Mathematical foundations of elasticity. Dover, New York

Materna D, Barthold FJ (2006a) Coherence of structural optimization and configurational mechanics. Proc Appl Math Mech 6(1):245–246. DOI 10.1002/pamm.200610103

Materna D, Barthold FJ (2006b) Relations between structural optimization and configurational mechanics with applications to mesh optimization. In: Sienz J, Querin OM, Toropov VV, Gosling PD (eds) Proceedings of the 6th ASMO-UK/ISSMO conference on engineering design optimization, St Edmund Hall, Oxford, UK, University of Leeds, UK, pp 173–181

Materna D, Barthold FJ (2007a) On variational sensitivity analysis and configurational mechanics. Comput Mech. Accepted for publication

Materna D, Barthold FJ (2007b) The use of singular value decomposition in sensitivity analysis and mesh optimization. To be submitted

Maugin GA (1993) Material inhomogeneities in elasticity. Chapman & Hall, London

McNeice GM, Marcal PE (1973) Optimization of finite element grids based on minimum potential energy. ASME J Eng Indust 95(1):186–190

Mosler J, Ortiz M (2006) On the numerical modeling of variational arbitrary lagrangian-eulerian (vale) formulations. Int J Numer Methods Eng 67:1272–1289

Mueller R, Maugin GA (2002) On material forces and finite element discretizations. Comput Mech 29:52–60

Nocedal J, Wright SJ (1999) Numerical optimization. Springer-Verlag

Noll W (1972) A new mathematical theory of simple materials. Arch Rational Mech Anal 48(1):1–50

Shield RT (1967) Inverse deformation results in finite elasticity. Z Angew Math Phys 18:490–500

Steinmann P (2000) Application of material forces to hyperelastic fracture mechanics. I. Continuum mechanical settings. Int J Solids Struct 37:7371–7391

Steinmann P, Maugin GA (eds) (2005) Mechanics of material forces. Springer-Verlag

Thoutireddy P (2003) Variational arbitrary lagrangian-eulerian method. Phd thesis, California Institute of Technology, Passadena, California

Truesdell C, Noll W (2004) The nonlinear field theories of mechanics, 3rd edn. Springer-Verlag, Berlin

Zhang S, Belegundu A (1993) Mesh distortion control in shape optimization. AIAA J 31(7):1360–1362

An anisotropic elastic formulation for configurational forces in stress space

Anurag Gupta · Xanthippi Markenscoff

Abstract A new variational principle for an anisotropic elastic formulation in stress space is constructed, the Euler–Lagrange equations of which are the equations of compatibility (in terms of stress), the equilibrium equations and the traction boundary condition. Such a principle can be used to extend recently obtained configurational balance laws in stress space to the case of anisotropy.

Keywords Configurational forces · Stress formulation · Anisotropy

1 Introduction

Configurational forces (in solid mechanics) provide us with the forces necessary for driving the dissipative mechanisms which are responsible for the kinetics of defect flow. In case of a thermodynamic equilibrium, vanishing of configurational forces are additional relations required for a complete determination of the system. Popular examples are dislocation motion, crack propagation, delamination etc. (Eshelby 1956; Maugin 1993; Gupta and Markenscoff 2007 and references therein). A systematic way of obtaining expressions for these forces follows from Noether's theorem (Noether 1918; Gelfand and Fomin 2000), where given a variational principle, conserved integrals (path independent in 2D) are obtained as necessary and sufficient conditions for satisfying respective symmetries of the variational principle. Configurational forces are then interpreted as these conserved integrals which vanish in case of a non-dissipative flow, but fail to vanish if dissipative mechanisms are active (i.e. when symmetries are broken) (Eshelby 1975).

In (Li et al. 2005), the authors use a variational principle of Pobedrya (Pobedrja 1980; Pobedria and Holmatov 1982) to obtain a class of conservation laws. The Euler–Lagrange equations of this principle were the Beltrami–Michell compatibility equations in the domain and stress equilibrium with traction boundary condition on the boundary. It was shown by Pobedrya (1980) and Kucher et al. (2004) that such a boundary value problem in terms of stress is well defined and it is sufficient to satisfy equilibrium on the boundary to satisfy it in the domain. The conservation laws were obtained assuming the case of linear, homogeneous and isotropic elasticity with vanishing body forces and incompatibility. In a sequel paper (Markenscoff and Gupta 2007) these were extended for non-vanishing incompatibility and body force distribution and applicability of such quantities was demonstrated using examples from dislocation theory and heat flow in a domain with a spherical cavity. More recently these

A. Gupta (✉)
Department of Mechanical Engineering, University of California, Berkeley, CA 94720, USA
e-mail: agupta@berkeley.edu

X. Markenscoff
Department of Mechanical and Aerospace Engineering, University of California, La Jolla, San Diego, CA 92093, USA
e-mail: xmarkens@ucsd.edu

Defect and Material Mechanics. C. Dascalu, G.A. Maugin & C. Stolz (eds.),
doi: 10.1007/978-1-4020-6929-1_14, © Springer Science+Business Media B.V. 2007

conserved quantities (or configurational forces) were interpreted as the necessary and sufficient dissipative mechanisms so as to maintain compatibility during the propagation of a inhomogeneity (or a defect) (Gupta and Markenscoff 2007).

The present paper aims at formulating a variational principle which would extend the earlier principle of Pobedrya (1980) to the case of anisotropy. Such a variational principle can then be used to obtain the configurations balance laws in stress space (Li et al. 2005; Markenscoff and Gupta 2007) for the case of anisotropic elasticity. Before constructing a variational principle in Sect. 3, we first discuss Pobedrya's formulation of the boundary value problem of anisotropic linear elasticity in stress space.

2 Formulation

The classical problem of linear and homogeneous elasticity in terms of stress involves equilibrium and compatibility equations in the bulk (bulk is denoted by $\Omega \in \mathbf{R}^3$) and a traction boundary condition (boundary is denoted by $\partial\Omega \equiv \Omega \cap \mathbf{R}^3/\Omega$). Assuming absence of body forces and inertial terms, the equilibrium equation is given as follows,

$$\sigma_{ij,j} = 0, \quad \forall x_k \in \Omega \tag{1}$$

with $\sigma_{ij}(x_k)$ denoting the stress. The compatibility relations are written in terms of strain $e_{ij}(x_k)$ as,

$$\eta_{ij} \equiv -\epsilon_{ikl}\epsilon_{jmn}e_{ln,km} = 0, \quad \forall x_k \in \Omega \tag{2}$$

where ϵ_{ikl} is the alternating tensor and η_{ij} is the Kröner's incompatibility tensor (Kröner 1981) (vanishing of which ensures compatibility). The compatibility relations in terms of stresses can be obtained by using an appropriate constitutive law in the framework of linear elasticity,

$$e_{ij} = D_{ijkl}\sigma_{kl}, \tag{3}$$

where D_{ijkl} is the (constant) elastic compliance tensor. Denote $\bar{\eta}_{ij}$ as the incompatibility tensor thus obtained as a function of the stress tensor. Therefore the compatibility relations in terms of stresses are,

$$\bar{\eta}_{ij} \equiv -D_{lnpq}\epsilon_{ikl}\epsilon_{jmn}\sigma_{pq,km} = 0, \quad \forall x_k \in \Omega. \tag{4}$$

Equations 1 and 4 are to be supplemented by prescribing traction on the boundary,

$$\sigma_{ij}n_j = p_i, \quad \forall x_k \in \partial\Omega \tag{5}$$

with $p_i(x_k)$ being the prescribed traction and $n_j(x_k)$, the unit normal to $\partial\Omega$. The system of Eqs. 1 and 4 is over-determined as there are nine equations for six unknowns. On the boundary, there is an under-determinacy by three conditions (Georgievskii and Pobedrya 2004). Pobedrya (1994) has introduced a set of equations to deal with this problem. His formulation transfers the equilibrium Eq. 1 to the boundary and therefore leaving only Eq. 4 to be solved in the bulk. The resulting system of equations is well defined over the whole domain (Pobedrja 1980; Kucher et al. 2004). An outline of his theory for anisotropic elasticity is now discussed.

Let $a_{ij}^{(\alpha)}$ be a set of tensors ($\alpha = 1, \ldots, N$) constructed from the invariant basis tensors of the symmetry group $\mathcal{G}$. These invariant basis tensors were originally obtained for the representation of anisotropic tensors and there are systematic procedures for their evaluation corresponding to the symmetry group $\mathcal{G}$ (see (Weyl 1946; Smith and Rivlin 1957; Smith 1970) for foundations and (Markenscoff 1976) for a possible application). The tensors $a_{ij}^{(\alpha)}$ are constructed such that they are pairwise orthogonal and add up to the unit tensor,

$$\frac{a_{ij}^{(\alpha)}a_{ij}^{(\beta)}}{a_{(\alpha)}a_{(\beta)}} = \delta_{\alpha\beta}, \quad \sum_{\alpha=1}^{N} a_{ij}^{(\alpha)} = \delta_{ij}, \tag{6}$$

where $a_{(\alpha)} = \sqrt{a_{ij}^{(\alpha)}a_{ij}^{(\alpha)}}$ (no summation implied under α). A set of linear invariants of the incompatibility tensor can then be constructed as,

$$\eta_{(\alpha)} = \eta_{ij}a_{ij}^{(\alpha)}. \tag{7}$$

Define a tensor H_{ij} as,

$$H_{ij} = \eta_{ij} + \sum_{\alpha=1}^{N} \xi^{(\alpha)}\eta_{(\alpha)}a_{ij}^{(\alpha)}, \tag{8}$$

where $\xi^{(\alpha)}$ are arbitrary scalars. The condition $H_{ij} = 0$ is equivalent to the compatibility condition (2), $\eta_{ij} = 0$ if $\xi^{(\alpha)} \neq -(a_{(\alpha)})^{-2}$ for all α. Indeed if $\eta_{ij} = 0$ then by definitions (7) and (8), $H_{ij} = 0$. If $H_{ij} = 0$, then by Eq. 8, $H_{ij}a_{ij}^{(\beta)} = \eta_{(\beta)} + \xi^{(\beta)}\eta_{(\beta)}(a_{(\beta)})^2 = 0$, where relations (6) and (7) have been used. If $\xi^{(\alpha)} \neq -(a_{(\alpha)})^{-2}$, then $\eta_{(\alpha)} = 0$ and consequently, using $H_{ij} = 0$ in (8) we obtain $\eta_{ij} = 0$.

Define a vector valued function,

$$A_i = R_{ij}\sigma_{jk,k}, \tag{9}$$

where R_{ij} is a positive definite operator. Therefore, condition $A_i = 0$ is equivalent to equilibrium relation (1). Construct a tensor $\mathrm{A}_{ij} = A_{i,j} + A_{j,i}$ whose linear invariants are denoted by $A_{(\alpha)} = \mathrm{A}_{ij}a_{ij}^{(\alpha)}$. Another tensor can then be defined as,

$$\bar{\mathrm{A}}_{ij} = \mathrm{A}_{ij} + \sum_{\alpha=1}^{N} \xi^{(\alpha)} A_{(\alpha)} a_{ij}^{(\alpha)}. \tag{10}$$

Reasoning along the lines of the paragraph following Eq. 8, we note that $\mathrm{A}_{ij} = 0$, if and only if $\bar{\mathrm{A}}_{ij} = 0$ given that $\xi^{(\alpha)} \neq -(a_{(\alpha)})^{-2}$.

Form a tensor,

$$\bar{H}_{ij} = H_{ij}^{\sigma} + \bar{\mathrm{A}}_{ij}, \tag{11}$$

where superscript σ in H_{ij}^{σ} indicates that H_{ij} is expressed in terms of stress rather than strain (using (6)). Therefore $H_{ij}^{\sigma} = \bar{\eta}_{ij} + \sum_{\alpha=1}^{N} \xi^{(\alpha)} \bar{\eta}_{(\alpha)} a_{ij}^{(\alpha)}$ where $\bar{\eta}_{ij}$ is as given in (4) and $\bar{\eta}_{(\alpha)} = \bar{\eta}_{ij} a_{ij}^{(\alpha)}$. The formulation of Pobedrya (Pobedrya 1994) involves satisfying equation,

$$\bar{H}_{ij} = 0, \quad \forall x_k \in \Omega \tag{12}$$

in the bulk and conditions,

$$\sigma_{ij} n_j = p_i, \quad \sigma_{ij,j} = 0, \quad \forall x_k \in \partial\Omega \tag{13}$$

on the boundary. The solution to Eqs. 12 and 13 satisfies equations of equilibrium and of compatibility as given in (1) and (4). From Eqs. 11 and 12 obtain, $\bar{H}_{ij} a_{ij}^{(\beta)} = (\bar{\eta}_{(\beta)} + A_{(\beta)})(1 + \xi^{(\beta)} (a_{(\beta)})^2) = 0$, and therefore if $\xi^{(\alpha)} \neq -(a_{(\alpha)})^{-2}$ is satisfied, we obtain,

$$\bar{\eta}_{(\beta)} + A_{(\beta)} = 0. \tag{14}$$

Differentiate Eq. 11 to get,

$$\bar{H}_{ij,j} = H_{ij,j}^{\sigma} + \bar{\mathrm{A}}_{ij,j} \tag{15}$$

which will now be evaluated term by term. Let $\bar{\eta}$ denote the trace of $\bar{\eta}_{ij}$. Therefore $\bar{\eta} = \bar{\eta}_{ij}\delta_{ij}$ and from (4), $\bar{\eta} = 2(D_{iijk}\sigma_{jk,pp} - D_{mnjk}\sigma_{jk,mn})$. It is easy to see that $\bar{\eta}_{ij,j} = \frac{1}{2}(\bar{\eta})_{,i}$. Also, note that $\bar{\eta} = \bar{\eta}_{ij}\delta_{ij} = \bar{\eta}_{ij} \sum_{\alpha=1}^{N} a_{ij}^{(\alpha)} = \sum_{\alpha=1}^{N} \bar{\eta}_{(\alpha)}$, where the first relation in (6) has been used. Using these results $H_{ij,j}^{\sigma}$ can then be evaluated as,

$$H_{ij,j}^{\sigma} = \sum_{\alpha=1}^{N} \left(\frac{1}{2}\delta_{ij} + \xi^{(\alpha)} a_{ij}^{(\alpha)} \right) \bar{\eta}_{(\alpha),j}. \tag{16}$$

For calculating $\bar{\mathrm{A}}_{ij,j}$, start from (10) and note that $\mathrm{A}_{ij,j} = \Delta A_i + A_{j,ji}$ and $A_{j,j} = \frac{1}{2}\mathrm{A}_{ij}\delta_{ij} = \frac{1}{2}\mathrm{A}_{ij} \sum_{\alpha=1}^{N} a_{ij}^{(\alpha)} = \frac{1}{2}\sum_{\alpha=1}^{N} A_{(\alpha)}$ to obtain,

$$\bar{\mathrm{A}}_{ij,j} = \Delta A_i + \sum_{\alpha=1}^{N} \left(\frac{1}{2}\delta_{ij} + \xi^{(\alpha)} a_{ij}^{(\alpha)} \right) A_{(\alpha),j} \tag{17}$$

and thereafter combine Eqs. 16 and 17 to write,

$$\bar{H}_{ij,j} = \Delta A_i + \sum_{\alpha=1}^{N} \left(\frac{1}{2}\delta_{ij} + \xi^{(\alpha)} a_{ij}^{(\alpha)} \right) \times (\bar{\eta}_{(\alpha)} + A_{(\alpha)})_{,j}. \tag{18}$$

Substitution of relations (12) and (14) into (18) then results into $\Delta A_i = 0$, i.e. A_i is harmonic. The vector A_i vanishes on the boundary (by $(13)_2$) and therefore it vanishes inside Ω (since A_i is harmonic). Therefore equilibrium relation (1) is satisfied in the bulk. As a consequence of this result and Eqs. 11 and 12, obtain $H_{ij}^{\sigma} = 0$. The compatibility relation (4) then follows from this equality. This is exactly the generalization to anisotropy of Pobedrya's method.

3 A New Variational Principle

Consider the following functional,

$$\Pi(\sigma_{ij}, \sigma_{ij,k}) = \int_{\Omega} E_{ijk}\sigma_{ij,k} dV - \int_{\partial\Omega} \left(E_{ijk}\sigma_{ij}n_k - \frac{1}{2}\sigma_{ij,j}\sigma_{ik,k} - \frac{1}{2}\sigma_{ij}n_j\sigma_{ik}n_k + p_i\sigma_{ij}n_j \right) dA, \tag{19}$$

where

$$E_{ijk} = - D_{lnpq}\epsilon_{ikl}\epsilon_{jmn}\sigma_{pq,m} + A_i\delta_{jk} + A_j\delta_{ik} + \sum_{\alpha=1}^{N} \xi^{(\alpha)} (-D_{lnpq}\epsilon_{rkl}\epsilon_{smn}\sigma_{pq,m} + A_r\delta_{sk} + A_s\delta_{rk}) a_{rs}^{(\alpha)} a_{ij}^{(\alpha)} \tag{20}$$

and A_i is as given in (9). Define $L_{\Omega} = E_{ijk}\sigma_{ij,k}$, then to obtain the Euler–Lagrange equations corresponding to the above functional, evaluate $\frac{\partial L_{\Omega}}{\partial \sigma_{uv,w}}$ using following formulae,

(i) $$\frac{\partial (D_{lnpq}\epsilon_{ikl}\epsilon_{jmn}\sigma_{pq,m}\sigma_{ij,k})}{\partial \sigma_{uv,w}} = Sym_{uv}[D_{lnpq}\epsilon_{ikl}\epsilon_{jmn}(\delta_{up}\delta_{vq}\delta_{wm}\sigma_{ij,k} + \delta_{ui}\delta_{vj}\delta_{wk}\sigma_{pq,m})] = D_{lnuv}\epsilon_{ikl}\epsilon_{jwn}\sigma_{ij,k} + Sym_{uv}(D_{lnpq}\epsilon_{uwl}\epsilon_{vmn}\sigma_{pq,m})$$

(ii) $\dfrac{\partial(D_{lnpq}\epsilon_{rkl}\epsilon_{smn}\sigma_{pq,m}\sigma_{ij,k})}{\partial\sigma_{uv,w}}$

$$= Sym_{uv}[D_{lnpq}\epsilon_{rkl}\epsilon_{smn}(\delta_{up}\delta_{vq}\delta_{wm}\sigma_{ij,k} + \delta_{ui}\delta_{vj}\delta_{wk}\sigma_{pq,m})]$$
$$= D_{lnuv}\epsilon_{rkl}\epsilon_{swn}\sigma_{ij,k} + Sym_{uv}(D_{lnpq}\epsilon_{rwl}\epsilon_{smn}\delta_{ui}\delta_{vj}\sigma_{pq,m})$$

(iii) $\dfrac{\partial(R_{ip}\delta_{qr}\delta_{jk}\sigma_{pq,r}\sigma_{ij,k})}{\partial\sigma_{uv,w}} = Sym_{uv}(R_{iu}\delta_{vw}\sigma_{ij,j}) + Sym_{uv}(R_{up}\delta_{vw}\sigma_{pq,q})$

(iv) $\dfrac{\partial(R_{rp}\delta_{qt}\delta_{sk}\sigma_{pq,t}\sigma_{ij,k})}{\partial\sigma_{uv,w}} = Sym_{uv}(R_{ru}\delta_{vw}\sigma_{ij,s}) + Sym_{uv}(R_{rp}\delta_{iu}\delta_{jv}\delta_{sw}\sigma_{pq,q})$,

where the notation $Sym_{uv}(\cdot)_{uv} = \frac{1}{2}[(\cdot)_{uv} + (\cdot)_{vu}]$ has been employed. Also note, that for a symmetric tensor σ_{ij}, $\frac{\partial\sigma_{ij}}{\partial\sigma_{uv}} = Sym_{uv}(\delta_{iu}\delta_{jv})$. Using these relations obtain,

$$\frac{\partial L_\Omega}{\partial\sigma_{uv,w}} = Sym_{uv}(E_{uvw}) - D_{lnuv}\epsilon_{ikl}\epsilon_{jwn}\sigma_{ij,k} + 2Sym_{uv}(R_{iu}\delta_{vw}\sigma_{ij,j}) + \sum_{\alpha=1}^{N}\xi^{(\alpha)}(-D_{lnuv}\epsilon_{rkl}\epsilon_{swn}\sigma_{ij,k} + 2Sym_{uv}Sym_{rs}(R_{ru}\delta_{vw}\sigma_{ij,s}))a_{rs}^{(\alpha)}a_{ij}^{(\alpha)}. \quad (21)$$

Taking a variation of the functional (19) with respect to σ_{uv}, we evaluate,

$$\delta\Pi = \int_\Omega -\left(\frac{d}{dx_w}\frac{\partial L_\Omega}{\partial\sigma_{uv,w}}\right)\delta\sigma_{uv}dV - \int_{\partial\Omega}\left(E_{uvw} - \frac{\partial L_\Omega}{\partial\sigma_{uv,w}}\right)\delta\sigma_{uv}n_w dA + \int_{\partial\Omega}\sigma_{ij,j}\delta\sigma_{ik,k}dA + \int_{\partial\Omega}(\sigma_{ik}n_k - p_i)\delta\sigma_{ij}n_j dA, \quad (22)$$

where the 'frozen condition' (Pobedrja 1980): $\delta E_{ijk}n_k = 0$ on $\partial\Omega$ has been assumed. The frozen condition ensures that variation $\delta\sigma_{uv}$ does not introduce a flux of incompatibility across the boundary into the bulk. Noting expression 21, Eqs. 12 and 13 are then recovered as the Euler–Lagrange equations if,

$$-D_{lnuv}\epsilon_{ikl}\epsilon_{jwn}\sigma_{ij,k} + 2(R_{iu}\delta_{vw}\sigma_{ij,j}) + \sum_{\alpha=1}^{N}\xi^{(\alpha)}\big(-D_{lnuv}\epsilon_{rkl}\epsilon_{swn}\sigma_{ij,k} + 2Sym_{rs}(R_{ru}\delta_{vw}\sigma_{ij,s})\big)a_{rs}^{(\alpha)}a_{ij}^{(\alpha)} = 0. \quad (23)$$

These equations give conditions on the arbitrary variables $\xi^{(\alpha)}$ and R_{ij} which make the variational principle suitable. Under such conditions Eq. 22 simplifies to,

$$\delta\Pi = -\int_\Omega E_{uvw,w}\delta\sigma_{uv}dV + \int_{\partial\Omega}\sigma_{ij,j}\delta\sigma_{ik,k}dA + \int_{\partial\Omega}(\sigma_{ik}n_k - p_i)\delta\sigma_{ij}n_j dA. \quad (24)$$

The equation $E_{uvw,w} = 0$ is equivalent to (12). Finally, note that Eq. 23 should be satisfied for arbitrary $\sigma_{ij,k}$ and therefore we obtain necessary and sufficient conditions,

$$-D_{lnuv}\epsilon_{ikl}\epsilon_{jwn} + 2R_{iu}\delta_{vw}\delta_{jk} + \sum_{\alpha=1}^{N}\xi^{(\alpha)}(-D_{lnuv}\epsilon_{rkl}\epsilon_{swn}a_{rs}^{(\alpha)} + 2Sym_{rk}(R_{ru}\delta_{vw}a_{rk}^{(\alpha)}))a_{ij}^{(\alpha)} = 0 \quad (25)$$

which are used as restrictions on the arbitrary variables $\xi^{(\alpha)}$ and R_{ij}.

Functional (19) can be used to derive conservation laws using the formalism of Noether's theorem (Noether 1918; Gelfand and Fomin 2000). Noether's theorem provides a systematic procedure to obtain divergence-free quantities from given symmetries of a variational problem such as one formulated in (19). The divergence-free quantities in an integral form provide conservation laws which in the presence of singularities and inhomogeneities will result into configurational forces acting on these defects. An extension of Noether's theorem to tensorial fields was already achieved (Li et al. 2005; Markenscoff and Gupta 2007) and conservation laws were obtained for translation, rotation, pre-stress and scaling symmetries, which are new and distinct (Gupta and Markenscoff 2007). Application of these conservation laws allowed for the determination of the incompatibility in the interior of the domain by surface data. The variational principle obtained in this paper extends previous work to the case of anisotropy.

Acknowledgements This research was supported by a SEGRF fellowship from Lawrence Livermore National Laboratory (AG) and by NSF Grant No. CMS-0555280 (XM).

References

Eshelby JD (1975) The elastic energy-momentum tensor. J Elasticity 5:321–335

Eshelby JD (1956) The continuum theory of lattice defects. In: Seitz F, Turnbull D (eds) Solid state physics. Academic Press, New York pp 79–144

Gelfand IM, Fomin SV (2000) Calculus of variations, Dover

Georgievskii DV, Pobedrya BYe (2004) The number of independent compatibility equations in the mechanics of deformable solids. Prikl Mat Mekh 68(6):1043–1048 (trans: J Appl Math Mech 68:941–946 (2004))

Gupta A, Markenscoff X (2007) Configurational forces as dissipative mechanisms: a revisit, Comptes Rendus Mecanique (In press)

Kröner E (1981) Continuum theory of defects. In: Balian R et al (eds) Les Houches, Session 35, 1980 – Physique des defauts. North-Holland, New York, pp 215–315

Kucher V, Markenscoff X, Paukshto M (2004) Some properties of the boundary value problem of linear elasticity in terms of stresses. J Elasticity 74(2):135–145

Li S, Gupta A, Markenscoff X (2005) Conservation laws of linear elasticity in stress formulations. Proc Roy Soc Lon A 461:99–116

Markenscoff X (1976) Independent 4th-order elastic coefficients for quartz. Appl Phys Lett 29(12):768–770

Markenscoff X, Gupta A (2007) Configurational balance laws for incompatibility in stress space. Proc R Soc Lond A, 463:1379–1392

Maugin GA (1993) Material inhomogeneities in elasticity. Chapman & Hall, London

Noether E (1918) Invariante variationsprobleme, Göttinger Nachrichten, (Mathematisch-physikalische Klasse) 2:235–257 (trans: Transport Theory and Statistical Physics 1:186–207 (1971))

Pobedrja BE (1980) A new formulation of the problem in mechanics of a deformable solid body under stress. Soviet Math Doklady 22(1):88–91

Pobedria BE, Holmatov T (1982) On the existence and uniqueness of solutions in the elasticity theory problem with stresses, Vestnik Moskovskogo Universiteta, Seriya 1 (Matematika Mekhanika), 1, pp 50–51

Pobedrya BYe (1994) The problem in terms of a stress tensor for an anisotropic medium. Prikl Mat Mekh 58(1):77–85. (trans: J Appl Math Mech 58(1):81–89 (1994))

Smith GF, Rivlin RS (1957) The anisotropic tensors. Quart Appl Math 15(3):308–314

Smith GF (1970) The crystallographic property tensors of orders 1 to 8. Ann N Y Acad Sci 172(5):57–106

Weyl H (1946) The classical groups, their invariants and representations. Princeton Univ Press, Princeton, NJ

Conservation laws, duality and symmetry loss in solid mechanics

Huy Duong Bui

Abstract The paper deals with conservation laws which are not of the pure divergence type and thus do not provide a path-independent integral for use in Fracture Mechanics. It is shown that Duality is the right tool to re-establish the symmetry between equations and to provide conservation laws of the pure divergence type. The loss of symmetry of some energetic expressions is exploited to derive a new method for solving some inverse problems. In particular, the earthquake inverse problem is solved analytically.

Keywords Conservation laws · Duality · Symmetry · Symmetry loss · Inverse problem

1 Introduction

This review paper summarizes two informal presentations at the Conference on "Problems in Solids Mechanics" 2006 at Symi, which is a paradisaical island of Greece, and at a workshop at Aussois 2007, which is a nice place of the French Alps, held in the memory of George Herrmann. Both conferences and the previous one, at Symi 2005, belong to a series of meetings originated from G. Herrmann's idea to gather many people involved in Conservation laws and Configurational Forces in Solid Mechanics.

My first contact with George Herrmann began in the middle of 70s, when he encouraged me to investigate on Conservation laws in thermoelasticity. At this time, most works concentrated on computational problems, for example on how to derive stress intensity factors from path-independent integrals, like Rice's J-Integral. I realized that George Herrmann's question for thermo-elasticity is rather difficult, because of the presence of an entropical "source" term in the corresponding conservation law which prevents us from deriving a path-independent integral. I was haunted by his question for a while, until the solution found for a true path-independent integral in linear thermoelasticity (Bui and Proix 1984). The method of solution to derive a path-independent integral from the one which is *not* a true path-independent integral, appears to be a quite general one. It is based on the notion of *Duality*.

The notion of duality and symmetry is closely linked to the concept of "Virtual Power". This was introduced in France in the Mechanics of continuous media by Paul Germain for an adequate representation of the action (forces, stress...) on a body (Germain 1973, 1978). In one of his paper, he wrote "*This concept is very seldom considered in the English scientific community, which directly made use of equations, for example the classical Newton law* ($\mathbf{f} = m\mathbf{a}$) *or Cauchy law* ($\text{div}\sigma = \rho\mathbf{a}$)". It originates from the mathematical concept of *spaces*

Dedicated to George Herrmann.

H. D. Bui (✉)
Department of Mechanics, Laboratory of Solid Mechanics, Ecole Polytechnique, Palaiseau, France
e-mail: hdbui37@yahoo.fr

H. D. Bui
Lamsid/CNRS Electricite de France, Clamart, France

doi: 10.1007/978-1-4020-6929-1_15,

and *dual spaces* of functions. As an example, to introduce the generalized functions, including the Dirac Delta function, Laurent Schwartz invented the Distribution theory (for which he was awarded the Fields medal, 1951). A distribution is a continuous linear form in some space of function F, equipped with some topology. It belongs to the *dual* space F′. In the mechanics of continuous media, solids or fluids, duality is always present in the formulation of mechanical problems. As Paul Germain liked to tell us that ≪force is the dual of the mobility≫, we kept in our mind that "force" is indeed a dual vector, i.e. an element of V′, the dual space of the space V of velocity fields. Stress is an element of the dual space D′, which is dual to the strain rate space D etc., so that the stress space S is identical to $D' \equiv S$. By an extension, stress and strain are often considered as dual variables. The interpretation of stress as the dual of an element in the strain rate space D leads to the abstract definition of stress as a linear form on D, called a virtual power. The duality becomes the bilinear form denoted by $\langle d', d\rangle$, i.e. the map $D' \times D \rightarrow \mathbb{R} : (d',d) \rightarrow \langle d', d\rangle$.

Let us consider an elementary example. We wish to find a force F equal to the prescribed one F^d. It is thus equivalent to require the equality between scalars $\langle F,v^*\rangle = \langle F^d,v^*\rangle$ for any $v^* \in V$. From the linearity of the form $\langle .,.\rangle$ we get $\langle F - F^d, v^*\rangle = 0, \forall v^* \in V$. This concept was known in analytical mechanics since Lagrange, two and half centuries ago. In modern computational mechanics, virtual motion and virtual displacement are known as *test functions*. Therefore, there are no new topics unfamiliar to every body, but only new interpretations and new applications allowed by the concepts of duality, virtual power and symmetry loss, as illustrated by many papers devoted to inverse problems.

2 Generalized Tonti's diagram

In the monograph of Tonti (1975) one can find duality in various domains of Physics (electromagnetism, gravitation, thermodynamics, electrostatics, quasi-static elasticity, rod, strings, etc.). When I generalized Tonti's diagram of elasto-statics to dynamic elasticity (Bui 1992), I saw the beautiful structure of the equations using dual variables. I discovered that the links between dual variables are always governed by differential operators and their adjoint ones (*) (Fig. 1).

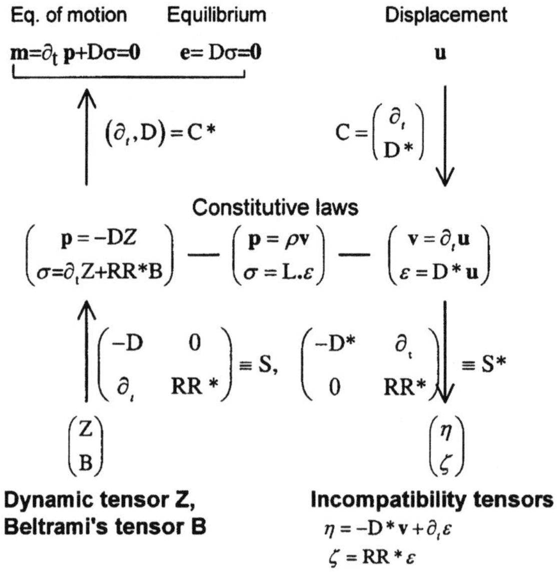

Fig. 1 Generalisation of Tonti's diagram to elastodynamics

In elasticity or plasticity theory, most works presented the field equations of equilibrium and constitutive laws with the *stress* space, ignoring another dual presentation of equations with the *strain* space. It is not simply an academic point of view, but sometimes it is a necessity. An example is given by the softening of elastic–plastic materials which convinces us that, to describe the decrease of the load and to avoid the ambiguity between elastic unloading and plastic loading with softening, it is necessary to consider the strain space (Nguyen and Bui 1974; Lubliner 1974). Another example of application of duality is obtained by completing the diagrams of Fig. 1 by the Helmholtz potentials φ, **H**, such that $\mathbf{u} = \text{grad}\ \varphi + \text{curl}\ \mathbf{H}$. The pairs ($\varphi$,**H**) and (B,Z), called *conjugate* potentials, enable the derivation of dynamic boundary integral equations, with *symmetrical* and *regular* kernels, while classical methods deal with unsymmetrical and singular kernels (Bui 1992).

(*) Tonti's diagram for elastic rods, beams and elastostatics is generalized to elastodynamics, using operators C, S* in the kinematic chart which are adjoint to C* and S of the dynamic chart. Operators D, D* are $D = -\text{div}$, $D^* = \frac{1}{2}(\nabla + \nabla^t)$. R is the left curl for symmetrical second order tensor, R* is the right curl for symmetrical second order tensor.

3 The DN and ND maps

When I was still a research student, I wondered how the Dirichlet boundary value problem in elasticity, with the prescribed datum $u_i^d(x)$, $x \in \partial\Omega$ on the boundary can be replaced by the corresponding Neumann boundary value problem with the prescribed datum $T_i^d(x)$, $x \in \partial\Omega$. I found the solution to my problem for a half plane but ignored at this time that the map $u_i^d(x) \rightarrow T_i^d(x)$, called the *Dirichlet-to-Neumann* map (DN) or its inverse the *Neumann-to-Dirichlet* map (ND), will have very important applications in the solution of some inverse problems in elasticity investigated in the 90s by mathematicians. This paper published in the sixties (Bui 1967), indicates that the concept of duality has been the driving concept of my works for many decades. The DNs are the key tools to solve the "Crack detection problem by a geometry approach" (Bui 1993).

4 Duality in fracture mechanics

In 1973, I found that Rice's J-integral is only one description of the energy release rate G, by a path-independent integral and that the dual I-integral is another possible one. This offers a great advantage in considering both descriptions with dual variables and spaces, conservation laws and dual laws since the minimum theorems for the potential energy $W(\varepsilon)$ and the complementary potential $U(\sigma)$, under certain conditions, provide us exact bounds of the J-integral (Bui 1974). Another important applications of the virtual power principle in Fracture Mechanics is provided by the notion of the *virtual crack propagation*. Classically, one deals with the energy release rate G as the *derivative* of the energy with respect to the crack length. Therefore, G in mixed modes I+II is well known as the quadratic form $G = (1-\nu^2)(K_I^2 + K_{II}^2)/E$. The question was raised on how to separately extract the stress-intensity factors. Many methods were proposed consisting in calculating the derivatives of the energy in the Ox_1 direction $J_1 = (1-\nu^2)(K_I^2 + K_{II}^2)/E$ (crack propagation along Ox_1) and in the Ox_2 direction $J_2 = -(1-\nu^2)K_I K_{II}/E$ (crack translation out of its plane). Such an unphysical method (for an actual derivative) was criticized by many authors. I tried to look at the virtual power method, with arbitrary adjoint fields u^* and discovered that the virtual power of the energy of a cracked body, in two-dimensions, is equal to the bilinear form $G(u, u^*) = 2(1-\nu^2)(K_I K_I^* + K_{II} K_{II}^*)/E$. By choosing a symmetric adjoint field ($K_I^* = 1, K_{II}^* = 0$) we extracted the SIF in mode I by $G(u, u^*) = 2(1-\nu^2)K_I/E$ and similarly by considering an anti-symmetric adjoint field ($K_I^* = 0, K_{II}^* = 1$) I obtained the stress-intensity factor in mode II. The virtual crack propagation is richer than the derivative of energy since it contains the classical result $1/2G(u, u) \equiv J_1$-integral. There is a profound difference between the virtual crack propagation method and the J_2-method which involves a crack translation out of its plane with the *same* loading. The virtual method $G(u, u^*)$ is based on a crack propagation in its direction, but under a *virtual load* giving rise to u^*, K_I^* and K_{II}^*.

In the 70s, I observed some intriguing results on the energy release rate in elastodynamics. As a student, I always learned that a formula describing physical phenomena must be independent of the motion of the frame reference in which measurements are made. This is the objectivity principle in Physics. The energy release rate formula in elasto-dynamics for a moving crack with the velocity V does not satisfy this principle since the velocity is *explicitly* present in its expression, in plane strain mode I (Achenbach and Bazant 1972)

$$G = \frac{1-\nu^2}{E} K_I^2 f_I(V) \tag{1}$$

where $f_I(V) = \beta_1(1-\beta_2^2)/\{(1-\nu)(4\beta_1\beta_2 - (1+\beta_2^2)^2\}$, $\beta_i = \sqrt{1-(V^2/c_i^2)}$, c_1 for P-wave, c_2 for S-wave. How to restore the objective formula for the dynamic G? My response to this question was *duality*.

Let me introduce for the mode I the same local definitions of *stress-intensity factor* and *crack displacement intensity factor* as known in quasi-statics, respectively

$$K_I^\sigma = \lim_{r\to 0} \sigma_{22}\sqrt{2\pi r},$$

$$K_I^u = \lim_{r\to 0} \frac{\mu}{4(1-\nu)} [[u_2]] \sqrt{\frac{2\pi}{r}} \tag{2}$$

In quasi-statics, both definitions provide the same SIF. In elastodynamics, I found a symmetrical formula for the energy release rate which is nothing but the duality between stress and strain rate near the moving crack tip (Bui 1977)

$$G = \frac{1-\nu^2}{E} K_I^\sigma K_I^u \tag{3}$$

This objective formula agrees with the traditional one since it can be proved that $K_I^u = K_I^\sigma f_I(\mathrm{V})$.

Another beautiful application of duality is about a conservation law in linear thermo-elasticity, which is George Herrmann's question to me. It is well known that classical conservation laws in thermo-elasticity involves a source term, namely in the form $\mathrm{div}\mathrm{A}(\mathbf{u}, \tau) = \mathrm{B}\mathbf{u}, \tau)$. Precisely because of the source term B that the thermo-elastic J-integral is not a purely path-independent integral, since it involves an area integral too. The symmetry is *lost* when we consider the pair $(\mathbf{u}, \tau)$ lone. We restore the symmetry by considering the dual pair $(\{\mathbf{u}, \tau\}, \{\mathrm{u}^*, \mathrm{w}^*\})$ and obtained a conservation law in the form

$\mathrm{div}\mathrm{A}(\mathbf{u}, \tau; \mathrm{u}^*, \mathrm{w}^*) = 0$ using dual variables, without a source term. The conservation law in linear thermoelasticity of the pure divergence form, for a line crack problem (along negative Ox_1) is given by

$$\frac{\partial}{\partial x_j}\left\{\frac{1}{2}u_i\sigma^*_{ij,1} - \frac{1}{2}u^*_{i,1}\sigma_{ij} - \gamma\tau(u^*_{1,j} - w^*_{,j}) + \gamma\tau_{,j}(u^*_1 - w^*)\right\} = 0 \tag{4}$$

where $\gamma = -\alpha\mu(3\lambda + 2\mu)/(\lambda + \mu)$, α is the thermal coefficient, λ and μ are Lamé's coefficients. The actual temperature field τ as well as the scalar adjoint field w^* are harmonic, while the field $\mathbf{u}^*$ is an elastic one. The stress free condition is assumed on the crack $\sigma \cdot \mathbf{n} = 0$ as well as the zero normal heat flux $\partial\tau/\partial \mathrm{n} = 0$. The adjoint fields $(\mathbf{u}^*, \mathrm{w}^*)$ is *not* a thermoelastic one. There is a small coupling between adjoint fields, by imposing the following condition on the crack faces $\partial(\mathrm{w}^* - \mathrm{u}_1^*)/\partial\mathrm{n} = 0$, which has no precise physical meaning $^{(*)}$. This boundary condition on the crack faces has only been introduced as a mathematical condition which enables the following *path independent* integral

$$T = \int_\Gamma \left\{\frac{1}{2}u_i\sigma^*_{ij,1}n_j - \frac{1}{2}u^*_{i,1}\sigma_{ij}n_j - \gamma\tau(u^*_{1,n} - w^*_{,n}) + \gamma(u^*_1 - w^*)\tau_{,n}\right\}\mathrm{d}s \tag{5}$$

The path independent T-integral in thermoelasticity (5) was presented at the Eshelby Symposium (*Fundamentals of deformation and Fracture*, Sheffield April, 1984), in honour of a great scientist who impinged on many works in Fracture Mechanics. Among applications of Eq. 5 to Fracture Mechanics, we recall the most important one, namely $T = (1 - \nu^2)(K_I K_I^* + K_{II} K_{II}^*)/E$ in mixed mode plane strain, whenever the adjoint field $\mathbf{u}^*$ is taken as the asymptotic field for a semi-infinite crack, with stress-intensity factors K_I^*, K_{II}^* and the other adjoint field is $w^* = (2K_I^*/\mu)(r/2\pi)^{1/2}(1 - \nu)\cos(\varphi/2)$. For a cracked body having a line crack along the negative part of Ox_1, the condition $\partial(\mathrm{w}^* - \mathrm{u}_1^*)/\partial\mathrm{n} = 0$ is satisfied by such explicit asymptotic solutions $\mathbf{u}^*$, w^* so that there is no need for a finite element computation of the adjoint fields which would equally introduce further difficulties in the estimation of near tip fields.

Does a symmetry exist in conservation laws in elastodynamics? It is clear that the conservation law $\mathrm{div}\sigma[\mathbf{u}] = \rho\ddot{\mathbf{u}}$ or the ones derived by Fletcher (1976) are not symmetric. To restore the symmetry, it is necessary to introduce adjoint fields $\mathbf{v}(\mathbf{x}, \mathrm{t}; \tau)$ satisfying the elastodynamic wave equations $\mathrm{div}\sigma[\mathbf{v}] = \rho\ddot{\mathbf{v}}$ such that $\mathbf{v}(\mathbf{x}, \mathrm{t}; \tau) \equiv 0$ for $\mathrm{t} > \tau$ where τ is an arbitrary constant. We obtained the symmetric conservation law in elastodynamics given in (Bui and Maigre 1988) as

$$\mathrm{div}\left\{\int_0^\tau (\mathbf{n} \cdot \sigma[\mathbf{u}] \cdot \mathbf{v} - \mathbf{n} \cdot \sigma[\mathbf{v}] \cdot \mathbf{u})\mathrm{d}t\right\} = 0 \tag{6}$$

This conservation law has been exploited in (Bui et al. 1992) to extract the stress-intensity factors in dynamic modes I and II by choosing appropriately the adjoint dynamical fields.

(*) The loss of physical significance of adjoint problems is often encountered. For example consider the heat diffusion equation $\mathrm{div}(\mathrm{kgradT}) - \partial_t\mathrm{T} = 0$. Its adjoint equation $\mathrm{div}(\mathrm{kgradT}^*) + \partial_t\mathrm{T}^* = 0$ is not physical, since it describes backwards diffusion, violating the second principle of thermodynamics.

5 Duality in plasticity

Dual variables are crucial in the thermodynamics of irreversible processes. The contributions of J.J. Moreau, Q.S. Nguyen, P. Germain, P. Suquet, A. Ehrlacher, C. Stolz and others in France, during the period 1960–1990 are very important to clarify the nature of dissipation in Plasticity and Fracture. Internal rate variables $\dot{\alpha}$, including the plastic strain rate $\dot{\varepsilon}^p, \dot{\alpha} = (\dot{\varepsilon}^p, \dot{\beta})$ describe the evolution of materials. The variable $\dot{\alpha}$ is the dual to the generalized force $\mathbf{A}$, so that $\mathbf{A} \cdot \dot{\alpha} \geq 0$ represents the *dissipation rate*. If one introduces the free energy per unit volume $\mathrm{W}(\varepsilon, \alpha)$ so that

$$\sigma = \frac{\partial W}{\partial \varepsilon}, \quad \mathbf{A} = -\frac{\partial W}{\partial \alpha} \tag{7}$$

then one obtains the state equation. One needs to introduce a complementary law by introducing a pseudo-potential $\Phi(\dot{\alpha})$ so that (Halphen and Nguyen 1975)

$$A = \frac{\partial \Phi}{\partial \dot{\alpha}} \tag{8}$$

The dual presentation consists in introducing the *conjugate* function $\psi(\mathbf{A})$ in the sense that conjugate functions $\Phi(\dot{\alpha})$ and $\psi(\mathbf{A})$ are linked by Fenchel–Legendre transform

$$\Phi(\dot{\alpha}) = \sup_{A \in V} \{\mathbf{A} \cdot \dot{\alpha} - \Psi(\mathbf{A})\} \tag{9}$$

where V is some convex of generalized forces. In the smooth convex case one has $\dot{\alpha} = \partial \Psi(\mathbf{A})/\partial \mathbf{A}$, while in the case of a non differentiable convex, one may use the notion of sub-differential $\dot{\alpha} \in \partial \Psi(\mathbf{A})$ (Moreau 1963, 1966; Rockafellar 1966).

If the crack length is the state variable of the cracked body, then $\dot{a}J_{tip}$ can be identified as the *dissipation rate* due to fracture at the crack tip! Here we have defined J_{tip} as the J-integral for a vanishing contour around the crack tip. In plasticity J_{tip} is equal to zero, which is the paradoxical result revealed by Rice (1968), so that the dissipation rate in a cracked body is essentially distributed by plastic heating inside the solid domain rather than concentrated at the crack tip. In elasticity, J_{tip} is not equal to zero and $\dot{a}J_{tip}$ represents the dissipation rate the crack tip even in an elastic body (which is a non dissipative medium in its volume!). The dissipative nature of crack propagation in an elastic body resulted in a new interpretation of the energy release rate and to the discovery of the *logarithmic singularity* of the temperature field T(r,θ) for a moving crack tip which behaves like a moving point heat source (k: coefficient of Fourier's law)

$$\mathrm{T} \simeq -\frac{\dot{a} J_{tip}}{2k\pi} \log r \tag{10}$$

(Bui et al. 1980). These are new aspects of Fracture Mechanics based on thermodynamical considerations. Such considerations have been introduced in Plasticity by my teacher, Professor Jean Mandel, with whom I wrote my first research paper on experimental Plasticity (1962). Later, I published another paper on the experimental verification of his plastic dissipation formula $D_p = \int (\mathbf{T} \cdot \dot{\mathbf{u}} - \dot{\mathbf{T}} \cdot \mathbf{u})dt$, which is nothing but a *symmetry loss* in Plasticity (Bui 1965).

6 Symmetry loss, dissipation and inverse problems

Virtual power is more general than the time derivative of the energy. Dual variables in continuum Mechanics are more general than the variables considered in the formulation of equations. Consider the expression of the energy release rate in Linear Fracture Mechanics as the derivative of the energy with respect to the crack length

$$G = \frac{1}{2} \int_{\partial\Omega} \left(\mathbf{u} \cdot \frac{\partial \mathbf{T}}{\partial a} - \mathbf{T} \cdot \frac{\partial \mathbf{u}}{\partial a} \right) ds \tag{11}$$

I remember a discussion with Paul Germain in which he questioned me about the *anti-symmetry* found in the above formula, in the sense that the interchanges u $\leftrightarrow$ du/da and T $\leftrightarrow$ dT/da change the sign of G. It seems that G looks like a Poisson's bracket! like Mandel's formula of plastic dissipation. At this time I had no correct answer to his question on the anti-symmetry. Today, I can say that this is simply a *symmetry loss*. In recent works with my colleagues in two research teams at Ecole Polytechnique and Electricité de France I discovered that symmetry loss is a fundamental notion in crack detection problems. The reciprocity theorem in classical elasticity, with the symmetry in the bilinear form of the strain energy density $a(\mathbf{u}, \mathbf{v}) = a(\mathbf{v}, \mathbf{u})$, expresses the symmetry between two states $(\mathbf{u}^1, \mathbf{u}^2)$. Consider a homogeneous linear elastic solid with the two states and the integral R defined as

$$\mathrm{R} = \int_{\partial\Omega} (\mathbf{u}^1 \cdot T(\mathbf{u}^2) - \mathbf{u}^2 \cdot T(\mathbf{u}^1)) ds \tag{12}$$

which expresses the Betti reciprocity theorem by R = 0, revealing the *symmetry* between the two states, which is the consequence of the symmetry of the elastic law. The symmetry is lost when R $\neq$ 0, for instance in the case where the solid is not homogeneous or contains cracks. One field corresponds to the non homogeneous or cracked body, another field for the homogeneous one.

Therefore, R is a *defect indicator* (also called a *reciprocity gap*):

R = 0 ⇔ no defect inside $\partial\Omega$,

R ≠ 0 ⇔ existence of a defect.

Therefore, defects detection reduces to the search of the zeros of a functional. A series of recent papers of my two research teams showed that the reciprocity gap functional provided a closed form solution to some inverse crack detection problems, for electrostatics (Andrieux and Ben Abda 1992) static elasticity (Andrieux et al. 1999; Andrieux and Ben Abda 1996), diffusion equation (Ben Abda and Bui 2003), transient acoustics (Bui et al. 1999), and elastodynamics with the exact solution to an earthquake inverse problem (Bui et al. 2005a, b). Closed form solutions can only be obtained in particular case as for a planar crack in linear material. The method of exact solutions does apply neither to inhomogeneities nor to non linear materials. In the particular case of small inhomogeneities, i.e. small perturbations of the homogeneous medium, closed form solutions can be obtained for the linearized inverse problems:

a. For the scalar equation $\mathrm{div}\{(1+h(\mathbf{x})\mathrm{grad}\varphi\}=0$ with unknown $h(\mathbf{x})$, with data φ and $\partial\varphi/\partial n$ on the boundary, see the papers by Bui (1994) and Calderon (1980).
b. For a small isotropic perturbation δL of homogeneous isotropic elastic medium, we can cite Ikehata (1998).

7 Solution to the earthquake inverse problem

As shown above, the reciprocity gap R is defined as the *external* boundary functional over S_{ext} which is known from the data $\mathbf{u}^1 = \mathbf{u}^d$ and $\mathbf{T}(\mathbf{u}^1) = \mathbf{T}^d$ and from the chosen adjoint functions $\{\mathbf{u}^*, \mathrm{T}(\mathbf{u}^*)\}$, denoted by $R(\mathbf{u}^d, \mathbf{T}^d; \mathbf{u}^*, \mathbf{T}(\mathbf{u}^*))$. In planar crack detection problems in 3D quasi-static elasticity, we can prove the following variational equation: Find $\Sigma(u)$, $[[\mathbf{u}]]$ such that

$$\int_{\Sigma(u)} [[\mathbf{u}]] \cdot \mathbf{T}(\mathbf{u}^*) ds = \int_{S_{ext}} \left\{\mathbf{u}^d \cdot \mathbf{T}(\mathbf{u}^*) - \mathbf{u}^* \cdot \mathbf{T}^d\right\} ds \doteq R(\mathbf{u}^d, \mathbf{T}^d; \mathbf{u}^*, \mathbf{T}(\mathbf{u}^*)), \forall \mathbf{u}^* \quad (13)$$

In the homogeneous body case (no crack), the left hand side of the above equation equals zero. By R = 0, we recover the symmetry between fields **u** and $\mathbf{u}^*$. In the *symmetry lost* case, the above equation provides a *non-linear* equation for determining the crack plane (containing Σ) as well as the displacement discontinuity $[[\mathbf{u}]]$. The non-linearity come from the fact that $\Sigma(u)$ is unknown. It is *impossible* to solve the non-linear inverse problem with classical methods based on the fields equations, because the crack support $\Sigma(u)$ depends on the unknown **u**. Now, the variational form makes it possible to solve the inverse problem in a closed form. The method of solution consists of two steps:

1. Find the normal **n** to the fault plane by adequate choice of $\mathbf{u}^*$, and the fault plane $\Sigma(u)$.
2. Once the fault plane having determined, find the discontinuity $[[\mathbf{u}]]$.

Remark that step 2 is a *linear* inverse problem which is incomparably simpler than the original problem.

An adequate choice of the adjoint function allows the invertibility of the above equation. We exploit here the arbitrariness of the choice of functions $\mathbf{u}^*$ to obtain the desired results. Traditional methods deal with the fields equations (elastic equilibrium equation, boundary conditions, with an *unknown* geometry). Therefore, the only possible method consists in finding the best fitting of measurements with predicted data corresponding to some guess geometry S

$$\{\Sigma \text{ and } [[\mathbf{u}]]\} = \arg \min_{S, [[\mathbf{v}]]} \left\{\left|\mathbf{v} - \mathbf{u}^d\right|^2 + \left|\mathbf{T}(\mathbf{v}) - \mathbf{T}^d\right|^2\right\} \quad (14)$$

where **v** is the solution of the boundary value problem with the geometry Ω/S and with *one* of the boundary condition, either $\mathbf{u}^d$ or $\mathbf{T}^d$ (two possible numerical solutions!). This classical method of solution is essentially a numerical one. It is well known that the above optimization procedure is mathematically an ill-posed problem (Tikhonov and Arsenine 1977). Undoubtedly, the reciprocity gap functional, based on the symmetry loss in Fracture Mechanics is the right tool to solve these inverse problems in closed form.

In practice, the earthquake inverse problem considered here is illustrated in Fig. 2, with measured acceleration data on the ground G and estimated far field data on the half-sphere H using a classical model of point source (Aki and Richards 1980). The near fields measurements, together with the far-fields and adequate adjoint fields are inserted in the elastodynamic reciprocity gap functional

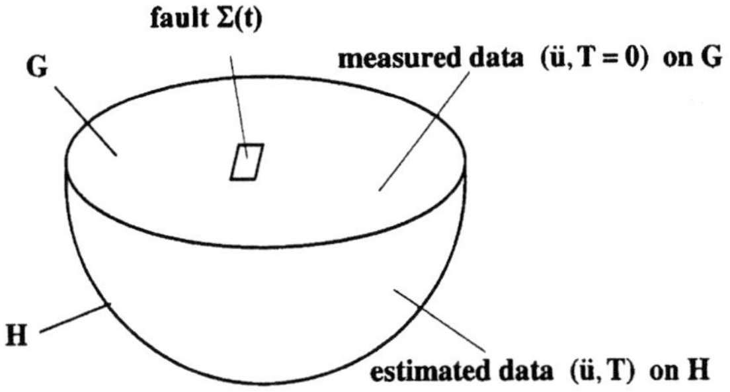

Fig. 2 The unknown moving fault Σ(t) in its plane near the free surface G. Acceleration data are measured on the ground G; Estimated data on the half-sphere H are provided by a model of a point source near the origin with measured seismic moments

$$\int_0^\infty \int_{\Sigma(u)} [[\mathbf{u}]] \cdot T(\mathbf{u}^*) \mathrm{d}s\mathrm{d}t$$
$$= \int_0^\infty \int_{S_{ext}} \left\{ \mathbf{u}^d \cdot \mathbf{T}(\mathbf{u}^*) - \mathbf{u}^* \cdot \mathbf{T}^d \right\} \mathrm{d}s\mathrm{d}t$$
$$\doteq R(\mathbf{u}^d, \mathbf{T}^d; \mathbf{u}^*, \mathbf{T}(\mathbf{u}^*)) \qquad (15)$$

with adequate adjoint fields $\mathbf{u}^*$, one adjoint field to determine first the fault plane, and another one to determine the fault geometry.

Step 1 can be solved by different choice of adjoint fields. One interesting method is provided by the "*instantaneous reciprocity gap functional*" (IRGF) which makes use of plane shear impulse waves $\mathbf{k}$ of propagation vectors $\mathbf{p}$. The adjoint displacement is given by

$$\mathbf{u}^*(x, t; \tau) = a\mathbf{k}Y(t - \mathbf{x} \cdot \mathbf{p}/c_s - \tau), \qquad (16)$$

where τ is a parameter defining the initial wave front as $\mathbf{x} \cdot \mathbf{p}/c_s + \tau = 0$ $(t = 0)$, and $Y(y)$ is the *down* step function $Y(y < 0) = 1$, $Y(y > 0) = 0$ defined in terms of the usual Heaviside function $H(y)$ by $Y(y) = 1 - H(y)$ (Y is understood as the limit for $h \to 0$ of the regularized step function $Y_h(y)$ which represents a smooth transition from 0 to 1 in a narrow band of width h, at $y = 0$, $Y_h(y > h/2) = 0$, $Y_h(y < -h/2) = 1$). Suppose that parameter τ s chosen so that the initial wave front is outside the crack, in the left as displayed in Fig. 3. As the adjoint wave propagates in direction $\mathbf{p}$, towards the region where $\mathbf{u}^* \neq 0$, the adjoint field vanishes $\mathbf{u}^* = 0$ on the crack for any $t > 0$. Therefore, IRGF vanishes for any time $t > 0$. If the initial wave front intersects the crack, then IRGF $\neq 0$. The adjoint field vanishes again when the wave front "arrives" at the left crack tip.

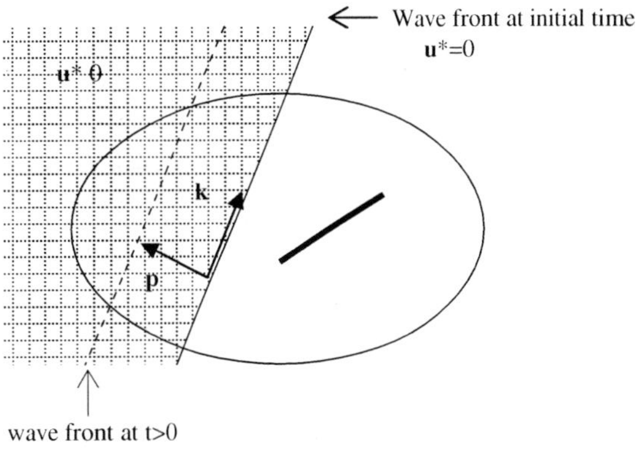

Fig. 3 Back propagation in direction $\mathbf{p}$ of shear adjoint wave $\mathbf{u}^*$ parallel to $\mathbf{k}$ ($\mathbf{k} \perp \mathbf{p}$)

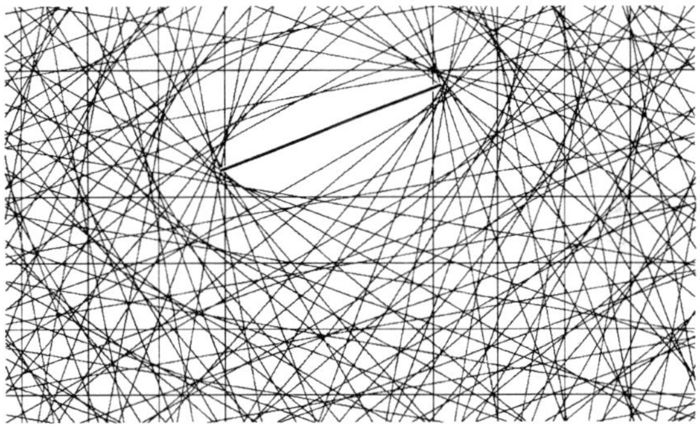

Fig. 4 Convex hull containing the crack defined by the positions of plane shock wave when $|\text{IRGF}| = 0(\varepsilon)$

If the initial wave front is still in the right of the right crack tip, the adjoint stress vector is equal to zero on the crack faces, as $\mathbf{u}^*$ is constant. The IRGF is equal to zero up to the moment when the wave front reaches the crack tip. By varying both the directions of propagation vectors $\mathbf{p}$ and the time parameters τ, one obtains a set of adjoint waves encompassing the crack. The condition IRGF = 0 of the set of adjoint fields defines a *convex hull* containing the crack. The smallest convex hull is obtained when all waves "arrive" at the crack tips (or the crack front in 3D). The numerical criterion for the arrival times of waves is IRGF = $O(\varepsilon)$, where ε is a given small number.

Figure 4 shows the computed wave front when the magnitude of IRGF is equal to a given small quantity $O(\varepsilon)$. In this case the adjoint stress is the travelling Dirac delta. The simulated earthquake is provided by a sudden release of geological stresses on the crack.

This is analogous to X-rays tomography. When the ray is outside a tumour, one does not have any attenua-

Table 1 Duality in solid mechanics

Variables & functions	Action or results	Dual variables & conjugate functions	Remarks
Displacement **u**	Work	Force **f**, traction vector **T**	
Virtual velocity **v***	Virtual power	Force **f**, traction vector **T**	
Deformation ε, strain rate $\dot{\varepsilon}$		Stress σ	
Potential energy P		Complementary potential Q	
Thermoelastic fields **u**, τ	Conservation law	Adjoint fields **u***, w*	Symmetry between (**u**, τ)
Internal variables α	Dissipation $A \cdot \dot{\alpha}$	Generalized forces A	and (**u***, w*)
$\Phi(\dot{\alpha})$ (pseudo-dissipation)		$\Psi(A)$ (Conjugate function)	$\Phi(\dot{\alpha}) = \sup_{A \in V} \{\mathbf{A} \cdot \dot{\alpha} - \Psi(\mathbf{A})\}$
J-integral	Derivative of the energy—dW/da	Dual I-integral	
J(u,u*)	Virtual power		J-integral $=\frac{1}{2}J(\mathbf{u}, \mathbf{u})$
$(\mathbf{v}, \varepsilon)^{\tau} = \mathbf{C}\mathbf{u}$	Tonti's diagram	$C^*(\mathbf{p}, \sigma)^{\tau} = (\mathbf{m}, \mathbf{e})$	
	Equation of motion	**m**=**0**	
	Equilibrium equation	**e**=**0**	
$S[\mathbf{v},\varepsilon]^{\tau} = [\eta, \zeta]^{\tau}$	Constitutive law	$S^*[Z,B]^{\tau} = [\mathbf{p}, \sigma]^{\tau}$	
$\eta = 0$	Compatibility $v \leftrightarrow \dot{\varepsilon}$		
$\zeta = 0$	Compatibility of ε		
Projection P	X-rays tomography	Back projection P*	The inverse Radon *trans-* form makes use of P and P*
Propagation	Scattering of waves	Back propagation	
Forward equations	Reciprocity gap	Time reversal mirror	
	Functional (RG)	Adjoint equations	D (space of C^{∞}compactly
Forward diffusion	RG functional	Backward diffusion	supported function), D′ (space of distributions)
State equation	Control theory	Adjoint state equation	S (rapidly decreasing
Primal problem	Convex analysis	Dual problem	functions f at infinity,
Functional spaces:	Mathematical	Dual spaces:	$f/\lvert x\rvert^{n} \to 0 \forall n$ positive
D, S	analysis	D′, S′ (Schwartz's spaces)	integer), S′ (tempered
$H^{m,p}$		$H^{-m,p'}$ (Sobolev's spaces)	distributions) (1/p)+(1/p′)=1.

tion of the X-ray signal A = 0. But an attenuation signal A ≠ 0 is detected when the ray intersects the tumour. This is why the terminology of "generalized tomography" is sometimes used in the literature of inverse problems (Bui 2006).

The IRGF provides both the normal and the fault plane. It does not determine the crack geometry when the crack is concave. To solve step 2, we make use of the following adjoint field

$$\mathbf{u}^* \equiv \mathbf{w}^{(s,q)}(\mathbf{x}, t) = \mathrm{curl}\{\psi(\mathbf{x}, t; \mathbf{s}, q)\mathbf{e}^3\} \tag{17}$$

where $\mathbf{x} = (x_1, x_2, x_3)$, with Ox_3 along the normal $\mathbf{n} = \mathbf{e}^3$ to the fault plane $x_3 = 0$. The adjoint field is parameterized by two variables $\mathbf{s} = (s_1, s_2, 0)$ and q a real number, and

$$\psi(\mathbf{x}, t; \mathbf{s}, q) = \exp(iqt - \eta t)\exp(i\mathbf{s} \cdot \mathbf{x})\exp[x_3\{|\mathbf{s}^2| + (iq - \eta)^2/c_2^2\}^{1/2}] \tag{18}$$

where η is a real vanishing positive number ($\eta \to 0^+$) which has been introduced for convergence reasons (in the space of tempered distributions). Let us introduce the vector $[[\mathbf{u}^{\perp}]] = ([[u_2]], -[[u_1]], 0)$ orthogonal to vector $[[\mathbf{u}]]$ in the fault plane ($x_3 = 0$). Vectors $[[\mathbf{u}]]$ and $[[\mathbf{u}^{\perp}]]$ as well as the field div $([[\mathbf{u}^{\perp}]])$ have the same spatial support in the fault plane. From Eq. 15, it can be proved that div $([[\mathbf{u}^{\perp}]])$ is explicitly determined by the data $R(\mathbf{u}^d, \mathbf{T}^d; \mathbf{w}^{(s,q)}, \mathbf{T}(\mathbf{w}^{(s,q)}))$, via an inverse *spatial* Fourier transform $(F_x)^{-1}$ with respect to variables (s_1,s_2) images of (x_1, x_2) and an inverse *time* Fourier transform, $(F_t)^{-1}$ with respect to q, with F_t defined as

$$F_t[[\mathbf{u}]](\mathbf{x}, q) \doteq \int_0^\infty [[\mathbf{u}(\mathbf{x}, t)]] \exp(iqt) \mathrm{d}t$$

$$\begin{aligned} \operatorname{div}([[\mathbf{u}^\perp]]) = & \frac{1}{2\mu}(F_\mathrm{t})^{-1}(F_\mathrm{x})^{-1} \\ & \times \mathrm{R}\left(\mathbf{u}^\mathrm{d}, \mathbf{T}^\mathrm{d}; \mathbf{w}^{(\mathrm{s,q})}, \mathbf{T}\left(\mathbf{w}^{(\mathrm{s,q})}\right)\right) \\ & \times \left\{|\mathrm{s}|^2 - (\mathrm{q} + \mathrm{i}\eta)^2/\mathrm{c}_1^2\right\}^{-1/2} \quad (\eta \to 0^+) \end{aligned} \tag{19}$$

Equation 19 *explicitly* solves the earthquake inverse problem for determining the *history* of the fault, since $\Sigma(\mathrm{t}) = \operatorname{supp}\{\operatorname{div}([[\mathbf{u}^\perp]])\}$.

8 Concluding remarks

I would like to mention first that *duality* is a very old philosophical principle in Asia. Duality is synonym of parallelism, or complementary things, sometimes an opposition between things: Yin and Yang in China, Am and Duong (the Vietnamese words for Female and Male, respectively), Positive and Negative, the Sky and the Earth, Water and Fire etc. In Sciences, duality is found in Solid Mechanics, X-rays Tomography and Mathematics (Schwartz 1978) etc., see Table 1.

References

Achenbach JD, Bazant ZP (1972) Elastodynamic near-tip stress and displacement fields for rapidly propagating cracks in orthotropic materials. J Appl Mech 97:183

Aki K, Richards PG (1980) Quantitative seismology, theory & methods, vol 1. W.H. Freeman and Cie, New York

Andrieux S, Ben Abda A (1992) Identification de fissures planes par une donnée de bord unique: un procédé direct de localisation et d'identification. C R Acad Sci Paris 315(I):1323–1328

Andrieux S, Ben Abda A (1996) Identification of planar cracks by complete overdetermined data inversion formulae. Inverse Probl 12:553–563

Andrieux S, Bui HD, Ben Abda A (1999) Reciprocity and crack identification. Inverse Probl 15:59–65

Ben Abda A, Bui HD (2003) Planar cracks identification for the transient heat equation. Inverse Ill-posed Probl 11(1):67–86

Bui HD (1965) Dissipation of energy in plasticity. Cahier Groupe Français de Rhéologie 1:15

Bui HD (1967) Transformation des données aux limites relatives au demi-plan élastique. C R Acad Sci Paris 265:862–865

Bui HD (1974) Dual path-independent integrals in the boundary-value problems of cracks. Eng Fract Mech 6:287–296

Bui HD (1977) Stress and crack displacement intensity factors in elastodynamics. In: Proceedings of 4th international conference fracture, vol 3. Waterloo, p 91

Bui HD (1984) A path-independent integral for mixed modes of fracture in linear thermo-elasticity. In: IUTAM symposium on fundamental of deformation and fracture. Sheffield, p 597, April 1984

Bui HD (1992) On the variational boundary integral equations in elastodynamics with the use of conjugate functions. J Elast 28:247

Bui HD (1993) Detection de fissure par une méthode géométrique, In: Horowitch J, Lions JL (eds) ≪A propos des grands Systèmes des Sciences et de la Technologie≫. Masson, Paris

Bui HD (1994) Inverse problems in the mechanics of materials: an introduction. CRC Press, Boca Raton

Bui HD (2006) Fracture mechanics: inverse problems and solutions. Springer

Bui HD, Maigre H (1988) Extraction of stress intensity factors from global mechanical quantities. C R Acad Sci Paris 306(II):1213

Bui HD, Proix JM (1984) Lois de conservation en thermoélasticité linéaire. C R Acad Sci Paris 298(II):325

Bui HD, Ehrlacher A, Nguyen QS (1980) Crack propagation in coupled thermo-elasticity. J Meca 19:697. Gauthier-Villars, Paris

Bui HD, Maigre H, Rittel D (1992) A new approach to the experimental determination of the dynamic stress intensity factors. Int J Solids Struct 29:2881–2895

Bui HD, Constantinescu A, Maigre H (1999) Inverse scattering of a planar crack in 3D acoustics: closed form solution for a bounded solid. C R Acad Sci Paris 327(II):971–976

Bui HD, Constantinescu A, Maigre H (2005a) An exact inversion formula for determining a planar fault from boundary measurements. Inv Ill-posed Probl 13(6):553–565

Bui HD, Constantinescu A, Maigre H (2005b) The reciprocity gap functional for identifying defects and cracks. In: Mroz Z, Stavroulakis GE (eds) In Parameter identification of materials and structures. CISM course and lecture, vol 469. Springer, Wien, New York

Calderon AP (1980) On an inverse boundary problem. In: Meyer WH, Raupp MA (eds) Seminar on numerical and applications in continuum physics. Brazilian Mathematical Society, Rio de Janeiro, pp 65–73

Fletcher DC (1976) Conservation laws in linear elastodynamics. Arch Rat Mech Anal 60:329

Germain P (1973) The method of virtual power in continuum mechanics. Part II. In: Domingos JJD et al (eds) Applications to continuum thermodynamics. J. Wiley, New York, pp 317–333

Germain P (1978) Duality and convection in continuum mechanics. In: Fichera G (ed) Trends in applications to mechanics. Pitman, London, pp 107–128

Halphen B, Nguyen QS (1975) Sur les matériaux standards généralisés. J Méca 14:39

Ikehata M (1998) Inversion for the linearized problem for an inverse boundary value problem in elastic prospection. SIAM J Math Anal 50(6):1635

Lubliner J (1974) A simple theory of plasticity. Int J Solids Struct 10:310

Mandel J (1965) Energie élastique et travail dissipé dans les modèles rhéologiques. Cahier Groupe Français de Rhéologie 1:1

Moreau JJ (1963) Fonctionnelles sous différentiables, C R Acad Sci Paris 257:4117–4119
Moreau JJ (1966) Fonctionnelles convexes. Séminaire Equations aux Dérivées Partielles. Collège de France
Nguyen QS, Bui HD (1974) Sur les matéraux à écrouissage positif ou négatif. J Meca 13(2):321–342
Rice JR (1968) Mathematical analysis in the Mechanics of Fracture. In: Liebowitz H (ed) Fracture. Academic Press, p 191
Rockafellar RT (1966) Characterization of the subdifferentials of convex functions. Pac J Math 17:497–510
Schwartz L (1978) Théorie des distributions. Hermann, Paris
Tikhonov A, Arsenine V (1977) Méthode de résolution de problèmes mal posés. Editions Mir, Moscou
Tonti E (1975) On the formal structure of physical theories. Cooperative Library Instituto di Polytechnico di Milano, Milano

Phase field simulation of domain structures in ferroelectric materials within the context of inhomogeneity evolution

Ralf Müller · Dietmar Gross · David Schrade · B. X. Xu

Abstract A phase field model for simulating the domain structures in ferroelectric materials is proposed. It takes mechanical and electric fields into account, thus allowing for switching processes due to mechanical and/or electrical loads. The central idea of the model is to take the spontaneous polarisation as an order parameter and to provide an evolution law for this parameter. The concept of evolving inhomogeneities (configuratioanl forces) can be used in this context, as the spatial distribution of the spontaneous polarisation describes the inhomogeneity of the system. The evolution is found to be in agreement with the second law of thermodynamics and to resemble the (classical) Ginzburg-Landau equation. Numerical simulations show the features of the model and the interaction of domain structures with defects.

Keywords Phase field · Configurational mechanics · Micromechanics · Ferroelectrics · Finite element method

1 Introduction

Ferroelectric materials are widely used in sensor and actuator applications. These materials allow for a conversion of electric signals into mechanical output and vice versa. The working principle is based on the atomic structure of these materials. Many piezoelectric materials have a so-called Perovskite structure. At a critical temperature, the Curie temperature, a spontaneous break of the symmetric distribution of charge carriers in unit cells occurs. This leads to a spontaneous polarisation, which is responsible for the multifunctional material behavior.

On the microscopic level the spontaneous polarisation arranges in areas in which it is almost constant. These areas are termed domains. In between these domains interfaces form, which are called domain walls. Due to the crystal structure a single crystal contains 90 and 180° domain walls, according to the change of the polarisation across a domain wall. Under external loads of mechanical or electric character these domain structures change, and cause a rather complex microstructural evolution, which influences the meso- and macroscopic material behaviour. The macroscopic material properties are often modeled by ferroelectric material models, which resemble plasticity models. The main shortcoming of these models is that they take the microstructure only in an averaged sense into account, neglecting the fine scale structures of the domains. The intention of this investigation is to contribute to the simulation of these ferroelectric materials on the mesoscopic level of domains. In the present approach a phase field model based on continuum physics is proposed and developed. The phase field model, also sometimes termed order parameter model, is derived and analysed using thermodynamic arguments within a

R. Müller (✉) · D. Gross · D. Schrade · B. X. Xu
Civil Engineering and Geodesy, TU Darmstadt, Hochschulstr.1, 64289 Darmstadt, Germany
e-mail: r.mueller@mechanik.tu-darmstadt.de

Defect and Material Mechanics. C. Dascalu, G.A. Maugin & C. Stolz (eds.),
doi: 10.1007/978-1-4020-6929-1_16,

configurational mechanics approach of evolving inhomogeneities. Similar approaches can be found in modelling of martensitic phase transitions, see for example McCormack et al. (1992), Müller (1998), Wang et al. (1993). In the context of ferroelectric materials phase field models can be found in Wang et al. (2004), Wang and Zhang (2006a, b), Zhang and Bhattacharya (2005a, b), Bhattacharya and Ravichandran (2003), Soh et al. (2006). In other models the domain walls are treated as sharp interfaces, at which jump conditions have to be satisfied. Within the context of ferroelectrics and in a variational setting of configurational forces this can be found in Mueller et al. (2005). A thermodynamic approach based on configurational forces on a point defect or a domain wall is given in Goy et al. (2006), Schrade et al. (2007). For the general theory of configurational forces the reader is referred to Maugin (1993), Gurtin (1996). The model presented here is solved by using finite element techniques. For a general treatment of configurational forces in a finite element context, we refer the reader to Mueller et al. (2002), Mueller and Maugin (2002) and the works cited in there. Within the context of ferroelectrics stationary solutions are given in Su and Landis (2006, 2007). Various examples demonstrate the main features and possible applications of the model.

2 Phase field model

2.1 Standard balance law

We will assume quasi-stationary conditions for both the mechanical and electric fields, thus the stress $\boldsymbol{\sigma}$ satisfies an equilibrium condition and the electric displacement $\boldsymbol{D}$ obeys Gauß' law

$$\operatorname{div}\boldsymbol{\sigma} + \boldsymbol{f} = \boldsymbol{0}, \quad \operatorname{div}\boldsymbol{D} = q, \tag{1}$$

where $\boldsymbol{f}$ represents a volume force and q a charge density. In addition there might be boundary conditions by tractions and surface charges, given by

$$\boldsymbol{\sigma n} = \boldsymbol{t}^* \quad \text{on } \partial\mathcal{B}_t, \quad \boldsymbol{D}\cdot\boldsymbol{n} = -Q^* \quad \text{on } \partial\mathcal{B}_q. \tag{2}$$

The strains $\boldsymbol{\varepsilon}$ and the electric field φ are defined by

$$\boldsymbol{\varepsilon} = \frac{1}{2}\left(\operatorname{grad}\boldsymbol{u} + \operatorname{grad}^{\mathrm{T}}\boldsymbol{u}\right), \quad \boldsymbol{E} = -\operatorname{grad}\varphi, \tag{3}$$

where $\boldsymbol{u}$ is the displacement and φ the electric potential. These fields are associated with appropriate boundary conditions, as

$$\boldsymbol{u} = \boldsymbol{u}^* \quad \text{on } \partial\mathcal{B}_u, \quad \varphi = \varphi^* \quad \text{on } \partial\mathcal{B}_\varphi. \tag{4}$$

To close the system constitutive relations are needed. For a piezoelectric material usually a linear relation between $\boldsymbol{\sigma}$ and $\boldsymbol{E}, \boldsymbol{\varepsilon}$ as well as $\boldsymbol{D}$ and $\boldsymbol{E}, \boldsymbol{\varepsilon}$ are postulated. These material laws neglect the spontaneous polarisation $\boldsymbol{P}$, which must to be considered on the mesoscopic level of domains.

2.2 Configurational force balance

Assuming that in the volume the following equation

$$\operatorname{div}\tilde{\boldsymbol{\Sigma}} + \tilde{\boldsymbol{g}} = \boldsymbol{0} \tag{5}$$

holds true, with appropriate boundary conditions, we will formulate the second law of thermodynamics in the following way

$$\begin{aligned}\int_{\partial\mathcal{B}} (\boldsymbol{t}^*\cdot\dot{\boldsymbol{u}} - Q^*\dot{\varphi})\,dA + \int_{\mathcal{B}} (\boldsymbol{f}\cdot\dot{\boldsymbol{u}} - q\dot{\varphi})\,dV \\ + \int_{\partial\mathcal{B}} (\tilde{\boldsymbol{\Sigma}}\boldsymbol{n})\cdot\dot{\boldsymbol{P}}\,dA - \int_{\mathcal{B}} \dot{H}(\boldsymbol{\varepsilon}, \boldsymbol{E}, \boldsymbol{P}, \operatorname{grad}\boldsymbol{P})\,dV \geq 0,\end{aligned} \tag{6}$$

where the work done by the forces and charges on the surface and in the volume are taken into account. Using the balance law (5) the second law can be put in the format

$$\begin{aligned}\int_{\mathcal{B}} \Bigg[\left(\boldsymbol{\sigma} - \frac{\partial H}{\partial\boldsymbol{\varepsilon}}\right) : \dot{\boldsymbol{\varepsilon}} - \left(\boldsymbol{D} + \frac{\partial H}{\partial\boldsymbol{E}}\right)\cdot\dot{\boldsymbol{E}} \\ + \left(\tilde{\boldsymbol{\Sigma}} - \frac{\partial H}{\partial\operatorname{grad}\boldsymbol{P}}\right) : \overset{\cdot}{\overline{\operatorname{grad}\boldsymbol{P}}} - \left(\tilde{\boldsymbol{g}} + \frac{\partial H}{\partial\boldsymbol{P}}\right)\cdot\dot{\boldsymbol{P}}\Bigg]\,dV \\ \geq 0.\end{aligned} \tag{7}$$

From this we can deduce the following constitutive relations

$$\boldsymbol{\sigma} = \frac{\partial H}{\partial\boldsymbol{\varepsilon}}, \quad \boldsymbol{D} = -\frac{\partial H}{\partial\boldsymbol{E}}, \quad \tilde{\boldsymbol{\Sigma}} = \frac{\partial H}{\partial\operatorname{grad}\boldsymbol{P}}. \tag{8}$$

It is worth mentioning, that besides the classical fluxes, i.e. stress and electric displacement, the polarisation forces appear as work conjugated quantities to the order parameter $\boldsymbol{P}$ and thus become 'constitutively' related to $\boldsymbol{P}$. The derivation presented here relies on ideas given in Gurtin (1996), which are extended to a vector valued order parameter $\boldsymbol{P}$, see also Su and Landis (2007).
In order to ensure a non-negative dissipation, the following relation has to be fulfilled in all processes,

$$\begin{aligned}-\left(\frac{\partial H}{\partial\boldsymbol{P}} + \tilde{\boldsymbol{g}}\right)\dot{\boldsymbol{P}} &= -\left(\frac{\partial H}{\partial\boldsymbol{P}} - \operatorname{div}\tilde{\boldsymbol{\Sigma}}\right)\dot{\boldsymbol{P}} \\ &= -\left(\frac{\partial H}{\partial\boldsymbol{P}} - \operatorname{div}\frac{\partial H}{\partial\operatorname{grad}\boldsymbol{P}}\right)\dot{\boldsymbol{P}} \\ &\geq 0,\end{aligned} \tag{9}$$

thus the evolution law for the order parameter $\boldsymbol{P}$ has to be in agreement with this inequality. Before specifying the form of the evolution equation, the choice of the phase field potential H is discussed briefly.

2.3 Phase field potential

The potential H of the presented phase field model consists of three contributions:

$$H = H^{\text{ent}} + H^{\text{sep}} + H^{\text{int}}. \tag{10}$$

Here the 'classical' electric enthalpy is given by

$$H^{\text{ent}} = \frac{1}{2}(\boldsymbol{\varepsilon} - \boldsymbol{\varepsilon}^0) : [\mathbb{C}(\boldsymbol{\varepsilon} - \boldsymbol{\varepsilon}^0)] - (\boldsymbol{\varepsilon} - \boldsymbol{\varepsilon}^0) : \mathbb{b}^{\mathrm{T}}\boldsymbol{E} - \frac{1}{2}\boldsymbol{E} \cdot \boldsymbol{A}\boldsymbol{E} - \boldsymbol{P} \cdot \boldsymbol{E}, \tag{11}$$

where $\mathbb{C}$ is the elastic tensor, $\mathbb{b}$ is the piezo-electric and $\boldsymbol{A}$ the dielectric tensor. The spontaneous strain and polarisation are denoted by $\boldsymbol{\varepsilon}^0$ and $\boldsymbol{P}$. It is crucial and physically senseful to let the spontaneous strain and the piezo-electric tensor $\mathbb{b}$ depend on the spontaneous polarisation. This is done by

$$\boldsymbol{\varepsilon}^0(\boldsymbol{P}) = \frac{3}{2}\varepsilon_0 \frac{|\boldsymbol{P}|}{P_0}\left(\boldsymbol{e} \otimes \boldsymbol{e} - \frac{1}{3}\mathbf{1}\right), \tag{12}$$

where $\boldsymbol{e}$ is the direction of $\boldsymbol{P}$, i.e. $\boldsymbol{e} = \boldsymbol{P}/|\boldsymbol{P}|$. For the piezo-electric coupling tensor $\mathbb{b}$ the dependency is given in index notation by

$$b_{kij} = \frac{|\boldsymbol{P}|}{P_0}\left\{b_{\|}e_i e_j e_k + b_{\perp}(\delta_{ij} - e_i e_j)e_k + b_{=}\frac{1}{2}[(\delta_{ki} - e_k e_i)e_j + (\delta_{kj} - e_k e_j)e_i]\right\}. \tag{13}$$

Details on this can also be found in Kamlah (2001). This introduces the material parameters $\boldsymbol{\varepsilon}_0, P_0, b_{\|}, b_{\perp}, b_{=}$. The separation energy forces the system to assume poled states in the absence of external loads. It is modelled by a forth order polynomial of the form

$$H^{\text{sep}} = a_1 + a_2(P_1^2 + P_2^2) + a_3(P_1^4 + P_2^2) + a_4 P_1^2 P_2^2, \tag{14}$$

which is applicable to a system with 90 and 180° domains in the $x_1 - x_2$ plane. A sketch of H^{sep} for this 2d setting is depicted in Fig. 1.
The 'domain wall energy' H^{int} is given by

$$H^{\text{int}} = \frac{1}{2}\lambda \parallel \operatorname{grad}\boldsymbol{P} \parallel^2, \tag{15}$$

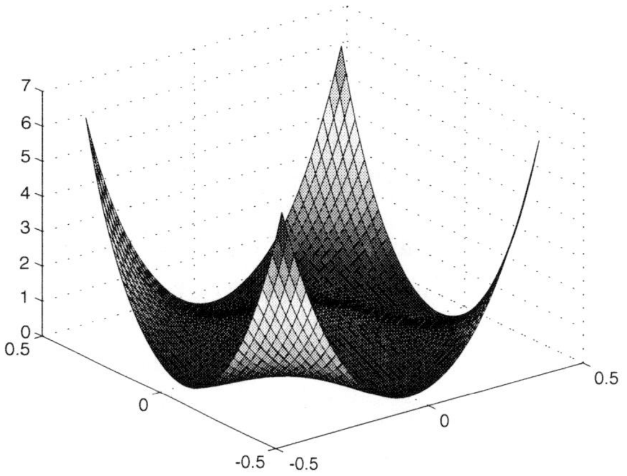

Fig. 1 Domain separation part in phase field potential with minima at $\boldsymbol{P} = (\pm P_0, 0)$ and $\boldsymbol{P} = (0, \pm P_0)$

and takes variations in the polarisation into account. Due to the thermodynamic analysis in the previous subsection, the following constitutive relations now can be derived

$$\boldsymbol{\sigma} = \frac{\partial H}{\partial \boldsymbol{\varepsilon}} = \mathbb{C}(\boldsymbol{\varepsilon} - \boldsymbol{\varepsilon}^0) - \mathbb{b}^{\mathrm{T}}\boldsymbol{E}.$$
$$\boldsymbol{D} = -\frac{\partial H}{\partial \boldsymbol{E}} = \mathbb{b}(\boldsymbol{\varepsilon} - \boldsymbol{\varepsilon}^0) + \boldsymbol{A}\boldsymbol{E} + \boldsymbol{P}. \tag{16}$$

The evolution equation for the polarisation $\boldsymbol{P}$ is chosen in agreement with the thermodynamic consideration using configurational forces.

$$\dot{\boldsymbol{P}} = -M\frac{\delta H}{\delta \boldsymbol{P}} = -M\left(\frac{\partial H}{\partial \boldsymbol{P}} - \operatorname{div}\frac{\partial H}{\partial \operatorname{grad}\boldsymbol{P}}\right) = -M\left(\frac{\partial H^{\text{ent}}}{\partial \boldsymbol{P}} + \frac{\partial H^{\text{sep}}}{\partial \boldsymbol{P}} - \lambda\Delta\boldsymbol{P}\right), \tag{17}$$

where M is the mobility parameter and $\delta H/\delta \boldsymbol{P}$ the variational derivative. This equation can be understood as the Ginzburg-Landau equation in order parameter models. The three terms in (17) describe the evolution of the order parameter $\boldsymbol{P}$ due to the electro-mechanical energy, due to the separation energy and due to a smoothing term. The parameter λ introduces the energy associated with areas of varying order parameter, i.e. domain walls. It can also be thought of as an internal length scale, which is introduced in the model. Eq. (17) is assumed in many theories with non-local internal variables or order parameters. Within the context of ferroelectric materials (17) can be understood as a special realisation of the theories proposed in Maugin and Pouget (1980), Maugin (1988).

3 Numerical implementation

The numerical implementation is done using a finite element formulation. For simplicity and to keep the numerical effort limited we restrict ourselves to 2d problems. The discretisation is done by 4-node bi-linear plane elements. Details on this can be found in any standard text book on finite elements, without claim of completeness we cite Hughes (2000). The degrees of freedom at each node I are the two displacements u_1^I, u_2^I, the electric potential φ^I and the spontaneous polarisation (order parameter) P_1^I, P_2^I. Thus each node has five degrees of freedom,

$$\underline{\mathbf{d}}^I = \left[u_1^I \;\; u_2^I \;\; \varphi^I \;\; P_1^I \;\; P_2^I\right]^{\mathrm{T}}. \tag{18}$$

The discretisation leads to a set of non-linear equations (residual) which depend on the nodal values and the rates of the nodal values, i.e.

$$\underline{\mathbf{R}}^I(\underline{\mathbf{d}}^J, \dot{\underline{\mathbf{d}}}^J) = \underline{\mathbf{0}}. \tag{19}$$

Discretising with a first order difference formula

$$\dot{\underline{\mathbf{d}}}^J \approx \frac{1}{\Delta t}\left(\underline{\mathbf{d}}_{n+1}^J - \underline{\mathbf{d}}_n^J\right), \tag{20}$$

where $_{n+1}$ and $_n$ denote the values at time t_{n+1} and t_n in conjunction with an evaluation of the residual at time t_{n+1} yields an implicit time integration scheme, where

$$\underline{\mathbf{R}}_{n+1}^I = \underline{\mathbf{R}}^I\left(\underline{\mathbf{d}}_{n+1}^J, \;\; \frac{\underline{\mathbf{d}}_{n+1}^J - \underline{\mathbf{d}}_n^J}{\Delta t}\right) = \underline{\mathbf{0}}. \tag{21}$$

Due to the non-linear character of the phase field model, this establishes a non-lienear system of equations for the unknown nodal values $\underline{\mathbf{d}}_{n+1}^J$. A Newton iteration is used to solve this system. For the numerical performance a consistent tangent matrix has to be derived. It has the following structure:

$$\underline{\mathbf{S}}^{IJ} = \underline{\mathbf{K}}^{IJ} + \frac{1}{\Delta t}\underline{\mathbf{D}}^{IJ} \quad \text{where}$$
$$\underline{\mathbf{K}}^{IJ} = \frac{\partial \underline{\mathbf{R}}^I}{\partial \underline{\mathbf{d}}^J}, \quad \underline{\mathbf{D}}^{IJ} = \frac{\partial \underline{\mathbf{R}}^I}{\partial \dot{\underline{\mathbf{d}}}^J}. \tag{22}$$

This establishes a robust and fast algorithm, which allows for sufficiently large time steps. The iteration matrix $\underline{\mathbf{S}}^{IJ}$ turns out to be symmetric, which allows for an efficient storage and solution. The symmetry of $\underline{\mathbf{S}}^{IJ}$ is due to the fact that all constitutive laws and the evolution law are derived from a potential.

4 Results

In all simulations we assume that there is no flux of polarisation across the boundary of the simulation, i.e. homogeneous Neumann boundary conditions are assumed for the phase field parameter $\boldsymbol{P}$. The boundary conditions on $\boldsymbol{P}$ arise as a mathematical necessity from the introduction of higher gradients on $\boldsymbol{P}$ to model domain walls. Using Voigt notation the material data are chosen as, which approximates a ferroelectric material.

$$\mathbb{C} = \begin{bmatrix} 12.0 & 7.5 & 0.0 \\ 7.5 & 12.0 & 0.0 \\ 0.0 & 0.0 & 2.6 \end{bmatrix} \cdot 10^{10}\mathrm{N/m^2},$$
$$\mathbb{b} = \begin{bmatrix} 0.0 & 0.0 & 13.0 \\ -5.2 & 15.0 & 0.0 \end{bmatrix} \mathrm{C/m^2},$$
$$A = \begin{bmatrix} 6.0 & 0.0 \\ 0.0 & 6.0 \end{bmatrix} \cdot 10^{-9}\mathrm{C/Vm}, \quad P_0 = 0.4\mathrm{C/m^2},$$
$$\varepsilon_0 = 0.003$$
$$M = 10\mathrm{A/Vm}, \quad \lambda = 10^{-6}\mathrm{Vm^3C}. \tag{23}$$

The coefficients $a_1, \ldots a_4$ in the separation part are chosen as

$$a_1 = 10^7\mathrm{N/m^2}, \; a_2 = -12.5 \cdot 10^7\mathrm{Nm^2/C^2},$$
$$a_3 = 39.0625 \cdot 10^7\mathrm{Nm^2/C^4},$$
$$a_4 = 234.375 \cdot 10^7\mathrm{Nm^2/C^6}. \tag{24}$$

This ensures a potential minimum at $\boldsymbol{P} = (0, \pm P_0)$ and at $\boldsymbol{P} = (\pm P_0, 0)$.

4.1 Microstructure evolution—self organisation

Starting with a random distribution of initial polarisations the development of subsequent microstructures is calculated. As we assume charge free boundaries, i.e. $\boldsymbol{D} \cdot \boldsymbol{n} = 0$, it is energetically very unfavourable to have polarisations with a normal component to the boundary. These components would require very strong electric fields to satisfy the boundary conditions. Therefore the system starts to arrange in such a manner, that it forms two vortices, with eight 90° domain walls. In Fig. 2 the evolution and self organisation of the system is shown. Due to non aligned polarisation vectors, strong internal fields develop and a rapid self organisation to structures with larger domains is the consequence. From a numerical point of view this initial period is very difficult and crucial with respect to the stability of the integration scheme.

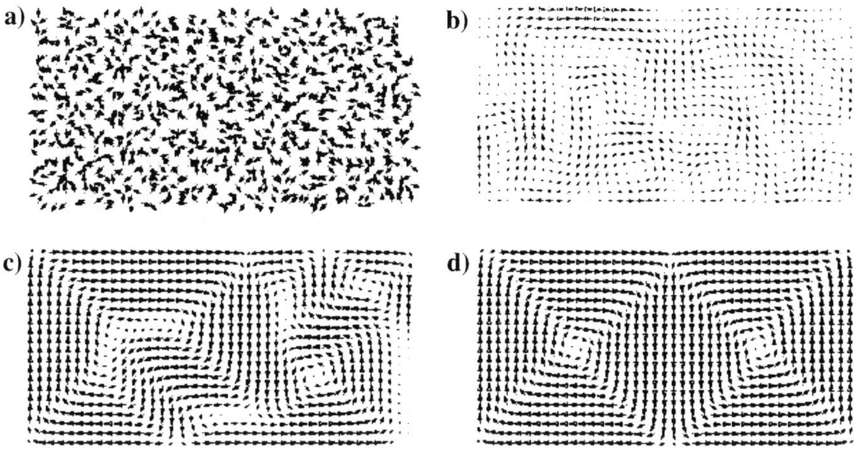

Fig. 2 Evolution of polarisation in absence of external loads or fields (**a**) random initial configuration, going through states (**b**) and (**c**), ending at (**d**) equilibrium state

4.2 Single domain wall

In the next example, the case of an isolated 180° domain wall is studied. The boundary is free of mechanical stresses, and charge free on the lateral surfaces, while on the top and the bottom boundary a zero electric potential is applied. This setting causes an isolated 180° domain wall to be stable. Starting with two blocks with opposite (up/down) spontaneous polarisation yields a smoothing of the spontaneous polarisation across the domain wall, see Fig. 3. The initially sharp domain wall becomes diffuse. The parameter λ can be used to calibrate the width of the domain wall. For numerical reasons the parameter has to be chosen, such that a smooth variation of the phase field parameter $\boldsymbol{P}$ can be resolved by the discretisation. This implies the use of a few elements (2–4 elements) in the domain wall region.

If a single domain wall is subjected to an external electric field, which is applied by a potential difference between the top and bottom electrode, it starts to move. The velocity depends on the magnitude of the external field. In Fig. 4 this relation is reported. It is observed, that any non zero external field will cause domain wall motion. There is no threshold value introduced in the present model. The apparent small threshold value in Fig. 4 is due to the numerical scheme. Refining time and space discretisation shows a vanishing offset value in the electric field. Defects in the material might make it necessary to introduce a critical value in the evolution equation (17). The relation between external field and velocity can be assumed to be almost linear for the range under consideration. A constitutive law of this type is for example discussed in Schrade et al. (2007) and experimentally measured in Flippen (1975).

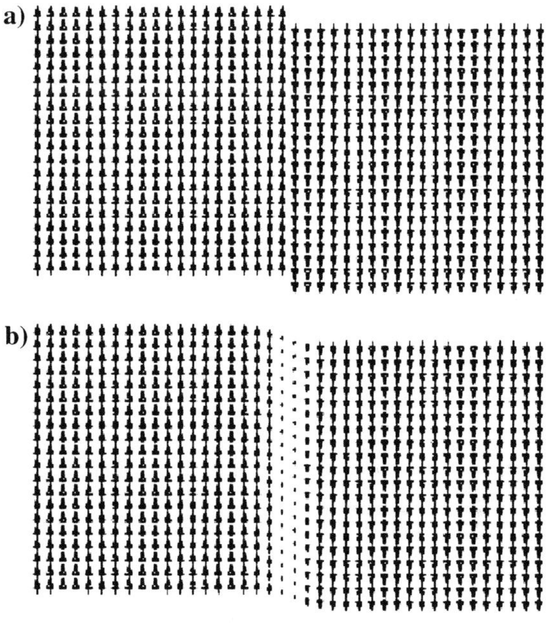

Fig. 3 Formation of a diffuse 180° domain wall (**a**) initial setup (**b**) stationary, diffuse domain wall

4.3 Defective electrode

To model a defective electrode, different boundary conditions are applied at parts of the top electrode, see Fig. 5. A defect is modeled by changing the boundary condition in the defect area to $\boldsymbol{D} \cdot \boldsymbol{n} = 0$. Figure 5a) shows the initial polarisation. The domain wall is driven towards the defect by an external electric field. If the field is below a critical value the domain wall stops at the defect position as can be seen in Fig. 5b). If the field is sufficiently strong, 90° domains form, see

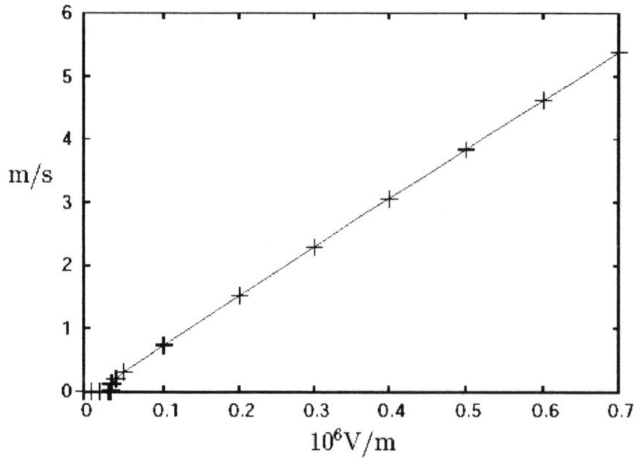

Fig. 4 Domain wall velocity as a function of the applied electric field E

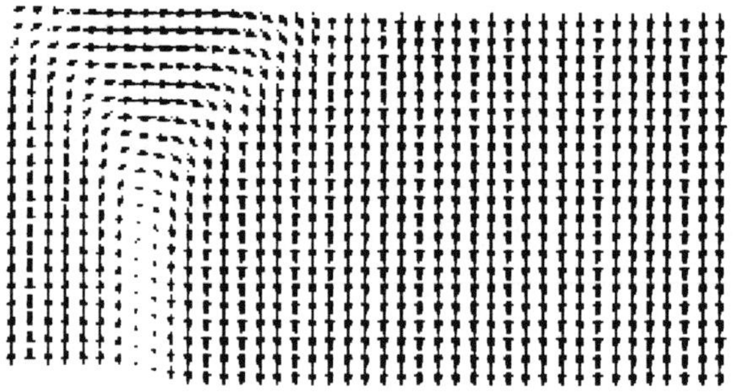

Fig. 6 Overrunning of an electrodefect by forming 90° domain walls

Fig. 6. In the final configuration the polarisation has switched completely. In Schrade et al. (2007) a sharp interface model was used to study this kind of defect for gadolinium molybdate. There, domain wall pinning was observed for low fields, whereas for high fields the domain wall could be moved over the defect.

4.4 Mechanical poling

Besides electrically induced domain switching the model is also capable of reproducing ferroelastic switching. Figure 7a shows the random initial distribution of the polarisation. The boundary conditions are $\varphi = 0$ on all edges, and bipolar normal stress loading is applied in the vertical direction. The series of pictures in Fig. 7 shows the polarisation at alternating compressive (b, d, f) and tensile (c, e) stresses. Pictures (e and f) reflect the states at maximum tensile and compressive load, respectively. These are the states the system would switch back and fourth under continued cyclic loading. If $\boldsymbol{D} \cdot \boldsymbol{n} = 0$ on the left and right boundary, the system behaves a lot stiffer when compressive loading is applied. This situation is illustrated in Fig. 8b, d, f where multiple horizontal domains form.

5 Summary

A continuum phase field model has been established. Using the concept of an order parameter in conjunction with the evolving inhomogeneities (configurational forces), a thermodynamically consistent evolution law was proposed. The polarisation balance is used to propose an evolution law for the order parameter of the systen, i.e. an evolution law for the spontaneous polarisation $\boldsymbol{P}$, which is in agreement with the Ginzburg-Landau approach. The model takes electrical and mechanical fields into account to model the microstructure evolution in ferroelectric materials.

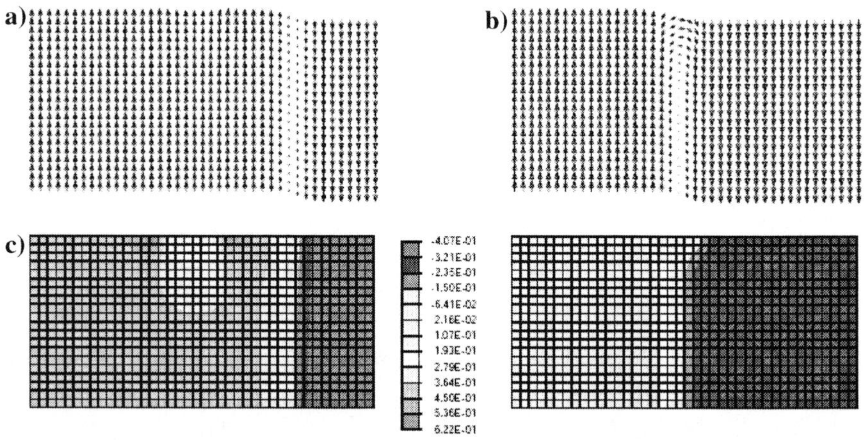

Fig. 5 Pinning of a domain wall in an electrode defect, (**a**) initial configuration, (**b**) stationary state (**c**, **d**) corresponding electric displacements D_2 to (**a**) and (**b**)

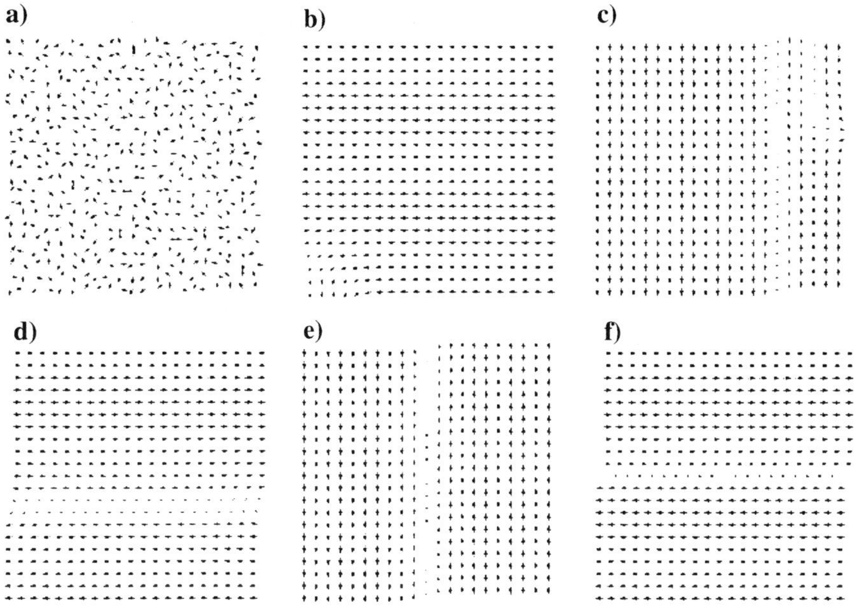

Fig. 7 Cyclic mechanilcal loading with zero electric potential on the boundary

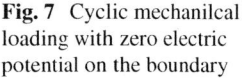

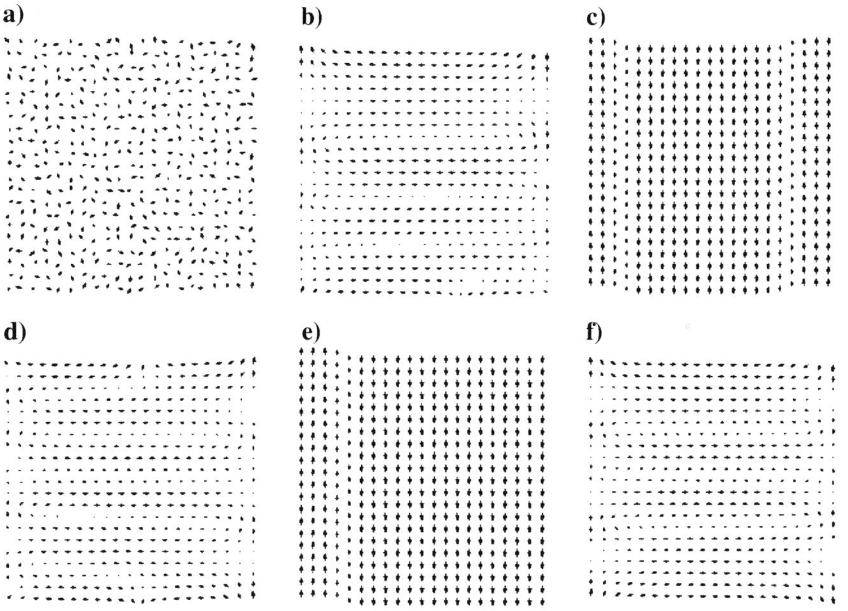

Fig. 8 Cyclic mechanical loading with zero electric potential on the top and bottom surface and zero charge on lateral boundaries

Central part in the numerical derivation is a finite element scheme, which is used to solve the non-linear partial differential equations. In extension to finite element formulations for piezoelectric materials, which use the mechanical displacement and the electric potential as nodal variable, the spontaneous polarisation $\boldsymbol{P}$ is introduced. This is also different from ferroelectric models, which use the remanent polarisation as an internal (Gaußpoint) variable. The present model is capable of reproducing the microstructures on the domain level. Crucial for the numerical Implementation is a robust time integration, which is achieved by an implicit time integration scheme together with a Newton iteration in each time step.

The microstructures predicted by the model agree with simple analytical models. Effects like domain wall pining, which are experimentally observed, can be reproduced. The effect of possible defect structures, such as point defects, can be included, but this is outside of the scope of this paper. Future works will include the simulation of mesoscopic material responses, i.e. the overall properties of single crystals with domain structure evolution during poling, as well as poly crystalline structures. The incorporation of defects (point defects) is necessary to describe effects like fatigue and ageing and will be pursued in future works.

References

Ahluwalja R, Cao W (2000) Influence of dipolar defects on switching behavior in ferroelectrics. Phys Rev B 63:012103

Ahluwalja R, Cao W (2001) Size dependence of domain patterns in a constrained ferroelectric system. J Appl Phys 89(12):8105–8109

Bhattacharya K, Ravichandran G (2003) Ferroelectric perovskites for electromechanical actuation. Acta Mater 51:5941–5960

Cao W, Cross L (1991) Theory of tetragonal twin structures in ferroelectric perovskites with a first-order phase transition. Phys Rev B 44(1):5–12

Flippen R (1975) Domain wall dynamics in ferroelectric/ferroelastic molybdates. J Appl Phys 46(3):1068–1071

Gross D, Kolling S, Mueller R, Schmidt I (2003) Configurational forces and their application in solid mechanics. Eur J Mech A/Solids 22:669–692

Goy O, Mueller R, Gross D (2006) Interaction of point defects in piezoelectric materials—numerical simulations in the context of electric fatigue. J Thero Appl Mech 44(4):819–836

Gurtin ME (1996) Generalized Ginzburg-Landau and Cahn-Hillirad equations based on a microforce balance. Physica D 92:178–192

Hughes T (2000) The finite element method. Dover, Mineola, New York

Kamlah M (2001) Ferroelectric and ferroelastic piezoceramics—modeling of electromechanical hysteresis phenomena. Continuum Mech Thermodyn 13:219–268

Maugin GA (1988) Continuum mechanics of electromagnetic solids. North-Holland, Amsterdam

Maugin GA (1993) Material inhomogeneities in elasticity. Chapman & Hall, London, Glasgow, New York, Tokyo, Melbourne, Madras

Maugin GA, Pouget J (1980) Electroacoustic equations for one-domain ferroelectric bodies. J Acoust Soc Am 68(2):575–587

McCormack M, Khachaturyan AG, Morris JW (1992) A two-dimensional analysis of the evolution of coherent precipitate in elastic media. Acta metall mater 40(2):325–336

Mueller R, Maugin GA (2002) On material forces and finite element discretizations. Comp Mech 29(1):52–60

Mueller R, Kolling St, Gross D (2002) On configurational forces in the context of the finite element method. Int J Numer Methods Eng 53:1557–1574

Mueller R, Gross D, Lupascu D (2005) Driving forces on domain walls in ferroelectric materials and interaction with defects. Comp Mat Sci 35:42–52

Müller WH (1998) Zur Simulation des Mikroverhaltens thermomechanisch fehlgepasster Verbundwerkstoffe. Fortschritt-Berichte VDI, Reihe 18, Nr. 234, Düsseldorf

Schrade D, Mueller R, Gross D, Utschig T, Shur V, Lupascu D (2007) Interaction of domain walls with defects in ferroelectric materials. Mech Mater 39:161–174

Soh A, Song Y, Ni Y (2006) Phase field simualtions of hysteresis and butterfly loops in ferroelectrics subjected to electromechanical coupled loading. J Am Ceram Soc 89:652–661

Su Y, Landis C (2006) A non-equlibrium thermodynamics framework for domain evolution; phase field models and finite element implementation. Proceeding to the SPIE

Su Y, Landis C (2007) Continuum thermodynamics of ferroelectric domain evolution: Theory, finite element implementation, and application to domain wall pinning. J Mech Phys Sol 55:280–305

Wang J, Zhang T-Y (2006a) Effect of long-range elastic interactions on the toroidal moment of polarisation in a ferroelectric nanoparticle. Appl Phys Lett 88:182904

Wang J, Zhang T-Y (2006b) Size effects in epitaxial ferroelectric islands and thin films. Phys Rev B 73:144107

Wang Y, Chen L, Khachaturyan AG (1993) Kinetics of strain-induced morphological transformation in cubic alloys with a miscibility gap. Acta metall mater 41(1):279–296

Wang J, Shi S-Q, Chen L-Q, Li Y, Zhang T-Y (2004) Phase field simulations of ferroelectric/ferroelastic polarisation switching. Acta mater 52:749–764

Xiao Y, Shenoy V, Bhattacharya K (2005) Depletion layers and domain walls in semiconducting ferroelectric thin films. Phys Rev Lett 95:247603

Zhang W, Bhattacharya K (2005a) A computational model of ferroelectric domains. Part I: model formulation and domain switching. Acta mater 53:185–198

Zhang W, Bhattacharya K (2005b) A computational model of ferroelectric domains. Part II: grain boundaries and defect pinning. Acta mater 53:199–209

An adaptive singular finite element in nonlinear fracture mechanics

Ralf Denzer · Michael Scherer · Paul Steinmann

Abstract In the case of nonlinear fracture mechanics the type of singularity induced by the crack tip is commonly not known. This results in a poor approximation of the near crack tip fields in a finite element setting and induces so called spurious—or residual—discrete material forces in the vicinity of the crack tip. Thus the numerical calculation of the crack driving material force in nonlinear fracture is often not that precise as in linear elasticity where we can use special crack tip elements and/or path independency. To overcome this problem we propose an adaptive singular element, which adapts automatically to the type of singularity. The adaption is based on an optimisation procedure using a variational principle.

Keywords Singular finite element · Singularity computation · Nonlinear fracture

1 Introduction

Problems with singular behaviour in the stress and/or strain fields at certain points, e.g. re-entrant corners or cracks, lead often to large inaccuracies of the numerical field approximations based on a finite element formulation in the vicinity of the singular points. This is caused by the fact that in finite element formulations typically Lagrangian or serendipity polynomials are used as shape function $N^n(\boldsymbol{\xi})$. It is obvious that polynomials are not adequate to represent singular behaviours. To overcome this problem different special finite elements with singular behaviour were introduced in the literature. Some of them are restricted to typical stress singularities like $1/\sqrt{r}$ or $1/r$ which occur in the case of cracks in linear elastic or perfect plastic materials, respectively, see e.g. Barsoum (1977). Other special finite elements have a variable, i.e. a user given, type of singularity, like Akin (1976), Tracey and Cook (1977), Stern (1978), Hughes and Akin (1980), Staab (1983) and Lim and Kim (1994). One of the essential problems while using these elements is the determination of the correct type of singularity of the given problem, which usually needs at least an asymptotic analytical solution of the problem at the position of the singularity. Especially in the case of nonlinear material behaviour and/or large strain analysis there is only a rare number of analytical solutions available in the literature, see e.g. Rice (1968), Knowles and Sternberg (1973, 1983), Herrmann (1989, 1992) and Le and Stumpf (1993). We therefore propose in this work an optimisation procedure based on a variational principle to determine the type of singularity numerically which minimises the (discrete) total potential energy of a system.

R. Denzer (✉) · M. Scherer · P. Steinmann
Applied Mechanis, University of Kaiserslautern,
67663 Kaiserslautern, Germany
e-mail: denzer@rhrk.uni-kl.de

Defect and Material Mechanics. C. Dascalu, G.A. Maugin & C. Stolz (eds.),
doi: 10.1007/978-1-4020-6929-1_17,

2 Spatial motion problem

2.1 Kinematics and kinetics

To set the stage and in order to introduce terminology and notation, we briefly reiterate some key issues pertaining to the geometrically nonlinear kinematics and kinetics of the quasi-static spatial motion problem.

Thereby, in order to introduce the relevant concepts, we merely consider a conservative mechanical system. In this case, the internal potential energy density W_τ per unit volume in $\mathcal{B}_\tau$ with $\tau = 0, t$ characterises the hyperelastic material response and is commonly denoted as stored energy density. Moreover, an external potential energy density V_τ characterises the conservative loading. Then, the conservative mechanical system is essentially characterised by the total potential energy density per unit volume $U_\tau = W_\tau + V_\tau$.

In the spatial motion problem in Fig. 1, the placement $\boldsymbol{x}$ of a 'physical particle' in the spatial configuration $\mathcal{B}_t$ is described by the nonlinear spatial motion deformation map

$$\boldsymbol{x} = \boldsymbol{\varphi}(\boldsymbol{X}) \tag{1}$$

in terms of the placement $\boldsymbol{X}$ of the same 'physical particle' in the material configuration $\mathcal{B}_0$. The spatial motion deformation gradient, i.e. the linear tangent map associated with the spatial motion deformation map, together with its determinant are then given by

$$\boldsymbol{F} = \nabla_X \boldsymbol{\varphi}(\boldsymbol{X}) \quad \text{and} \quad J = \det \boldsymbol{F} \tag{2}$$

and its inverse by

$$\boldsymbol{f} = \boldsymbol{F}^{-1} \quad \text{and} \quad j = \det \boldsymbol{f} \tag{3}$$

which will be discussed more precisely later.

The quasi-static balance of momentum for the spatial motion problem reads

$$-\operatorname{Div} \boldsymbol{P} = \boldsymbol{b}_0 \quad \Longrightarrow \quad -\operatorname{div} \boldsymbol{\sigma} = \boldsymbol{b}_t \tag{4}$$

The two-point description stress $\boldsymbol{P}$ and the spatial description stress $\boldsymbol{\sigma}$, see Fig. 1, which are called the spatial motion Piola and Cauchy stresses, have been introduced here. For the present case of a conservative mechanical system, they follow from the potential energy density as

$$\boldsymbol{P} = \partial_{\boldsymbol{F}} U_0 \quad \Longrightarrow \quad \boldsymbol{\sigma} = j\boldsymbol{P} \cdot \boldsymbol{F}^t = U_t \boldsymbol{I} - \boldsymbol{f}^t \cdot \partial_{\boldsymbol{f}} U_t \tag{5}$$

The second expression in Eq. 5 thereby denotes the energy-momentum format of the spatial motion Cauchy stress. For the sake of conciseness and without danger of confusion, we omit the explicit indication of the spatial or material parametrisation.

Moreover, distributed volume forces $\boldsymbol{b}_\tau$ per unit volume follow from the explicit spatial gradient of the total potential energy density

$$\boldsymbol{b}_0 = -\partial_{\boldsymbol{\varphi}} U_0 \Longrightarrow \boldsymbol{b}_t = j\boldsymbol{b}_0 \tag{6}$$

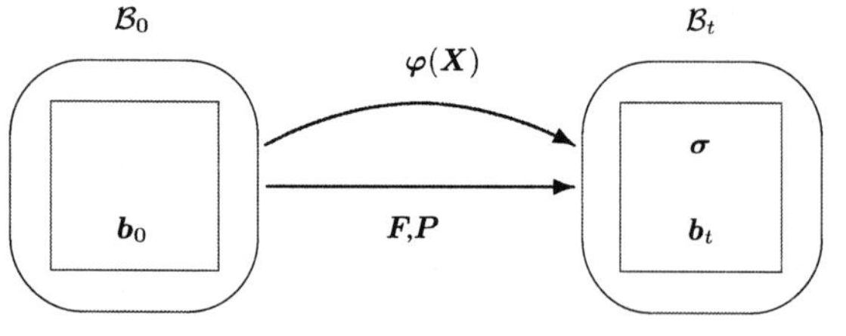

Fig. 1 Kinematics and kinetics of the spatial motion problem

2.2 Virtual work and discretisation

The pointwise statement in Eq. 4 for the solution of the spatial motion problem is multiplied by a test function (spatial virtual displacement) $\boldsymbol{w}$ under the necessary smoothness and boundary assumptions to render the virtual work expression

$$\underbrace{\int_{\partial\mathcal{B}_t} \boldsymbol{w} \cdot \boldsymbol{\sigma} \cdot \boldsymbol{n} \,\mathrm{d}a}_{\mathfrak{w}^{\text{sur}}} = \underbrace{\int_{\mathcal{B}_t} \nabla_x \boldsymbol{w} : \boldsymbol{\sigma} \,\mathrm{d}v}_{\mathfrak{w}^{\text{int}}} - \underbrace{\int_{\mathcal{B}_t} \boldsymbol{w} \cdot \boldsymbol{b}_t \,\mathrm{d}v}_{\mathfrak{w}^{\text{vol}}} \quad \forall\, \boldsymbol{w} \tag{7}$$

whereby $\mathfrak{w}^{\text{sur}}$ denotes the spatial variation of the total bulk potential energy due to its complete dependence on the spatial position, whereas the contributions $\mathfrak{w}^{\text{int}}$ and $\mathfrak{w}^{\text{vol}}$ denote the spatial variations of the total bulk potential energy due to its implicit and explicit dependence on the spatial position, respectively.

Obviously, the quasi-static equilibrium of spatial forces is recovered, if arbitrary uniform spatial virtual displacements $\boldsymbol{w} = \boldsymbol{\theta} = \text{const.}$ are selected for the evaluation of Eq. 7.

$$\boldsymbol{\theta} \cdot \left[\int_{\partial\mathcal{B}_t} \boldsymbol{\sigma} \cdot \boldsymbol{n} \,\mathrm{d}a + \int_{\mathcal{B}_t} \boldsymbol{b}_t \,\mathrm{d}v \right] = 0 \quad \forall\, \boldsymbol{\theta} \tag{8}$$

The domain is next discretised in n_{el} elements with $\mathcal{B}_0^h = \cup_{e=1}^{n_{\text{el}}} \mathcal{B}_0^e$ and $\mathcal{B}_t^h = \cup_{e=1}^{n_{\text{el}}} \mathcal{B}_t^e$. The geometry in $\mathcal{B}_t$ and $\mathcal{B}_0$ is interpolated from the positions $\boldsymbol{\varphi}_n$ and $\boldsymbol{X}_n$ of the n_{en} nodes by shape functions $N^n(\boldsymbol{\xi})$ on each element, with $n \in [1, n_{\text{en}}]$ denoting the local node numbering

$$\boldsymbol{\varphi}^h|_{\mathcal{B}^e_\square} = \sum_{n=1}^{n_{\text{en}}} N^n(\boldsymbol{\xi})\boldsymbol{\varphi}_n \quad \text{and}$$

$$\boldsymbol{X}^h|_{\mathcal{B}^e_\square} = \sum_{n=1}^{n_{\text{en}}} N^n(\boldsymbol{\xi})\boldsymbol{X}_n \tag{9}$$

whereby we introduced the isoparametric domain $\square = [-1, 1]^{n_{\text{dim}}}$ with n_{dim} the dimension of the problem.

Thus, the elementwise discretisation of the virtual spatial displacement field $\boldsymbol{w}$ into nodal values $\boldsymbol{w}_n$ which are interpolated as well by the shape functions N^n in the spirit of an isoparametric expansion, renders the representation

$$\boldsymbol{w}^h|_{\mathcal{B}^e_\square} = \sum_{n=1}^{n_{\text{en}}} N^n(\boldsymbol{\xi})\boldsymbol{w}_n \tag{10}$$

Corresponding gradients of the virtual spatial displacement field are given in each element by

$$\nabla_X \boldsymbol{w}^h|_{\mathcal{B}^e_\square} = \sum_{n=1}^{n_{\text{en}}} \boldsymbol{w}_n \otimes \nabla_X N^n(\boldsymbol{\xi}) \tag{11}$$

$$\nabla_x \boldsymbol{w}^h|_{\mathcal{B}^e_\square} = \sum_{n=1}^{n_{\text{en}}} \boldsymbol{w}_n \otimes \nabla_x N^n(\boldsymbol{\xi}) \tag{12}$$

Lastly, based on the above discretisations, the corresponding deformation gradient $\boldsymbol{F}$ takes the elementwise format

$$\boldsymbol{F}^h|_{\mathcal{B}^e_\square} = \sum_{n}^{n_{\text{en}}} \boldsymbol{\varphi}_n \otimes \nabla_X N^n(\boldsymbol{\xi}) \tag{13}$$

The elementwise expansions for the internal and the volume contributions therefore read

$$\mathfrak{w}_e^{\text{int}} = \sum_{n=1}^{n_{\text{en}}} \boldsymbol{w}_n \cdot \int_{\mathcal{B}^e_t} \boldsymbol{\sigma} \cdot \nabla_x N^n \,\mathrm{d}v \tag{14}$$

$$\mathfrak{w}_e^{\text{vol}} = \sum_{n=1}^{n_{\text{en}}} \boldsymbol{w}_n \cdot \int_{\mathcal{B}^e_t} \boldsymbol{b}_t N^n \,\mathrm{d}v \tag{15}$$

Finally, considering the arbitrariness of the spatial virtual node point displacements $\boldsymbol{w}_n$, the global discrete spatial node point forces characterising external spatial surface loads are computed as

$$\mathfrak{f}^h_{\text{sur}} = \mathop{\mathsf{A}}_{e=1}^{n_{\text{el}}} \sum_{n=1}^{n_{\text{en}}} \int_{\mathcal{B}^e_t} \boldsymbol{\sigma} \cdot \nabla_x N^n - \boldsymbol{b}_t N^n \,\mathrm{d}v \tag{16}$$

where by $\mathop{\mathsf{A}}_{e=1}^{n_{\text{el}}}$ denotes the assembly operator for all n_{el} finite elements. In conclusion of these considerations, the discrete spatial node point (surface) forces are thus energetically conjugated to variations of the spatial node point positions.

3 Variable power singular element

In the sequel we will follow the generation of a 2D triangular element introduced by Hughes and Akin (1980) with a variable type of singularity. In the next section we then formulate a numerical scheme based on a variational principle to compute the type of singularity for a discretised problem.

We start with a one-dimensional 3-node singular element, as depicted in Fig. 2a, with the following shape functions:

$$\begin{aligned} B^1(r,\kappa) &= 1 - 2r + \frac{r^\kappa - 2\left[\frac{1}{2}\right]^\kappa r}{1 - 2\left[\frac{1}{2}\right]^\kappa} \\ B^2(r,\kappa) &= 2r - 2\left[\frac{r^\kappa - 2\left[\frac{1}{2}\right]^\kappa r}{1 - 2\left[\frac{1}{2}\right]^\kappa}\right] \\ B^3(r,\kappa) &= \frac{r^\kappa - 2\left[\frac{1}{2}\right]^\kappa r}{1 - 2\left[\frac{1}{2}\right]^\kappa} \end{aligned} \tag{17}$$

whereby $r \in [0, 1]$. Please note that these shape functions are capable to represent a point singularity of order r^κ. Even though this one-dimensional element has little practical interest, it is possible to construct a two-dimensional 9-node element, see Fig. 2b, by a tensorial expansion $\tilde{N}^i(r,s,\kappa) = B^j C^k$ with a standard 3-node element $C^i(s) = B^i(r = s, \kappa = 2)$, i.e. a 2nd order Lagrange polynomial. The corresponding shape functions

$$\begin{array}{lll} \tilde{N}^1 = B^1C^1 & \tilde{N}^4 = B^1C^3 & \tilde{N}^7 = B^2C^3 \\ \tilde{N}^2 = B^3C^1 & \tilde{N}^5 = B^2C^1 & \tilde{N}^8 = B^1C^2 \\ \tilde{N}^3 = B^3C^3 & \tilde{N}^6 = B^3C^2 & \tilde{N}^9 = B^2C^2 \end{array} \tag{18}$$

are capable of representing a line singularity of the order r^κ exactly. By degenerating this 9-node quadrilateral element with

$$N^1(r,s,\kappa) = \tilde{N}^1 + \tilde{N}^4 + \tilde{N}^8 \tag{19}$$

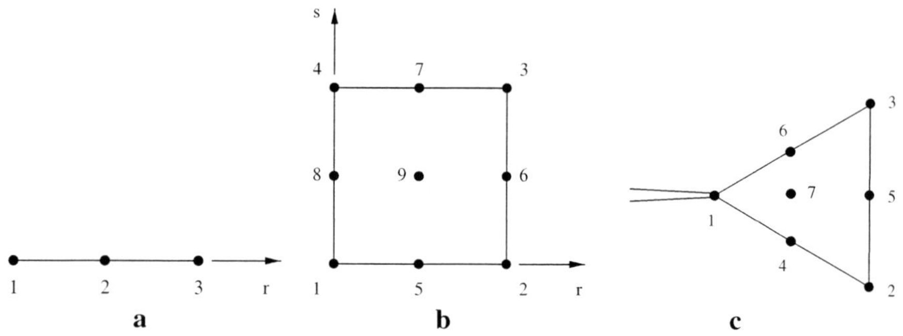

Fig. 2 Generation of a 7-node triangular singular finite element. (**a**) 1D 3-noded element; (**b**) 2D 9-noded element; (**c**) 7-noded triangular element

and renumbering the rest of the nodes appropriately, we end up with a 7-node triangular element, Fig. 2c, which captures a point singularity of the order r^κ and is therefore an appropriate finite element for e.g. crack problems.

If we use these elements in the vicinity of a crack tip, then the obvious numerical representation of the displacement field and e.g. a linearised strain is

$$\boldsymbol{u}(r,s,\kappa) = \sum_{i=1}^{7} N^i(r,s,\kappa)\boldsymbol{u}_i \tag{20}$$

$$\text{i.e.}\quad \boldsymbol{u} \propto r^\kappa \quad\text{and}\quad \boldsymbol{\varepsilon} \propto r^{[\kappa-1]}. \tag{21}$$

Now the problem arises how to determine κ for non-linear problems, where we do not know the value of κ in advance by e.g. an analytical solution. A solution will be proposed in the next section.

4 Adaptive singular elements

4.1 Dirichlet principle

By means of the total potential energy of a continuous system

$$\Pi(\boldsymbol{\varphi}) = \int_{\mathcal{B}_0} W_0 - \boldsymbol{\varphi}\cdot\boldsymbol{b}_0\,\mathrm{d}V - \int_{\partial\mathcal{B}_0^t} \boldsymbol{\varphi}\cdot\boldsymbol{t}_0^p\,\mathrm{d}A \to \min \tag{22}$$

and Dirichlet boundary condition $\boldsymbol{\varphi} = \boldsymbol{\varphi}^p$ on $\partial\mathcal{B}_0^\varphi$ the state of static equilibrium could be expressed as the minimum of this Dirichlet functional. If we restrict ourselves to convex problems, a necessary condition for the minimum is

$$\mathrm{D}_\delta\Pi = \int_{\mathcal{B}_0} \mathrm{D}_\delta\boldsymbol{F} : \partial_{\boldsymbol{F}} W_0 - \mathrm{D}_\delta\boldsymbol{\varphi}\cdot\boldsymbol{b}_0\,\mathrm{d}V - \int_{\partial\mathcal{B}_0^t} \mathrm{D}_\delta\boldsymbol{\varphi}\cdot\boldsymbol{t}_0^p\,\mathrm{d}A = 0 \tag{23}$$

Hereby the admissible variations $\mathrm{D}_\delta\boldsymbol{\varphi}$ are given by $\nabla_X\,\mathrm{D}_\delta\boldsymbol{\varphi} = \mathrm{D}_\delta\boldsymbol{F}$ in $\mathcal{B}_0$ and $\mathrm{D}_\delta\boldsymbol{\varphi} = \boldsymbol{0}$ on $\partial\mathcal{B}_0^\varphi$

4.2 Discrete Dirichlet functional

If we reformulate the Dirichlet functional with the discrete global spatial deformation map

$$\boldsymbol{\varphi}^h(\boldsymbol{X},\boldsymbol{\varphi}_n,\kappa) = \sum_{n=1}^{n_{\mathrm{np}}} N^n(\boldsymbol{X},\kappa)\boldsymbol{\varphi}_n(\kappa) \tag{24}$$

whereby n_{np} is the total number of node points in the discretised system, we get the discrete version of the Dirichlet functional

$$\Pi^h = \Pi^h(\boldsymbol{\varphi}_n(\kappa),\kappa) \to \min \tag{25}$$

Hereby, we use in the vicinity of a singular point, i.e. a crack tip, the previously described singular elements with variable power r^κ. This functional depends explicitly and implicitly, via $\boldsymbol{\varphi}_i$ on the unknown singularity type r^κ. We now propose, that the type of singularity adjusts in such a way, that it minimises the total potential energy of the system. Therefore we are able to formulate the necessary condition to minimise Π^h with respect to κ by

$$\left.\frac{\partial\Pi^h}{\partial\kappa}\right|_{\mathrm{impl}} = \left.\frac{\partial\Pi^h}{\partial\boldsymbol{\varphi}_n}\right|_{\mathrm{expl}}\cdot\frac{\partial\boldsymbol{\varphi}_n}{\partial\kappa} + \left.\frac{\partial\Pi^h}{\partial\kappa}\right|_{\mathrm{expl}} = \boldsymbol{0} \tag{26}$$

If we use a staggered scheme for the minimisation, we can incorporate the spatial equilibrium condition

$$\boldsymbol{r}_n = \left.\frac{\partial\Pi^h}{\partial\boldsymbol{\varphi}_n}\right|_{\mathrm{expl}} = \boldsymbol{0} \tag{27}$$

and end up with a minimisation problem on the spatial equilibrium surface

$$r_\kappa = \left.\frac{\partial\Pi^h}{\partial\kappa}\right|_{\mathrm{impl}} = \left.\frac{\partial\Pi^h}{\partial\kappa}\right|_{\mathrm{expl}} = 0 \tag{28}$$

5 Minimisation procedure

For simplicity we restrict ourselves to problems without body forces, i.e. $\boldsymbol{b}_0 = \boldsymbol{0}$, and traction free boundaries, i.e. $\boldsymbol{t}_0^p = \boldsymbol{0}$. Therefore the only loads applied on the system are Dirichlet boundary conditions $\boldsymbol{\varphi}^p$ on $\partial\mathcal{B}_0^{\varphi}$. This reduces our discrete Dirichlet functional to

$$\Pi^h(\boldsymbol{\varphi}^h) = \int_{\mathcal{B}_0} W_0(\boldsymbol{F}^h(\kappa))\,\mathrm{d}V \rightarrow \min \tag{29}$$

Incorporating the discrete global deformation map from Eq. 24 leads to

$$\Pi^h(\boldsymbol{\varphi}_n, \kappa) = \mathop{\mathbf{A}}_{e=1}^{n_{\mathrm{el}}} \sum_{q=1}^{n_{\mathrm{qp}}} \left[W_0(\boldsymbol{\varphi}_n, \kappa) \det \boldsymbol{J}^h(\boldsymbol{\varphi}_n, \kappa) w_q \right] \tag{30}$$

whereby we approximate the integral over the body $\mathcal{B}_0$ as a numerical integration scheme with n_{qp} quadrature points in each element with weights w_q and the Jacobian

$$\boldsymbol{J}^h(\kappa) = \frac{\partial \boldsymbol{X}_n^h}{\partial \boldsymbol{\xi}} = \sum_{n=1}^{n_{\mathrm{en}}} \boldsymbol{X}_n \otimes \nabla_{\xi} N^n(\kappa) \tag{31}$$

of the map between interpolated material positions $\boldsymbol{X}_n$ and the isoparametric coordinates $\boldsymbol{\xi} = (r, s)$. Next we need the explicit derivative of Eq. 30 with respect to κ which is given by

$$\frac{\partial \Pi^h}{\partial \kappa} = \mathop{\mathbf{A}}_{e=1}^{n_{\mathrm{el}}} \sum_{q=1}^{n_{\mathrm{qp}}} \left[\frac{\partial W_0^h}{\partial \kappa} \det \boldsymbol{J}^h(\kappa) \right] w_q \tag{32}$$

where we already incorporated

$$\frac{\partial \boldsymbol{J}^h}{\partial \kappa} = \sum_{n=1}^{n_{\mathrm{en}}} \boldsymbol{X}_n \otimes \frac{\partial}{\partial \kappa} \nabla_{\xi} N^n(\kappa) = \boldsymbol{0} \tag{33}$$

which holds for the variable power shape functions presented in Eqs. 18 and 19. The remaining term in Eq. 32 is the stored energy density W_0 of the material, which can be formulated as

$$\frac{\partial W_0^h}{\partial \kappa} = \frac{\partial W_0^h}{\partial \boldsymbol{F}^h} : \frac{\partial \boldsymbol{F}^h}{\partial \kappa} = \boldsymbol{P}^h : \frac{\partial \boldsymbol{F}^h}{\partial \kappa} \tag{34}$$

In the next section we will discuss this term in the case of a compressible neo-Hookean material and for comparison linear elastic material.

Another key point is the numerical integration, given by Eq. 32, of the singular elements. This leads to indefinite integrals with a singularity at one end point, e.g. the crack tip. Therefore we apply the so called "double exponential integration scheme" by Muhammad and Mori (2003).

5.1 Compressible neo-Hookean material

For a compressible neo-Hookean material we assume a stored energy density as

$$W_0^h(\boldsymbol{F}^h(\kappa), \kappa) = \frac{1}{2}\mu\left[I_{\boldsymbol{b}(\kappa)}^h - 3 \right] - \mu \ln J^h(\kappa) + \frac{1}{2}\lambda\left[\ln J^h(\kappa) \right]^2 \tag{35}$$

Here μ and λ are the Lame parameters, I_b is the first invariant, i.e. the trace, of the Finger tensor $\boldsymbol{b} = \boldsymbol{F} \cdot \boldsymbol{F}^t$ and J the determinant of the deformation gradient $\boldsymbol{F}$. In Eq. 34 we need the Piola stress tensor which reads in this case

$$\boldsymbol{P} = [\lambda \ln J - \mu]\boldsymbol{F}^{-t} + \mu \boldsymbol{F} \tag{36}$$

as well as the derivative of the (discrete) deformation gradient

$$\boldsymbol{F}^h(\kappa) = \sum_{n=1}^{n_{\mathrm{en}}} \boldsymbol{\varphi}_n \otimes \nabla_X N^n(\kappa) \tag{37}$$

which is calculated as

$$\frac{\partial \boldsymbol{F}^h}{\partial \kappa} = \sum_{n=1}^{n_{\mathrm{en}}} \boldsymbol{\varphi}_n \otimes \left[[J^h]^{-t}(\kappa) \cdot \frac{\partial \nabla_{\xi} N^n}{\partial \kappa} \right] \tag{38}$$

where we took already Eq. 33 into account.

5.2 Linear elasticity

The case of a linear elastic material is characterised by the stored energy density

$$W_0^h(\varepsilon^h(\kappa), \kappa) = \frac{1}{2}\boldsymbol{\varepsilon}^h(\kappa) : \mathbb{E} : \boldsymbol{\varepsilon}^h(\kappa) \tag{39}$$

with $\boldsymbol{\varepsilon}$ the linearised strain measure and $\mathbb{E}$ the fourth order elasticity tensor. If we take the definition of the linearised strain measure $\boldsymbol{\varepsilon} = \frac{1}{2}[\boldsymbol{h} + \boldsymbol{h}^t]$ and the (discrete) displacement gradient into account

$$\boldsymbol{h}^h(\kappa) = \sum_{n=1}^{n_{\mathrm{en}}} \boldsymbol{u}_n \otimes \nabla_x N^n(\kappa) \tag{40}$$

we are able to compute the required derivative as

$$\frac{\partial \boldsymbol{h}^h}{\partial \kappa} = \sum_{n=1}^{n_{\mathrm{en}}} \boldsymbol{u}_n \otimes \left[[J^h]^{-t}(\kappa) \cdot \frac{\partial \nabla_{\xi} N^n}{\partial \kappa} \right] \tag{41}$$

Thus the format of the linearised case is very similar to the geometrically nonlinear case which is an advantage for the numerical implementation.

5.3 Numerical minimisation

At this point we have all necessary information together to minimise the discrete total potential energy of the system, Eq. 30, reformulated as a minimisation problem on the spatial equilibrium surface, Eq. 28, with respect to the unknown singularity type represented by the power r^{κ}. This is a nonlinear one-dimensional equation, so as a minimisation algorithm we use a golden section search with parabolic interpolation as described in Forsythe et al. (1977) which converges very robust but sometimes—depending on the example—slow. As an alternative method we have also implemented a quasi-Newton BFGS method as described in Bertsekas (1995) which converges often much faster.

6 V-shaped notched specimen

As a model problem for the determination of the type of singularity we choose a V-shaped notched specimen under Mode-I load in plain strain conditions. The re-entrant corner, as depicted in Fig. 3, leads to a singularity in the stress and strain field whereby the order of the singularity r^{κ} varies with the opening angle 2β or its counterpart α. In the case of linear elasticity Williams (1952) has analytically derived an eigenvalue problem

$$\sin 2\kappa\alpha + \kappa \sin 2\alpha = 0 \quad (42)$$

to determine κ. In the numerical model, the vicinity of the notch is discretised by elements with a variable power singularity and the rest of the specimen is discretised by standard biquadratic Lagrange finite elements. In the numerical model we use our proposed minimisation procedure for the discrete total potential energy of the system with respect to κ. As shown in Fig. 3, the numerical solution coincides very good with the analytical solution. The difference between both solution is of the order 10^{-4}. In the case of the large strain problem, where we use the above given neo-Hookean material for the V-shape notch problem, the calculated types of singularity differs from the linear elastic case. But in this case an analytical solution is not known in the literature. We should mention here, that in the geometrically nonlinear case the deformation gradient $\boldsymbol{F} \propto R^{[\kappa-1]}$ within the complete singular finite element, whereby R is the radius in the material configuration. Therefore the presented solution may be not directly comparable with an asymptotic analytic solution of the problem, in the case it is known.

7 Material motion problem

In the next sections we like to use our proposed numerical minimisation procedure to study numerically the crack driving force, i.e. a material or configurational force acting on the crack tip, in the geometrically nonlinear case. Therefore we shortly review the underlying basic geometrically nonlinear kinematics of the quasi-static material motion problem and based on this the Material Force Method is introduced to calculate the discrete material forces in an elegant and straightforward way, see also Steinmann (2000, 2001) and Denzer et al. (2003) for a more detailed discussion.

7.1 Kinematics and kinetics

In the material motion problem $\mathcal{B}_t$ denotes the spatial configuration occupied by the body of interest at time t. Then $\boldsymbol{\Phi}(\boldsymbol{x})$ denotes the nonlinear deformation

Fig. 3 V-shape notched specimen under mode I loading

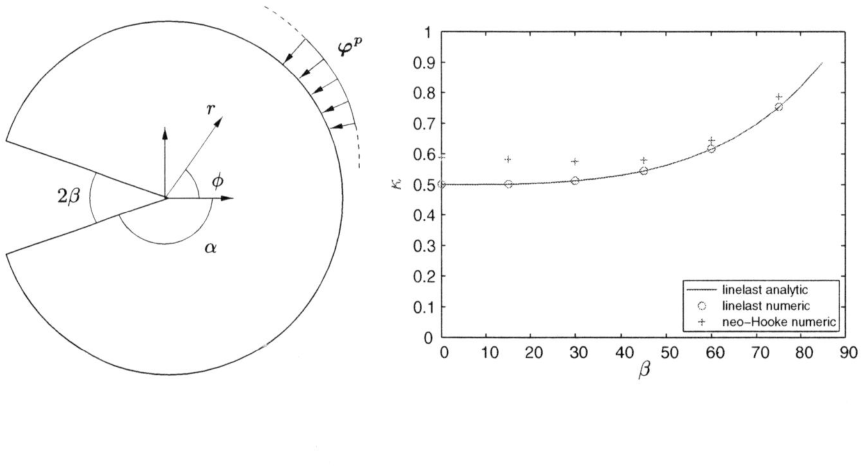

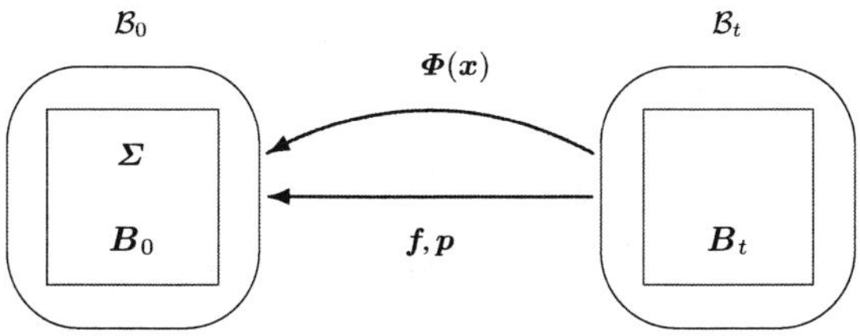

Fig. 4 Kinematics and kinetics of the material motion problem

map assigning the spatial placements $\boldsymbol{x} \in \mathcal{B}_t$ of a 'physical particle' to the material placements $\boldsymbol{X} = \boldsymbol{\Phi}(\boldsymbol{x}) \in \mathcal{B}_0$ of the same 'physical particle'. Thus, the 'physical particle' are followed through the ambient material at fixed spatial position, i.e. the observer takes essentially the Eulerian viewpoint.

Next, the material motion linear tangent map is given by the deformation gradient $\boldsymbol{f} = \nabla_x \boldsymbol{\Phi}$ transforming line elements from the tangent space $T\mathcal{B}_t$ to line elements from the tangent space $T\mathcal{B}_0$, see Fig. 4. The spatial Jacobian, i.e. the determinant of $\boldsymbol{f}$ is denoted by $j = \det \boldsymbol{f}$ and relates volume elements $\mathrm{d}v \in \mathcal{B}_t$ to volume elements $\mathrm{d}V \in \mathcal{B}_0$.

For the material motion problem the quasi-static balance of momentum reads

$$-\operatorname{div} \boldsymbol{p} = \boldsymbol{B}_t \Longrightarrow -\operatorname{Div} \boldsymbol{\Sigma} = \boldsymbol{B}_0 \tag{43}$$

It involves the momentum flux $\boldsymbol{p}$, a two-point tensor that we shall call the material motion Piola stress, see Fig. 4, and the momentum source $\boldsymbol{B}_t$, a vector in material description with spatial reference called the material motion volume force density.

The Piola transformation of $\boldsymbol{p}$ is called the Eshelby stress $\boldsymbol{\Sigma} = J\, \boldsymbol{p} \cdot \boldsymbol{f}^t$, alternatively the terminology energy-momentum tensor or configurational stress tensor is frequently adopted. The spatial motion volume force density with material reference is given by $\boldsymbol{B}_0 = J\, \boldsymbol{B}_t$.

8 Constitutive equation

For the material motion problem the free energy density $W_t = j\, W_0$ with spatial reference is expressed in terms of the material motion deformation gradient $\boldsymbol{f}$ (or its inverse $\boldsymbol{F}$). The explicit dependence on the material placement is captured by the field $\boldsymbol{X} = \boldsymbol{\Phi}(\boldsymbol{x})$

$$W_t = W_t(\boldsymbol{f}, \boldsymbol{\Phi}(\boldsymbol{x})) \tag{44}$$

Then the familiar constitutive equations for the so-called Eshelby stress in $\mathcal{B}_0$ are given as $\boldsymbol{\Sigma} = j\, \boldsymbol{p} \cdot \boldsymbol{f}^t = W_0 \boldsymbol{I} - \boldsymbol{F}^t \cdot \boldsymbol{P}$ with $\boldsymbol{P} = \partial_{\boldsymbol{F}} W_0$. Note that the distributed volume forces now take the following particular format $\boldsymbol{B}_t = -\partial_{\boldsymbol{\Phi}} W_t + \boldsymbol{B}_t^{\text{ext}}$ where the first part is related to material inhomogeneities.

9 Weak form and discretisation

Here, the pointwise statement of the material balance of momentum, see Eq. 43, is tested by material virtual displacements $\delta\boldsymbol{\Phi} = \boldsymbol{W}$ under the necessary smoothness and boundary assumptions to render the virtual work expression

$$\underbrace{\int_{\partial\mathcal{B}_0} \boldsymbol{W} \cdot \boldsymbol{\Sigma} \cdot \boldsymbol{N}\, \mathrm{d}A}_{\mathfrak{W}^{\text{sur}}} - \underbrace{\int_{\mathcal{B}_0} \nabla_X \boldsymbol{W} : \boldsymbol{\Sigma}\, \mathrm{d}V}_{\mathfrak{W}^{\text{int}}} + \underbrace{\int_{\mathcal{B}_0} \boldsymbol{W} \cdot \boldsymbol{B}_0\, \mathrm{d}V}_{\mathfrak{W}^{\text{vol}}} = \boldsymbol{0} \quad \forall \boldsymbol{W} \tag{45}$$

For a conservative system the different energetic terms $\mathfrak{W}^{\text{sur}}$, $\mathfrak{W}^{\text{int}}$ and $\mathfrak{W}^{\text{vol}}$ may be interpreted by considering the material variation at fixed $\boldsymbol{x}$ of the free energy density W_t. As a result, the contribution $\mathfrak{W}^{\text{sur}}$ denotes the material variation of W_t due to its complete dependence on the material position, whereas the contributions $\mathfrak{W}^{\text{int}}$ and $\mathfrak{W}^{\text{vol}}$ denote the material variations of W_t due to its implicit and explicit dependence on the material position, respectively.

By expanding the geometry $\boldsymbol{x}$ elementwise with shape functions N^n in terms of the positions $\boldsymbol{x}_n$ of the node points and by using a Bubnov-Galerkin finite element method based on the iso-parametric concept, we end up with discrete material node point (surface) forces at the global node point K given by

$$\mathfrak{F}^h_{\text{sur}} = \mathop{\mathbf{A}}_{e=1}^{n_{\text{el}}} \sum_{n=1}^{n_{\text{en}}} \int_{\mathcal{B}_0^e} \boldsymbol{\Sigma} \cdot \nabla_X N^n - N^n \boldsymbol{B}_0\, \mathrm{d}V \tag{46}$$

Based on these results we advocate the Material Force Method with the notion of global discrete material node point (surface) forces, that (in the sense of Eshelby) are generated by variations relative to the ambient material at fixed spatial positions. Such forces corresponding to the material motion problem are trivially computable once the spatial motion problem has been solved.

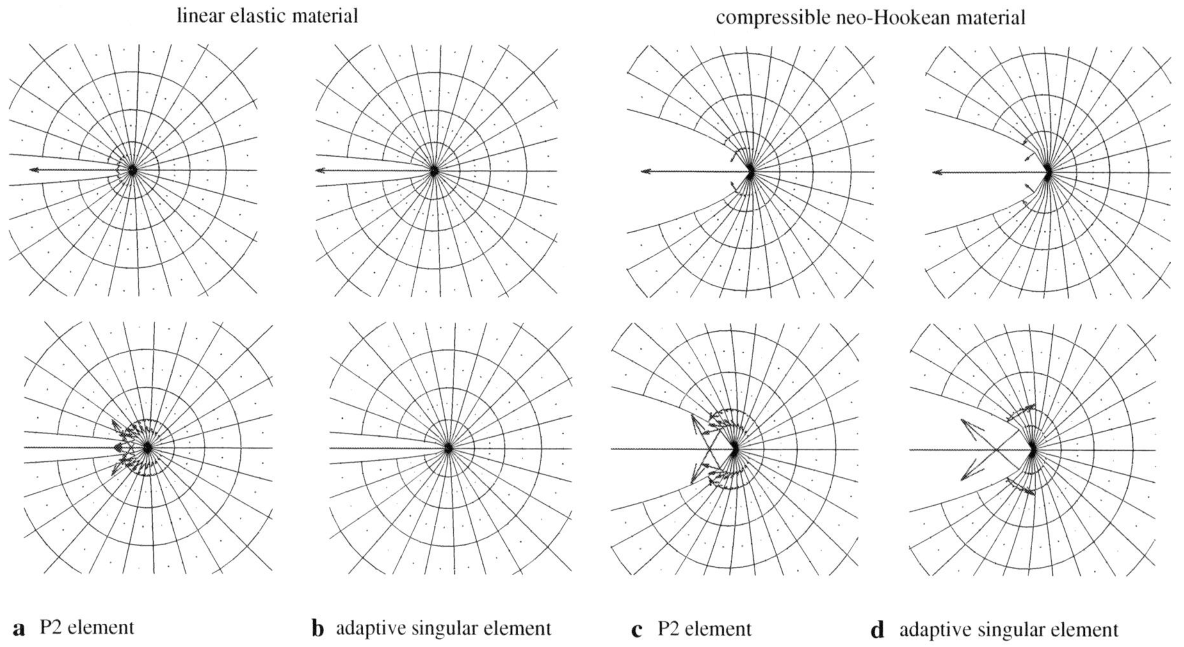

Fig. 5 Material forces in the vicinity of the crack tip

10 Cracked specimen

In this section we discuss the numerical results for a cracked specimen under mode I loading. The problem is similar to that depicted in Fig. 3 with an opening angle of $2\beta = 0$. For comparison reason we consider a so called "Modified Boundary Layer"-formulation (Rice 1974), where we apply Dirichlet boundary conditions in the displacement field at the considered region given by

$$\boldsymbol{u}^p = \frac{K_I}{2G}\sqrt{r/2\pi}\, \boldsymbol{f}_I(\phi) \tag{47}$$

whereby K_I is the mode I stress intensity factor, G the shear modulus and $\boldsymbol{f}_I$ are given functions depending only on the angle ϕ. Under the condition, that the overall deformations are small, the linear elastic relation for the J-integral, i.e. the material force acting on the crack tip, $J_{\text{appl}} = K_I^2/E'$ with $E' = E/[1-\nu]$ holds. Nevertheless, the deformation in the vicinity of the crack are often large and therefore a geometrically nonlinear analysis maybe necessary. With this model problem we have performed our optimisation procedure to evaluate the type of singularity in the vicinity of the crack tip and additionally have calculate the numerically resulting crack driving forces.

Figure 5a and b depict the calculated material forces in the vicinity of the crack for the linear elastic case. The first row of elements directly connected to the crack tip node have a length of $l_{\text{el}} = 0.01$ whereas the overall radius of the considered region is $L = 100$. In Fig. 5a, we used standard biquadratic triangular elements connected to the crack tip and in Fig. 5b we use the proposed adaptive singular elements. All next rows of elements are standard biquadratic elements based on Lagrangian polynomials. In this example, we have applied $J_{\text{appl}} = 1.0$ at the outer boundary of the problem. The type of singularity is calculated numerically with the proposed minimisation procedure as $\kappa = 0.5002$ which agrees very well with the analytical result $\kappa = 0.5$. This results in a very high precision in the calculation of the crack driving force. In this example we got a numerical value of $\mathfrak{F}_{\text{sur}}^{\text{cracktip}} = 1.0004$ which is very close to the applied value of $J_{\text{appl}} = 1.0$. In the lower row of Fig. 5a and b the calculated discrete material forces are scaled by a factor 20 in comparison to the upper row. Whereas the standard finite element (P2) shows a lot of numerical "noise" in the calculated discrete material forces the adaptive singular element reduces these inaccuracies considerably. The crack driving force at the crack tip node in the case of the standard P2 element has a value of only $\mathfrak{F}_{\text{sur}}^{\text{cracktip}} = 0.8346$.

The second example, shown in Fig. 5c and d, a straight crack in a compressible neo-Hookean material is discussed in the sequel. The geometry of the specimen is the same as in the linear elastic case, but in this case we applied a load of $J_{\text{appl}} = 16$. This results in much larger deformations in the vicinity of the crack tip, but the overall deformations remains still small, so that we can approximately use the linear elastic relation $J_{\text{appl}} = K_I^2/E'$. Also in this case the adaptive singular element performs better than the standard P2 element. In the lower row of fig. 5c and d, where the discrete material forces are scaled by a factor of 20 in comparison to the upper row, the numerical "noise", the so called spurious or residual material forces, in the case of the adaptive singular element are lower in comparison to the P2 element. The resulting (discrete) crack driving forces are $\mathfrak{F}_{\text{sur}}^{\text{cracktip}}/J_{\text{appl}} = 0.8393$ for the P2 element and $\mathfrak{F}_{\text{sur}}^{\text{cracktip}}/J_{\text{appl}} = 0.8878$ for the proposed element. In this example we calculate $\kappa = 0.6949$ as the order of the singularity. It should be noted, that the calculation of the crack driving force by a domain integral method, which is in the sense of discrete material forces simply the sum of all discrete material forces in a given subdomain including the crack tip, see Denzer et al. (2003), the "exact" value of $\mathfrak{F}_{\text{sur}}^{\text{cracktip}}/J_{\text{appl}} = 1.0$ is regained. We should state here once more, that the calculated order of singularity is not directly comparable to an asymptotic analytic solution, because κ is calculated for a finite length $\mathfrak{l}_{\text{el}}$ of the adaptive finite element. This is indicated by further computational experiments, where we observe that the numerically calculated order of singularity κ in the geometrically nonlinear case depends on the element length $\mathfrak{l}_{\text{el}}$ and the applied load level. Here, further numerical studies are necessary. Additionally, it may be possible, that the singularity is not of a power type and may be calculated by the fact that the Eshelby stress $\boldsymbol{\Sigma} \propto 1/R$, see Maugin (1995), using W_0 for a specific material.

11 Summary

We introduced an optimisation procedure based on a variational principle to determine adaptively the order of singularity in problems even in geometrically nonlinear cases, where we normally do not know the type of singularity at re-entrant corners or cracks in advance by (asymptotic) analytical solutions. In the case of linear elasticity, the proposed method shows a very high accuracy in comparison to known analytic solutions and the precision of calculation of material forces, e.g. crack driving forces, are significantly improved. Although the proposed method improves also the precision of numerically calculated material forces in the geometrically nonlinear case, it turns out, that the order of singularity depends on the element length and the load level.

References

Akin JE (1976) The generation of elements with singularities. Int J Numer Methods Eng 10:1249–1259

Barsoum RS (1977) Triangular quarter-point elements as elastic and perfectly-plastic crack tip elements. Int J Numer Methods Eng 11:85–98

Bertsekas DP (1995) Nonlinear programming. Athena Scientific

Denzer R, Barth FJ, Steinmann P (2003) Studies in elastic fracture mechanics based on the material force method. Int J Numer Methods Eng 58:1817–1835

Forsythe GE, Malcolm MA, Moler CB (1977) Computer methods for mathematical computations. Prentice-Hall

Herrmann JM (1989) An asymptotic analysis of finite deformations near the tip of an interface-crack. J Elasticity 21:227–269

Herrmann JM (1992) An asymptotic analysis of finite deformations near the tip of an interface-crack: part i. J Elasticity 29:203–241

Hughes TJR, Akin JE (1980) Techniques for developing 'special' finite element shape functions with particular reference to singularities. Int J Numer Methods Eng 15:733–751

Knowles JK, Sternberg E (1973) An asymptotic finite-deformation analysis of the elastostatic field near the tip of a crack. J Elasticity 3:67–107

Knowles JK, Sternberg E (1983) Large deformations near a tip of an interface-crack between two neo-Hookean sheets. J Elasticity 13:257–293

Le KCh, Stumpf H (1993) The singular elastostatic field due to a crack in rubberlike materials. J Elasticity 32:183–222

Lim W-K, Kim S-C (1994) Further study to obtain a variable power singularity using quadratic isoparametric elements. Eng Fract Mech 47(2):223–228

Maugin GA (1995) Material forces: concepts and applications. Appl Mech Rev 48:213–245

Muhammad M, Mori M (2003) Double exponential formulas for numerical indefinite integration. J Comput Appl Math 161:431–448

Rice JR (1968) A path independent integral and the approximate analysis of strain concentration by notches and cracks. J Appl Mech 35:379–386

Rice JR (1974) Limitations to the small scale yielding approximation for crack tip plasticity. J Mech Phys Solids 22:17–26

Staab GH (1983) A variable power singular element for analysis of fracture mechanics problems. Comput Struct 17(3):449–457

Steinmann P (2000) Application of material forces to hyperelastostatic fracture mechanics. i. Continuum mechanical setting. Int J Solids Struct 37:7371–7391

Steinmann P, Ackermann D, Barth FJ (2001) Application of material forces to hyperelastostatic fracture mechanics. ii. Computational setting. Int J Solids Struct 38:5509–5526

Stern M (1978) Families of consistent conforming elements with singular derivative fields. Int J Numer Methods Eng 14:409–421

Tracey DM, Cook TS (1977) Analysis of power type singularities using finite elements. Int J Numer Methods Eng 11:1225–1233

Williams ML (1952) Stress singularities resulting from various boundary conditions in angular corners of plates in extension. J Appl Mech 19:526–528

Moving singularities in thermoelastic solids

Arkadi Berezovski · Gérard A. Maugin

Abstract The solution of the evolution problem of a discontinuity requires the formulation of a kinetic law of the progress relating the driving force and the velocity of the singularity. In the case of a crack, the energy-release rate can be computed (in quasi-statics and in the absence of thermal and intrinsic dissipations) by means of the celebrated J-integral of fracture that is known to be path-independent and, therefore, provides a very convenient estimation of the driving force once the field solution is known. However, the velocity at the crack tip remains undetermined. A similar situation holds for a displacive phase-transition front propagation. The driving force acting on the phase boundary can be determined, but not the velocity of the displacive phase-transition front. From the thermodynamic point of view, both the phase transition and the crack propagation are non-equilibrium processes; entropy is produced at the evolving discontinuity. Therefore, stress jumps are determined by means of non-equilibrium jump relations at the discontinuity. Then the kinetic relations can be obtained depending on the choice of excess stress behavior. The procedure is illustrated on the example of a phase-transition front propagation in a shape-memory alloy bar.

Keywords Moving discontinuity · Phase-transition front · Jump relations · Kinetic relation

A. Berezovski (✉)
Institute of Cybernetics, Tallinn University of Technology, Akadeemia tee 21, 12618 Tallinn, Estonia
e-mail: Arkadi.Berezovski@cs.ioc.ee

G. A. Maugin
Institut Jean Le Rond d'Alembert, Université Pierre et Marie Curie, UMR 7190, 4 Place Jussieu, 75252 Paris Cedex 05, France

1 Introduction

Phase-transition front and crack propagation are most known examples of moving singularities in solids. Both the phase-transition front and the crack represent quasi-inhomogeneities, and, therefore, these phenomena can be considered under one umbrella in the framework of the canonical thermomechanical theory (Maugin 2000). At the same time, it is well known that the material (driving) force in each case is computed differently. The main problem in both cases is the establishment of the kinetic relation between the driving force and the velocity of the singularity (Abeyaratne and Knowles 2006). The product of this velocity and the corresponding driving force determines the entropy production at the evolving discontinuity (Maugin 2000).

Recently (Berezovski and Maugin 2007), a kinetic relation for the straight brittle crack was discussed. It was shown that the equation of motion for a crack following from the energy balance at the crack tip in the framework of the linear elasticity theory (Freund 1990; Fineberg and Marder 1999) corresponds to the local equilibrium approximation. A more general non-equilibrium description allows to achieve a better

doi: 10.1007/978-1-4020-6929-1_18,

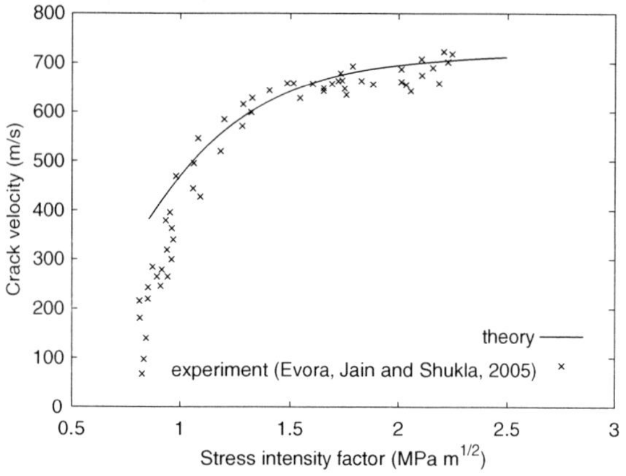

Fig. 1 Crack growth toughness in Polyester/TiO_2 nanocomposite

interpretation of experimental data. A typical example of the comparison of the theory (Berezovski and Maugin 2007) with experimental data for the crack propagation in Polyester/TiO_2 nanocomposite (Evora et al. 2005) is shown in Fig. 1, where the theoretical curve is calculated using the measured value of the limiting velocity $V_T = 724\,\mathrm{m/s}$.

It should be noted that the theoretical prediction is applied only for the values of the stress intensity factor $K_I > K_{Ic}$, and its critical value is $K_{Ic} = 0.85\,\mathrm{MPa\,m^{1/2}}$ (Evora et al. 2005).

Here a similar description is kept in the derivation of the kinetic relation for the phase-transition front. We start with the material setting of thermoelasticity. Then we discuss the jump relations at a discontinuity. Due to the irreversibility of the considered process, we apply non-equilibrium jump relations at the singularity (Berezovski and Maugin 2004). These jump relations allow us to determine the stress jump in terms of the driving force. Having the value of the stress jump, we are able to derive the kinetic relation in the case of a displacive phase-transition front. This is demonstrated on the one-dimensional example of phase-transition front propagation in a bar.

2 Material setting

We express the governing equations on the material manifold. This gives rise to the notion of material forces, which are made apparent through a canonical projection of the thermomechanical problem onto the material manifold.

In the Piola-Kirchhoff formulation (Maugin 1993), the balance of mass, linear momentum, and energy with no external supply of energy and body force can be written at any regular material point $\mathbf{X}$ in a continuous body per unit reference volume as follows (Maugin 1997):

$$\left.\frac{\partial}{\partial t}\rho_0\right|_{\mathbf{X}} = 0, \tag{1}$$

$$\left.\frac{\partial}{\partial t}\mathbf{p}\right|_{\mathbf{X}} - \nabla_R \cdot \mathbf{T} = 0, \tag{2}$$

$$\left.\frac{\partial}{\partial t}\mathcal{H}\right|_{\mathbf{X}} - \nabla_R \cdot (\mathbf{T}\cdot\mathbf{v} - \mathbf{Q}) = 0, \tag{3}$$

where t is time, $\rho_0(\mathbf{X})$ is the matter density in the reference configuration, $\mathbf{v}$ is the physical velocity, $\mathbf{T}$ is the first Piola-Kirchhoff stress, $\mathbf{p} = \rho_0(\mathbf{X})\mathbf{v}(\mathbf{X},t)$ is the linear momentum, S is the entropy per unit volume, $\mathcal{H} = K + E$, $K(\mathbf{v},\mathbf{X}) = \rho_0\mathbf{v}^2/2$ is the kinetic energy per unit volume in the reference configuration, $E(\mathbf{F},\theta;\mathbf{X})$ is the corresponding internal energy, $\mathbf{Q}$ is the material heat flux.

The second law of thermodynamics reads

$$\left.\frac{\partial S}{\partial t}\right|_{\mathbf{X}} + \nabla_R \cdot (\mathbf{Q}/\theta) \geq 0, \tag{4}$$

where θ is the absolute temperature.

The above set of Eqs. 1–4 is valid in any continuously inhomogeneous material. It is clear that the existence of a discontinuity surface $\mathbb{S}$ breaks the symmetry of the problem and that, in the case of phase transition, it may be viewed as breaking the translational invariance of the whole physical system on the material manifold. Therefore, the presence of $\mathbb{S}$ manifests a lack of the material homogeneity for the whole system under study. Accordingly, the equation associated with this lack of invariance must play a prominent role in further considerations concerning $\mathbb{S}$. This equation is the balance of pseudo-momentum (Maugin 1993)

$$\left.\frac{\partial \mathcal{P}}{\partial t}\right|_{\mathbf{X}} - \nabla_R \cdot \mathbf{b} = \mathbf{f}^{int} + \mathbf{f}^{inh}, \tag{5}$$

where $\mathcal{P} = -\mathbf{p}\cdot\mathbf{F}$ is the pseudo-momentum, $\mathbf{b} = -(\mathcal{L}\mathbf{1_R} + \mathbf{T}\cdot\mathbf{F})$ is the dynamical Eshelby stress tensor, $\mathcal{L} = K - W$ is the Lagrangian density, $\mathbf{F}$ is the deformation gradient,

$$\mathbf{f}^{inh} := \left.\frac{\partial \mathcal{L}}{\partial \mathbf{X}}\right|_{expl} \equiv \left.\frac{\partial \mathcal{L}}{\partial \mathbf{X}}\right|_{fixed\ fields} = \left(\frac{1}{2}\mathbf{v}^2\right)\nabla_R \rho_0 - \left.\frac{\partial \overline{W}}{\partial \mathbf{X}}\right|_{expl}, \tag{6}$$

$$\mathbf{f}^{int} = \mathbf{T} : (\nabla_R \mathbf{F})^T - \nabla_R W|_{impl}\,. \tag{7}$$

Here the subscript notations *expl* and *impl* mean, respectively, the material gradient keeping the fields fixed (and thus extracting the explicit dependence on $\mathbf{X}$), and taking the material gradient only through the fields present in the function.

The energy equation in the bulk (Eq. 3) of each phase can also be written as

$$\frac{\partial (S\theta)}{\partial t} + \nabla_R \cdot \mathbf{Q} = h^{int}, \quad h^{int} := \mathbf{T} : \dot{\mathbf{F}} - \frac{\partial W}{\partial t}, \tag{8}$$

where the right-hand side of Eq. 8_1 is formally an internal heat source.

The corresponding jump relations across a homothermal ($[\theta] = 0$) and coherent ($[\mathbf{V}] = 0$) front $\mathbb{S}$ are the Rankine-Hugoniot jump relations (Maugin 1997)

$$\bar{V}_N[\rho_0] = 0, \tag{9}$$

$$\bar{V}_N[\mathbf{p}] + \mathbf{N}\cdot[\mathbf{T}] = 0, \tag{10}$$

$$\bar{V}_N[\mathcal{H}] + \mathbf{N}\cdot[\mathbf{T}\cdot\mathbf{v} - \mathbf{Q}] = 0, \tag{11}$$

and

$$\bar{V}_N[S] - \mathbf{N}\cdot[\mathbf{Q}/\theta] = \sigma_{\mathbb{S}} \geq 0, \tag{12}$$

where square brackets denote jumps, $\mathbf{V}$ is the material velocity, and $\bar{V}_N$ is its normal component at the points of the discontinuity surface $\mathbb{S}$.

Both Eqs. 5 and 8 are nonconservative. Therefore, the corresponding jump relations across $\mathbb{S}$ should exhibit source terms to be jointly determined by the thermodynamic study:

$$\bar{V}_N[\mathcal{P}] + \mathbf{N}\cdot[\mathbf{b}] = -\mathbf{f}, \tag{13}$$

$$\bar{V}_N[\theta S] - \mathbf{N}\cdot[\mathbf{Q}] = q_{\mathbb{S}}, \tag{14}$$

where $q_{\mathbb{S}}$ and $\sigma_{\mathbb{S}}$ are unknown scalars, and $\mathbf{f}$ is an unknown material co-vector. These three quantities are surface sources and are collectively constrained to satisfy the second law of thermodynamics at the front $\mathbb{S}$ such that

$$\mathbf{f}\cdot\mathbf{V} = \theta_{\mathbb{S}}\sigma_{\mathbb{S}} = q_{\mathbb{S}} \geq 0. \tag{15}$$

The above-mentioned formulation for a problem displays clearly the need of the kinetic relation, because the jump relations (Eqs. 9–14) are useless until the velocity of the front is determined.

3 Jump relations at a front

Classical jump relations at the interface between two distinct media follow from thermodynamic equilibrium conditions which consist in zero jumps of temperature, pressure, and chemical potentials for fluid-like systems. In the thermomechanics of solids they are expressed as (cf. Cermelli and Sellers 2000)

$$[\theta] = 0, \quad [\mathbf{T}]\cdot\mathbf{N} = 0, \quad [\mu] = 0, \tag{16}$$

where μ is the chemical potential.

The conditions (16) represent thermal, mechanical, and chemical equilibrium, respectively. It is clear that these equilibrium conditions are insufficient in the case of an irreversible motion of crack or phase-transition front.

In practice, the uniformity of the chemical potential is substituted by the non-zero driving force, which in the case of homothermal phase transition can be represented as (Maugin 1997; Maugin and Trimarco 1995)

$$f_{\mathbb{S}} = -[W] + \mathbf{N}\cdot\langle\mathbf{T}\rangle\cdot[\mathbf{F}\cdot\mathbf{N}], \tag{17}$$

where W is the free energy per unit reference volume and $\langle\ldots\rangle$ denotes the mean value across a discontinuity surface.

Let us consider the mechanical equilibrium condition $(16)_2$ in more detail. It is equivalent, for fluid-like system, to the uniformity of pressures, which can be represented in terms of thermodynamic derivatives by

$$[p] = \left[\left(\frac{\partial U}{\partial V}\right)_{S,M}\right] = 0, \tag{18}$$

where p is pressure, U is the internal energy, V is volume, and M denotes mass.

It should be emphasized that the thermodynamic derivative in Eq. 18 is computed for a fixed entropy value. However, an irreversible progress of crack front or phase boundary is accompanied by entropy production. Therefore, we have proposed to exploit a more convenient thermodynamic derivative (Berezovski and Maugin 2004), namely

$$\left[\left(\frac{\partial U}{\partial V}\right)_{p,M}\right] = 0, \tag{19}$$

In the thermomechanical case, the latter is equivalent to (Berezovski and Maugin 2004)

$$\left[\theta\left(\frac{\partial S}{\partial \mathbf{F}}\right)_{\mathbf{T}} + \mathbf{T}\right] \cdot \mathbf{N} = 0. \tag{20}$$

This means that we can determine the value of the jump of stress tensor if we know the behavior of entropy in the vicinity of a discontinuity.

4 Stress jumps

As shown (Abeyaratne et al. 2001), in the case of phase transformations there are two sources of entropy production: heat conduction and phase transformation. In the isothermal case ($\theta = const$), the entropy production rate due to heat conduction vanishes. This means that the entropy jump is determined only by the driving force and the temperature at the phase boundary,

$$[S] = \frac{f_{\mathbb{S}}}{\theta_{\mathbb{S}}}. \tag{21}$$

In the case of phase-transition front, we may expect that the entropy behaves like

$$S = \frac{f}{\theta}, \tag{22}$$

where the function f is defined similarly to Eq. 17

$$f = -W + \mathbf{N} \cdot \langle \mathbf{T} \rangle \cdot \mathbf{F} \cdot \mathbf{N}. \tag{23}$$

This formally defined quantity—a generator function—acquires a physical meaning only when evaluated at a point $\mathbf{X}$ belonging to an oriented surface of unit normal $\mathbf{N}$ and with an operation such as that of taking the discontinuity applied to it.

Then the derivative of entropy with respect to deformation gradient can be computed by

$$\left(\frac{\partial S}{\partial \mathbf{F}}\right)_{\mathbf{T}} = \frac{1}{\theta}\left(\frac{\partial f}{\partial \mathbf{F}}\right)_{\mathbf{T}} - \frac{f}{\theta^2}\left(\frac{\partial \theta}{\partial \mathbf{F}}\right)_{\mathbf{T}}, \tag{24}$$

and one can determine the stress jump at the discontinuity in terms of the driving force as follows (Berezovski and Maugin 2005a)

$$[\mathbf{T}] \cdot \mathbf{N} = \left(\frac{f_{\mathbb{S}}}{\theta_{\mathbb{S}}}\left\langle\left(\frac{\partial \theta}{\partial \mathbf{F}}\right)_{\mathbf{T}}\right\rangle - \left[\left(\frac{\partial f}{\partial \mathbf{F}}\right)_{\mathbf{T}}\right]\right) \cdot \mathbf{N}. \tag{25}$$

This relation gives us the possibility to determine the velocity of the phase boundary.

5 Phase-transition fronts in a bar

In the case of phase-transition front propagation, we consider the simplest possible one-dimensional formulation. This is the boundary value problem of the tensile loading of a shape memory alloy bar that has uniform cross-sectional area A_0 and temperature θ_0 (Abeyaratne et al. 2001). The bar occupies the interval $0 < x < L$ in a reference configuration and the boundary $x = 0$ is subjected to the tensile loading

$$\sigma(0, t) = \hat{\sigma}(t) \quad \text{for} \quad t > 0. \tag{26}$$

The bar is assumed to be long compared to its diameter so it is under a uniaxial stress state and the stress $\sigma(x, t)$ depends only on the axial position and time. Supposing the temperature is constant during the process, it is characterized by the displacement field $u(x, t)$, where x denotes the location of a particle in the reference configuration and t is time. Linearized strain is further assumed so the axial component of the strain $\varepsilon(x, t)$ and the particle velocity $v(x, t)$ are related to the displacement by

$$\varepsilon = \frac{\partial u}{\partial x}, \quad v = \frac{\partial u}{\partial t}. \tag{27}$$

The density of the material ρ_0 is assumed constant. All field variables are averaged over the cross-section of the bar. The geometry of the problem is shown in Fig. 2.

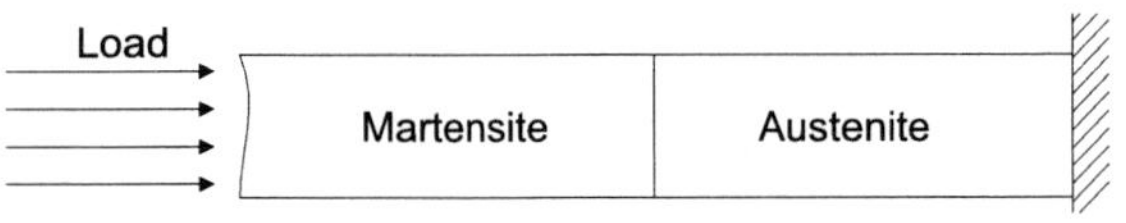

Fig. 2 Test problem for one-dimensional front propagation

Limiting ourselves to the isothermal case, we need to keep in mind the complete thermomechanical theory, since we exploit the entropy production at the front.

At each instant t during a process, the strain $\varepsilon(x, t)$ varies smoothly within the bar except at phase boundaries; across a phase boundary, it suffers jump discontinuity. The displacement field is assumed to remain continuous throughout the bar. Away from a phase boundary, the balance of linear momentum and the kinematic compatibility require that

$$\rho_0 \frac{\partial v}{\partial t} = \frac{\partial \sigma}{\partial x}, \tag{28}$$

$$\frac{\partial \varepsilon}{\partial t} = \frac{\partial v}{\partial x}. \tag{29}$$

Suppose that at time t there is a moving discontinuity in strain or particle velocity at $x = \mathbb{S}(t)$. Then one also has the corresponding jump conditions (cf. Abeyaratne et al. 2001)

$$\rho_0 V_{\mathbb{S}}[v] + [\sigma] = 0, \tag{30}$$

$$V_{\mathbb{S}}[\varepsilon] + [v] = 0, \tag{31}$$

$$V_{\mathbb{S}}\theta[S] = f_{\mathbb{S}} V_{\mathbb{S}}, \tag{32}$$

where $V_{\mathbb{S}}$ is the velocity of the phase boundary and the driving traction $f_{\mathbb{S}}(t)$ at the discontinuity is defined by (cf. Truskinovsky 1987; Abeyaratne and Knowles 1990)

$$f_{\mathbb{S}} = -[W] + \langle \sigma \rangle [\varepsilon], \tag{33}$$

where W is the free energy per unit volume.

The second law of thermodynamics requires that

$$f_{\mathbb{S}} V_{\mathbb{S}} \geq 0 \tag{34}$$

at strain discontinuities. If $f_{\mathbb{S}}$ is not zero, the sign of $V_{\mathbb{S}}$, and hence the direction of motion of discontinuity, is determined by the sign of $f_{\mathbb{S}}$.

The difficulties relate to an unknown motion of the phase boundary.

6 Velocity of a phase boundary

In the case of phase transition front propagation, we propose the satisfaction of the non-equilibrium jump relation at the phase boundary (20) (Berezovski and Maugin 2005a), which in the uniaxial case is reduced to

$$\left[\theta \left(\frac{\partial S}{\partial \varepsilon}\right)_{\sigma} + \sigma\right] = 0. \tag{35}$$

Moreover, we assume the continuity of the excess stresses at the phase boundary (Berezovski and Maugin 2005b). This continuity means that the full stress jump is equal to the jump in local equilibrium stress values.

The derivative of entropy (24) in the considered case is given by

$$\left(\frac{\partial S}{\partial \varepsilon}\right)_{\sigma} = \frac{1}{\theta}\left(\frac{\partial f}{\partial \varepsilon}\right)_{\sigma} - \frac{f}{\theta^2}\left(\frac{\partial \theta}{\partial \varepsilon}\right)_{\sigma}, \tag{36}$$

where

$$f = -W + \langle \sigma \rangle \varepsilon. \tag{37}$$

It follows from Eq. 37 that the jump of the first term in the right hand side of Eq. 36 is

$$\left[\frac{1}{\theta}\left(\frac{\partial f}{\partial \varepsilon}\right)_{\sigma}\right] = -\frac{1}{2}[\sigma], \tag{38}$$

and the relation (35) can be represented as

$$\frac{1}{2}[\sigma] = \left[\frac{f}{\theta}\left(\frac{\partial \theta}{\partial \varepsilon}\right)_{\sigma}\right] = f_{\mathbb{S}}\left\langle \frac{1}{\theta}\left(\frac{\partial \theta}{\partial \varepsilon}\right)_{\sigma}\right\rangle + \left[\frac{1}{\theta}\left(\frac{\partial \theta}{\partial \varepsilon}\right)_{\sigma}\right]\langle f \rangle. \tag{39}$$

In general, the function f is determined up to an arbitrary constant, which in the isothermal case can be chosen such that $\langle f \rangle = 0$ (Berezovski and Maugin 2005a). In the latter case, we obtain the value of the stress jump at the phase boundary in the form (Berezovski and Maugin 2005b)

$$[\sigma] = 2 f_{\mathbb{S}} \left\langle \frac{1}{\theta} \left(\frac{\partial \theta}{\partial \varepsilon} \right)_{\sigma} \right\rangle. \tag{40}$$

Having the value of the stress jump, we can determine the material velocity at the moving phase boundary by means of the jump relation for linear momentum (30) rewritten in terms of averaged quantities because of the continuity of excess quantities at the phase boundary (Berezovski and Maugin 2005b)

$$V_{\mathbb{S}}[\rho_0 \bar{v}] + [\bar{\sigma}] = 0. \tag{41}$$

Here overbars denote averaged (local equilibrium) quantities.

The kinematic compatibility condition (31) reads

$$[\bar{v}] = -[\bar{\varepsilon}] V_{\mathbb{S}}, \tag{42}$$

and the jump relation for linear momentum (41) can be rewritten in a form which is more convenient for the calculation of the velocity at discontinuity

$$\rho_0 V_{\mathbb{S}}^2 [\bar{\varepsilon}] = [\bar{\sigma}]. \tag{43}$$

The direction of the front propagation is determined by the positivity of the entropy production

$$f_{\mathbb{S}} V_{\mathbb{S}} \geq 0. \tag{44}$$

Relations (43) and (44) fully determine the velocity of the phase transition front propagation. However, we can express them in a more explicit form. In the isothermal case we have for the strain jump

$$[\bar{\varepsilon}] = \left[\frac{\bar{\sigma}}{\lambda + 2\mu} \right] - \varepsilon_{tr}, \tag{45}$$

or

$$[\bar{\varepsilon}] = [\bar{\sigma}] \left\langle \frac{1}{\lambda + 2\mu} \right\rangle + \langle \bar{\sigma} \rangle \left[\frac{1}{\lambda + 2\mu} \right] - \varepsilon_{tr}. \tag{46}$$

Here ε_{tr} denotes transformation strain.

Representing the mean value of the stress as

$$\langle \bar{\sigma} \rangle = \bar{\sigma}_A - [\bar{\sigma}]/2, \tag{47}$$

where $\bar{\sigma}_A$ is the averaged stress in austenite, we can rewrite the expression for the stress jump at the phase boundary as follows

$$[\bar{\varepsilon}] = A[\bar{\sigma}] - B, \tag{48}$$

with

$$A = \left\langle \frac{1}{\lambda + 2\mu} \right\rangle - \frac{1}{2} \left[\frac{1}{\lambda + 2\mu} \right], \tag{49}$$

$$B = \varepsilon_{tr} - \bar{\sigma}_A \left[\frac{1}{\lambda + 2\mu} \right]. \tag{50}$$

Therefore, Eq. 42 can be represented in terms of the stress jump

$$\rho_0 V_{\mathbb{S}}^2 = \frac{[\bar{\sigma}]}{A[\bar{\sigma}] - B}. \tag{51}$$

As follows from Eq. 40, the stress jump is connected to the driving force by

$$[\bar{\sigma}] = D f_{\mathbb{S}}, \tag{52}$$

where $f_{\mathbb{S}}$ is defined in Eq. 33 and we have set

$$D = 2 \left\langle \frac{\lambda + 2\mu}{\alpha \theta (3\lambda + 2\mu)} \right\rangle, \tag{53}$$

where α is the thermal expansion coefficient.

This leads to the kinetic relation in the form

$$\rho_0 V_{\mathbb{S}}^2 = \frac{D f_{\mathbb{S}}}{A D f_{\mathbb{S}} - B}. \tag{54}$$

It should be noted that the phase transformation process begins only if the value of the driving force is over a certain critical value. The critical value of the driving force is expressed in the isothermal case as (Berezovski and Maugin 2005b)

$$f_{\mathbf{cr}} = \theta_{\mathbb{S}}^2 [\alpha(3\lambda + 2\mu)] \left\langle \frac{(\lambda + 2\mu)}{\alpha(3\lambda + 2\mu)} \right\rangle^{-1}. \tag{55}$$

Therefore, we exploit $f_{\mathbb{S}} - f_{\mathbf{cr}}$ instead of $f_{\mathbb{S}}$ in the previous expression. However, the strain jump should be calculated by $f_{\mathbb{S}}$. This means that the final expression

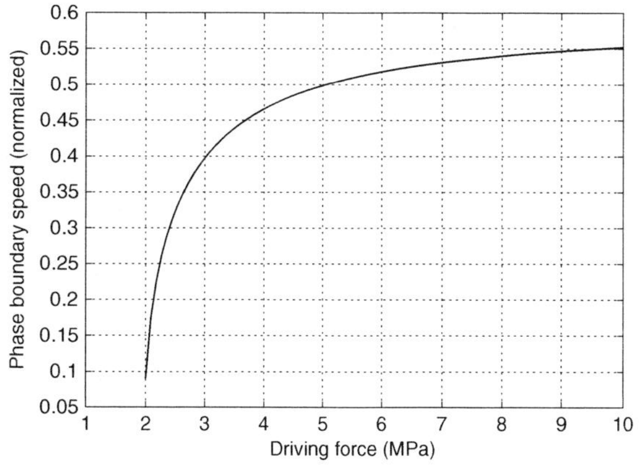

Fig. 3 Kinetic behavior of Ni–Ti

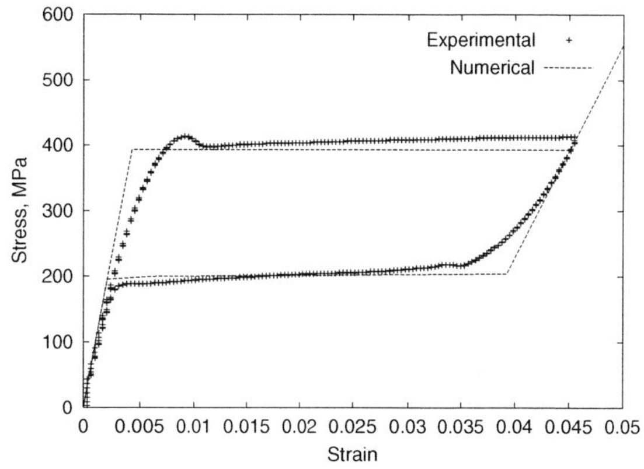

Fig. 4 Stress–strain relation at a fixed point

for kinetic relation has the form

$$\rho_0 V_{\mathbb{S}}^2 = \frac{D(f_{\mathbb{S}} - f_{\mathbf{cr}})}{AD(f_{\mathbb{S}} - f_{\mathbf{cr}}) + ADf_{\mathbf{cr}} - B}. \tag{56}$$

The obtained kinetic behavior is illustrated in Fig. 3, where the coefficients are calculated for the particular case of a Ni–Ti shape memory alloy.

Material properties for the Ni–Ti shape memory alloy were extracted from the paper by McKelvey and Ritchie (2000), where all the details of experimental procedure are well explained. The Young's moduli are 62 and 22 GPa for austenite and martensite phases, respectively, Poisson's ratio is the same for both phases and is equal to 0.33, the density for both phases is 6450 kg/m^3.

The corresponding stress in austenite is 393 MPa (McKelvey and Ritchie 2000), and the critical value of the driving force is 2.25 MPa. The values of coefficients are the following: $D = 414$, $AD = 0.0127\,\text{MPa}^{-1}$, $B = 0.0124$. The values of the velocity of the phase boundary are normalized by the longitudinal wave velocity in austenite (3774 m/s).

To validate the obtained results, the local stress–strain relation at a fixed point which was initially in austenitic state was calculated numerically. The best fitting of experimental observations (McKelvey and Ritchie 2000) corresponds to the value for the transformation strain 3.3%. Just this value is used for the comparison with the experimental data. After unloading, the reverse phase transformation occurs. It is supposed that the reverse transformation begins immediately after unloading. The calculated local stress–strain relation is plotted in Fig. 4 together with experimental data for the quasi-static loading (McKelvey and Ritchie 2000).

7 Conclusions

A kinetic relation for phase-transition front propagation is obtained by means of the procedure similar to that for moving straight-brittle crack (Berezovski and Maugin 2007), exploiting the driving force concept and non-equilibrium jump relations at the discontinuity. The driving force is determined in the framework of the material setting of thermoelasticity. Stress jump across the discontinuity is related to the driving force due to the non-equilibrium jump relation. The value of the stress jump is used then in the jump relation for linear momentum, resulting in an explicit form of the kinetic relation. Considered example represents the moving phase boundary in a simplified form, but it opens a possibility to generalizations and more detailed numerical simulations.

Acknowledgements Support of the Estonian Science Foundation (A.B.) is gratefully acknowledged. G.A.M. benefits from a Max Planck Award for International Cooperation (2001–2005).

References

Abeyaratne R, Knowles JK (1990) On the driving traction acting on a surface of strain discontinuity in a continuum. J Mech Phys Solids 38:345–360

Abeyaratne R, Knowles JK (2006) Evolution of phase transitions: a continuum theory. Cambridge University Press, Cambridge

Abeyaratne R, Bhattacharya K, Knowles JK (2001) Strain-energy functions with local minima: modeling phase transformations using finite thermoelasticity. In: Fu Y, Ogden RW (eds) Nonlinear elasticity: theory and application. Cambridge University Press, Cambridge, pp 433–490

Berezovski A, Maugin GA (2004) On the thermodynamic conditions at moving phase-transition fronts in thermoelastic solids. J Non-Equilib Thermodyn 29:37–51

Berezovski A, Maugin GA (2005a) On the velocity of moving phase boundary in solids. Acta Mech 179:187–196

Berezovski A, Maugin GA (2005b) Stress-induced phase-transition front propagation in thermoelastic solids. Eur J Mech – A/Solids 24:1–21

Berezovski A, Maugin GA (2007) On the propagation velocity of a straight brittle crack. Int J Fract 143:135–142

Cermelli P, Sellers S (2000) Multi-phase equilibrium of crystalline solids. J Mech Phys Solids 48:765–796

Evora VMF, Jain N, Shukla A (2005) Stress intensity factor and crack velocity relationship for polyester/TiO_2 nanocomposites. Exp Mech 45:153–159

Fineberg J, Marder M (1999) Instability in dynamic fracture. Phys Rep 313:1–108

Freund LB (1990) Dynamic fracture mechanics. Cambridge University Press, Cambridge

Maugin GA (1993) Material inhomogeneities in elasticity. Chapman and Hall, London

Maugin GA (1997) Thermomechanics of inhomogeneous – heterogeneous systems: application to the irreversible progress of two- and three-dimensional defects. ARI – Int J Phys Eng Sci 50:41–56

Maugin GA (2000) On the universality of the thermomechanics of forces driving singular sets. Arch Appl Mech 70:31–45

Maugin GA, Trimarco C (1995) The dynamics of configurational forces at phase-transition fronts. Meccanica 30:605–619

McKelvey AL, Ritchie RO (2000) On the temperature dependence of the superelastic strength and the prediction of the theoretical uniaxial transformation strain in Nitinol. Philos Mag A 80:1759–1768

Truskinovsky L (1987) Dynamics of nonequilibrium phase boundaries in a heat conducting nonlinear elastic medium. J Appl Math Mech (PMM) 51:777–784

Dislocation tri-material solution in the analysis of bridged crack in anisotropic bimaterial half-space

T. Profant · O. Ševeček · M. Kotoul ·
T. Vysloužil

Abstract The problem of an edge-bridged crack terminating perpendicular to a bimaterial interface in a half-space is analyzed for a general case of elastic anisotropic bimaterials and specialized for the case of orthotropic bimaterials. The edge crack lies in the surface layer of thickness h bonded to semi-infinite substrate. It is assumed that long fibres bridge the crack. Bridging model follows from the assumption of "large" slip lengths adjacent to the crack faces and neglect of initial stresses. The crack is modelled by means of continuous distribution of dislocations, which is assumed to be singular at the crack tip. With respect to the bridged crack problems in finite dissimilar bodies, the reciprocal theorem (Ψ-integral) is demonstrated as to compute, in the present context, the generalized stress intensity factor through the remote stress and displacement field for a particular specimen geometry and boundary conditions using FEM. Also the application of the configurational force mechanics is discussed within the context of the investigated problem.

Keywords Anisotropic bimaterials · Generalized stress intensity factor · FEM · Reciprocal theorem · Distributed dislocations technique · Bridged crack

T. Profant · O. Ševeček · M. Kotoul (✉)
Faculty of Mechanical Engineering, Brno University of Technology, 616 69 Brno, Czech Republic
e-mail: kotoul@fme.vutbr.cz

T. Vysloužil
Jan E. Purkyne University, Na okraji 1001, 400 96 Usti nad Labem, Czech Republic

1 Introduction

It is well-known that in the case of a matrix crack impinging on the interface, a differential energy analysis is unsuitable due to the discontinuity in the elastic properties: finite crack extensions are to be considered and the relevant criterion depends on the ratio of these crack extensions. This approach forms a basis of so called *finite fracture mechanics* (Martin et al. 2001).

Understanding the mechanism of crack deflection by an interface and/or crack penetration through the interface is required to determine the suitable interlayer and the optimum interface fracture energy in layered composites. Various attempts have been made to attain this objective. With the help of asymptotic analysis, He and Hutchinson (1989) derived a deflection criterion which compares the ratio of the interfacial toughness G_{ci} over the toughness of the penetrated material G_{c_1}. The approach was extended by taking anisotropy into account (Martinez and Gupta 1994) and was confirmed by Tullock et al. (1994). Martin et al. (2001) improved the criterion which does not require any assumption concerning the crack extension ratio has been established.

The toughening mechanisms induced by multilayered structure, defined as 1st-level toughening mechanisms include crack deflection mechanism, bridging mechanism of interlocking layers, frictional sliding mechanism, and so on. The toughening mechanisms induced by fibres in multilayer ceramics, namely 2nd-level toughening mechanisms, are similar to the

doi: 10.1007/978-1-4020-6929-1_19,

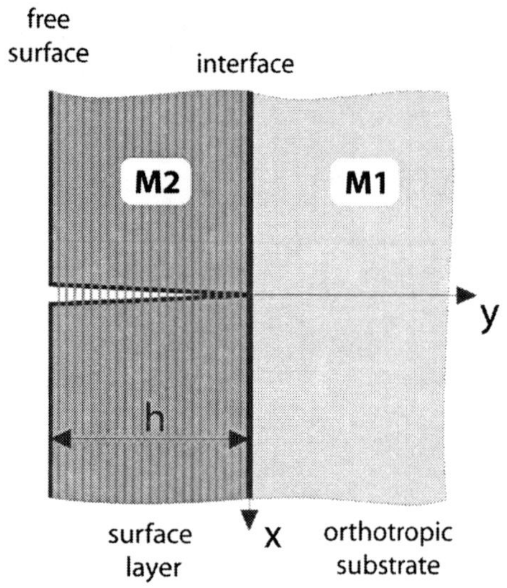

Fig. 1 Scheme of bridged crack impinging the layer interface

toughening mechanisms in fibre reinforced ceramic composites and they include crack deflection mechanism, crack trapping mechanism, fibre pull-out mechanism, bridging mechanism, micro-crack mechanism, and so on. In authors' opinion, a synergy between the 1st-level toughening mechanism such as the crack deflection at the interface between layers and the 2nd-level toughening mechanism such as the crack bridging by fibres has not been investigated yet.

A generic problem whose solution is a prerequisite for the investigation of the aforementioned synergy of toughening mechanisms of the 1st and 2nd level may be specified as follows: find the local generalized stress intensity factor H_{tip} for a bridged crack impinging the layer interface. The crack is initiated from the surface defect and extends through the surface layer of thickness h, see Fig. 1.

It is assumed that a great number of fibres are bridging the crack hence the bridging effect of fibres is treated in an average "smeared" sense. The bridging stress gives rise to the generalized bridging stress intensity factor, H_{br}, and as a result, the local generalized stress intensity $H_{tip} = H_{appl} - H_{br}$ acting in the very crack tip is lower than the remote applied stress intensity H_{appl}. Although the FE analysis is capable of capturing the singular stress behaviour near a corner or a crack tip in homogeneous regions with a refined mesh of conventional elements, this traditional FE approach fails to *accurately* capture the appropriate singular behaviour near a corner or a crack tip at the junction of dissimilar materials.

A very promising approach to an accurate calculation of the near crack tip fields consists in the application of the two-state (or mutual) conservation integrals. Among these two-state conservation laws, the reciprocal theorem (Ψ-integral) has been widely employed for obtaining stress intensities and elastic T-stresses for cracks, as well for finding dislocation strength. The success of the mutual integral methods is crucially linked to the existence of the auxiliary solution in the form of the complementary eigenfunction. A simple modification of the Ψ-integral consisting in the incorporation of a pair of body forces which acts on the crack faces at a general point allows to set up weight functions which can be used further to calculate the generalized bridging stress intensity factor, H_{br}.

The bridging crack problems can be efficiently solved using the distributed dislocation technique. This technique leads to a Fredholm equation that can be solved very accurately using e.g. polynomial-base Galerkin method. However, the application of this method requires determining the solution for a dislocation in a complicated domain. Such an approach is not economical, but there are strategies, which may be employed to overcome this problem. For a bimaterial half-space, see Fig. 1, the solution can be worked out due to recent findings by Choi and Earmme (2002), who studied singularities in anisotropic trimaterials. For the case of *unbridged* crack, the solution based upon the findings of Choi and Earmme (2002) has been worked out in the accompanying paper by Profant et al. (2007). First, Profant et al. (2007) have tested in details the validity of the semi-analytical solution of Choi and Earmme (2002) using FEM. Afterwards, this solution has been applied in modelling of unbridged crack and both the generalized stress intensity factor (GSIF) and the T-stress have been evaluated. The results obtained have been compared with the evaluation of GSIF based upon the reciprocal theorem (Ψ-integral). The evaluation of T-stress based upon the dislocation arrays technique has been compared with FEM computations using a detailed analysis of the near crack tip stress field.

It should be noted that the configurational force mechanics can be applied to verify the results obtained by former method. It is known that the configurational forces approach has some fundamental advantages for studying fracture: first, in addition to crack growth, it can model diffusion, martensitic and diffusional phase transformations, thin film growth, etc. (Eshelby 1970; Maugin 1995; Gurtin 2000; Kienzler and Herrmann 2000; Simha 2000), and hence provides a comprehensive framework for studying the influence of these processes on fracture. Second, based on the configurational balance laws computational errors due to discretization and approximate numerical solution methods can be quantified (Braun 1997; Mueller et al. 2002). Consequently, the configurational forces approach allows evaluating the crack shielding/anti-shielding due to sharp interface.

2 Determination of eigenvalues and eigenvectors in Williams-like asymptotic expansion

The procedure originally developed by Ting (1997) is an efficient tool for the singular characterization of non-degenerate anisotropic multimaterial corners, of which a crack impinging bimaterial interface is a special case. The ith material wedge occupies the polar sector $\omega_{i-1} < \theta < \omega_i$, $i = 1, \ldots, N$. Perfect bonding is considered between material wedges. Fixed or free boundary conditions are considered at the external faces. The solution can be written in the condensed form using the complex variable $z_\alpha = x_1 + p_\alpha x_2 = r(\cos\theta + p_\alpha \sin\theta) = r\zeta_\alpha(\theta)$:

$$\mathbf{w}(r, \theta) = r^{1-\lambda} \mathbf{X} \mathbf{Z}^{1-\lambda}(\theta)\, \mathbf{q}, \tag{1}$$

where $\mathbf{w}^T(r, \theta) = [\mathbf{u}(r, \theta), \mathbf{\Phi}(r, \theta)]^T$, $\mathbf{u}$ stands for the displacement vector and $\mathbf{\Phi}$ is the stress function vector, see Appendix A. p_α are three distinct complex numbers with positive imaginary parts, which are obtained as the roots of the characteristic equation (A1). The matrices $\mathbf{X}$ and $\mathbf{Z}$ are defined as

$$\mathbf{X} = \begin{bmatrix} \mathbf{A} & \bar{\mathbf{A}} \\ \mathbf{L} & \bar{\mathbf{L}} \end{bmatrix}, \quad \mathbf{Z}^{1-\lambda} = \begin{bmatrix} \langle \zeta_*^{1-\lambda} \rangle & 0 \\ 0 & \langle \bar{\zeta}_*^{1-\lambda} \rangle \end{bmatrix}, \tag{2}$$

where the matrices $\mathbf{A}$ and $\mathbf{L}$ are defined in Appendix A, the overbar denotes the complex conjugate. $\langle \zeta_*^{1-\lambda} \rangle = \mathrm{diag}[\zeta_1^{1-\lambda}, \zeta_2^{1-\lambda}, \zeta_3^{1-\lambda}]$. $\lambda \in (0, 1)$ is the stress singularity exponent, $\mathbf{q}^T = [\mathbf{v}, \tilde{\mathbf{v}}]^T$ is the corresponding eigenvector which can be determined up to a multiplicative constant. If λ is a real number, then $\tilde{\mathbf{v}} = \bar{\mathbf{v}}$. Ting's procedure makes use of a transfer matrix, which transfers the displacements and stress function vector components from one edge of the material wedge to the other. Using the continuity conditions introduced by the hypothesis of perfect bonding between the wedges, $\mathbf{w}_i(r, \omega_i) = \mathbf{w}_{i+1}(r, \omega_i)$ $(i = 1, \ldots, N-1)$, and the transfer matrix for each wedge, it is easy to arrive at an expression for the whole multimaterial corner, as it relates the variables between its external faces (ω_0 and ω_N):

$$\begin{bmatrix} \mathbf{u}_N(r, \omega_N) \\ \mathbf{\Phi}_N(r, \omega_N) \end{bmatrix} = \begin{bmatrix} \mathbf{K}_N^{(1)} & \mathbf{K}_N^{(2)} \\ \mathbf{K}_N^{(3)} & \mathbf{K}_N^{(4)} \end{bmatrix} \begin{bmatrix} \mathbf{u}_1(r, \omega_0) \\ \mathbf{\Phi}_1(r, \omega_0) \end{bmatrix}, \tag{3}$$

where $\mathbf{K}_N$ is obtained by the product of the sequence of the successive transfer matrices $\mathbf{E}_i$ of all the wedges in the corner:

$$\begin{aligned} \mathbf{K}_N &= \mathbf{E}_N \cdot \mathbf{E}_{N-1} \cdots \mathbf{E}_2 \cdot \mathbf{E}_1, \\ \mathbf{E}_i &= \mathbf{X} \mathbf{Z}^{1-\lambda}(\omega_i) \left[\mathbf{Z}^{1-\lambda}(\omega_{i-1}) \right]^{-1} \mathbf{X}^{-1}. \end{aligned} \tag{4}$$

This is worthy of note that Ting's procedure directly yields a linear system whose size is 3×3 or 6×6, irrespective of the number of materials N, contrary to traditional analytical procedures leading a linear system of $(6N \times 6N)$.

The eigenpairs λ, $\mathbf{q}$ can also be evaluated using the method developed by Papadakis and Babuska (1995). Their method can be used with multi-material wedges, with anisotropic materials and general boundary conditions under the assumption of plane strain. Along the interfaces at $\theta = \omega_i$, the following continuity conditions are assumed

$$[\mathbf{u}]_i = 0, \quad [\mathbf{t}]_i = 0, \tag{5}$$

where $\mathbf{u}$ is the displacement vector, $\mathbf{t}$ is the traction vector and the brackets denote a jump along $\theta = \omega_i$. The problem of finding the characteristic exponent λ can be viewed as the following eigenvalue problem: Find the characteristic exponent λ such that there exists $\mathbf{F} \neq 0$ such that

$$\begin{aligned} &\frac{\mathrm{d}\mathbf{F}(\theta)}{\mathrm{d}\theta} = \mathcal{H}(\lambda; \theta)\, \mathbf{F}(\theta) \quad \text{in } \omega_i \le \theta \le \omega_{i+1}, \\ &\mathbf{O}_1(\lambda)\, [\mathbf{F}(\omega_i)] = 0 \quad \text{for } \theta = \omega_i, \\ &\mathbf{O}_2(\lambda) \left[\mathbf{F}(\omega_{i+1}) \right] = 0 \quad \text{for } \theta = \omega_{i+1}, \end{aligned} \tag{6}$$

$\mathcal{H}$ is a 4×4 matrix whose elements depend in a complicated manner upon elastic constants and the angle θ, $\mathbf{O}_1$, $\mathbf{O}_2$ are 2×4 matrices and $\mathbf{F}(\theta)$ is 4×1 vector $\mathbf{F}(\theta) = [u_r, u_\theta, u_r', u_\theta']^T$, where u_r and u_θ stand for radial and tangential displacement component respectively. The general idea in solving the above problem is as follows: First construct two initial value problems using the matrix $\mathcal{H}(\lambda, \theta)$ and start with two independent initial vectors that satisfy the left boundary conditions

$$\frac{d\mathbf{F}(\theta)}{d\theta} = \mathcal{H}(\lambda;\theta)\,\mathbf{F}(\theta), \quad \mathbf{F}(0) = \boldsymbol{\zeta}_0, \quad \text{and}$$
$$\frac{d\mathbf{F}(\theta)}{d\theta} = \mathcal{H}(\lambda;\theta)\,\mathbf{F}(\theta), \quad \mathbf{F}(0) = \boldsymbol{\psi}_0, \tag{7}$$

where $\boldsymbol{\zeta}_0\,\boldsymbol{\psi}_0$ are two linearly independent vectors which satisfy the boundary conditions $\mathbf{O}_1(\lambda)\boldsymbol{\zeta}_0=0$, $\mathbf{O}_2(\lambda)\boldsymbol{\psi}_0=0$; $\boldsymbol{\zeta}_0\boldsymbol{\psi}_0$ can be determined a priori. Then the fact that a linear combination of the solution of the two initial value problems $k_1\boldsymbol{\zeta}(\theta) + k_2\boldsymbol{\psi}(\theta)$ will be a solution of Eq. 7 only if it satisfies the right-hand side boundary conditions, leads to the formation of the determinant of a matrix which depends on λ. Specifically, in each material, two initial value problems are solved and the interface conditions are used to calculate two independent vectors which will be used as the initial vectors for the initial value problem in the next material. Finally, $k_1\boldsymbol{\zeta}(\theta)+k_2\boldsymbol{\psi}(\theta)$ solves Eq. 6 if k_1, k_2 are chosen to satisfy

$$\mathbf{O}_2(\lambda)\left(k_1\zeta(\omega_N) + k_2\psi(\omega_N)\right) = 0, \quad \text{or equivalently}$$
$$\mathbf{D}(\lambda)\begin{bmatrix} k_1 \\ k_2 \end{bmatrix} = 0 \tag{8}$$

with $\mathbf{D}$ a 2×2 matrix. Then for non-zero k_1, k_2 satisfying Eq. 9 the determinant of $\mathbf{D}$, $W(\lambda) = \mathrm{Det}[\mathbf{D}(\lambda)]$ must vanish. A special iterative procedure named Shoot was developed to solve the problem Eq. 6–8 in the MATLAB 7.1. For comparison, the results are shown in Table 1 and referred to as the Shoot method.

Also an array of continuously distributed edge dislocations is an efficient tool in modelling of special cases of multi-material wedge like the crack impinging bimaterial interface, see Appendix B. The crack is modelled by means of continuous distribution of dislocations, which is assumed to be singular at the crack tip. A system of simultaneous functional equations is obtained that enables to find the singularity exponent λ.

3 Model of crack bridging

For simple sliding with a constant sliding resistance τ between the matrix and fibres, the following model of bridging fibres represented by a continuous distribution of bridging springs obeying the quadratic bridging law; was suggested (Budiansky et al. 1986; Budiansky and Amazigo 1989; Marshall and Cox 1988)

$$\delta(y) = 2\left(\frac{\sigma_{br}(y)}{\beta}\right)^2, \tag{9}$$

where $\delta(x)$ is the crack opening displacement and the constant β is defined as follows

$$\beta = \left[\frac{4c_f^2 E_f E^2 \tau}{R\left(1 - c_f\right)^2 E_m^2}\right]^{1/2} \tag{10}$$

where R is the fibre radius, $E = E_f c_f + (1 - c_f)E_m$, E_f and E_m stands for the fibre Young modulus and the matrix Young modulus respectively, τ is the frictional shear stress between the matrix and fibres, and c_f is the volume fraction of fibres. Relation (10) follows from an estimate of the extra elastic elongation of a long bridging fibre that occurs in regions on the two sides of a matrix crack wherein frictionally constrained sliding occurs.

Under the assumption that the strength of the fibres, σ_{0f}, has a single, deterministic value, failure occurs when the bridging spring stress at the original crack tips reaches $\sigma = c_f\sigma_{0f}$. Because of fibres/matrix slip, the fibre stress decays linearly from the crack mid-plane. Since the stress on the fibres has a maximum value in the plane of the matrix crack, the assumption of a single strength value of fibres leads to the conclusion, that fibres break in the plane of the crack. Consequently, the prediction of composite toughness and strength may be unduly conservative. The reason is that with a dispersion in the fibre tensile strength, fibres may fracture within the matrix rather than at bridged faces of the matrix crack, thereby leading to frictionally constrained fibre pullout before final failure occurs, and so leading to enhanced composite strength. Apparently, fractured fibres still contribute to the bridging stresses as they have to be pulled out from the matrix. The relative contribution of intact fibres, which act as elastic ligaments between the crack faces, and broken ones within the matrix, which are eventually pulled out, is analysed assuming that the fibre strength follows the Weibull statistics (Thouless and Evans 1988). This gave an explicit expression for the average stress transferred by the fibres across crack given by

$$\hat{\sigma}_{br} = \sigma_{br}\exp\left[-\left(\frac{\sigma_{br}}{c_f\Sigma}\right)^{m+1}\right] + \sigma_P\left\{1 - \exp\left[-\left(\frac{\sigma_{br}}{c_f\Sigma}\right)^{m+1}\right]\right\}, \tag{11}$$

where σ_P is the average stress exerted by the broken fibres pulled out from the matrix, and $\exp[-(\sigma_{br}/c_f\Sigma)^{m+1}]$ stands for the fraction of intact fibres in the crack wake. The fibre strength distribution is introduced through the parameter $\Sigma = (m+1)^{1/(m+1)}\sigma_{0f}$, which

Table 1 Singularity exponent calculations for a specified configuration

Materials	$\frac{M1}{M2}$	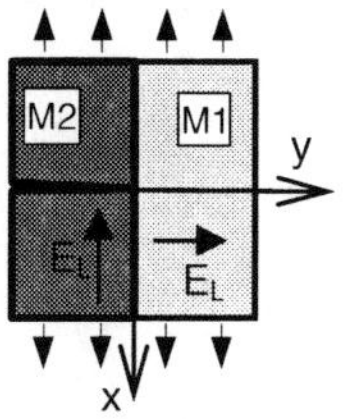	**Graphite/Epoxy T300/5208** $E_L = 137$ GPa, $E_T = E_Z = 10.8$ GPa, $G_{ZT} = 3.36$ GPa, $G_{ZL} = G_{TL} = 5.65$ GPa, $\nu_{TZ} = 0.49$, $\nu_{ZL} = \nu_{TL} = 0.238$
Singularity exponent	λ	0.6716824	Computed using the dislocation approach Eq. B16
		0.671825	Computed using the Shoot method
Auxiliary singularity exponent	$s = 2 - \lambda$	1.3283176	Computed using the dislocation approach Eq. B16
		1.328175	Computed using the Shoot method

includes the information on the fibre tensile properties given by the Weibull modulus m and the fibre characteristic strength σ_{0f}. Physically, there is typically one flaw of strength σ_{0f} in a length l_c of fibre and $l_c = R\sigma_{0f}/\tau$ is twice the fibre slip length at an applied stress of σ_{0f}.

Using a simple shear-lag approach, the stress transferred by the broken fibres as they are pulled out from the matrix can be expressed as

$$\sigma_P = \frac{2c_f\tau}{R}\left(\langle d\rangle - \frac{\delta}{2}\right), \tag{12}$$

where $\langle d\rangle$ is the average distance from the fibre failure position to the crack plane for the broken fibres which was computed in (Thouless and Evans 1988) as

$$\langle d\rangle = \frac{R}{2\tau}\frac{\Sigma}{m+1}\Gamma\left(\frac{m+2}{m+1}\right) = \frac{l_c}{2}\frac{1}{(m+1)^{m/(m+1)}}\Gamma\left(\frac{m+2}{m+1}\right), \tag{13}$$

where Γ is the Gamma function. Fibre pullout, thus, scales directly with the characteristic length l_c. Figures 2 and 3 show several courses of the bridging law (11) for selected values of m and τ. The fibre volume fracture $c_f = 0.4$ and the fibre radius $R = 7\,\mu$m. In brittle matrix composites, the fibres used to date have been almost exlusively the Nicalon fibres. Their in-situ strength parameter σ_{0f} typically ranges from 2200 MPa to 2800 MPa while the Weibull modulus m typically ranges from 2 to 7, see e.g. Curtin (2000).

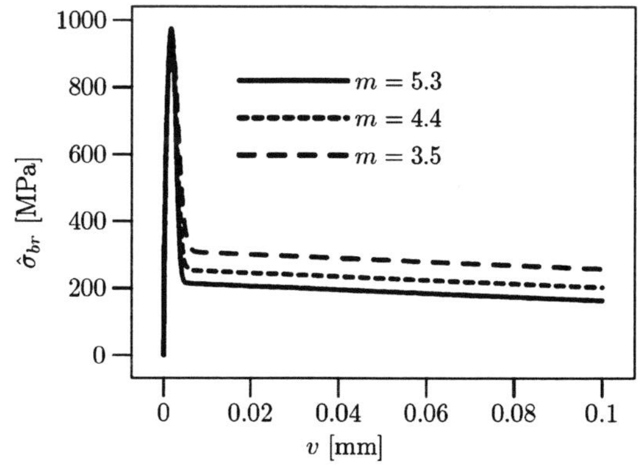

Fig. 2 Bridging stress according to Eq. 11 as a function of $v = \delta/2$ for several values of Weibull modulus m, $\tau = 6$ MPa, $\sigma_{0f} = 2750$ MPa

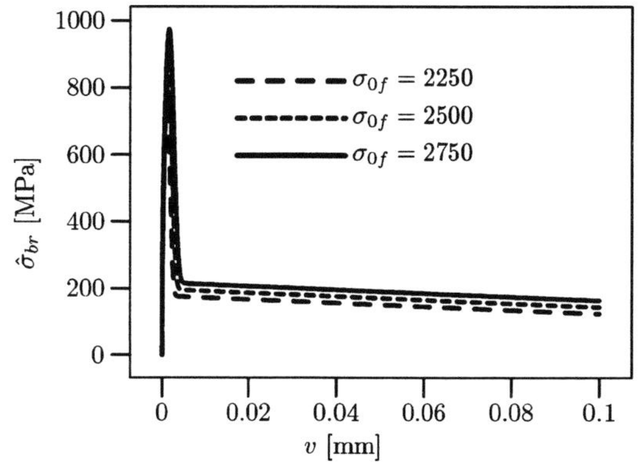

Fig. 3 Bridging stress according to Eq. 11 as a function of $v = \delta/2$ for several values of the characteristic fibre strength σ_{0f}, $m = 5.3$, $\tau = 6$ MPa

4 Crack modeling by distributed dislocation technique

Choi and Earmme (2002) used the so-called alternating technique that generalizes the formulas (B5) and (B6), and for the case in Fig. 4, it gives the following relations for potentials

Fig. 4 Scheme of the bimaterial half-plane

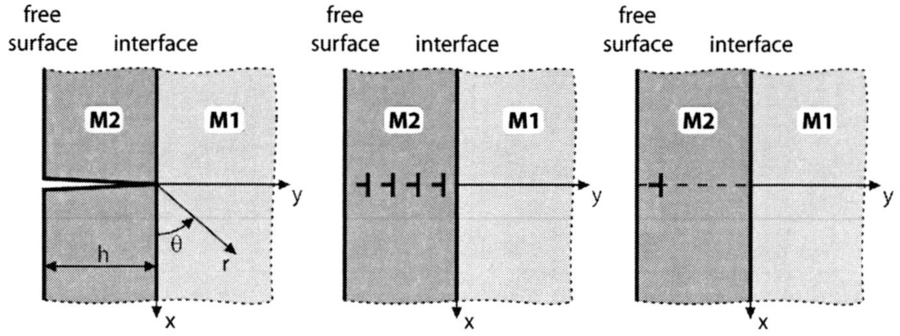

$$\Phi_\alpha(z) = \begin{cases} \sum_{n=1}^{\infty}\left[\Phi_\alpha^n(z) + \sum_\beta \left(\mathbf{M}^{II}\bar{\mathbf{L}}^{II}\right)_{\alpha\beta}\Phi_\beta^n\left(z - p_\alpha^{II}h + p_\beta^{II}h\right)\right], & z \in 2 \\ C_{\alpha\beta}\bar{\Phi}_{\beta o}(z) + \sum_\beta\sum_\gamma \left(\mathbf{C}\mathbf{M}^{II}\right)_{\alpha\beta}\bar{L}_{\beta\gamma}^{II}\sum_{n=1}^{\infty}\Phi_\beta^n\left(z - p_\beta^{II}h + p_\gamma^{II}h\right), & z \in 1, \end{cases} \tag{14}$$

in which the recurrence formula for $\Phi_\alpha^n(z)$ is

$$\Phi_\alpha^{n+1}(z) = \begin{cases} \Phi_{\alpha o}(z) + \sum_\beta G_{\alpha\beta}\bar{\Phi}_{\beta o}(z), & \text{if } n = 0 \\ \sum_\beta G_{\alpha\beta}\left(\bar{\mathbf{M}}^{II}\mathbf{L}^{II}\right)_{\beta\gamma}\Phi_\gamma^n\left(z - \bar{p}_\beta^{II}h + \bar{p}_\gamma^{II}h\right), & \text{if } n = 1, 2, 3, \ldots \end{cases} \tag{15}$$

Using the preceding formula, the stress σ_{xx} was calculated along the dislocation plane. The comparison was made with FEM calculations performed for a sufficiently large specimen, made of two layers M1 and M2 of composite such as FP/Al system with elastic constants: $E_L = 225\,\text{GPa}$, $E_T = 150\,\text{GPa}$, $G_L = 58\,\text{GPa}$, $\nu_L = 0.28$ (fibres in the substrate 1 are parallel with the y axis). The geometric parameters of the specimen were as follows: thickness of the layer M2 $h = 4\,\text{mm}$, thickness of the layer M1 $L = 40\,\text{mm}$, the specimen height 200 mm. Further increase of specimen size did not lead to the change of numerical results. As it can be seen from the Fig. 5, FEM results are in excellent agreement with semi-analytical solution based on Eqs. 14 and 15, see also Profant et al. (2007).

An integral equation is obtained by choosing the dislocation distribution to meet the traction conditions along of the line of the crack and within the crack bridging zone. Since the material interface and the crack plane correspond to the material symmetry planes, and the specimen is subjected to simple tensile loading conditions, the Burgers vector component b_y is equal to zero. The integral equation then reads

$$\sigma_{xx}^{appl}(y) + \hat{\sigma}_{br}(\delta(y)) + \frac{1}{2\pi}$$

$$\times \mathrm{Re}\left\{\sum_{\beta=1}^{2} L_{1\beta}^{II}\sum_{\alpha=1}^{2} M_{\beta\alpha}^{II}\left(B_{\alpha 1}^{II}\right)^{-1}\right\}\int_{-h}^{0}\frac{b_x(y_0)}{y_0 - y}dy_0$$

$$+ \int_{-h}^{0} b_x(y_0)\,K_{xx}(y, y_0)\,dy_0 = 0. \tag{16}$$

$\sigma_{1i}^{appl}(y)$ denotes the negated stresses in $x = 0$ produced by the given boundary loads, acting on a specimen with boundary $\partial\Omega$, but without cracks and dislocations,

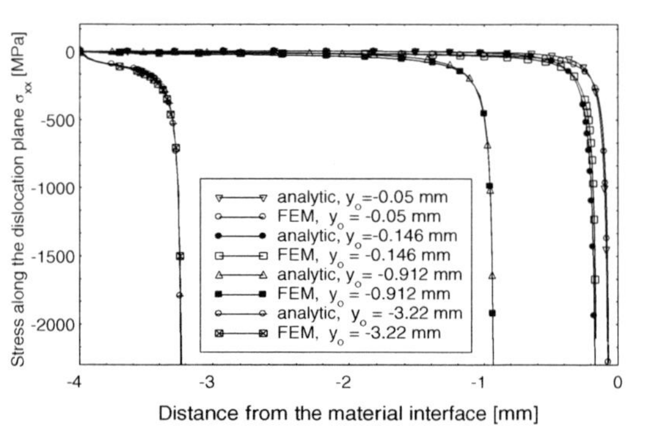

Fig. 5 Calculations of σ_{xx} along the plane of dislocation with Burgers vector $\mathbf{b} = (0.01, 0)$ [mm] using either FEM and the alternating technique for several locations y_o of dislocation core

and $\hat{\sigma}_{br}(\delta(y))$ is the bridging stress from Eq. 11. The integral equation (16) must be complemented with the condition

$$\delta(y) = \int_{y}^{0} b_x(y_0)\,\mathrm{d}y_0, \tag{17}$$

which relates the crack opening displacement δ to the dislocation density b_x. The regular kernel $K_{xx}(y, y_0)$ was obtained from the truncated semi-analytical solution of Choi and Earmme (2002). The regular kernel describes the interaction of a dislocation with the bimaterial interface and with the free surface as well. $K_{xx}(y, y_0)$ possesses a complicated structure and depends on elastic constants of both materials and on the layer thickness. It can expressed as the truncated series

$$K_{xx}(y, y_0) = \sum_{n=1}^{N_K} \frac{k'_{1,n}}{k'_{2,n}y_0 - k'_{3,n}y - k'_{4,n}}. \tag{18}$$

The $k'_{i,n}$ are the constants developed from the alternating technique discussed above. The substitutions

$$s = 2\frac{y_0}{h} + 1, \quad t = 2\frac{y}{h} + 1 \tag{19}$$

allow to reduce the integral equation (16) to the form

$$\sigma_{xx}^{appl}(y) + \hat{\sigma}_{br}(\delta(y)) + \frac{1}{2\pi}\mathrm{Re}\left\{\sum_{\beta=1}^{2} L_{1\beta}^{II} \sum_{\alpha=1}^{2} M_{\beta\alpha}^{II}\left(B_{\alpha 1}^{II}\right)^{-1}\right\}\int_{-1}^{1}\frac{b_x(s)}{s-t}\mathrm{d}s + \int_{-1}^{1} b_x(s) K_{xx}(t, s)\,\mathrm{d}s = 0, \tag{20}$$

where

$$K_{xx}(t, s) = \sum_{n=1}^{N_K} \frac{k_{1,n}}{k_{2,n}s - k_{3,n}t - k_{4,n}}. \tag{21}$$

The procedure involves the reduction of the integral equation and constraints to a system of algebraic equations using the collocation technique.

The dislocation density is sought in the form $b_x(s) = (1-s)^{-\lambda}(1+s)^{\lambda} g(s)$, where $g(s)$ is a bounded function. As mentioned elsewhere (Hills et al. 1996), this choice means that $b_x(-1)$ must vanish, i.e. that crack faces at the mouth are forced to be parallel and the solution is over-constrained. Nevertheless, this incorrect end-point behaviour at the crack mouth had a negligible effect on the calculated stress intensity factor.

Remark A more acceptable form of the density function can be found as (Dewynne et al. 1992)

$$b_x^*(s) = \frac{g(s)}{(1-s)^{\lambda}} \tag{*}$$

A quadrature method has to be adapted to the singularities of $b_x^*(s)$ using Jacobi polynomials $P_n^{(-\lambda,0)}(s)$ at all. The form of the density function $b_x^*(s)$ disables to receive the closed form solution of the regular kernel $K_{xx}(s, t)$, which has to be approximated (Profant et al. 2007). To avoid the application of $P_n^{(-\lambda,0)}(s)$ polynomials, the following approximation of the density function can be used (Dewynne et al. 1992)

$$b_x^{**}(s) = b_{-1}g_{-1}(s) + \left(\frac{1+s}{1-s}\right)^{\lambda} g(s)$$

where $g_{-1}(s)$ is some known, bounded function on $[-1, 1]$ such that $g_{-1}(s) = -1$ and b_{-1} is an unknown constant, which equals $b_x^{**}(-1)$. This equality serves as an additional consistency condition. This form of the density function corrects the crack opening at the crack mouth without the influence on the stress intensity factor. Because the objective of the paper is to find the stress intensity factor, the crack mouth opening correction will be omitted below. Observe that while the value of the function $g(s)$ at the crack tip obtained by using the density function $b_x(s)$ is $g(1) = 0.021$, the density function $b_x^*(s)$ leads to the result g(1)=0.020. The last value corresponds to the approximation of the regular kernel $N_f = 20$, the approximation of the density of the Burgers vector $N_B = 10$ and the chosen collocation points

$$t_i = 2\cos\left(\frac{\pi}{4}\frac{2i+1}{N_B}\right) - 1.$$

The integral equation may be solved using the Gauss-Jacobi quadrature. The function g(s) is sought in the form of linear combination of Jacobi polynomials $P_n^{(-\lambda,\lambda)}(s)$

$$g(s) = \sum_{n=0}^{\infty} c_n P_n^{(-\lambda,\lambda)}(s) \cong \sum_{n=0}^{N_B} c_n P_n^{(-\lambda,\lambda)}(s). \tag{22}$$

This allows to express the integral containing the regular kernel $K_{xx}(t, s)$ in the closed form by integrating each component of the truncated series in Eq. 18. It is useful to apply the theory of the curve complex integrals developed by Muschelishvili (1953) because of the ambiguous behaviour of the $(1-s)^{-\lambda}(1+s)^{\lambda}$ around the points $s = \pm 1$. Hence, after the integration, the regular part of the integral equation (20) can be written as

$$\int_{-1}^{1} b_x(s)K_{xx}(t,s)\mathrm{d}s = \sum_{n=1}^{N_K}\int_{-1}^{1}\frac{k_{1,n}b_x(s)\mathrm{d}s}{k_{2,n}s-k_{3,n}t-k_{4,n}}$$
$$= \sum_{n=1}^{N_K}\frac{k_{1,n}}{k_{2,n}}\int_{-1}^{1}\left(\frac{1+s}{1-s}\right)^{\lambda}\frac{k_{1,n}g(s)\mathrm{d}s}{s-(k_{3,n}t+k_{4,n})/k_{2,n}}$$
$$= \sum_{n=1}^{N_K}\frac{k_{1,n}}{k_{2,n}}\frac{2\pi i e^{i\lambda\pi}}{1-e^{i\lambda 2\pi}}\left[\left|\frac{(k_{3,n}t+k_{4,n})+k_{2,n}}{(k_{3,n}t+k_{4,n})-k_{2,n}}\right|^{\lambda}\right.$$
$$\times g((k_{3,n}t+k_{4,n})/k_{2,n})$$
$$\left.-\left(\alpha_n s^n+\cdots+\alpha_0\right)\Big|_{s=(k_{3,n}t+k_{4,n})/k_{2,n}}\right], \tag{23}$$

where α_n are the coefficients of the pole at the infinity of the function $b_x(z) = (1-z)^{-\lambda}(1+z)^{\lambda}g(z)$.

The boundary conditions along crack faces, $\sigma_{xx} = \sigma_{xy} = 0$, are controlled at the collocation points $t = t_i$, $i = 0, 1, \ldots, N_B - 1$ given by

$$t_i = \cos\left(\frac{\pi}{2}\frac{2i+1}{N_B}\right). \tag{24}$$

Using Eqs. 21 and 22 in the integral equation (20) and employing the integral relations given in (Erdogan et al. 1974) one obtains the system of algebraic equations through which the unknown coefficients c_n can be evaluated:

$$\sigma_{xx}^{appl}(t_i) + \frac{1}{2}\mathrm{Re}\left\{\sum_{\beta=1}^{2}L_{1\beta}^{II}\sum_{\alpha=1}^{2}M_{\beta\alpha}^{II}\left(B_{\alpha 1}^{II}\right)^{-1}\right\}$$
$$\times\sum_{n=0}^{N_B}c_n\left[\cot(-\pi\lambda)(1-t_i)^{-\lambda}(1+t_i)^{\lambda}P_n^{(-\lambda,\lambda)}(t_i)\right.$$
$$\left.-\frac{\Gamma(-\lambda)\Gamma(n+\lambda+1)}{\Gamma(n+1)}\mathrm{F}\left(n+1,-n;1+\lambda;\frac{1-t_i}{2}\right)\right]$$
$$+\sum_{m=1}^{N_k}\frac{k_{1,m}}{k_{2,m}}\frac{2\pi i e^{i\lambda\pi}}{1-e^{i\lambda 2\pi}}\left[\left|\frac{(k_{3,m}t_i+k_{4,m})+k_{2,m}}{(k_{3,m}t_i+k_{4,m})-k_{2,m}}\right|^{\lambda}\right.$$
$$\times\sum_{n=0}^{N_B}c_n P_n^{(-\lambda,\lambda)}((k_{3,m}t_i+k_{4,m})/k_{2,m})$$
$$\left.-\left(\alpha_m s^m+\cdots+\alpha_0\right)\Big|_{s=(k_{3,m}t_i+k_{4,m})/k_{2,m}}\right] = 0, \tag{25}$$

where $\mathrm{F}(n_1, n_2; n_3; x)$ stands for the hypergeometric function, $\Gamma(n)$ is the Gamma function and $i = 0, 1, \ldots, N_B - 1$. The strength of the singularity in stress may be quantified in the usual way by defining the local generalized SIF H_{tip}. Using the function-theoretic methods (Erdogan et al. 1974; Hills et al. 1996) one obtains

$$H_{tip} = \lim_{r\to 0}\sqrt{2\pi}r^{\lambda}\sigma_{xx}(r,\theta=\pi/2)$$
$$= \sqrt{2\pi}h^{\lambda}g(1)\mathrm{Re}\left\{\frac{ie^{\lambda i\pi}}{1-e^{2\pi i\lambda}}\left[L_{11}^{I}\left(C_{11}\left(M_{11}^{II}\right.\right.\right.\right.$$
$$\left.\times\left(B_{11}^{II}\right)^{-1}+M_{12}^{II}\left(B_{21}^{II}\right)^{-1}\right)\left|\frac{p_1^{II}}{p_1^{I}}\right|^{\lambda-1}+C_{12}$$
$$\left.\times\left(M_{21}^{II}\left(B_{11}^{II}\right)^{-1}+M_{22}^{II}\left(B_{21}^{II}\right)^{-1}\right)\left|\frac{p_2^{II}}{p_1^{I}}\right|^{\lambda-1}\right)+L_{12}^{I}$$
$$\times\left(C_{21}\left(M_{11}^{II}\left(B_{11}^{II}\right)^{-1}+M_{12}^{II}\left(B_{21}^{II}\right)^{-1}\right)\left|\frac{p_1^{II}}{p_2^{I}}\right|^{\lambda-1}\right.$$
$$+C_{22}\left(M_{21}^{II}\left(B_{11}^{II}\right)^{-1}+M_{22}^{II}\left(B_{21}^{II}\right)^{-1}\right)$$
$$\left.\left.\left.\times\left|\frac{p_2^{II}}{p_2^{I}}\right|^{\lambda-1}\right)\right]\right\}, \tag{26}$$

where $g(1)$ denotes a value of the function $g(s)$, see Eq. 22, at $s=1$; other quantities were already defined above.

5 Bridged crack modelling using weight function method

It should be noted that an important task in the analysis of a component with a bridged crack is the calculation of the bridging stress intensity factor for a specified bridging stress-crack opening displacement relationship. There are a great number of methods available for the determination of stress intensity factors such as e.g. the finite element method with contact elements, the boundary element method, the boundary collocation method, or the weight function method. High efficiency of the weight function method consists in that once the weight function(s) are known the bridging intensity factor can be easily calculated for any bridging stress distribution by evaluating the integral of the form of Eq. 40. Moreover, it allows setting up a bridging stress-crack opening displacement relationship by analysing the experimental crack opening displacement data and solving an integral equation. The weight function method has been extensively used to the modelling of bridged crack problems (Fett and Munz 1997). For a complicated domain, the weight function has to be obtained numerically, e.g. from FEM calculations (Sarrafi et al. 1998). As to a crack impinging on the bimaterial interface, such calculations have not been reported yet. The weight function is obtained numerically by performing a number of calculations of the generalised stress intensity factor due to unit line load applied to the crack face at arbitrary points. To this end, an application of the reciprocal theorem seems to be very efficient.

5.1 Application of the Ψ-integral

In the absence of body forces the reciprocal theorem states that the following integral is path independent (Stern and Soni 1976)

$$\Psi(\mathbf{u}, \mathbf{v}) = \int_{\Gamma} \left[\sigma_{ij}(\mathbf{u}) n_i v_j - \sigma_{ij}(\mathbf{v}) n_i u_j\right] \mathrm{d}s, \tag{27}$$

where Γ is any contour surrounding the crack tip and **u**, **v** are two admissible displacement fields. If the following displacement fields are considered $\mathbf{u} = \mathcal{U}_i(x) = r^{1-\lambda_i}\bar{\mathbf{u}}_i(\theta)$, $\mathbf{v} = \mathcal{U}_j(x) = r^{1-\lambda_j}\bar{\mathbf{u}}_j(\theta)$, one can show that the contour integral Ψ is equal to zero for $-(1-\lambda_i) \neq 1-\lambda_j$ and non-zero if $-(1-\lambda_i) = 1-\lambda_j$. Thus, for the set of basis functions $\{\mathcal{U}_i(x) = r^{\frac{i}{|i|}(1-\lambda_{|i|})}\bar{\mathbf{u}}_i, i = \pm 1, \pm 2, \ldots\}$, there is a kind of orthogonality with respect to the "scalar" product $\Psi(\cdot,\cdot)$.

Since the basis function corresponding to coefficient $f_1 = H$ (the generalized stress intensity factor) in the asymptotic expansion for **u** is $r^{1-\lambda_1}\bar{\mathbf{u}}_1(\theta)$, due to the former orthogonality conditions it holds

$$\Psi\left(\mathbf{u}, r^{-(1-\lambda_1)}\bar{\mathbf{u}}_{-1}\right) = \sum_{i=1}^{\infty} f_i \Psi\left(r^{1-\lambda_i}\bar{\mathbf{u}}_i, r^{-(1-\lambda_1)}\bar{\mathbf{u}}_{-1}\right) = f_1 \Psi\left(r^{1-\lambda_1}\bar{\mathbf{u}}_1, r^{-(1-\lambda_1)}\bar{\mathbf{u}}_{-1}\right). \tag{28}$$

Thus, the generalized stress intensity factor $H = f_1$ can be computed as follows:

$$H = \frac{\Psi\left(\mathbf{u}, r^{-(1-\lambda_1)}\bar{\mathbf{u}}_{-1}\right)}{\Psi\left(r^{1-\lambda_1}\bar{\mathbf{u}}_1, r^{-(1-\lambda_1)}\bar{\mathbf{u}}_{-1}\right)}. \tag{29}$$

On the other hand, since the exact solution **u** is not known, a finite element solution $\mathbf{u}^h$ can be used as an approximation for **u** so to obtain an approximation for H, (Profant et al. 2007). Due to the path independence, the Ψ-integral standing in the denominator of Eq. 29 is evaluated along an infinitesimal path that shrinks to the crack tip. The solution for stresses and displacements corresponding to the basis function $r^{-(1-\lambda_1)}\bar{\mathbf{u}}_{-1}$ are sometimes referred to as auxiliary (dual) fields which satisfy the same equilibrium and constitutive equations as the actual fields σ_{ij} and u_i and the dual tractions vanish on crack faces. These dual fields are more singular than the actual fields and they are not physically meaningful.

Remark For the crack making a right angle with the interface between two different orthotropic materials whose axes of material symmetry are parallel with or perpendicular to the interface, the most singular stress fields at the tip can be written as

$$\sigma_{ij} = H_I r^{-\lambda_1}\sigma_{ij}^I(\theta) + H_{II} r^{-\lambda_1}\sigma_{ij}^{II}(\theta). \tag{30}$$

In this case, the eigenvalue problem for the exponent λ_1 has a double root yielding two linearly independent fields σ_{ij}^I and σ_{ij}^{II} which can be taken to be symmetric and anti-symmetric relative to the crack plane. In the case where λ_1 is a multiple eigenvalue, then so is $2-\lambda_1$, and one has to solve a system of equations for the corresponding generalized intensity factors. For example, consider the case where λ_1 is a double eigenvalue, corresponding to two eigenfunctions $\bar{\mathbf{u}}_1^I(\theta)$ and $\bar{\mathbf{u}}_1^{II}(\theta)$. The displacement field **u** has the following expansion

$$\mathbf{u} = H_I r^{1-\lambda_1}\bar{\mathbf{u}}_1^I(\theta) + H_{II} r^{1-\lambda_1}\bar{\mathbf{u}}_1^{II}(\theta) + h.o.t. \ldots \tag{31}$$

The orthogonality condition (28) then leads to the following system to solve for the generalized intensity factors H_I and H_{II}

$$\begin{bmatrix} \Psi\left(r^{1-\lambda_1}\bar{\mathbf{u}}_1^I, r^{-(1-\lambda_1)}\bar{\mathbf{u}}_{-1}^I\right) & \Psi\left(r^{1-\lambda_1}\bar{\mathbf{u}}_1^I, r^{-(1-\lambda_1)}\bar{\mathbf{u}}_{-1}^{II}\right) \\ \Psi\left(r^{1-\lambda_1}\bar{\mathbf{u}}_1^{II}, r^{-(1-\lambda_1)}\bar{\mathbf{u}}_{-1}^I\right) & \Psi\left(r^{1-\lambda_1}\bar{\mathbf{u}}_1^{II}, r^{-(1-\lambda_1)}\bar{\mathbf{u}}_{-1}^{II}\right) \end{bmatrix} \begin{bmatrix} H_I \\ H_{II} \end{bmatrix} = \begin{bmatrix} \Psi\left(\mathbf{u}, r^{-(1-\lambda_1)}\bar{\mathbf{u}}_{-1}^I\right) \\ \Psi\left(\mathbf{u}, r^{-(1-\lambda_1)}\bar{\mathbf{u}}_{-1}^{II}\right) \end{bmatrix}. \tag{32}$$

For the crack impinging interface at an oblique angle and/or for an arbitrary orientation of the axes of material symmetry with respect to the interface, the eigenvalue problem no longer has double roots. Instead of (32), the two most singular fields of the interest are

$$\sigma_{ij} = H_1 r^{-\lambda_1}\sigma_{ij}^{(1)}(\theta) + H_2 r^{-\lambda_2}\sigma_{ij}^{(2)}(\theta) \tag{33}$$

Corresponding to each eigenvalue is only one eigenfunction instead of two. Thus, the generalized intensity factors H_1 and H_2 in (33) are calculated using the formula (29).

The dual fields can be easily calculated using the dislocation approach described in the Appendix B assuming the distribution of dislocations in the form

$$\hat{f}_k(r) = \frac{w_k}{(-r)^{2-\lambda}}, \quad \infty < r < 0, \tag{34}$$

where w_k is the eigenvector of matrix **D** corresponding to the eigenvalue $s = 2-\lambda$. After a tedious algebra, the Ψ-integral along the infinitesimal contour follows as

$$\Psi\left(r^{1-\lambda_1}\bar{\mathbf{u}}_1, r^{-(1-\lambda_1)}\bar{\mathbf{u}}_{-1}\right) = c_1 - c_2. \tag{35}$$

where

$$c_1 = \frac{1}{4(\lambda-1)}\left\{\int_{-\frac{\pi}{2}}^{0} \mathrm{Re}\left\{\sum_\alpha L_{i\alpha}^{II}\left(\sin\theta - p_\alpha^{II}\cos\theta\right)\right.\right.$$
$$\times\sum_\beta\left[G_{\alpha\beta}\bar{M}_{\beta k}^{II}\left(\bar{p}_\beta^{II}\right)^{\lambda-1} + \delta_{ik}\left(p_\alpha^{II}\right)^{\lambda-1}\right]$$
$$\left.\times\frac{(-1)^\lambda}{\tau_\alpha^{II}(\theta)^\lambda}\left(\cot(\pi\lambda) - i\right)v_k\right\} \times \mathrm{Re}\left\{\sum_\alpha A_{i\alpha}\right.$$
$$\times\sum_\beta\left[G_{\alpha\beta}\bar{M}_{\beta k}^{II}\left(\bar{p}_\beta^{II}\right)^{\lambda-1} + \delta_{ik}\left(p_\alpha^{II}\right)^{\lambda-1}\right]$$
$$\left.\times\frac{(-1)^\lambda}{\tau_\alpha^{II}(\theta)^{1-\lambda}}\left(\cot(\pi\lambda) - i\right)w_k\right\}\mathrm{d}\theta + \int_0^\pi \mathrm{Re}$$
$$\times\left\{\sum_\alpha L_{i\alpha}^{I}\left(\sin\theta - p_\alpha^{I}\cos\theta\right)\sum_\beta C_{\alpha\beta}M_{\beta k}^{II}\right.$$
$$\left.\times\left(p_\beta^{II}\right)^{\lambda-1}\frac{(-1)^\lambda}{\tau_\alpha^{I}(\theta)^\lambda}\left(\cot(\pi\lambda) - i\right)v_k\right\} \times \mathrm{Re}$$
$$\times\left\{\sum_\alpha A_{i\alpha}\sum_\beta C_{\alpha\beta}M_{\beta k}^{II}\left(p_\beta^{II}\right)^{\lambda-1}\frac{(-1)^\lambda}{\tau_\alpha^{I}(\theta)^{1-\lambda}}\right.$$
$$\left.\times\left(\cot(\pi\lambda) - i\right)w_k\right\}\mathrm{d}\theta + \int_\pi^{\frac{3\pi}{2}}\mathrm{Re}\left\{\sum_\alpha L_{i\alpha}^{II}\right.$$
$$\times\left(\sin\theta - p_\alpha^{II}\cos\theta\right)\sum_\beta\left[G_{\alpha\beta}\bar{M}_{\beta k}^{II}\frac{1}{p_\alpha^{II}}\left(\frac{\bar{p}_\beta^{II}}{p_\alpha^{II}}\right)^{\lambda-1}\right.$$
$$\left.\left.+\frac{\delta_{ik}}{p_\alpha^{II}}\right]\frac{(-1)^\lambda}{\tau_\alpha^{II}(\theta)^\lambda}v_k\right\} \times \mathrm{Re}\left\{\sum_\alpha A_{i\alpha}\right.$$
$$\times\sum_\beta\left[G_{\alpha\beta}\bar{M}_{\beta k}^{II}\left(\bar{p}_\beta^{II}\right)^{\lambda-1} + \delta_{ik}\left(p_\alpha^{II}\right)^{\lambda-1}\right]$$
$$\left.\left.\times\frac{(-1)^\lambda}{\tau_\alpha^{II}(\theta)^{1-\lambda}}\left(\cot(\pi\lambda) - i\right)w_k\right\}d\theta\right\}, \tag{36}$$

$$c_2 = \frac{1}{4(\lambda-1)}\left\{\int_{-\frac{\pi}{2}}^{0} \mathrm{Re}\left\{\sum_\alpha L_{i\alpha}^{II}\left(\sin\theta - p_\alpha^{II}\cos\theta\right)\right.\right.$$
$$\times\sum_\beta\left[G_{\alpha\beta}\bar{M}_{\beta k}^{II}\left(\bar{p}_\beta^{II}\right)^{\lambda-1} + \delta_{ik}\left(p_\alpha^{II}\right)^{\lambda-1}\right]$$
$$\left.\times\frac{(-1)^\lambda}{\tau_\alpha^{II}(\theta)^{2-\lambda}}\left(\cot(\pi\lambda) - i\right)w_k\right\} \times \mathrm{Re}\left\{\sum_\alpha A_{i\alpha}\right.$$
$$\times\sum_\beta\left[G_{\alpha\beta}\bar{M}_{\beta k}^{II}\left(\bar{p}_\beta^{II}\right)^{\lambda-1} + \delta_{ik}\left(p_\alpha^{II}\right)^{\lambda-1}\right]$$
$$\left.\times\frac{(-1)^\lambda}{\tau_\alpha^{II}(\theta)^{\lambda}}\left(\cot(\pi\lambda) - i\right)v_k\right\}\mathrm{d}\theta + \int_0^\pi \mathrm{Re}$$
$$\times\left\{\sum_\alpha L_{i\alpha}^{I}\left(\sin\theta - p_\alpha^{I}\cos\theta\right)\sum_\beta C_{\alpha\beta}M_{\beta k}^{II}\right.$$
$$\left.\times\left(p_\beta^{II}\right)^{\lambda-1}\frac{(-1)^\lambda}{\tau_\alpha^{I}(\theta)^{2-\lambda}}\left(\cot(\pi\lambda) - i\right)w_k\right\} \times \mathrm{Re}$$
$$\times\left\{\sum_\alpha A_{i\alpha}\sum_\beta C_{\alpha\beta}M_{\beta k}^{II}\left(p_\beta^{II}\right)^{\lambda-1}\frac{(-1)^\lambda}{\tau_\alpha^{I}(\theta)^{\lambda}}\right.$$
$$\left.\times\left(\cot(\pi\lambda) - i\right)v_k\right\}d\theta + \int_\pi^{\frac{3\pi}{2}}\left\{\int_{-\frac{\pi}{2}}^{0}\mathrm{Re}\left\{\sum_\alpha L_{i\alpha}^{II}\right.\right.$$
$$\times\left(\sin\theta - p_\alpha^{II}\cos\theta\right)\sum_\beta\left[G_{\alpha\beta}\bar{M}_{\beta k}^{II}\left(\bar{p}_\beta^{II}\right)^{\lambda-1} + \delta_{ik}\right.$$
$$\left.\left.\times\left(p_\alpha^{II}\right)^{\lambda-1}\right]\frac{(-1)^\lambda}{\tau_\alpha^{II}(\theta)^{2-\lambda}}\left(\cot(\pi\lambda) - i\right)w_k\right\} \times \mathrm{Re}$$
$$\times\left\{\sum_\alpha A_{i\alpha}\sum_\beta\left[G_{\alpha\beta}\bar{M}_{\beta k}^{II}\left(\bar{p}_\beta^{II}\right)^{\lambda-1} + \delta_{ik}\left(p_\alpha^{II}\right)^{\lambda-1}\right]\right.$$
$$\left.\left.\times\frac{(-1)^\lambda}{\tau_\alpha^{II}(\theta)^{\lambda}}\left(\cot(\pi\lambda) - i\right)v_k\right\}d\theta\right\}. \tag{37}$$

Now assume that a pair of line forces acts on the crack faces at a point y_b, see Fig. 6. Other loading is absent. Equation 27 modifies with help of Eq. 28 as

$$\int_{\Gamma_3}\left[\sigma_{ij}(\mathbf{u})\,n_i r^{-(1-\lambda_1)}\bar{u}_{-1_j} - \sigma_{ij}\left(r^{-(1-\lambda_1)}\bar{\mathbf{u}}_{-1}\right)n_i u_j\right]$$
$$\mathrm{d}s + 2\mathbf{F}r^{-(1-\lambda_1)}\bar{\mathbf{u}}_{-1} = H\Psi\left(r^{1-\lambda_1}\bar{\mathbf{u}}_1, r^{-(1-\lambda_1)}\bar{\mathbf{u}}_{-1}\right). \tag{38}$$

Γ_3 is an arbitrary contour enclosing a domain containing both the crack tip and the pair of line forces. By definition, the weight function $W(y_b, h)$ follows as

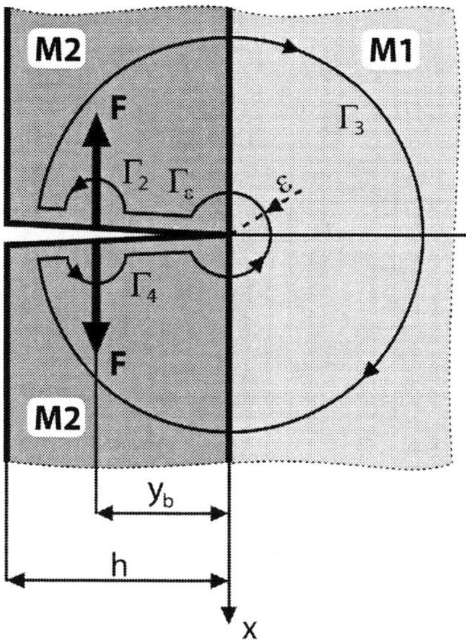

Fig. 6 A pair of line forces acting on the crack faces and the integration path

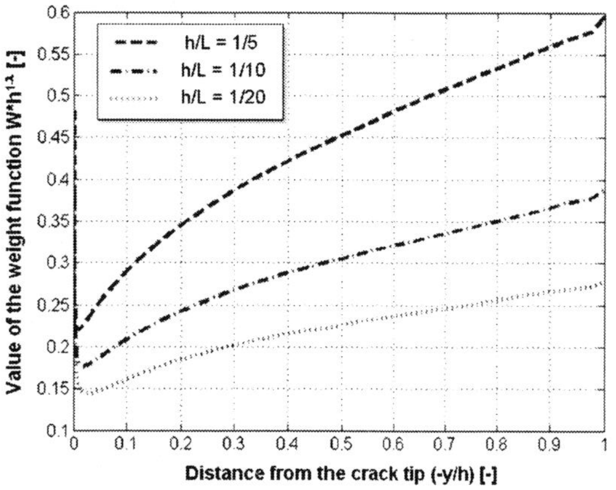

Fig. 7 Bimaterial normalized weight function against the dimensionless distance from the crack tip for several values of the ratio h/L

$$W \equiv \frac{H}{|\mathbf{F}|} = \frac{1}{|\mathbf{F}|} \frac{\int\limits_{\Gamma_3} \left[\sigma_{ij}(\mathbf{u})\, n_i r^{-(1-\lambda_1)} \bar{u}_{-1_j} - \sigma_{ij}\left(r^{-(1-\lambda_1)} \bar{\mathbf{u}}_{-1}\right) n_i u_j\right] \mathrm{d}s + 2\mathbf{F} r^{-(1-\lambda_1)} \bar{\mathbf{u}}_{-1}}{\Psi\left(r^{1-\lambda_1}\bar{\mathbf{u}}_1, r^{-(1-\lambda_1)}\bar{\mathbf{u}}_{-1}\right)} \tag{39}$$

A finite element solution $\mathbf{u}^h$ was used as an approximation for $\mathbf{u}$ in Eq. 39. Having calculated a value of the weight function W for sufficiently large number of line force positions, the generalized bridging stress intensity factor, H_{br}, can be obtained for an arbitrary bridging stress distribution $\sigma_{br}(y)$ as

$$H_{br} = \int\limits_{-h}^{0} W(y, h)\, \sigma_{br}(y)\, \mathrm{d}y \tag{40}$$

With elastic constants of two layers M1 and M2 specified in the preceding section, the weight function were calculated for several ratio of the layer thicknesses h/L. Observe that for the given material combination the stress singularity exponent $\lambda \equiv \lambda_1 = 0.67168$. The results are presented in Fig. 7 in dimensionless form such that the product $W.h^{1-\lambda}$ is plotted against the dimensionless distance from the crack tip $-y/h$. The influence of the longitudinal modulus E_L is demonstrated in Fig. 8.

5.2 Determination of the generalized bridging stress intensity factor

The evaluation of the generalized bridging stress intensity factor, H_{br}, from Eq. 40 requires knowledge of the bridging stress along the crack faces. Since the bridging stress is prescribed as a function of the crack opening displacement, first the total displacements of the crack surface $\delta = \delta_{appl} + \delta_{br}$ (note that $\delta_{br} < 0$) has to be derived. A special procedure to achieve this goal was devised and it is described in Appendix C. The approach is based on numerical data obtained from FEM which are used as an input for a recurrent procedure designed in the programme system MAPLE. The procedure requires considering two sets of loading in the FEM modelling:

(1) the specimen was loaded by unit tensile loading σ_0 and the displacement of the upper crack face $v_{appl} = \delta_{appl}/2$ was recorded;
(2) the specimen was loaded by the unit line force P_0 acting at the distance $-y$ from the crack tip. $v_{br} = \delta_{br}/2$ was calculated for a wide range of crack-notch geometries. The results were recorded into files which were used as input files for recurrent

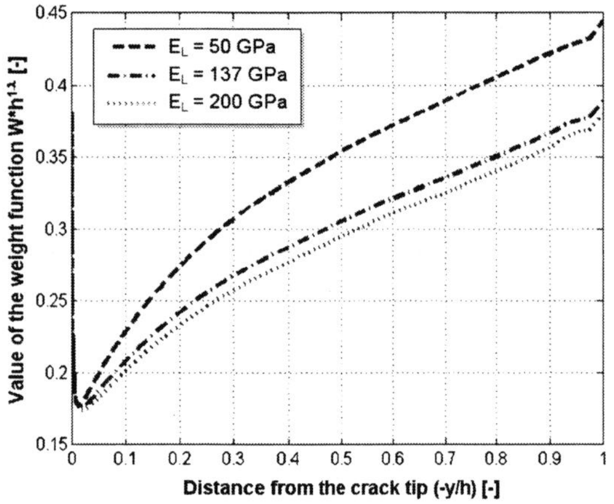

Fig. 8 Bimaterial normalized weight function against the dimensionless distance from the crack tip for several values of the longitudinal modulus E_L

calculations performed in the system MAPLE, see Appendix C.

6 Numerical results

At first the results of numerical analysis based upon the weight function method are presented. The bridging model described in the Sect. 3 was applied with the fibre volume fraction $c_f = 0.4$, the fibre radius $R = 7\,\mu$m, the sliding resistance $\tau = 6$ MPa, the fibre Young modulus $E_f = 228,000$ MPa, and the matrix Young modulus $E_m = 76,000$ MPa. A parametric study was performed in order to examine an influence of the Weibull modulus m and the fibre characteristic strength σ_{0f}. As stated in the Sect. 5.2, the total displacements of the crack surface $\delta = \delta_{appl} + \delta_{br}$ has to be derived. Figure 9 reveals the influence of the Weibull modulus upon the crack opening displacement. As expected, the lower value of m leads to the lower crack opening due to higher bridging stress, see Fig. 2. Similarly, higher value of the fibre characteristic strength σ_{0f} leads to higher bridging stress, and as a consequence, the crack opening displacement is reduced, see Fig. 10.

Having calculated the total displacements of the crack surface, the bridging stress distribution can be obtained and, consequently, the generalized bridging stress intensity factor, H_{br} can be evaluated from Eq. 40. The results of these calculations are presented in Fig. 11, where the remote, bridging, and local generalized stress intensity factors are plotted as functions of the applied tensile loading σ_0 for several values of the Weibull modulus and the fibre characteristic strength σ_{0f}.

It is evident that the local generalized SIF decreases with the decrease of the Weibull modulus m and with the increase of the characteristic fibre strength σ_{0f}. Also observe that the bridging generalized SIF (grey lines) begins to decrease with loading when the broken fibres are pulled out from the matrix. As a consequence, the resulting local generalized SIF begins to increase more rapidly with loading.

It is a matter of interest to compare the calculations based upon the weight functions method with the results obtained using the distribution dislocation technique (DDT) according to Eq. 26. So far there are avail-

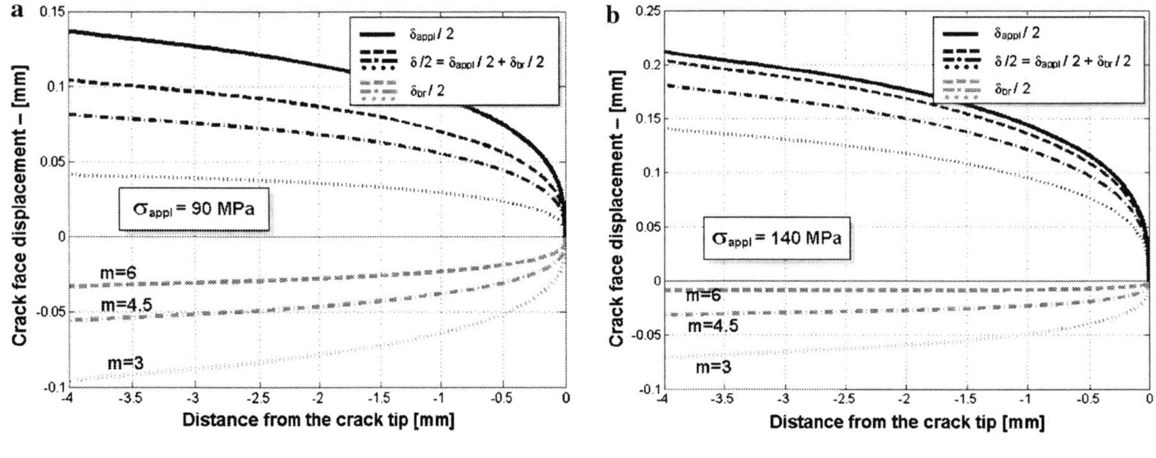

Fig. 9 Applied, closure and total crack opening displacement for several values of the Weibull modulus m, $\sigma_{0f} = 2300$ MPa. The applied tensile loading (**a**) $\sigma_0 = 90$ MPa, (**b**) $\sigma_0 = 140$ MPa

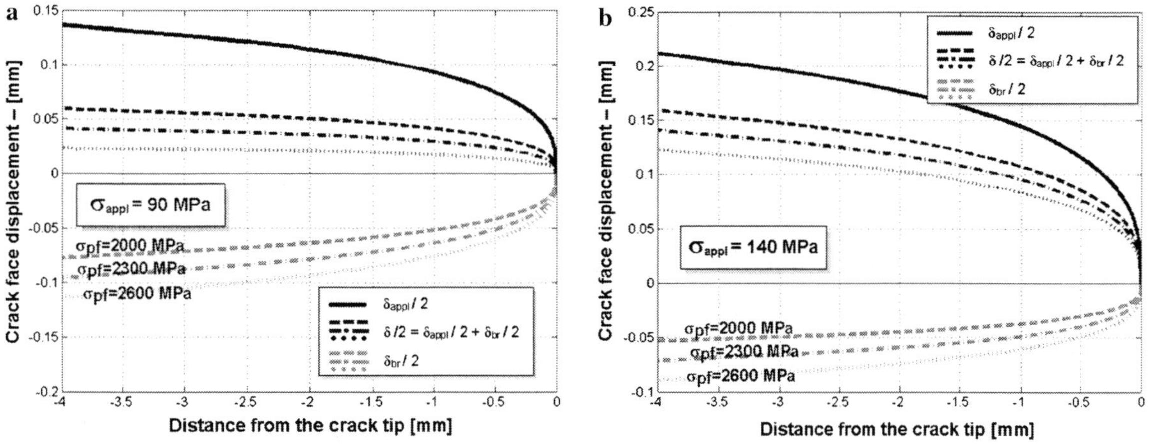

Fig. 10 Applied, closure and total crack opening displacement for several values of the Weibull modulus m, $\sigma_{0f} = 2300\,\text{MPa}$. The applied tensile loading (**a**) $\sigma_0 = 90\,\text{MPa}$, (**b**) $\sigma_0 = 140\,\text{MPa}$

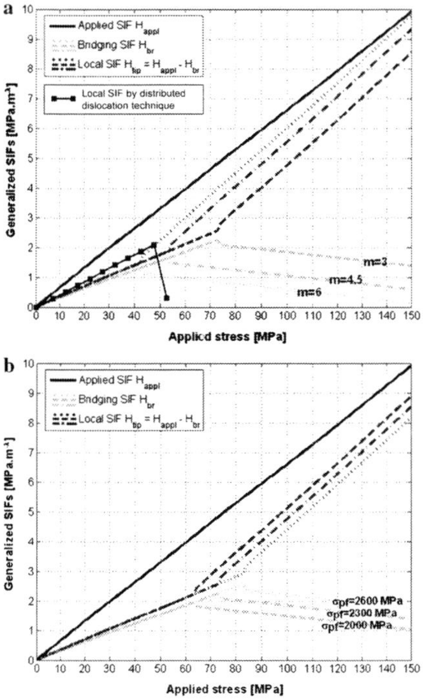

Fig. 11 Remote, bridging, and local generalized stress intensity factors plotted as functions of the applied tensile loading σ_0 for (**a**) several values of the Weibull modulus, (**b**) several values of the fibre characteristic strength σ_{0f}

able only numerical data for the first stage of loading when the broken fibres are not massively pulled out from the matrix. There exist certain numerical problems for the subsequent stage of loading which were not resolved satisfactorily yet. Nevertheless, Fig. 11 shows that the results obtained via DDT in the first stage of loading are in a good accordance with the results obtained via weight functions method.

Remark A method of solution should be sought for the situation when the preferred directions of the orthotropic material may not coincide with the reference axes in addition to having the crack and/or the material interface with an arbitrary orientation. Apparently, the concept of generalized anisotropic bimaterial applies to such situations. In the case of the generalized anisotropic bimaterial some aspects of the solution take place. Because of the participation of the all components of the displacement vector and stress tensor, the potentials describe the stress and displacement field must be extended to three ones, as well as the number of the eigenvalues characterizing each material. The characteristic equation (4) is usually in complex form, provided that media on both sides of the interface are generally anisotropic. Hence, it is usually difficult to investigate the general features of the characteristic roots for such anisotropic media. The determined eigenvalue λ can be complex, which leads to the oscillation with the increasing frequency of the expression r^{λ} as $r \to 0$. If a complex λ is an eigenvalue, so is its complex conjugate $\bar{\lambda}$. One can superimpose two solutions associated with λ and $\bar{\lambda}$ to obtain real values for the expressions describing the displacement and stress field. When the eigenvalues change from real to complex at some range of bimaterial elastic properties and crack/interface orientation respectively, multiple eigenvalues corresponding to the same independent eigenfunction may occur. The power series solution breaks down and exhibits very low numerical accuracy for a specific range of opening angles around the critical values. The analytical solution of such a special case cannot be expressed as a single function of the radial coordinate multiplied by a single function of the angular coordinate. When the number of the independent eigenvectors is less than the multiplicity of the eigenvalues, the missing eigenfunction can be determined by the differentiation of the stress and displacement functions.

7 Discussion

As it was already mentioned, a detailed quantification of the strength of singularity in stress and displacement field around the crack terminating at the bimaterial interface is a prerequisite for an assessment of fracture behaviour of many modern composites. The computational model developed in this paper provides a basis for an application of the criterion of fracture at an interface between anisotropic materials when a competition between deflection and penetration at the interface has to be analysed by seeking the path, which maximises the additional energy released by the fracture process for a finite and small crack extension. If crack deflection occurs preferentially to penetration at the interface, the following condition must be satisfied: $\Delta W_d = \delta W_d - G_{c_i} l_d > \Delta W_p = \delta W_p - G_{c1} l_p$, where G_{ci} is the interface toughness, G_{c1} is the toughness of the material M1 and δW is a change of the potential energy between the original and new crack position. Matched asymptotic procedure can be used to derive the change of potential energy.

Consider a perturbation of the domain Ω with crack impinging on the interface between materials M2 and M1 as shown in Fig. 12; the perturbation is a deflected (double) crack extension of length l_d or penetrating crack extension of length l_d with the small perturbation parameter ε defined as

$$\delta = \frac{l}{L} \ll 1, \quad l = l_p, l_d, \tag{41}$$

where L is the characteristic length of Ω. A second scale to the problem can be introduced, represented by the scaled-up coordinates

$$z = \frac{x}{\varepsilon}, \quad \text{or } (z_1, z_2) = \left(\frac{x_1}{\varepsilon}, \frac{x_2}{\varepsilon}\right), \tag{42}$$

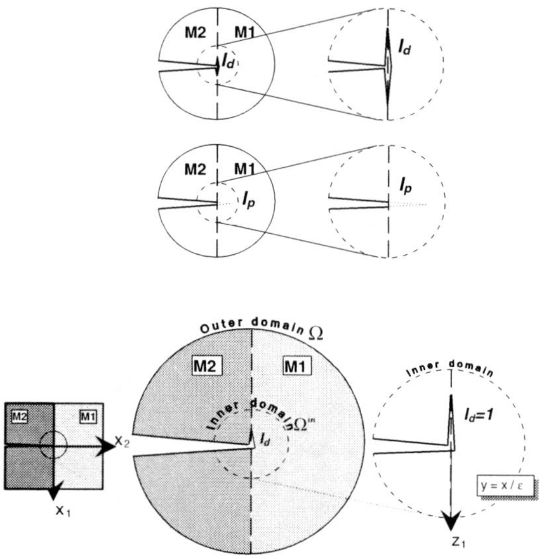

Fig. 12 Zoomed-in view of crack neighbourhood perturbated by a small crack extension

which provides a zoomed-in view into the region surrounding the crack. The displacement $\mathbf{u}^\varepsilon$ of the perturbed elasticity problem due to the presence of the crack can now be expressed in terms of the regular coordinate x and the scaled-up coordinate z as follows:

$$\mathbf{u}^\varepsilon(x) = \mathbf{u}^\varepsilon(\varepsilon z) = \mathbf{V}^\varepsilon(z), \tag{43}$$

where the definition of the function $\mathbf{V}^\varepsilon$ has been introduced, simply by a change of variable from x to z. Note that $\mathbf{V}^\varepsilon(z)$ is viewed here as exactly the same displacement as denoted by $\mathbf{u}^\varepsilon(x)$, i.e., the solution of the perturbed elasticity problem, but in terms of the scaled-up variable y, instead of the regular variable x. The change in the potential energy δW between the solutions in the unperturbed $\mathbf{u}^0(x_1, x_2)$ and perturbed $\mathbf{u}^\varepsilon(x_1, x_2)$ situations for unchanged boundary conditions (unchanged applied loads) is by use of Betti's theorem as follows:

$$\begin{aligned} \delta W &= \frac{1}{2}\int_\Gamma \left(\sigma_{ij}\left(\mathbf{u}^\varepsilon\right) n_i u_j^0 - \sigma_{ij}\left(\mathbf{u}^0\right) n_i u_j^\varepsilon\right) \mathrm{d}s \\ &= K_1 H^2 \varepsilon^{2(1-\lambda_1)} + \cdots \\ &= K_1 H^2 \left(\frac{l}{L}\right)^{2(1-\lambda_1)} + \cdots, \quad \varepsilon \ll 1, \end{aligned} \tag{44}$$

where Γ is any contour surrounding the crack tip and n_i its normal pointing toward the origin, see also the Sect. 5.1. The inner domain asymptotics approximates δW with K_1 depending on the perturbation due to a small crack extension l through inner expansion functions which can be computed once for all by FEM in a similar fashion as for H using the Ψ-integral.

The release rate G is easily obtained by taking a derivative of δW with respect to l as

$$G = \frac{1}{L} K_1 H^2 2 (1 - \lambda_1) \left(\frac{l}{L} \right)^{1-2\lambda_1}. \tag{45}$$

Apparently, G is singular for $l \to 0$ when $\lambda_1 > 0.5$ and zero when $\lambda_1 < 0.5$.

Recent finding of the configurational force mechanics (Simha et al. 2005) shows that the relation between the *unbridged* crack tip and far field J-integrals reads as

$$J_{tip} = J_{far} + C_{inh}, \tag{46}$$

where, in the case of sharp interface Σ between homogeneous materials, the inhomogeneity term C_{inh} is the projection of the total configurational interface force in the direction of the crack growth $\hat{\mathbf{e}}$:

$$C_{inh} = -\frac{1}{2} \mathbf{n} \cdot \hat{\mathbf{e}} \int_{\Sigma} [\![c_{ijkl}]\!] \, \varepsilon_{ij}^{+} \left(\mathbf{u}^{\varepsilon} \right) \varepsilon_{kl}^{-} \left(\mathbf{u}^{\varepsilon} \right) \mathrm{d}s, \tag{47}$$

where $\mathbf{n}$ is the interface unit normal, c_{ijkl} is the elastic moduli tensor, $\boldsymbol{\varepsilon}^{+}$ and $\boldsymbol{\varepsilon}^{-}$ are the strain values on the right side and left side of the interface respectively, $[\![c_{ijkl}]\!] = c_{ijkl}^{+} - c_{ijkl}^{-}$. C_{inh} quantifies the crack tip shielding or antishielding and originates from material inhomogeneity in the direction of crack growth. Equating the crack tip J-integral in Eq. 46 to the release rate G in Eq. 45 allows scaling of the far field J-integral with the generalized stress intensity factor H. Such scaling could be helpful in experimental study of crack growth across the bimaterial interface. Observe that C_{inh} equals to zero when crack propagates along the interface. In the case of a *bridged* crack terminating at the bimaterial interface, Eq. 46 modifies as (see Fig. 13)

$$\frac{1}{L} K_1 H_{tip}^2 2 (1 - \lambda_1) \left(\frac{l}{L} \right)^{1-2\lambda_1} + J_{br} = J_{far} + C_{inh}, \tag{48}$$

where Eq. 45 was used. J_{br} represents the resisting force arising from the bridging tractions and in the present geometry is given by

$$J_{br} = \int \frac{\partial \delta}{\partial y} \sigma_{br} (y) \, \mathrm{d}y = \int_{0}^{\delta^*} \sigma_{br} (\delta) \, \mathrm{d}\delta, \tag{49}$$

where δ^* is the crack opening at the end of the bridged region.

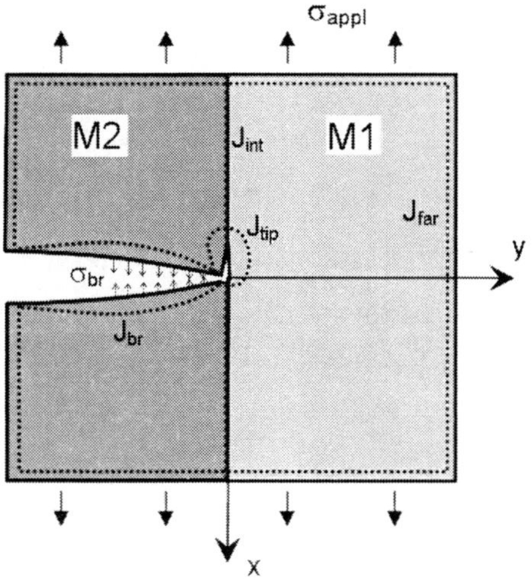

Fig. 13 J integration contours around an edge bridged crack terminating at the sharp interface

In terms of the configurational mechanics, the correct physical interpretation of J_{br} is following—it is the total force on the dipole distribution (per unit length of crack) and the quantity $-J_{br} t \delta a$ is the Gibbs free energy change when the bridging traction are shifted along the crack faces by δa in the crack direction keeping the crack tip fixed in position (t is the thickness of the material).

8 Summary

- Dislocation trimaterial solution was found to hold also for the considered finite specimen.
- GSIF was calculated both by CDD technique and FEM combined with the Ψ-integral. The results obtained were found to be in good agreement.
- The weight functions for a crack in the surface layer terminating at the layer/substrate interface were calculated using FEM combined with the Ψ-integral. The weight functions were further used in the calculation of the bridging GSIF.
- The influence of the bridging fibres was analyzed using the advanced model, involving a statistical distribution of fibre strength and pull out of fibres from the matrix.
- The relationship with the configurational mechanics approach was discussed.

Acknowledgements The support through the grants GACR 101/05/0320, 106/05/H008, and 101/05/P290 is gratefully acknowledged.

Appendix A

Following Lekhnitskii (1963), for plane deformation, the elastic field can be represented in terms of complex potential functions $\Phi_1(z_1)$, $\Phi_2(z_2)$, $\Phi_3(z_3)$, each of which is holomorphic in its arguments $z_\alpha = x + p_\alpha y$. Here, p_α are three distinct complex numbers with positive imaginary parts, which are obtained as the roots of the characteristic equation

$$\det\left[c_{i1k1} + p\left(c_{i1k2} + c_{i2k1}\right) + p^2 c_{i2k2}\right] = 0, \quad \text{(A1)}$$

where c_{ijkl} is the tensor of elastic constants, i.e. $\sigma_{ij} = c_{ijkl}u_{k,l}$, which satisfies the symmetry conditions

$$c_{ijkl} = c_{ijlk} = c_{jikl} = c_{klij}. \quad \text{(A2)}$$

With these holomorphic functions, the representation for the displacements u_i and stresses σ_{ij} is

$$u_i = 2\text{Re}\left[\sum_{\alpha=1}^{3} A_{i\alpha}\Phi_\alpha(z_\alpha)\right],$$

$$\sigma_{2i} = 2\text{Re}\left[\sum_{\alpha=1}^{3} L_{i\alpha}\Phi'_\alpha(z_\alpha)\right],$$

$$\sigma_{1i} = -2\text{Re}\left[\sum_{\alpha=1}^{3} L_{i\alpha}p_\alpha\Phi'_\alpha(z_\alpha)\right], \quad \text{(A3)}$$

Here, ()' designates the derivative with respect to the associated arguments, and **A** and **L** are matrices given by

$$L_{i\alpha} = A_{k\alpha}\left(c_{i2k1} + p_\alpha c_{i2k2}\right), \quad \text{(A4)}$$

where $A_{k\alpha}$ denotes the eigenvector corresponding to the eigenvalue p_α above.

Appendix B

The singularity at the crack tip can be analysed using the distributed edge dislocations method. The semi-infinite crack is modelled as an array of continuously distributed edge dislocations along the negative y-axis, see Fig. B1. The potential functions for an isolated dislocation located at the point (x_o, y_o) in an infinite homogeneous anisotropic medium is

$$\Phi_{\alpha o}(z) = q_\alpha \ln\left(z - \varsigma_\alpha\right), \quad \text{(B1)}$$

where

$$\varsigma = x_o + p_\alpha y_o, \quad \alpha = 1, \ldots, 3 \quad \text{(B2)}$$

and

$$q_\alpha = \frac{1}{4\pi}M_{\alpha k}d_k, \quad \text{(B3)}$$

where the vector d_k is related to the Burgers vector b_i through the equation

$$b_i = B_{ik}d_k, \quad \text{with } B_{ik} = \frac{i}{2}\sum_\alpha\left(A_{i\alpha}M_{\alpha k} - \bar{A}_{i\alpha}\bar{M}_{\alpha k}\right)$$

$$= -\text{Im}\left(\sum_\alpha A_{i\alpha}M_{\alpha k}\right), \quad \text{(B4)}$$

where the matrix $M_{\alpha k}$ is defined as the inverse of $L_{i\alpha}$, $M_{\alpha k}\, L_{k\beta} = \delta_{\alpha\beta}$. The quantities p_α, $A_{i\alpha}$, $L_{i\alpha}$ are given in Appendix A.

The potentials for the interaction of an edge dislocation with the interface of two anisotropic materials can be obtained in terms of $\mathbf{\Phi}_o(z)$ by invoking the standard analytical continuation arguments along the interface, as described by Suo (1990). The solution for the two media can be written as

$$\mathbf{\Phi}(z) = \mathbf{\Phi}^I(z), \quad z \in 1,$$
$$\mathbf{\Phi}(z) = \mathbf{\Phi}^{II}(z) + \mathbf{\Phi}_o(z), \quad z \in 2, \quad \text{(B5)}$$

$$\mathbf{\Phi}^I(z) = \mathbf{C}\mathbf{\Phi}_o(z),$$
$$\mathbf{C} = i\mathbf{M}^I\mathbf{H}^{-1}\left(\mathbf{A}^{II}\mathbf{M}^{II} - \bar{\mathbf{A}}^{II}\bar{\mathbf{M}}^{II}\right)\mathbf{L}^{II}, \quad z \in 1,$$
$$\mathbf{H} = i\left(\mathbf{A}^I\mathbf{M}^I - \bar{\mathbf{A}}^{II}\bar{\mathbf{M}}^{II}\right),$$
$$\mathbf{\Phi}^{II}(z) = \mathbf{G}\bar{\mathbf{\Phi}}_o(z),$$
$$\mathbf{G} = -i\mathbf{M}^{II}\bar{\mathbf{H}}^{-1}\left(\bar{\mathbf{A}}^{II}\bar{\mathbf{M}}^{II} - \bar{\mathbf{A}}^I\bar{\mathbf{M}}^I\right)\bar{\mathbf{L}}^{II}, z \in 2. \quad \text{(B6)}$$

The solution for the stress field produced by an isolated dislocation located at point (x_o, y_o) with the Burgers vector b_i in an infinite anisotropic bimaterial follows from (A3), (B5) and (B6) as

$$\sigma_{1i}(x, y) = -\frac{1}{4\pi}\sum_\alpha L^{II}_{i\alpha}p^{II}_\alpha$$
$$\left[\sum_\beta\left(G_{\alpha\beta}\bar{M}^{II}_{\beta k}\frac{d_k}{z_\alpha - \bar{\varsigma}_\beta}\right) + M^{II}_{\alpha k}\frac{d_k}{z_\alpha - \varsigma_\alpha}\right]$$
$$+C.C., \quad z \in 2, \quad \text{(B7)}$$

$$\sigma_{2i}(x, y) = \frac{1}{4\pi}\sum_\alpha L^{II}_{i\alpha}$$
$$\left[\sum_\beta\left(G_{\alpha\beta}\bar{M}^{II}_{\beta k}\frac{d_k}{z_\alpha - \bar{\varsigma}_\beta}\right) + M^{II}_{\alpha k}\frac{d_k}{z_\alpha - \varsigma_\alpha}\right]$$
$$+C.C., \quad z \in 2, \quad \text{(B8)}$$

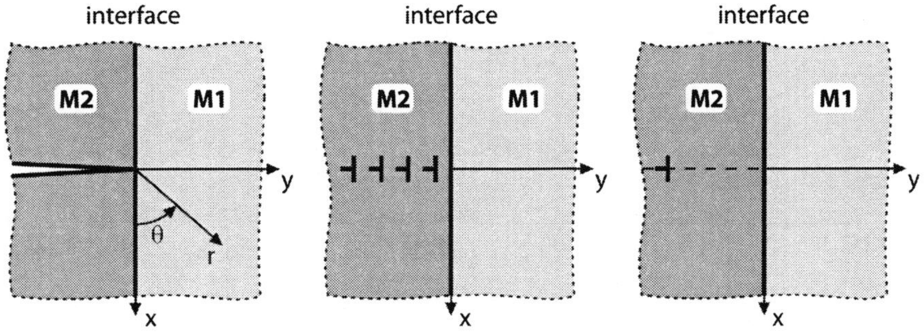

Fig. B1 Semi-infinite crack terminating perpendicular to the interface between two anisotropic materials

$$\sigma_{1i}(x,y) = -\frac{1}{4\pi}\sum_{\alpha} L^{I}_{i\alpha} p^{I}_{\alpha} \sum_{\beta}\left(C_{\alpha\beta} M^{II}_{\beta k}\frac{d_k}{z_\alpha - \varsigma_\beta}\right) + C.C., \quad z \in 1, \tag{B9}$$

$$\sigma_{2i}(x,y) = \frac{1}{4\pi}\sum_{\alpha} L^{I}_{i\alpha} \sum_{\beta}\left(C_{\alpha\beta} M^{II}_{\beta k}\frac{d_k}{z_\alpha - \varsigma_\beta}\right) + C.C., \quad z \in 1, \tag{B10}$$

where $C.C.$ denotes the complex conjugate of the preceding expression, superscript I and II refers to the material 1 and 2 respectively, and the convention of summing over repeated Latin indices is used. Introduce a function f_k at a point on the crack $(x = 0, y)$ which relates to the elemental Burgers vector δb_i between y_o and $y_0 + \delta y_0$ as

$$\delta b_i = B_{ik}\delta d_k = B_{ik} f_k(y_o)\,\delta y_o, \tag{B11}$$

and integrate (B7) along the whole crack. The tractions produced at a point $(0, y)$ by the density function f_k can be expressed as

$$\begin{aligned}\sigma_{1i}(y) = -\frac{1}{4\pi}\Bigg\{ & \sum_{\alpha} L^{II}_{i\alpha}\left[\sum_{\beta}\left(G_{\alpha\beta}\bar{M}^{II}_{\beta k}\frac{p^{II}_{\alpha}}{\bar{p}^{II}_{\beta}}\int_{-\infty}^{0}\frac{f_k(y_o)\,\mathrm{d}y_o}{\frac{p^{II}_{\alpha}}{\bar{p}^{II}_{\beta}}y - y_o}\right) + M^{II}_{\alpha k}\int_{-\infty}^{0}\frac{f_k(y_o)\,\mathrm{d}y_o}{y - y_o}\right] \\ & + \sum_{\alpha}\bar{L}^{II}_{i\alpha}\left[\sum_{\beta}\left(\bar{G}_{\alpha\beta}M^{II}_{\beta k}\frac{\bar{p}^{II}_{\alpha}}{p^{II}_{\beta}}\int_{-\infty}^{0}\frac{f_k(y_o)\,\mathrm{d}y_o}{\frac{\bar{p}^{II}_{\alpha}}{p^{II}_{\beta}}y - y_o}\right) + \bar{M}^{II}_{\alpha k}\int_{-\infty}^{0}\frac{f_k(y_o)\,\mathrm{d}y_o}{y - y_o}\right]\Bigg\}.\end{aligned} \tag{B12}$$

The asymptotic stress field near the crack tip is modelled as a continuous distribution of dislocations with density function

$$f_k(y_o) = H v_k(-y_o)^{-\lambda}, \quad y_o < 0, \tag{B13}$$

where λ is the stress singularity exponent, which is yet unknown, v_k are the components of corresponding eigenvector, and H is the generalized stress intensity factor (GSIF). Substitute Eq. B13 in Eq. B12, integrate and apply the traction-free condition on the plane of the crack to obtain

$$\mathrm{Re}\left\{\left[\left[\sum_{\alpha}\sum_{\beta} L^{II}_{i\alpha}G_{\alpha\beta}\bar{M}^{II}_{\beta k}\left(-\frac{\bar{p}^{II}_{\beta}}{p^{II}_{\alpha}}\right)^{\lambda-1} \csc(\pi\lambda) - \delta_{ik}\cot(\pi\lambda)\right]\right]\right\} v_k = 0. \tag{B14}$$

Eq. B14 can be briefly written as

$\mathbf{D}(\lambda)\,\mathbf{v} = 0$, where

$$D_{ik}(\lambda) = \mathrm{Re}\left\{\left[\left[\sum_{\alpha}\sum_{\beta} L^{II}_{i\alpha}G_{\alpha\beta}\bar{M}^{II}_{\beta k}\left(-\frac{\bar{p}^{II}_{\beta}}{p^{II}_{\alpha}}\right)^{\lambda-1} \csc(\pi\lambda) - \delta_{ik}\cot(\pi\lambda)\right]\right]\right\}. \tag{B15}$$

The parameter λ is calculated from the characteristic equation

$$\mathrm{Det}\,[\mathbf{D}(\lambda)] = 0 \tag{B16}$$

and the eigenvector $\mathbf{v}$ is determined from Eq. B15 up to a multiplicative constant.

Appendix C

Recurrent calculations of bridging stress as a function of position along the crack face consist of the following steps:

(1) In the first step, the displacement of the upper crack face for unbridged crack in a bimaterial body due to tensile loading σ_1 is calculated using FEM:

$$v_{appl(1)}(y) = v_{appl0}(y)\,\sigma_1 \tag{C1}$$

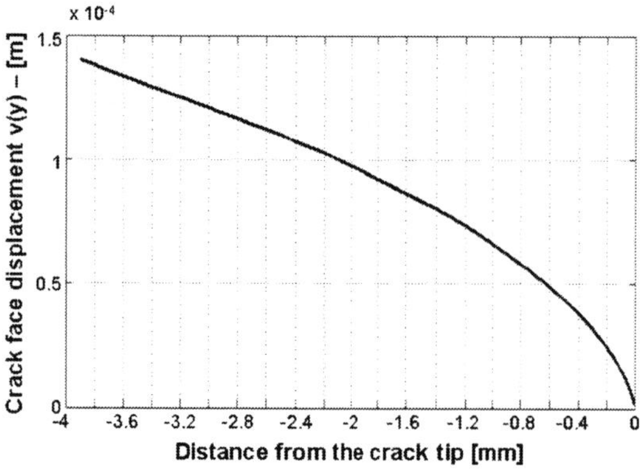

Fig. C1 Displacement of the upper crack face caused by the unit loading σ_0

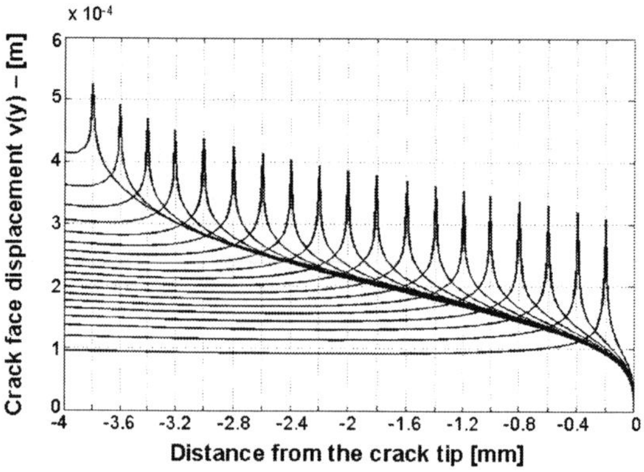

Fig. C2 Displacement of the upper crack face for a number of positions of the applied line load

where v_{appl0} is the displacement of the upper crack face caused by the unit loading σ_0, see Fig. C1.

(2) In the second step, the total displacement of the upper crack face is recurrently refined in several sub-steps. In the first sub-step of the recurrent calculation the total displacement, $v_{n,1}$, $n = 1$, is set equal to $v_{appl(n)}$, $n = 1$. Then the bridging stress is computed via

$$\hat{\sigma}_{br(n,1)} = \sigma_{br(n,1)} \underbrace{\exp\left[-\left(\frac{\sigma_{br(n,1)}}{c_f \Sigma}\right)^{m+1}\right]}_{\text{fraction of intact fibres}} + \sigma_{P(n,1)} \underbrace{\left\{1 - \exp\left[-\left(\frac{\sigma_{br(n,1)}}{c_f \Sigma}\right)^{m+1}\right]\right\}}_{\text{fraction of broken fibres}}, \tag{C2}$$

where $\sigma_{br(n,1)}(y) = \beta\sqrt{v_{n,1}(y)}$, $n = 1$, $\sigma_{P(n,1)}(x) = \frac{2c_f\tau}{R}(\langle h\rangle - v_{n,1}(y))$, $n = 1$. This makes possible to estimate the corresponding crack face displacement $v_{br(n,1)}$ using the FEM solution for the crack face displacement due to the unit line force from the relation

$$v_{br(n,1)}(y) = \sum_i v_{br0}(y, y_i)\, \hat{\sigma}_{br(n,1)}(y_i)\, S(y_i) \tag{C3}$$

where $v_{br0}(y, y_i)$ is the crack face displacement due to the unit line force acting at the point y_i, see Fig. C2. $S(y_i)$ is the area per node at the point y_i and the summation is performed over all node rows behind the crack tip.

In next sub-steps the total crack face displacement is refined as follows:

$$v_{n,m_n+1}(y) = v_{n,m_n} \frac{v_{appl(n)}(y)}{v_{n,m_n}(y) - v_{br(n,m_n)}(y)}, \quad n=1 \tag{C4}$$

where $v_{br(n,m_n)} < 0$ is the crack face displacement due to the bridging stress $\hat{\sigma}_{br(n,m_n)}(y) = \hat{\sigma}_{br(n,m_n)}[v_{n,m_n}(y)]$. Note that for the exact value of $v_{br}(x)$ the ratio in Eq. C4 equals to one. The recurrent calculation stops when

$$\left(\frac{v_{n,m_n+1}(x) - v_{n,m_n}(x)}{v_{n,m_n+1}(x)}\right)^2 < TO1 \tag{C5}$$

where $TO1$ is a prescribed tolerance.

References

Braun M (1997) Configurational forces induced by finite-element discretization. Proc Estonian Acad Sci Phys Math 46:24–31

Budiansky B, Amazigo J (1989) Toughening by aligned, frictionally constrained fibers. J Mech Phys Solids 37:93–109

Budiansky B, Hutchinson JW, Evans AG (1986) Matrix fracture in fiber-reinforced ceramics. J Mech Phys Solids 34:167–189

Choi ST, Earmme YY (2002) Elastic study on singularities interacting with interfaces using alternating technique part. Part I. Anisotropic trimaterial. Int J Solid Struct 39:943–957

Curtin WA (2000) Stress-strain behaviour of brittle matrix composites. In: Kelly A, Zweben C (eds) Comprehensive composite materials, vol 4. Pergamon Press, pp 47–76

Dewynne JN, Hills DA, Nowell D (1992) Calculation of the opening displacement of surface-breaking plane cracks. Comp Methods Appl Mech Eng 97:321–331

Erdogan F, Gupta GD, Ratwani M (1974) Interaction between a circular inclusion and an arbitrary oriented crack. J Appl Mech 12:1007–1013

Eshelby JD (1970) Energy relations and the energy-momentum tensor in continuum mechanics. In: Kanninen M et al (eds) Inelastic behavior of solids. McGraw Hill, New York pp 77–115

Fett T, Munz D (1997) Stress intensity factors and weight functions. Computational Mechanics, Inc.

Gurtin ME (2000) Configurational forces as basic concepts of continuum physics. Springer, New York

He MY, Hutchinson JW (1989) Crack deflection at an interface between dissimilar elastic materials. Int J Solids Struct 25:1053–1067

Hills DA, Kelly PA, Dai DN, Korsunsky AM (1996) Solution of crack problems. Kluwer Academic Publishers, Dordrecht

Kienzler R, Herrmann G (2000) Mechanics in material space. Springer, Berlin

Lekhnitskii SG (1963) Theory of elasticity of an anisotropic body. Holden-Day, San Francisco

Marshall DB, Cox BN (1988) J-integral method for calculating steady-state matrix cracking in composites. Mech Mater 7:127–133

Martin E, Leguillon D, Lacroix C (2001) A revisited criterion for crack deflection at an interface in a brittle bimaterial. Compos Sci Technol 61:1671–1679

Martinez D, Gupta V (1994) Energy criterion for crack deflection at an interface between two orthotropic media. J Mech Phys Solids 42:1247–1271

Maugin GA (1995) Material forces: concepts and applications. ASME Appl Mech Rev 48:213–245

Mueller R, Kolling S, Gross D (2002) On configurational forces in the context of the finite element method. Int J Numer Meth Eng 53:1557–1574

Muschelishvili NI (1953) Some basic problems in the mathematical theory of elasticity. Noordhoff, Holland

Papadakis PJ, Babuska I (1995) A numerical procedure for the determination of certain quantities related to the stress intensity factors in two-dimensional elasticity. Comp Methods Appl Mech Eng 122:69–92

Profant T, Ševeček O, Kotoul M (2007) Calculation of K-factor and T-stress for cracks in anisotropic bimaterials. accepted in Eng Fract Mech

Sarrafi-Nour GR, Coyle TW, Fett T (1998) A weight function for the crack surface tractions in chevron-notched specimens. Eng Fract Mech 59:439–445

Simha NK (2000) Toughening by phase boundary propagation. J Elast 59:195–211

Simha NK, Fischer FD, Kolednik O, Predan J, Shan GX (2005) Crack tip shielding or anti-shielding due to smooth and discontinuous material inhomogeneities. Int J Fract 135:73–93

Stern M, Soni ML (1976) On the computation of stress intensities at fixed-free corners. Int J Solid Struct 12:331–337

Suo Z (1990) Singularities, interfaces and cracks in dissimilar anisotr. media. Proc R Soc Lond A 427:331–358

Thouless MD, Evans AG (1988) Effects of pull-out on the mechanical properties of ceramic matrix composites. Acta Metall 36:517–522

Ting TCT (1997) Stress singularities at the tip of interfaces in polycrystals. In: Rossmanith HP (ed) Proceedings of first international conference on damage and failure of interfaces. Balkema Publishers, Rotterdam, Netherlands, pp. 45–82

Tullock DL, Reimanis IE, Graham AL, Petrovic JJ (1994) Deflection and penetration of cracks at an interface between two dissimilar materials. Acta Metall Mater 42:3245–3252

Study of the simple extension tear test sample for rubber with Configurational Mechanics

Erwan Verron

Abstract The simple extension tear test-piece also referred to as the trousers sample is widely used to study crack propagation in rubber. The corresponding energy release rate, called tearing energy for rubber materials, was first established by Rivlin and Thomas (*J Polym Sci*, 10:291–318, 1953); a second derivation was proposed later by Eshelby (In G.C. Sih, H. C. van Elst, and D. Broek, editors, *Prospects of Fracture Mechanics*, 69-84, Leyden, 1975). We show here that the derivation of this result can be advantageously revisited through the scope of Configurational Mechanics. Our approach is based on the rigorous definition of the configurations of the body and on the physical significance of the configurational stress tensor. More precisely, it is demonstrated that the change in energy due to crack growth, and then the tearing energy, is directly related to the components of the configurational stress tensor in the body.

Keywords Rubber · Tearing energy · Simple extension tear test · Configurational Mechanics

E. Verron (✉)
Institut de Recherche en Génie Civil et Mécanique, UMR CNRS 6183, École Centrale de Nantes, BP 92101, 44321 Nantes cedex 3, France
e-mail: erwan.verron@ec-nantes.fr

1 Introduction

In his seminal paper, Griffith (1920) proposed a criterion to determine the amount of energy involved during crack propagation in brittle materials. Denoting dU (<0) the change of total energy (change of strain energy and work of external forces) and dA the increase in crack surface during crack growth in the body, the energy release rate G is defined by

$$G = -\frac{dU}{dA}, \tag{1}$$

and the crack growth criterion can be simply written as

$$G < G_c, \tag{2}$$

where G_c is a critical value of the energy release rate directly related to the surface free energy of the material. It should be measured in experiments.

More than 30 years later, Rivlin and Thomas (1953) extended the Griffith theory to rubber-like materials considering the following assumptions:

(a) the approach of Griffith is valid for large strain (in fact, no restriction was formulated in the original paper of Griffith),
(b) irreversible changes in energy due to crack growth take place only in the neighbourhood of the crack tip,
(c) the change in energy is independent of the shape and dimensions of the body.

Defect and Material Mechanics. C. Dascalu, G.A. Maugin & C. Stolz (eds.),
doi: 10.1007/978-1-4020-6929-1_20, © Springer Science+Business Media B.V. 2007

Then, authors defined the tearing energy T, i.e. the counterpart of G for rubberlike materials, by

$$T = -\left.\frac{\partial w}{\partial A}\right|_l, \tag{3}$$

where w is the strain energy, A is the crack surface and the suffix $\cdot_l$ denotes differentiation with constant displacement of the boundaries over which forces are applied. Considering thin samples (uniform thickness h_0) and denoting c the length of the crack, the tearing energy reduces to

$$T = -\frac{1}{h_0}\left.\frac{\partial w}{\partial c}\right|_l, \tag{4}$$

and the corresponding crack growth criterion is simply $T > T_c$, T_c being the critical value of the tearing energy that only depends on the material. Nevertheless, due to assumption (b), T_c cannot be directly related to the surface free energy of the elastomer. Moreover, the correctness of the above-mentioned assumption (c) was investigated by several authors by making tearing measurement on thin test pieces of different shapes but of the same material, and examining the constancy of T_c values obtained (see Thomas (1994) and the references herein).

In their paper published in 1953, Rivlin and Thomas proposed several experimental samples to perform crack propagation experiments in rubber. One of these samples is the simple extension tear test-piece also referred to as the trousers sample. Authors calculated the corresponding expression for the tearing energy and used this sample to measure the critical value of the tearing energy. Later, Eshelby (1975b) considered the same test-piece to illustrate the relevance of path-independent integrals for the calculation of configurational forces that drive the evolution of singularities. In his paper, Eshelby proposed a second derivation for the tearing energy of the trousers sample. The present paper discusses a third derivation of this result. It is demonstrated that the tearing energy of the simple extension tear test sample can be easily recovered using the general theory of Configurational Mechanics and more precisely the definition of the configurational stress tensor. Previous derivations are first recalled. Then, our proposal is detailed.

2 Description of the problem

The sample is a rectangular thin sheet of length b and width $2a$; the thickness h_0 is uniform. The test piece

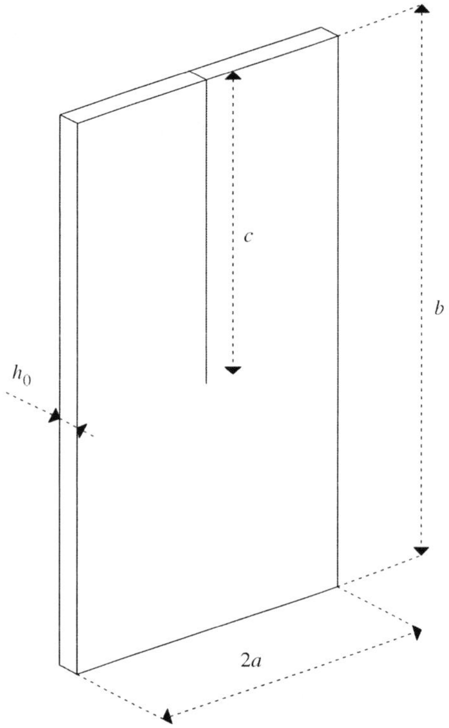

Fig. 1 Trousers sample for rubber tearing

contains a cut of length c (with $c \gg a$) parallel to the edges of length b. The sample geometry is presented in Fig. 1.

The description of the problem is based on the Fig. 2. Both the notations proposed by Rivlin and Thomas (1953, p. 302) and the three-dimensional sketch of Greensmith and Thomas (1956, p. 373) (considered later by Eshelby (1975b, p. 76)) are adopted. In the following, the emphasize is laid on the rigorous definition of body configurations. The initial position of the sample shown in Fig. 1 is adopted as the reference configuration ($\mathcal{C}_0$) (see Fig. 2a). During the experiments, the 'legs of the trousers' are first spread; then, the force F is applied to the legs to produce tearing. It leads to the definition of the deformed configuration ($\mathcal{C}$) as shown in Fig. 2b. In this configuration, the separation between legs extremities is l and the thickness h of the piece is no longer uniform. We consider now that the crack length increases by dc in the deformed configuration, the force F being kept constant. Thus, after unloading, i.e. $F = 0$, the sample occupies a new reference configuration ($\mathcal{C}_0'$) depicted in Fig. 2c. The motion between ($\mathcal{C}$) and ($\mathcal{C}_0'$) is defined by its deformation gradient $\mathbf{f}$.

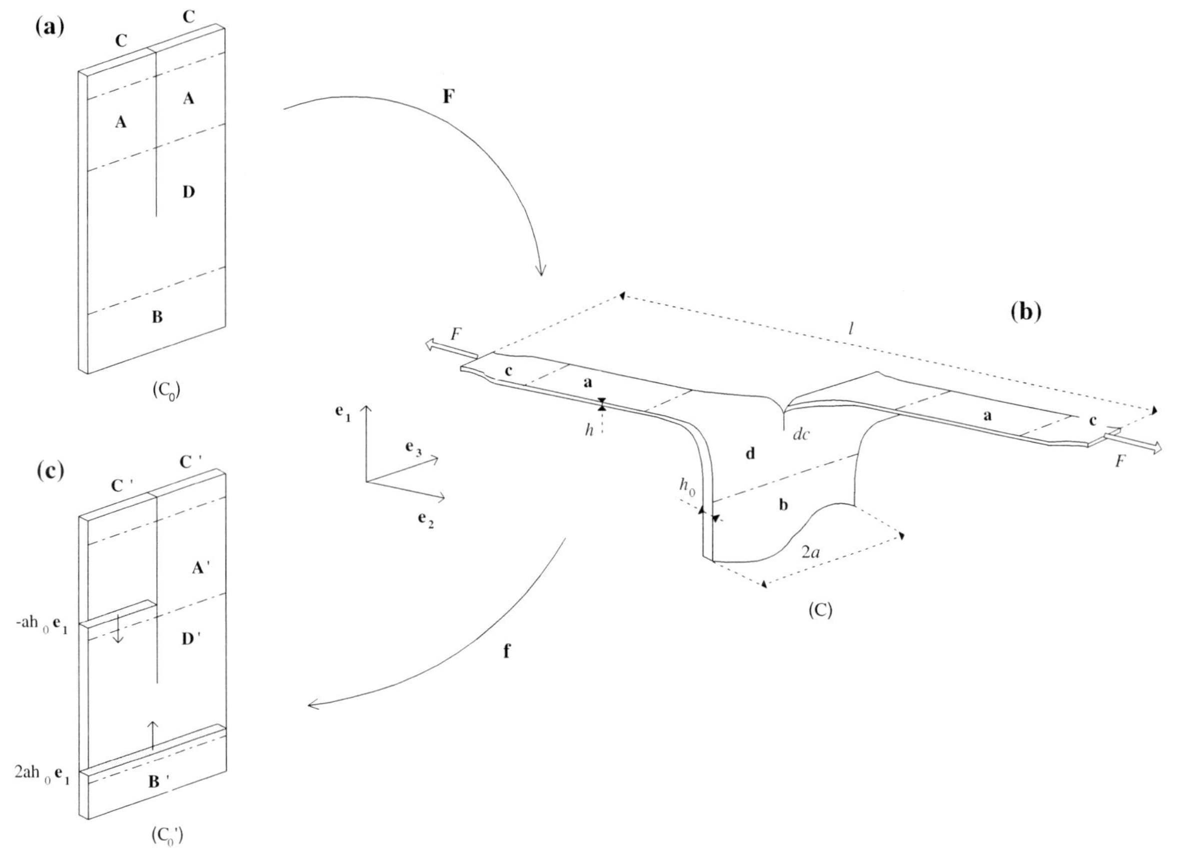

Fig. 2 Definition of body configurations

Obviously, **f** is not exactly equal to $\mathbf{F}^{-1}$ because the crack has grown and $(\mathcal{C}_0')$ differs from $(\mathcal{C}_0)$.

3 Previous derivations

3.1 Derivation of Rivlin and Thomas (1953)

The determination of the tearing energy as proposed by Rivlin and Thomas (1953) is based on an energetic analysis. For this purpose, authors considered Fig. 2a and b.

The sample can be separated into four different regions in the deformed configuration (Fig. 2b):

- the region **a** of each of the legs which is substantially in uniaxial extension, the corresponding stretch ratio being denoted λ,
- the region **b** which is substantially undeformed if the uncut part of the sample is sufficiently long,
- the regions **c** and the region **d** in which strain and stress distributions are complicated, the former correspond to the neighbourhood of the points in which forces apply and the latter corresponds to the neighbourhood of the tip.

When the crack length increases from c to $c + dc$, authors consider that:

- the size of regions **a** increases and the size of the region **b** decreases of the same expense. In the reference configuration (Fig. 2a), the increase in volume of each region **A** (which are transformed into regions **a** by the motion) is $a\, h_0\, dc$; thus, the decrease in volume of the region **B** (which is transformed into region **b** by the motion) is equal to $2a\, h_0\, dc$,
- the regions **c** and the region **d** are only moved and their respective sizes (or the sizes of their undeformed counterparts **C** and **D**) do not change,

- moreover, after crack growth the region **a** is still in uniaxial extension with the same stretch ratio λ because F does not change, the region **b** is still undeformed, and the strain and stress distributions in regions **c** and **d** remain unchanged.

So, the length between legs extremities increases by $dl = 2\lambda\, dc$, and then,

$$\left.\frac{\partial l}{\partial c}\right|_F = 2\lambda. \tag{5}$$

Considering that the strain energy of the sample is a function of the two quantities l and c, its change dw is related to the changes of crack length dc and overall length dl by

$$dw = \left.\frac{\partial w}{\partial l}\right|_c dl + \left.\frac{\partial w}{\partial c}\right|_l dc. \tag{6}$$

Recalling that crack growth is due to the applied force F which remains constant, the previous equation can be written as

$$\left.\frac{\partial w}{\partial c}\right|_F = \left.\frac{\partial w}{\partial l}\right|_c \left.\frac{\partial l}{\partial c}\right|_F + \left.\frac{\partial w}{\partial c}\right|_l . \tag{7}$$

With respect to the previous statement about the increase in volume of the regions **a** and the decrease in volume of the region **b**, the change in energy can be determined in the reference configuration $(\mathcal{C}_0)$,

$$\left.\frac{\partial w}{\partial c}\right|_F = W\, 2a\, h_0, \tag{8}$$

where W stands for the strain energy density per unit of undeformed volume and $2a\, h_0$ is the cross-section of the undeformed sample. Moreover, the force F is related to the change in energy by

$$\left.\frac{\partial w}{\partial l}\right|_c = F. \tag{9}$$

Finally, recalling Eqs. 5, 8 and 9, Eq. 7 becomes

$$\left.\frac{\partial w}{\partial c}\right|_l = W\, 2a\, h_0 - F\, 2\lambda, \tag{10}$$

the tearing energy being obviously deduced from Eq. 4.

3.2 Derivation of Eshelby (1975b)

In order to highlight the connection between the configurational stress tensor (also called the elastic energy momentum tensor or the Eshelby stress tensor) and the path-independent integral for energy release rate, Eshelby (1975b) proposed a new derivation of the tearing energy for the trousers sample.

In this way, following Knowles and Sternberg (1972) he calculated the following surface integral defined in the reference configuration $(\mathcal{C}_0)$ shown in Fig. 2a

$$\gamma_1 = \int_S \left(W - P_{i1}\, u_{i,1}\right) dS \tag{11}$$

where S is a surface embracing the tip of the crack, **P** is the first Piola-Kirchhoff stress tensor and **u** is the displacement vector. In fact, this scalar quantity corresponds to the first component of the configurational force

$$\boldsymbol{\gamma} = \int_S \overline{\boldsymbol{\Sigma}}\mathbf{n}\, dS \tag{12}$$

in which **n** is the unit vector normal to the surface dS and $\overline{\boldsymbol{\Sigma}}$ is the energy momentum (configurational stress) tensor proposed by Eshelby (1951, 1975a)

$$\overline{\boldsymbol{\Sigma}} = W\mathbf{I} - \nabla^t_{\mathbf{X}}\mathbf{u}\,\mathbf{P}, \tag{13}$$

where **I** is the 3×3 identity tensor and the superscript $\cdot^t$ denotes the transposition. γ_1 being path-independent, the author considers a surface S made up of parts of the specimen surface with normal $\mathbf{e}_2$ and $\mathbf{e}_3$ in regions **A**, **D** and **B**; a cross-section of the region **B** (which is not deformed); and cross-sections of both legs (in regions **A**). The two first contributions are equal to zero; then γ_1 reduces to

$$\gamma_1 = 2\left(W - P_{21}\,\frac{\partial u_2}{\partial X_1}\right) a\, h_0. \tag{14}$$

A given point with initial coordinates $(X_1, X_2 = 0)$ being first swung round then stretched, the displacement in the $\mathbf{e}_2$-direction is

$$u_2 = \pm\lambda\, X_1 - X_2, \tag{15}$$

where the sign depends on the leg: plus sign for the right-hand side leg and minus sign for the left-hand side leg (see Fig. 2). The stress is given by $P_{21} = \pm F/a\, h_0$, with the same remark for the signs. Thus, γ_1 reduces to

$$\gamma_1 = 2\,(W\, a\, h_0 - F\,\lambda)\,. \tag{16}$$

Recalling that the total energy release rate (tearing energy) is related to γ_1 by

$$T = -\frac{1}{h_0}\gamma_1, \tag{17}$$

in which the minus sign merely indicates that the tear will run to the bottom, the result of Rivlin and Thomas given by Eqs. 4 and 10 is recovered.

4 A new derivation

Our method to determine the tearing energy of the trousers test sample is based on the general theory of Configurational Mechanics. Before presenting our approach, some basic results concerning the configurational stress tensor are recalled. Only results necessary for our derivation are given, for more details the reader can refer to Maugin (1993, 1995).

4.1 Basic definitions

The configurational stress tensor permits to quantify the evolution of the reference configuration of a given body subject to deformation. This tensor, denoted $\boldsymbol{\Sigma}$ through the rest of the paper, is defined as

$$\boldsymbol{\Sigma} = W\mathbf{I} - \mathbf{F}^t\,\mathbf{P}, \tag{18}$$

where $\mathbf{F}$ is the deformation gradient (equal to $\mathbf{I}+\nabla_{\mathbf{X}}\mathbf{u}$). Note that the expression of the configurational stress tensor is different than the one considered by Eshelby (1975b) and given in Eq. 13, the displacement gradient being replaced by the deformation gradient (see Sect. 4.3 for comments).

The derivation will be essentially based on the physical significance of the components of $\boldsymbol{\Sigma}$. Considering the material space $\mathcal{M}^3$ in which the body is defined as a set of particles (a reference configuration), the scalar $d\mathbf{U}\cdot\boldsymbol{\Sigma}\,dS_0\mathbf{N}_0$ is the change in energy due to a material displacement, i.e. a displacement in $\mathcal{M}^3$, $d\mathbf{U}$ of the material surface $dS_0\mathbf{N}_0$. So, as previously established by Kienzler and Herrmann (1997) for small strain, Σ_{ij} is the change in energy due to a unit material translation in the direction of the vector $\mathbf{e_i}$ of a unit material surface which normal vector is $\mathbf{e_j}$.

4.2 Application to the trousers test sample

As recalled above, the tearing energy is related to the change in energy between a body with a crack of length c and the same body with a crack of length $c+dc$ for a given motion. In order to determine the tearing energy for the trousers sample, let us examine the Fig. 2. In this figure, the configurations $(\mathcal{C}_0)$ (Fig. 2a) and $(\mathcal{C}'_0)$ (Fig. 2c) are two reference configurations; then, with the help of the configurational stress tensor, it is possible to calculate the change in energy between them. In this way, only the material transformation, i.e. defined in $\mathcal{M}^3$, between $(\mathcal{C}_0)$ and $(\mathcal{C}'_0)$ has to be considered, the forces applied to the sample being known (Fig. 2b).

The analysis of the problem proposed by Rivlin and Thomas (1953) and recalled in Sect. 3.1 is adopted; regions **A**, **B**, **C** and **D** of $(\mathcal{C}_0)$ are transformed into **A**′, **B**′, **C**′ and **D**′ of $(\mathcal{C}'_0)$ in the following manner:

- the two regions **A** are transformed into the two regions **A**′ by the material translation $-dc\,\mathbf{e_1}$ of the material surface $a\,h_0\,(-\mathbf{e_1})$ (grey cross-section in Fig. 2c),
- the region **B** is transformed into the region **B**′ by the material translation $-dc\,\mathbf{e_1}$ of the material surface $2a\,h_0\,\mathbf{e_1}$ (grey cross-section in Fig. 2c),
- the regions **C** and **D** are transformed into regions **C**′ et **D**′ by a 'rigid body material motion'.

Then, the change in energy between configurations $(\mathcal{C}_0)$ and $(\mathcal{C}'_0)$ is

$$dw = dw_{\mathbf{A}\to\mathbf{A}'} + dw_{\mathbf{B}\to\mathbf{B}'} + dw_{\mathbf{C}\to\mathbf{C}'} + dw_{\mathbf{D}\to\mathbf{D}'} \tag{19}$$

with

$$\begin{aligned} dw_{\mathbf{A}\to\mathbf{A}'} &= 2\left[-dc\,\mathbf{e_1}\cdot\boldsymbol{\Sigma}^{\mathbf{A}}\,a\,h_0\,(-\mathbf{e_1})\right] \\ &= 2\,dc\,a\,h_0\,\Sigma^{\mathbf{A}}_{11}, \end{aligned} \tag{20}$$

$$\begin{aligned} dw_{\mathbf{B}\to\mathbf{B}'} &= -dc\,\mathbf{e_1}\cdot\boldsymbol{\Sigma}^{\mathbf{B}}\,2a\,h_0\,\mathbf{e_1} \\ &= -2\,dc\,a\,h_0\,\Sigma^{\mathbf{B}}_{11}, \end{aligned} \tag{21}$$

and

$$dw_{\mathbf{C}\to\mathbf{C}'} = 0 \quad \text{and} \quad dw_{\mathbf{D}\to\mathbf{D}'} = 0. \tag{22}$$

In these equations $\boldsymbol{\Sigma}^{\mathbf{A}}$ and $\boldsymbol{\Sigma}^{\mathbf{B}}$ stand for the values of the configurational stress tensor in regions **A** and **B**, respectively. Equation 22 is obvious because regions **C** and **D** are statically moved in $\mathcal{M}^3$ and consequently their energies are unchanged. Moreover, the region **B** being undeformed and stress-free, $\boldsymbol{\Sigma}^{\mathbf{B}} = \mathbf{0}$, and

$$dw_{\mathbf{B}\to\mathbf{B}'} = 0. \tag{23}$$

Finally, the change in energy of the body is only due to the transformation of the regions **A** into the regions **A**′. These regions being in simple extension and assuming that the material is incompressible, the deformation gradient and the first Piola-Kirchhoff stress tensor are:

- for the right-hand side leg of the trousers

$$\mathbf{F} = \lambda\, \mathbf{e_2} \otimes \mathbf{e_1} - \lambda^{-\frac{1}{2}} \mathbf{e_1} \otimes \mathbf{e_2} + \lambda^{-\frac{1}{2}} \mathbf{e_3} \otimes \mathbf{e_3} \quad (24)$$

and

$$\mathbf{P} = \frac{F}{a\, h_0}\, \mathbf{e_2} \otimes \mathbf{e_1}, \quad (25)$$

- for the left-hand side leg of the trousers

$$\mathbf{F} = -\lambda\, \mathbf{e_2} \otimes \mathbf{e_1} + \lambda^{-\frac{1}{2}} \mathbf{e_1} \otimes \mathbf{e_2} + \lambda^{-\frac{1}{2}} \mathbf{e_3} \otimes \mathbf{e_3} \quad (26)$$

and

$$\mathbf{P} = -\frac{F}{a\, h_0}\, \mathbf{e_2} \otimes \mathbf{e_1}. \quad (27)$$

For both legs, the first component of $\mathbf{\Sigma}^{\mathbf{A}}$ reduces to

$$\Sigma_{11}^{\mathbf{A}} = W - F_{j1}\, P_{j1} = W - \lambda \frac{F}{a\, h_0}. \quad (28)$$

So, the change in energy of the regions $\mathbf{A}$ is

$$dw_{\mathbf{A}\to\mathbf{A}'} = 2dc \left(W - \lambda \frac{F}{a\, h_0} \right) a\, h_0$$
$$= (W\, 2a\, h_0 - F\, 2\lambda)\, dc. \quad (29)$$

So considering Eqs. 19, 22 and 23, it is also equal to the total change in energy between configurations $(\mathcal{C}_0)$ and $(\mathcal{C}_0')$. It can be written as

$$\frac{\partial w}{\partial c} = W\, 2a\, h_0 - F\, 2\lambda. \quad (30)$$

So the result due to Rivlin and Thomas, and Eshelby is recovered.

4.3 Remark on the definition of the configurational stress tensor

It should be noted that the derivation proposed above does not depend on the choice of the reference configuration. As an example, consider the configuration depicted in Fig. 3. During experiments, the legs of the trousers are first spread then extended. Thus this configuration is undeformed and stress-free because the motion between the configuration $(\mathcal{C}_0)$ shown in Fig. 2a and this one reduces to a rigid body motion of a part of the sample. So, it can be adopted as a reference configuration. In this case, the change in energy between regions $\mathbf{A}$ and $\mathbf{A}'$ is simply due to a material unit translation $-dc\, \mathbf{e_2}$ of the material surface $a\, h_0\, (-\mathbf{e_2})$ for the right-hand side leg and to a material unit translation

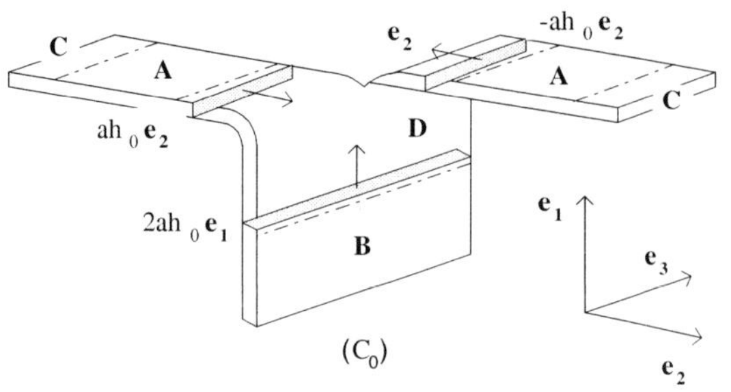

Fig. 3 A new reference configuration for the trousers sample

$dc\, \mathbf{e_2}$ of the material surface $a\, h_0\, \mathbf{e_2}$ for the right-hand side leg (see the grey cross-sections in Fig. 3)

$$dw_{\mathbf{A}\to\mathbf{A}'} = dc\, \mathbf{e_2} \cdot \mathbf{\Sigma}^{\mathbf{A}}\, a\, h_0\, \mathbf{e_2}$$
$$+ dc\, (-\mathbf{e_2}) \cdot \mathbf{\Sigma}^{\mathbf{A}}\, a\, h_0\, (-\mathbf{e_2})$$
$$= 2\, dc\, a\, h_0\, \Sigma_{22}^{\mathbf{A}}. \quad (31)$$

The changes in energy due to the other regions remain equal to zero. For both legs of the sample, the deformation gradient and the first Piola-Kirchhoff stress tensor are identical and respectively given by

$$\mathbf{F} = \lambda\, \mathbf{e_2} \otimes \mathbf{e_2} + \lambda^{-\frac{1}{2}}\, \mathbf{e_1} \otimes \mathbf{e_1} + \lambda^{-\frac{1}{2}}\, \mathbf{e_3} \otimes \mathbf{e_3} \quad (32)$$

and

$$\mathbf{P} = \frac{F}{a\, h_0}\, \mathbf{e_2} \otimes \mathbf{e_2}. \quad (33)$$

With these expressions, the result Eq. 30 is easily recovered.

As noted above, Eshelby (1975b) considered the configurational stress tensor $\overline{\mathbf{\Sigma}}$ defined in terms of the displacement gradient (Eq. 13) rather than $\mathbf{\Sigma}$ defined in terms of the deformation gradient (Eq. 18). It is to note that Knowles and Sternberg (1972) defined several J-type invariant integrals for elastostatic finite deformations and especially the integrals based on both configurational stress tensors. The former definition of the configurational stress tensor (Eq. 13) was first proposed by Eshelby (1951) for small strain but its use for large strain problems is revealed inappropriate (Maugin, 1993). In fact, this can be illustrated by the present example. Indeed, adopting the tensor $\overline{\mathbf{\Sigma}}$ and making the derivation with the reference configuration shown in Fig. 2a leads to the right expression of the tearing energy. Nevertheless, making it with the reference configuration of Fig. 3 fails.

5 Conclusion

In this paper, the derivation of the energy release rate of the simple extension tear sample was revisited in the framework of Configurational Mechanics. Obviously, such a derivation is possible because of the simplicity of the mechanical fields in the different regions of the sample: the tearing energy reduces to the value of one of the component of the tensor and does not involve a surface integral. In more general cases, the use of path-integrals, i.e. J-type integrals, is mandatory because the configurational stress tensor, i.e. the integrant, is not uniform. Nevertheless, such a study is useful to illustrate the relevance of the configurational mechanics quantities to determine changes in energy between different reference configurations of a given body. Moreover, this type of approach can be considered to design new test samples for fracture mechanics: similarly to the simple tests for the mechanical response of materials (uniaxial extension, simple shear, equibiaxial extension ...), new tear samples can be proposed by studying the configurational stress field and the possibility to calculate it analytically. Moreover, it has been shown here that the components of $\mathbf{\Sigma}$ are appropriate quantities for the study of macroscopic cracks in rubberlike materials. Motivated by this small study, the tensor can be considered as an efficient tool for the study of the evolution of microscopic defects in elastomers; such an approach was recently proposed by Verron et al., 2006, Verron and Andriyana, In Press, Andriyana and Verron, 2007) to predict fatigue crack initiation in rubber under multiaxial loading conditions.

References

Andriyana A, Verron E (2007) Prediction of fatigue life improvement in natural rubber using configurational stress. Int J Solids Struct 44:2079–2092

Eshelby JD (1951) The force on an elastic singularity. Phil Trans R Soc Lond A 244:87–112

Eshelby JD (1975a) The elastic energy-momentum tensor. J Elast 5(3-4):321–335

Eshelby JD (1975b) The calculation of energy release rates. In: Sih GC, van Elst HC, Broek D (eds) Prospects of fracture mechanics. Noordhoff, Leyden pp 69–84

Greensmith HW, Thomas AG (1956) Rupture of rubber - III - Determination of tear properties. Rubber Chem Technol 29:372–381

Griffith AA (1920) The phenomena of rupture and flow in solids. Phil Trans R Soc Lond A 221:163–198

Kienzler R, Herrmann G (1997) On the properties of the Eshelby tensor. Acta Mech 125:73–91

Knowles JK, Sternberg E (1972) On a class of conservation laws in linearized and finite elastostatics. Arch Rat Mech Anal 44:187–211

Maugin GA (1993) Material inhomogeneities in elasticity. Chapman and Hall, London

Maugin GA (1995) Material forces: concepts and applications. Appl Mech Rev 48:213–245

Rivlin RS, Thomas AG (1953) Rupture of rubber. I. Characteristic energy for tearing. J Polym Sci 10:291–318

Thomas AG (1994) The development of fracture mechanics for elastomers. Rubber Chem Technol 67:G50–G60

Verron E, Andriyana A (In Press) Definition of a new predictor for multiaxial fatigue crack nucleation in rubber. J Mech Phys Solids. doi:10.1016/j.jmps.2007.05.019

Verron E, Le Cam J-B, Gornet L (2006) A multiaxial criterion for crack nucleation in rubber. Mech Res Commun 33:493–498

Stress-driven diffusion in a deforming and evolving elastic circular tube of single component solid with vacancies

Chien H. Wu

Abstract The title problem is considered for an elastic circular tube of inner radius A and outer radius B. The tube is made of a single component solid with vacancies as its second component. The mole fraction of the massive species is denoted by $\mathbf{x}^1$, while that of the vacancies by $\mathbf{x}^0 = 1 - \mathbf{x}^1$. The tube is completely surrounded by vacuum, serving as a reservoir of vacancies. One of the standard elasticity boundary conditions is applied at time $t = 0$, when the composition is uniform. The ensuing coupled deformation and diffusion leads to the evolving of $A(t)$, $B(t)$ and $\mathbf{x}^1(R, t)$ as functions of time. Since the single component solid is not in contact with its vapor or liquid, the diffusion boundary condition is always tied to the elasticity problem through a surface condition that involves the normal configurational traction. Our chemical potential has an energy density term that serves as a source in the interior and the boundary conditions for the diffusion problem are such that the time rates of boundary accretion $\dot{A}(t)$ and $\dot{B}(t)$ must simultaneously satisfy two dissipative inequalities, one governed by the gradient of the internal chemical potential and the other by the normal configurational traction.

Keywords Energy momentum tensor · Configurational stress · Chemical potential · Diffusion

C. H. Wu (✉)
Department of Civil and Materials Engineering (MC 246), University of Illinois at Chicago, 842 West Taylor Street, Chicago IL 60607-7023, USA
e-mail: cwu@uic.edu

1 Introduction

Consider a single component solid, with vacancies as its second component, and let $\mathbf{x}^1$ be the mole fraction of the massive species. We denote by $\underline{V}\left(T, \mathbf{x}^1\right)$ the molar volume of the solid at zero pressure, temperature T, and composition $\mathbf{x}^1$. For an isotropic material, $\underline{V}\left(T, \mathbf{x}^1\right)$ may be thought of as a defining parameter for the *configuration* of a continuum material particle of composition $\mathbf{x}^1$. The change of a continuum particle's configuration from a reference $\underline{V}\left(T, \mathbf{x}^1_{\mathrm{ave}}\right)$, which is defined by a constant volume-average composition $\mathbf{x}^1_{\mathrm{ave}}$, to an arbitrary $\underline{V}\left(T, \mathbf{x}^1\right)$, via a change in composition, may be accomplished by a suitably defined *eigentransformation*, $\boldsymbol{F}^*$. If the uniform reference state is defined in a fixed reference coordinate system $\boldsymbol{X}$, then the effect of an eigentransformation, $\boldsymbol{F}^*$, is to convert a differential element $d\boldsymbol{X}$ into a new differential element $d\boldsymbol{X}^{SF} = \boldsymbol{F}^* d\boldsymbol{X}$. An eigentransformation field is, in general, incompatible and, as a result, must be accompanied by an elastic transformation, $\boldsymbol{F}^e$, so that the combined effect is an integrable deformation-gradient field, $\boldsymbol{F} = \boldsymbol{F}^e\boldsymbol{F}^*$, and $d\boldsymbol{x} = \boldsymbol{F}d\boldsymbol{X}$ is the differential element in the spatial coordinate system $\boldsymbol{x}$. In general, neither $d\boldsymbol{X}^{SF}$ nor $d\boldsymbol{X}^{SF} = \boldsymbol{F}^*\boldsymbol{F}^{-1}d\boldsymbol{x}$ is integrable but $\boldsymbol{F}^e = \boldsymbol{F}\boldsymbol{F}^{*-1}$ ties the stress-free element $d\boldsymbol{X}^{SF}$ to the spatial element $d\boldsymbol{x}$. We exploit this last observation by insisting that the Helmholtz free energy involved in $d\boldsymbol{X} \to d\boldsymbol{x}$ may be obtained in terms of the enthalpy of mixing in $d\boldsymbol{X} \to d\boldsymbol{X}^{SF}$ and the elastic

doi: 10.1007/978-1-4020-6929-1_21,

strain energy in $dX^{SF} \to dx$. The derivative of this free energy with respect to the eigentransformation is a generalized configurational stress, which, in the limit as the eigentransformation tends to the identity transformation, tends to the classical energy momentum tensor of Eshelby (1970). Moreover, the generalized configurational stress may be directly linked to the chemical potential, a result that was first reported by Wu (2001).

The eigentransformation in a single-component solid with vacancies as well as the three-way transformations among the three sets of coordinates are explicitly defined in Sect. 2 in an axially symmetric setting. The implication of a time dependent material surface is presented in Sect. 3 where the desired boundary condition is obtained. This condition requires the simultaneous satisfaction of the surface and bulk dissipation inequalities. It involves the linear combination of the chemical potential and its gradient and is, therefore, of a radiation type.

2 Elasticity and diffusion in the bulk

2.1 The eigentransformation, elastic transformation and deformation gradient

Let a single-component solid be defined by a molar concentration (mol/m^3) c^1 and a vacancy concentration c^0, so that the total molar concentration c, which is the number of available lattice sites per unit volume, and the associated mole fractions $\mathbf{x}^1 = c^1/c$ and $\mathbf{x}^0 = c^0/c$ satisfy

$$c = c^1 + c^0, \quad 1 = \mathbf{x}^1 + \mathbf{x}^0. \tag{2.1}$$

Since the two mole fractions are not independent, we use $\mathbf{x}^1$ to define the composition.

Let $\underline{V}(\mathbf{x}^1_{\text{ave}})$ be the molar volume (m^3/mol) of the solid at zero pressure, and mole fraction $\mathbf{x}^1 = \mathbf{x}^1_{\text{ave}}$. Isothermal conditions are presumed throughout this paper. We associate $\underline{V}(\mathbf{x}^1_{ave})$ with a uniform state of a solid body occupying a region $\mathcal{V}$ in a chosen reference cylindrical coordinate system $\boldsymbol{X} : (R, \Theta, Z)$. Axial symmetry is implied throughout the paper. If $\mathbf{x}^1 = \mathbf{x}^1(R, t)$ is radially nonuniform, the associated $\underline{V}(\mathbf{x}^1)$ may be used to compute the associated eigentransformation $\boldsymbol{F}^*$ defined in terms of the Jacobian of transformation $J^*(R, t) = \underline{V}(\mathbf{x}^1(R, t))/\underline{V}(\mathbf{x}^1_{\text{ave}})$. In terms of a linear elastic theory to be used in the sequel, this Jacobian is approximately

$$\begin{aligned} J^* &= \frac{\underline{V}\left(\mathbf{x}^1\right)}{\underline{V}\left(\mathbf{x}^1_{\text{ave}}\right)} = 1 + 3\varepsilon^*, \quad \varepsilon^* = \eta^1\left(\mathbf{x}^1 - \mathbf{x}^1_{\text{ave}}\right) \\ &= \eta^1\left(c^1 - c^1_{\text{ave}}\right)/c \end{aligned} \tag{2.2}$$

where

$$3\eta^1 = \frac{\underline{V}'\left(\mathbf{x}^1_{\text{ave}}\right)}{\underline{V}\left(\mathbf{x}^1_{\text{ave}}\right)} = \frac{\bar{V}^1\left(\mathbf{x}^1_{\text{ave}}\right) - \bar{V}^0\left(\mathbf{x}^1_{\text{ave}}\right)}{\underline{V}\left(\mathbf{x}^1_{\text{ave}}\right)},$$

$$\underline{V}'\left(\mathbf{x}^1_{\text{ave}}\right) = \frac{\partial}{\partial \mathbf{x}^1}\underline{V}\left(\mathbf{x}^1_{\text{ave}}\right). \tag{2.3}$$

Since $\bar{V}^1$ and $\bar{V}^0$ are the partial molar volumes, the above identity indicates that η^1 is the linear strain in the lattice due to replacing a vacancy with an atom. The principal eigentranasformation in cylindrical coordinates now becomes

$$\begin{aligned} \boldsymbol{F}^* &= \begin{bmatrix} \Lambda^*_r & 0 & 0 \\ 0 & \Lambda^*_\theta & 0 \\ 0 & 0 & \Lambda^*_z \end{bmatrix} = \begin{bmatrix} 1+\varepsilon^*_r & 0 & 0 \\ 0 & 1+\varepsilon^*_\theta & 0 \\ 0 & 0 & 1+\varepsilon^*_z \end{bmatrix}, \\ &\Lambda^*_r = \Lambda^*_\theta = \Lambda^*_z = \Lambda^* \\ &\varepsilon^*_r = \varepsilon^*_\theta = \varepsilon^*_z = \varepsilon^* \end{aligned} \tag{2.4}$$

The solid body, which occupies a cylindrical region $\mathcal{V}$ in $\boldsymbol{X}$, is deformed into a new cylindrical region v in $\mathbf{x} : (r, \theta, z)$. The associated principal deformation gradient in cylindrical coordinates is

$$\boldsymbol{F} = \begin{bmatrix} \Lambda_r & 0 & 0 \\ 0 & \Lambda_\theta & 0 \\ 0 & 0 & \Lambda_z \end{bmatrix}$$

$$\begin{aligned} \Lambda_r &= \frac{\partial r}{\partial R} = 1 + \varepsilon_r = 1 + \partial u/\partial r \\ \text{and } \Lambda_\theta &= \frac{r}{R} = 1 + \varepsilon_\theta = 1 + u/r \\ \Lambda_z &= \frac{\partial z}{\partial Z} = 1 + \varepsilon_z = 1 + \text{constant} \end{aligned} \tag{2.5}$$

where $u = u(r)$ is the radial displacement and the condition of a generalized plane strain is enforced. As a matter of practice the spatial radial coordinate r, instead of the reference R, will be used as the independent variable in all linear descriptions of the problem. Finally, the combination of $\boldsymbol{F}$ and $\boldsymbol{F}^*$ is termed the elastic transformation $\boldsymbol{F}^e = \boldsymbol{F}\boldsymbol{F}^{*-1}$, which has the principal form

$$\boldsymbol{F}^e = \begin{bmatrix} \Lambda^e_r & 0 & 0 \\ 0 & \Lambda^e_\theta & 0 \\ 0 & 0 & \Lambda^e_z \end{bmatrix} = \begin{bmatrix} 1+e_r & 0 & 0 \\ 0 & 1+e_\theta & 0 \\ 0 & 0 & 1+e_z \end{bmatrix}$$

$$\begin{aligned} e_r &= \varepsilon_r - \varepsilon^* \\ \text{and} \quad e_\theta &= \varepsilon_\theta - \varepsilon^* \\ e_z &= \varepsilon_z - \varepsilon^* \end{aligned} \tag{2.6}$$

This completes the kinematical description for axial symmetric problems.

2.2 Elasticity

Let E and ν be, respectively, Young's modulus and Poisson's ratio, and let τ_r, τ_θ and τ_z be the principal stresses in a linear elasticity setting. Applying Hooke's law to the elastic strain of (2.6) and simplifying, we obtain for the case $\varepsilon_z = 0$

$$\tau_r = \frac{(1-\nu)E}{(1+\nu)(1-2\nu)}\left[\varepsilon_r + \frac{\nu}{1-\nu}\varepsilon_\theta - \frac{1+\nu}{1-\nu}\varepsilon^*\right], \tag{2.7}$$

$$\tau_\theta = \frac{(1-\nu)E}{(1+\nu)(1-2\nu)}\left[\varepsilon_\theta + \frac{\nu}{1-\nu}\varepsilon_r - \frac{1+\nu}{1-\nu}\varepsilon^*\right], \tag{2.8}$$

$$\tau_z = \frac{\nu E}{(1+\nu)(1-2\nu)}(\varepsilon_r + \varepsilon_\theta) - \frac{E}{1-2\nu}\varepsilon^*. \tag{2.9}$$

The stress and displacement equations of equilibrium are, respectively,

$$\frac{\partial}{\partial r}\tau_r + \frac{\tau_r - \tau_\theta}{r} = 0, \tag{2.10}$$

$$\frac{\partial}{\partial r}\left[\frac{1}{r}\frac{\partial}{\partial r}(ru)\right] - \frac{1+\nu}{1-\nu}\frac{\partial}{\partial r}\varepsilon^* = 0, \tag{2.11}$$

where $\varepsilon^*(r,t)$ is the eigenstrain defined by (2.2). The most general solution of (2.11) is

$$u(r,t) = C_1 r + \frac{C_2}{r} + \frac{1+\nu}{1-\nu}\frac{1}{r}\int_A^r \varepsilon^*(\rho,t)\rho \mathrm{d}\rho \quad (A(t) \le r \le B(t)) \tag{2.12}$$

where C_1 and C_2 are arbitrary functions of time in case the boundaries $r = A$ and $r = B$ are time dependent. Since c^1_{ave} of (2.2) is the average of the concentration c^1 over the region $A \le r \le B$,

$$\int_A^B c^1(\rho,t)\rho \mathrm{d}\rho = c^1_{\text{ave}}\int_A^B \rho \mathrm{d}\rho \text{ and hence}$$

$$\int_A^B \varepsilon^*(\rho,t)\rho \mathrm{d}\rho = 0. \tag{2.13}$$

The sum of the principal stresses will be needed. It is

$$3\tau \equiv \tau_r + \tau_\theta + \tau_z = \frac{2E}{1-2\nu}C_1 - \frac{2E}{1-\nu}\varepsilon^*, \tag{2.14}$$

which satisfies

$$\nabla^2 3\tau = \nabla^2(\tau_r + \tau_\theta + \tau_z) = -\frac{2E}{1-\nu}\nabla^2\varepsilon^* = -\frac{2E\eta^1}{(1-\nu)c}\nabla^2 c^1. \tag{2.15}$$

It is clear from (2.12) that the problem of finding an elasticity solution is completely separated from that of finding a diffusion solution in that nonuniform concentration only affects the particular solution. In fact, for the traction boundary-value problem: $\tau_r(A,t) = \tau_{rA}$ and $\tau_r(B,t) = \tau_{rB}$, the constants C_1 and C_2 are just

$$C_1 = \frac{(1+\nu)(1-2\nu)}{E}\frac{B^2\tau_{rB} - A^2\tau_{rA}}{B^2 - A^2},$$

$$C_2 = \frac{(1+\nu)}{E}\frac{A^2B^2(\tau_{rB} - \tau_{rA})}{B^2 - A^2}, \tag{2.16}$$

so that the boundary conditions are not affected by the inclusion of the self-stress generated by a nonuniform concentration. However, one of our main interests in this paper is to examine the effect of applied mechanical tractions on diffusion. Before proceeding, we list the formulas for computing the elastic strain-energy density function. Let τ_{ij} and e_{ij} be, respectively, the linear stress and strain tensors. In terms of the decompositions

$$\tau_{ij} = \tau'_{ij} + \tau\delta_{ij}\left(\tau'_{ii} = 0, \tau_{ii} = 3\tau\right),$$

$$e_{ij} = e'_{ij} + \frac{e}{3}\delta_{ij}\left(e'_{ii} = 0, e_{ii} = e\right), \tag{2.17}$$

the strain-energy density W may be given as

$$W = \frac{1}{2}\tau_{ij}e_{ij} = \frac{1}{2}\tau e + \frac{1}{2}\tau'_{ij}e'_{ij}, \quad \tau'_{ij} = \frac{E}{1+\nu}e'_{ij},$$

$$\tau = \kappa e, \quad \text{and} \quad \kappa = \frac{E}{3(1-2\nu)}. \tag{2.18}$$

The diffusion aspect of the problem will be considered next.

2.3 Diffusion

The desired chemical potential can be more conveniently derived in a nonlinear setting. Let ψ be the Helmholtz free energy. In terms of the molar concentrations c^1 and c^0 and the principal stretch ratios of (2.5), we have

$$\dot{\psi} = P_r\dot{\Lambda}_r + P_\theta\dot{\Lambda}_\theta + P_z\dot{\Lambda}_z + \mu^1\dot{c}^1 + \mu^0\dot{c}^0 \tag{2.19}$$

where P_r, P_θ and P_z are the principal Piola stresses, which are related to the Cauchy stresses σ_r, σ_θ and σ_z by

$$P_r = \Lambda_z\Lambda_\theta\sigma_r, \quad P_\theta = \Lambda_z\Lambda_r\sigma_\theta, \quad P_z = \Lambda_r\Lambda_\theta\sigma_z. \tag{2.20}$$

We note that the symbol τ has been reserved for linear elasticity stresses to avoid confusion. The Piola counterpart of (2.10) is

$$\frac{\partial}{\partial R} P_r + \frac{P_r - P_\theta}{R} = 0, \tag{2.21}$$

where the use of the reference coordinate R as the independent variable is introduced for clarity. Since the molar concentrations are related by (2.1), (2.19) may be written

$$\dot{\psi} = P_r \dot{\Lambda}_r + P_\theta \dot{\Lambda}_\theta + P_z \dot{\Lambda}_z + \mu^{10} \dot{c}^1, \mu^{10} = \mu^1 - \mu^0, \tag{2.22}$$

where μ^{10} is a relative chemical potential between μ^1 and μ^0 in that the process merely indicates the replacing of a vacancy with an atom. Let g be the Gibbs energy and ω the grand potential. Then, for isothermal conditions,

$$\begin{aligned} g &= \psi - P_r \Lambda_r - P_\theta \Lambda_\theta - P_z \Lambda_z, \\ \dot{g} &= -\Lambda_r \dot{P}_r - \Lambda_\theta \dot{P}_\theta - \Lambda_z \dot{P}_z + \mu^{10} \dot{c}^1, \end{aligned} \tag{2.23}$$

$$\omega = \psi - \mu^{10} c^1, \dot{\omega} = P_r \dot{\Lambda}_r + P_\theta \dot{\Lambda}_\theta + P_z \dot{\Lambda}_z - c^1 \dot{\mu}^{10}. \tag{2.24}$$

The desired free energy is constructed as follows:

$$\begin{aligned} \psi(\Lambda_r, \Lambda_\theta, \Lambda_z, c^1, c^0) &= G(c^1, c^0) \\ &\quad + \hat{W}_\varepsilon(\Lambda_r, \Lambda_\theta, \Lambda_z, \Lambda_r^*(\mathbf{x}^1), \\ &\quad \Lambda_\theta^*(\mathbf{x}^1), \Lambda_z^*(\mathbf{x}^1)) \end{aligned} \tag{2.25}$$

in which

$$\begin{aligned} G &= c\underline{G}\left(\mathbf{x}^1\right), \underline{G}\left(\mathbf{x}^1\right) \\ &= \mathcal{R}\mathcal{T}\left[\mathbf{x}^1 \ln \mathbf{x}^1 + \left(1 - \mathbf{x}^1\right) \ln\left(1 - \mathbf{x}^1\right)\right] \\ &\quad + \Omega_{\text{int}} x^1 \left(1 - \mathbf{x}^1\right), \end{aligned} \tag{2.26}$$

$$\begin{aligned} &\hat{W}_\varepsilon\left(\Lambda_r, \Lambda_\theta, \Lambda_z, \Lambda_r^*\left(\mathbf{x}^1\right), \Lambda_\theta^*\left(\mathbf{x}^1\right), \Lambda_z^*\left(\mathbf{x}^1\right)\right) \\ &\quad = J^* \hat{W}_\varepsilon^{SF}\left(\Lambda_r^e, \Lambda_\theta^e, \Lambda_z^e\right), \end{aligned} \tag{2.27}$$

$$\Lambda_r^e = \frac{\Lambda_r}{\Lambda_r^*}, \quad \Lambda_\theta^e = \frac{\Lambda_\theta}{\Lambda_\theta^*}, \quad \Lambda_z^e = \frac{\Lambda_z}{\Lambda_z^*}. \tag{2.28}$$

In (2.26), Ω_{int} is the regular solution interaction energy, $\mathcal{R}$ the gas constant and $\mathcal{T}$ the absolute temperature. It is emphasized that the coupling between concentration and elastic deformation is completely captured in $\hat{W}_\varepsilon$. The principal Piola stresses are now giving by

$$\begin{aligned} P_r &= \frac{\partial \psi}{\partial \Lambda_r} = \frac{\partial \hat{W}_\varepsilon}{\partial \Lambda_r} = \frac{J^*}{\Lambda_r^*} \frac{\partial \hat{W}_\varepsilon^{SF}}{\partial \Lambda_r^e} = \frac{J^*}{\Lambda_r^*} P_r^e, \\ P_r^e &= \frac{\partial \hat{W}_\varepsilon^{SF}}{\partial \Lambda_r^e}, \end{aligned} \tag{2.29}$$

$$\begin{aligned} P_\theta &= \frac{\partial \psi}{\partial \Lambda_\theta} = \frac{\partial \hat{W}_\varepsilon}{\partial \Lambda_\theta} = \frac{J^*}{\Lambda_\theta^*} \frac{\partial \hat{W}_\varepsilon^{SF}}{\partial \Lambda_\theta^e} = \frac{J^*}{\Lambda_\theta^*} P_\theta^e, \\ P_\theta^e &= \frac{\partial \hat{W}_\varepsilon^{SF}}{\partial \Lambda_\theta^e}, \end{aligned} \tag{2.30}$$

$$\begin{aligned} P_z &= \frac{\partial \psi}{\partial \Lambda_z} = \frac{\partial \hat{W}_\varepsilon}{\partial \Lambda_z} = \frac{J^*}{\Lambda_z^*} \frac{\partial \hat{W}_\varepsilon^{SF}}{\partial \Lambda_z^e} = \frac{J^*}{\Lambda_z^*} P_z^e, \\ P_z^e &= \frac{\partial \hat{W}_\varepsilon^{SF}}{\partial \Lambda_z^e}. \end{aligned} \tag{2.31}$$

Differentiating $\hat{W}_\varepsilon$ with respect to Λ_r^*, Λ_θ^* and Λ_z^*, we get the principal components of the energy-momentum tensor Σ_r, Σ_θ and Σ_z

$$\begin{aligned} P_r^* &= \frac{\partial \hat{W}_\varepsilon}{\partial \Lambda_r^*} = \frac{J^*}{\Lambda_r^*}\left(W^{SF} - P_r^e \Lambda_r^e\right) = \frac{\Sigma_r}{\Lambda_r^*}, \\ \Sigma_r &\equiv \Lambda_r^* P_r^* = W - P_r \Lambda_r, \end{aligned} \tag{2.32}$$

$$\begin{aligned} P_\theta^* &= \frac{\partial \hat{W}_\varepsilon}{\partial \Lambda_\theta^*} = \frac{J^*}{\Lambda_\theta^*}\left(W^{SF} - P_\theta^e \Lambda_\theta^e\right) = \frac{\Sigma_\theta}{\Lambda_\theta^*}, \\ \Sigma_\theta &\equiv \Lambda_\theta^* P_\theta^* = W - P_\theta \Lambda_\theta, \end{aligned} \tag{2.33}$$

$$\begin{aligned} P_z^* &= \frac{\partial \hat{W}_\varepsilon}{\partial \Lambda_z^*} = \frac{J^*}{\Lambda_z^*}\left(W^{SF} - P_z^e \Lambda_z^e\right) = \frac{\Sigma_z}{\Lambda_z^*}, \\ \Sigma_z &\equiv \Lambda_z^* P_z^* = W - P_z \Lambda_z, \end{aligned} \tag{2.34}$$

in which W and W^{SF} are, respectively, $\hat{W}_\varepsilon$ and $\hat{W}_\varepsilon^{SF}$ evaluated at R. The chemical potentials μ^1 and μ^0 may now be calculated from (2.25)–(2.27) and (2.19). They are

$$\begin{aligned} \mu^1 &= \frac{\partial \psi}{\partial c^1} = \underline{G} + \mathbf{x}^0 \underline{G}' + \frac{\mathrm{d}\hat{W}_\varepsilon}{\mathrm{d}\mathbf{x}^1} \frac{1 - \mathbf{x}^1}{c}, \\ \mu^0 &= \frac{\partial \psi}{\partial c^0} = \underline{G} - \mathbf{x}^1 \underline{G}' + \frac{\mathrm{d}\hat{W}_\varepsilon}{\mathrm{d}\mathbf{x}^1} \frac{-\mathbf{x}^1}{c} \end{aligned} \tag{2.35}$$

where

$$\begin{aligned} \frac{\mathrm{d}\hat{W}_\varepsilon}{\mathrm{d}\mathbf{x}^1} &= \frac{\partial \hat{W}_\varepsilon}{\partial \Lambda_r^*} \frac{\mathrm{d}\Lambda_r^*}{\mathrm{d}\mathbf{x}^1} + \frac{\partial \hat{W}_\varepsilon}{\partial \Lambda_\theta^*} \frac{\mathrm{d}\Lambda_\theta^*}{\mathrm{d}\mathbf{x}^1} + \frac{\partial \hat{W}_\varepsilon}{\partial \Lambda_z^*} \frac{\mathrm{d}\Lambda_z^*}{\mathrm{d}\mathbf{x}^1} \\ &= \left(P_r^* + P_\theta^* + P_z^*\right) \frac{\mathrm{d}\Lambda^*}{\mathrm{d}\mathbf{x}^1} \end{aligned} \tag{2.36}$$

and

$$\frac{d\Lambda^*}{d\mathbf{x}^1} = \frac{d\varepsilon^*}{d\mathbf{x}^1} = \eta^1 \tag{2.37}$$

by (2.2). Making use of (2.26) and (2.32)–(2.37), we finally get

$$\mu^1 = \underline{G} + \left(1 - \mathbf{x}^1\right) \underline{G}' + 3P^*\eta^1 \frac{1 - \mathbf{x}^1}{c}, \tag{2.38}$$

$$\mu^0 = \underline{G} - \mathbf{x}^1 \underline{G}' - 3P^*\eta^1 \frac{\mathbf{x}^1}{c}, \tag{2.39}$$

$$\mu^{10} = \underline{G}' + 3P^* \frac{\eta^1}{c}. \tag{2.40}$$

where, after applying (2.4), (2.32)–(2.34) and (2.20),

$$P^* \equiv \frac{1}{3}\left(P_r^* + P_\theta^* + P_z^*\right) = \frac{1}{\Lambda^*}\left(W - J\sigma\right). \tag{2.41}$$

The above is an exact expression in a nonlinear elasticity formulation. In the linear elasticity setting of (2.7)–(2.18), this expression becomes, for plane strain,

$$P^* = \frac{1}{\Lambda^*}\left[W - (1 + \varepsilon_r + \varepsilon_\theta)\tau\right] \simeq W - \varepsilon\tau - \tau,$$
$$\varepsilon = \varepsilon_r + \varepsilon_\theta. \tag{2.42}$$

The grand potential ω of (2.24) is a functional of the principal stretches and the relative chemical potential μ^{10}, i.e., $\omega = \hat{\omega}(\Lambda_r, \Lambda_\theta, \Lambda_z, \mu^{10})$. The principal configurational stresses C_r, C_θ and C_z are now defined in terms of ω, following the definition of the energy momentum tensor (2.32)–(2.34), viz.

$$C_r = \omega - P_r\Lambda_r = \left(\psi - \mu^{10}\mathrm{c}^1\right) - P_r\Lambda_r,\ P_r = \frac{\partial\hat{\omega}}{\partial\Lambda_r}, \tag{2.43}$$

$$C_\theta = \omega - P_\theta\Lambda_\theta = \left(\psi - \mu^{10}\mathrm{c}^1\right) - P_\theta\Lambda_\theta,\ P_\theta = \frac{\partial\hat{\omega}}{\partial\Lambda_\theta}, \tag{2.44}$$

$$C_z = \omega - P_z\Lambda_z = \left(\psi - \mu^{10}\mathrm{c}^1\right) - P_z\Lambda_z,\ P_z = \frac{\partial\hat{\omega}}{\partial\Lambda_z}. \tag{2.45}$$

It follows from the identities

$$\frac{\partial\omega}{\partial R} = \frac{\partial}{\partial R}\hat{\omega}(\Lambda_r(R,t), \Lambda_\theta(R,t), \Lambda_z(R,t), \mu^{10}(R,t)),$$
$$c^1 = -\frac{\partial\hat{\omega}}{\partial\mu^{10}}, \tag{2.46}$$

and the equilibrium condition, (2.21), that

$$\frac{\partial}{\partial R}C_r + \frac{C_r - C_\theta}{R} = -c^1\frac{\partial\mu^{10}}{\partial R}, \tag{2.47}$$

which is the equilibrium condition for the configurational force system. Once again, the use of R as the independent variable is introduced for clarity. It is noted that the configurational body force on the right-hand side is proportional to the driving force for diffusion. It is related to the energy dissipation rate per unit volume and unit time, $\dot{d}_{\text{dissipation}}$, or the entropy production rate, $\dot{\eta}_{\text{production}}$, by

$$\dot{d}_{\text{dissipation}} = \mathcal{T}\dot{\eta}_{\text{production}} = -J_r^1\frac{\partial\mu^{10}}{\partial R} > 0. \tag{2.48}$$

By way of comparison, the energy-momentum tensor satisfies

$$\frac{\partial}{\partial R}\Sigma_r + \frac{\Sigma_r - \Sigma_\theta}{R} = \frac{1}{c}\frac{d\hat{W}_\varepsilon}{d\mathbf{x}^1}\frac{\partial c^1}{\partial R} = \left(\mu^{10} - \underline{G}'\right)\frac{\partial c^1}{\partial R}, \tag{2.49}$$

where the "body force" cannot be directly identified as a material driving force.

In connection with our earlier linear elastic result, the chemical potential, (2.40), in conjunction with (2.26) and (2.42), is further simplified to become

$$\mu^{10} = \mathcal{RT}\left[\ln\mathbf{x}^1 - \ln\left(1 - \mathbf{x}^1\right)\right] + \Omega_{\text{int}}\left(1 - 2\mathbf{x}^1\right)$$
$$-\frac{\eta^1}{c}3\tau + \frac{3\eta^1}{c}\left(W - \varepsilon\tau\right) \tag{2.50}$$

or, after the substitution of the general elasticity solution (2.14),

$$\mu^{10} = \mathcal{RT}\left[\ln\mathbf{x}^1 - \ln\left(1 - \mathbf{x}^1\right)\right] + \Omega_{\text{int}}\left(1 - 2\mathbf{x}^1\right)$$
$$-\frac{2E}{1-2\nu}\frac{\eta^1}{c}C_1 + \frac{2E}{1-\nu}\frac{(\eta^1)^2}{c}\left(\mathbf{x}^1 - \mathbf{x}^1_{\text{ave}}\right) \tag{2.51}$$
$$+\frac{3\eta^1}{c}\left(W - \varepsilon\tau\right)$$

where the constant C_1 depends on the boundary conditions of the elasticity problem, c.f. (2.16). It is also noted that the last term of (2.51), which involves the energy density, has rarely been included as a part of the full chemical potential. Applying (2.14) and (2.2), we obtain from the above

$$\frac{\partial}{\partial R}\mu^{10} = \frac{1}{c}\left[\frac{\mathcal{RT}}{\mathbf{x}^1\left(1 - \mathbf{x}^1\right)} - 2\Omega_{\text{int}} + \frac{2E\left(\eta^1\right)^2}{(1-\nu)c}\right]\frac{\partial c^1}{\partial R}$$
$$+\frac{3\eta^1}{c}\frac{\partial}{\partial R}\left(W - \varepsilon\tau\right). \tag{2.52}$$

The radial flux of the massive species $J_r^1\,(R,t)$ is

$$J_r^1 = c^1 v_r^1 = -c^1 \mathcal{M}\frac{\partial \mu^{10}}{\partial R} = -\frac{\mathcal{D}}{\mathcal{RT}}c^1\frac{\partial \mu^{10}}{\partial R} \tag{2.53}$$

where v_r^1 is the radial drift velocity of the massive component, $\mathcal{M}$ the mobility and $\mathcal{D}$ the diffusivity. Combining (2.52) and (2.53), we get

$$J_r^1 = -\mathcal{D}_{\text{eff}}\frac{\partial c^1}{\partial R} - \frac{\mathcal{D}3\eta^1}{\mathcal{RT}}\mathbf{x}^1\frac{\partial}{\partial R}\left(W - \varepsilon\tau\right), \tag{2.54}$$

$$\mathcal{D}_{\text{eff}} = \mathcal{D}\left\{\mathbf{x}^1\left[\frac{1}{\mathbf{x}^1\left(1-\mathbf{x}^1\right)} - \frac{2\Omega_{\text{int}}}{\mathcal{RT}} + \frac{2E\left(\eta^1\right)^2}{(1-\nu)\,\mathcal{RT}c}\right]\right\}, \tag{2.55}$$

where $\mathcal{D}_{\text{eff}}$ is the effective diffusion coefficient, a function of R. The effect of self-stress on $\mathcal{D}_{\text{eff}}$ in an axially symmetric setting has been studied by Wang et al. (2002). Once again, the appearance of the energy-density term, which acts like a source, appears to be new. The diffusion equation governing the process is

$$\frac{\partial c^1}{\partial t} = -\frac{1}{R}\frac{\partial}{\partial R}\left(RJ_r^1\right) = \frac{\mathcal{D}}{\mathcal{RT}}\frac{1}{R}\frac{\partial}{\partial R}\left(Rc^1\frac{\partial \mu^{10}}{\partial R}\right). \tag{2.56}$$

Combining the above with (2.54), we obtain

$$\frac{\partial c^1}{\partial t} = \frac{1}{R}\frac{\partial}{\partial R}\left(\mathcal{D}_{\text{eff}}R\frac{\partial c^1}{\partial R}\right) + \frac{\mathcal{D}3\eta^1}{\mathcal{RT}}\frac{1}{R}\frac{\partial}{\partial R}\left(\mathbf{x}^1 R\frac{\partial}{\partial R}\left(W-\varepsilon\tau\right)\right), \tag{2.57}$$

which is the governing equation for the concentration. Without the inclusion of the energy-density term, this equation is completely uncoupled from the elastic field even though the presence of Young's modulus in $\mathcal{D}_{\text{eff}}$ remains. If, in addition, the solid is also in equilibrium with its vapor/liquid then the boundary condition for diffusion becomes unrelated to the underlying mechanical equilibrium problem. We have thus ended up with a chemical version of the thermal stress problem where temperature is unaffected by mechanical stresses. Most of the published results fall into this category (Larche and Cahn 1978a,b, 1982). Thus, to examine the effect of applied mechanical tractions on diffusion when the elastic body is not in contact and equilibrium with its own vapor/liquid, boundary conditions for the diffusion problem must be reexamined.

We shall concentrate from now on situations where $\mathbf{x}^0 = 1 - \mathbf{x}^1 \ll 1$. The potentials μ^1, μ^0 and μ^{10} may be approximated by

$$\mu^{10} \simeq -\mu^0 \simeq -\left[\mathcal{RT}\ln \mathbf{x}^0 + \Omega_{\text{int}} - \frac{3\eta^1}{c}\tau + \frac{3\eta^1}{c}\left(W-\varepsilon\tau\right)\right], \tag{2.58}$$

$$\mu^1 \simeq \mathbf{x}^0\left[-\mathcal{RT} + \Omega_{\text{int}}\mathbf{x}^0 - \frac{3\eta^1}{c}\tau + \frac{3\eta^1}{c}\left(W-\varepsilon\tau\right)\right]\left(\left|\mu^1\right| \ll \left|\mu^0\right|\right). \tag{2.59}$$

Ignoring the products of the derivatives of μ^{10} and $\mathbf{x}^1$, we get from (2.56)

$$\frac{\partial \mathbf{x}^1}{\partial t} = \frac{\mathcal{D}}{\mathcal{RT}}\frac{1}{R}\frac{\partial}{\partial R}\left(R\frac{\partial \mu^{10}}{\partial R}\right) \quad \text{or} \quad \frac{\partial \mathbf{x}^0}{\partial t} = \frac{\mathcal{D}}{\mathcal{RT}}\frac{1}{R}\frac{\partial}{\partial R}\left(R\frac{\partial \mu^{0}}{\partial R}\right), \tag{2.60}$$

and from (2.57)

$$\frac{\partial c^1}{\partial t} = -\frac{\partial c^0}{\partial t} = \frac{\mathcal{D}_{\text{ave}}}{R}\frac{\partial}{\partial R}\left(R\frac{\partial c^1}{\partial R}\right) + \frac{\mathcal{D}3\eta^1}{\mathcal{RT}}\frac{1}{R}\frac{\partial}{\partial R}\left(R\frac{\partial}{\partial R}\left(W-\varepsilon\tau\right)\right) \tag{2.61}$$

where

$$\mathcal{D}_{\text{ave}} = \mathcal{D}\left[\frac{1}{\mathbf{x}^0_{\text{ave}}} - \frac{2\Omega_{\text{int}}}{\mathcal{RT}} + \frac{2E\left(\eta^1\right)^2}{(1-\nu)\,\mathcal{RT}c}\right]. \tag{2.62}$$

The boundary conditions for the diffusion equations will now be examined in conjunction with the evolution of a phase boundary, a dividing cylindrical surface between the solid cylinder containing vacancies and the vacuum surrounding taken to be a vacancy reservoir.

3 A time dependent isotropic surface

The inner cylindrical surface $R = A$ and outer surface $R = B$ will be considered while the end surfaces will not be treated. Both are assumed to be time dependent, i.e. $A(t)$ and $B(t)$. We shall first concentrate on the outer surface $B(t)$. For a generic function $F(R,t)$, the following notation will be used:

$$\overset{\circ}{F} = \frac{\mathrm{d}}{\mathrm{d}t} F(B(t),t) = \dot{F}(B(t),t) + F'(B(t),t)\,\dot{B}. \tag{3.1}$$

The quantity $\dot{B}$ is the radial velocity of the bounding surface $R = B$.

3.1 Balance of atoms

Recalling that $c^1(R,t)$ is the molar concentration of the massive species in the bulk, we denote by $\tilde{c}^1(B,t)$ the corresponding surface molar concentration on the bounding surface $B(t)$. Then, for an arbitrary region $\mathcal{B}_\varepsilon : B^-(\varepsilon) \le R \le B^+(\varepsilon)$ with $B^-(\varepsilon) \le B(t) \le B^+(\varepsilon)$ and $B^{\pm}_{\varepsilon=0} \to B(t)$, the integral form of the balance law is

$$\frac{\mathrm{d}}{\mathrm{d}t}\left\{\int_{\mathcal{B}_\varepsilon} c^1(R,t)\,2\pi R\mathrm{d}R + 2\pi B(t)\,\tilde{c}^1(B(t),t)\right\} = 2\pi B^-(\varepsilon)\,J_r^1\left(B^-,t\right), \tag{3.2}$$

where $J_r^1(B^-,t)$ is an inward flux to the region $\mathcal{B}_\varepsilon$. Since $B^+(\varepsilon)$ is in the vacuum, $J_r^1(B^+,t) = 0$. Taking the time derivative and then letting $\varepsilon \to 0$, we get

$$\dot{B}c^1(B,t) + \frac{\dot{B}}{B}\tilde{c}^1(B,t) + \overset{\circ}{\tilde{c}^1}(B,t) = J_r^1(B,t) \tag{3.3}$$

The vacancy counterpart of (3.2) is

$$\frac{\mathrm{d}}{\mathrm{d}t}\left\{\int_{\mathcal{B}_\varepsilon} c^0(R,t)\,2\pi R\mathrm{d}R + 2\pi B(t)\,\tilde{c}^0(B(t),t)\right\} = 2\pi B^-(\varepsilon)\,J_r^0\left(B^-,t\right). \tag{3.4}$$

Keeping in mind that $\mathcal{B}_\varepsilon$ is filled with vacancies everywhere, we obtain the vacancy counterpart of (3.3):

$$\dot{B}c^0(B,t) - \dot{B}c + \frac{\dot{B}}{B}\tilde{c}^0(B,t) + \overset{\circ}{\tilde{c}^0}(B,t) = J_r^0(B,t). \tag{3.5}$$

Unless the thickness of the physical surface layer is comparable to the surface radius, all the $\tilde{c}$ terms may be ignored from (3.3) and (3.5). We thus have

$$\dot{B}c^1(B,t) = J_r^1(B,t) = c^1 v_r^1(B,t) \Leftrightarrow \dot{B}\left[c^0(B,t) - c\right] = J_r^0(B,t) \tag{3.6}$$

which indicates that $\dot{B}$ is just the radial drift velocity $v_r^1(B,t)$.

3.2 Superficial dissipation inequality

The governing condition for a surface can be derived from the second law in the form of a dissipation inequality (Gurtin 2000). In terms of a radial thermodynamic force $\mathcal{G}_{rB}$, it is

$$-\mathcal{G}_{rB}\cdot\dot{B} \ge 0 \text{ and } \mathcal{G}_{rB} = \left[C_r + \frac{1}{B}\tilde{C}_\theta\right]_{R=B} \tag{3.7}$$

where C_r is the radial configurational stress (2.43) and $\tilde{C}_\theta$ a surface configurational stress. In terms of the grand potential ω, $\mathcal{G}_{rB}$ becomes

$$\mathcal{G}_{rB} = \left[C_r + \frac{1}{B}\tilde{C}_\theta\right]_{R=B} = \left[(\omega - \Lambda_r P_r) + \frac{1}{B}\left(\tilde{\omega} - \tilde{\Lambda}_\theta\tilde{P}_\theta\right)\right]_{R=B}, \tag{3.8}$$

where the second term is the surface counterpart of the first. In particular, $\tilde{P}_\theta$ is a surface stress and $\tilde{\Lambda}_\theta$ a surface principal stretch. Simplifying the above, we get

$$\mathcal{G}_{rB} = \left[\left(\psi - \mu^{10}c^1 - \Lambda_r P_r\right) + \frac{1}{B}\left(\tilde{\psi} - \tilde{\mu}^1\tilde{c}^1 - \tilde{\Lambda}_\theta\tilde{P}_\theta\right)\right]_{R=B} \tag{3.9}$$

where $\tilde{\psi}$ is the surface free energy, the surface counterpart of ψ. We shall from now on adopt the assumptions:

$$\frac{\tilde{c}^1}{B} \ll c^1, \quad \frac{\tilde{\sigma}_\theta}{B} \ll \sigma_r, \quad \tilde{\psi} = \Gamma \text{ (constant)} \tag{3.10}$$

so that

$$\mathcal{G}_{rB} = \left[C_r + \frac{\Gamma}{B}\right]_{R=B} = \left[(\omega - \Lambda_r P_r) + \frac{\Gamma}{B}\right]_{R=B} = \left[\left(\psi - \mu^{10}c^1 - \Lambda_r P_r\right) + \frac{\Gamma}{B}\right]_{R=B} \tag{3.11}$$

which, after applying (2.25)–(2.42), may be written

$$\mathcal{G}_{rB} = \left[c\mu^0 - J\sigma_r + W + \frac{\Gamma}{B}\right]_B = c\mu_B^0 - (1+\varepsilon_B)\,\tau_{rB} + W_B + \frac{\Gamma}{B}. \tag{3.12}$$

where a subscript B indicates evaluation at $R = B$.

It is now necessary to find a way to satisfy the dissipation inequality (3.7). In view of (3.6), $\dot{B}$ is just the radial drift velocity $v^1_{rB} = v^1_r(B, t)$ which satisfies the interior dissipation inequality (2.48) via (2.53). Thus, to satisfy both (3.7) and (2.48) in a consistent manner, we simply enforce the identities at $R = B$:

$$-\frac{\mathcal{D}}{\mathcal{R}T}\frac{\partial\mu^{10}}{\partial R}\bigg|_B = v^1_{rB} = \dot{B} = -\frac{\mathcal{D}}{\mathcal{R}T}\frac{1}{c\delta}\mathcal{G}_{rB} \qquad (3.13)$$

where δ is the thickness of the surface, which can at most be of a few atomic layer thick. For $\mathbf{x}^0 = 1 - \mathbf{x}^1 \ll 1$ and using (2.58) and (3.12), we deduce from the above

$$\delta c\,\frac{\partial\mu^{10}}{\partial R}\bigg|_B = \mathcal{G}_{rB} \quad \text{or} \quad -\delta c\,\frac{\partial\mu^{0}}{\partial R}\bigg|_B = \mathcal{G}_{rB}. \qquad (3.14)$$

While there is nothing new with the proposition behind (3.13), the actual form of (3.14) appears to be novel in the study of nonequilibrium phase-boundary motion. Since $\mathcal{G}_{rB}$ is a function of μ^0_B, the boundary condition is of the radiation type. The appearance of δ in the above indicates that the left-hand side is on the order of δ/B relative to the right-hand side. Thus, the equilibrium condition $\mathcal{G}_{rB} = 0$ is merely a first order approximation of the actual nonequilibrium requirement. The explicit form of this condition, written now in terms of the independent variable r for a linear-theory description, is

$$\mu^0_B + \delta\,\frac{\partial\mu^0}{\partial r}\bigg|_B = \phi_B \equiv \frac{1}{c}\left[(1+\varepsilon_B)\tau_{rB} - W_B - \frac{\Gamma}{B}\right],$$
$$\varepsilon_B = \varepsilon_{rB} + \varepsilon_{\theta B}. \qquad (3.15)$$

For the boundary at $r = A (A < B)$, the needed thermodynamic force is denoted by $\mathcal{G}_{rA}$ and the A-counterparts of (3.7) and the above are

$$-\mathcal{G}_{rA}\cdot\dot{A} \ge 0 \quad \text{and} \quad \mathcal{G}_{rA} = \left[-C_r + \frac{1}{A}\tilde{C}_\theta\right]_{R=A}, \qquad (3.16)$$

$$\mu^0_A - \delta\,\frac{\partial\mu^0}{\partial r}\bigg|_A = \phi_A \equiv \frac{1}{c}\left[(1+\varepsilon_A)\,\tau_{rA} - W_A + \frac{\Gamma}{A}\right],$$
$$\varepsilon_A = \varepsilon_{rA} + \varepsilon_{\theta A}. \qquad (3.17)$$

This completes the boundary conditions to be satisfied by the underlying diffusion problem.

4 Concluding remarks

Diffusion affected shape changes are usually small, for most practical problems, and the involved processes are slow. It is perhaps for these reasons that this class of problems is not widely studied. One exception is impression creep, which has been persistently pursued after by Li since 1977 (Li 2002; Chu and Li 1977). Even there the effect of shape change on the elastic response is deemed to be negligible. If such changes are to be rigorously followed the needed calculations can only be incremental, as the underlying elasticity solution has to be continuously updated. Moreover, to the best of our knowledge, none of the known results have made use of the configurational equilibrium equations to follow the process of energy dissipation. It just seems that the axial symmetric equations do provide us with a simpler setting to explore and/or understand some of the open questions further.

Acknowledgements The research support of the National Science Foundation, CMS-0010077, is gratefully acknowledged.

References

Chu SNG, Li JCM (1977) Impression creep; a new creep test. J Mater Sci 12:2200–2208

Eshelby JD (1970) Energy relations and the energy-momentum tensor in continuum mechanics. In: Kanninen MF, Adler WF, Rosenfeld AR, Jaffee RI (eds) Inelastic behavior of solids. McGraw-Hill, NY, pp 77–114

Gurtin ME (2000) Configurational forces as basic concepts of continuum physics. Appl Math Sci 137. Springer

Larche F, Cahn JW (1978a) A nonlinear theory of thermochemical equilibrium of solids under stress. Acta metall 26:53–60

Larche F, Cahn JW (1978b) Thermochemical equilibrium of multiphase solids under stress. Acta metall 26:1579–1589

Larche F, Cahn JW (1982) The effect of self-stress on diffusion in solids. Acta metall 30:1835–1845

Li JCM (2002) Impression creep and other localized tests. Mater Sci Eng A 322:23–42

Wang WL, Lee S, Chen JR (2002) Effect of chemical stress on diffusion in a hollow cylinder. J Appl Phys 91:9584–9590

Wu CH (2001) The role of Eshelby stress in composition-generated and stress-assisted diffusion. J Mech Phys Solids 49:1771–1794

Mode II intersonic crack propagation in poroelastic media

Enrico Radi · Benjamin Loret

Abstract A crack is steadily running in an elastic isotropic fluid-saturated porous solid at an intersonic constant speed c. The crack tip speeds of interest are bounded below by the slower between the slow longitudinal wave-speed and the shear wave-speed, and above by the fast longitudinal wave-speed. Biot's theory of poroelasticity with inertia forces governs the motion of the mixture. The poroelastic moduli depend on the porosity, and the complete range of porosities $n \in [0, 1]$ is investigated. Solids are obtained as the limit case $n = 0$, and the continuity of the energy release rate as the porosity vanishes is addressed. Three characteristic regions in the plane (n, c) are delineated, depending on the relative order of the body wave-speeds. Mode II loading conditions are considered, with a permeable crack surface. Cracks with and without process zones are envisaged. In each region, the analytical solution to a Riemann–Hilbert problem provides the stress, pore pressure and velocity fields near the tip of the crack. For subsonic propagation, the asymptotic crack tip fields are known to be continuous in the body [Loret and Radi (2001) J Mech Phys Solids 49(5):995–1020]. In contrast, for intersonic crack propagation without a process zone, the asymptotic stress and pore pressure might display a discontinuity across two or four symmetric rays emanating from the moving crack tip. Under Mode II loading condition, the singularity exponent for energetically admissible tip speeds turns out to be weaker than 1/2, except at a special point and along special curves of the (n, c)-plane. The introduction of a finite length process zone is required so that 1. the energy release rate at the crack tip is strictly positive and finite; 2. the relative sliding of the crack surfaces has the same direction as the applied loading. The presence of the process zone is shown to wipe out possible first order discontinuities.

E. Radi (✉)
Dipartimento di Scienze e Metodi dell'Ingegneria, Università di Modena e Reggio Emilia, via G. Amendola 2, 42100 Reggio Emilia, Italy
e-mail: eradi@unimore.it

B. Loret
Laboratoire Sols, Solides, Structures, Institut National Polytechnique de Grenoble, BP 53X, 38041 Grenoble Cedex, France

Keywords Dynamic fracture · First order discontinuity · Poroelastic material · Asymptotic analysis · Process zone

1 Introduction

Super-Rayleigh fracture propagation, and a fortiori intersonic and supersonic fracture propagation, have long been questioned. Laboratory experiments on metals, and in situ observations on the earth crust have been reported, but they were too isolated for their validity to be accepted. For example, super-shear fracture over a long portion of the 1979 Imperial Valley fault has been advocated by Archuleta (1984) and Spudich and Cranswick (1984). Mode I supersonic crack

Defect and Material Mechanics. C. Dascalu, G.A. Maugin & C. Stolz (eds.),
doi: 10.1007/978-1-4020-6929-1_22,

propagation along weak crystallographic planes in potassium chloride has been realized by Winkler et al. (1970) and Curran et al. (1970). Still, the argument that led to doubt the reality of the phenomenon and the validity of the theoretical justifications was as follows. Even if energy considerations would not exclude super-Rayleigh fracture and intersonic fracture propagations, the phenomenon should be unstable to branching. Such fractures were thought to branch quickly, unless they were guided by a plane of weakness.

In fact, the argument is subtle, as it turns out that, for isotropic solids, the range of fracture speeds (c_R, c_E) between the Rayleigh wave-speed c_R, and the Eshelby speed $c_E = \sqrt{2}\, c_S$, with c_S shear wave-speed, is indeed unstable with respect to perturbations of the fracture speed under most loading conditions. This is why observations of repetitive and controlled intersonic fracture correspond in fact to crack tip speeds higher than $\sqrt{2}\, c_S$.

The first laboratory experiments that seem to clearly witness the existence of intersonic fracturing are quite recent, Rosakis et al. (1999). Two plates of polyester resin in contact along a plane of weakness were subjected to Mode II shear loading, a projectile fired by a gas gun impacting one of the plates. Events were recorded through a ultra high speed camera. The publication of the laboratory experiments of the CalTech group has triggered several in situ observations of the phenomenon to be reported, and helped in structuring their interpretation. Bouchon and Vallée (2003) analyzed the Kunlunshan earthquake of magnitude 8.1. They report that fracture ran over 400 km with an average speed of 3.8 km/s, which is higher than the shear wave-speed $c_S \simeq 3.5$ km/s of the rocks at the sites. The fracture speed c was sub-Rayleigh in the first hundred kilometers, that is $c \leq c_R \simeq 0.92\, c_S \simeq 3.2$ km/s, and it later became equal to 5 km/s, which is close to $\sqrt{2}\, c_S$. These in situ observations agree with the succession of events captured by the CalTech group. Similar conclusions were drawn from observations on earthquakes that took place in Turkey in 1999, Bouchon et al. (2001). In the same vein, Dunham and Archuleta (2004) view the 2002 Denali earthquake as displaying evidence of a transition from sub-Rayleigh to supershear rupture: records show two simultaneous pulses, parallel and normal to the fault (supershear rupture), followed by a normal pulse (propagating at the Rayleigh wave-speed).

Mode I intersonic or supersonic speeds require the loading to be applied on the crack faces or at the crack tip. Indeed, experimental setups where the energy required for fracture to proceed is furnished by laser-pulsed expanding plasma indicate that supersonic Mode I crack propagation can be realized, Winkler et al. (1970), Curran et al. (1970). However, totally supersonic crack propagation is out of the realm of this analysis which is concerned with intersonic propagation only.

Indeed, this work aims at extending available theoretical analyzes on intersonic crack propagation in solids and addresses the phenomenon in fluid-infiltrated porous materials. The existence of cracks dynamically propagating at high speeds in such materials is expected to be relevant in geophysical and environmental problems, including the rupture of tectonic faults, as just pointed out, as well as in engineering applications such as hydraulic fracturing, and general mining operations.

The analysis is expected to be more complex than for solids for at least two reasons, namely inherent fluid-solid couplings and the existence of three body wave-speeds. Indeed, a prominent characteristics of the mechanical response of fluid-infiltrated geomaterials like soils, rocks, and biomaterials, like articular cartilages and intervertebral discs, lies in the significant coupling of deformation with diffusion. This feature gives rise to quite distinct short- and long-term responses.

Crack propagation in fluid saturated poroelastic media has been initially analyzed under quasistatic conditions, e.g., for shear loading by Rice and Cleary (1976); Rice and Simons (1976) and Rice (1985) and later for tensile loading by Atkinson and Craster (1991). Under these circumstances, the region close to the crack tip is practically drained, even for rapid crack propagation. Inclusion of inertia of the sole solid phase does not alter the above observation as shown in dynamic thermoelasticity by Atkinson and Craster (1992).

As for fluid-infiltrated elastic-plastic media, Radi et al. (2002) performed an asymptotic analysis for the problem of quasi-static crack propagation. The behavior was shown to be asymptotically drained at the crack tip as well. Moreover, the coupling between plastic dilatancy and fluid diffusion was shown to influence strongly the distribution of pore water pressure: by increasing the flux of water towards the crack tip, this coupling contributes to dissipate the amount of supplied energy, and thus reduces the energy available to fracture the material.

The analysis of crack propagation in poroelastic materials without a process zone has been extended to a dynamic context by Loret and Radi (2001). These authors obtained a closed-form asymptotic solution for the stress, pore pressure and displacement fields near the tip of a crack, rapidly propagating in a poroelastic material at a subsonic speed, that is, at a speed smaller than any of the three elastic wave-speeds. Indeed, three elastic body waves propagate in poroelastic materials, two longitudinal waves, affecting both solid and fluid phases, and one shear wave affecting the sole solid phase. The fast longitudinal wave-speed c_1 is the largest, whereas the order of the slow longitudinal wave-speed c_2 (also called Biot's wave-speed, or second longitudinal wave-speed) and of the shear wave-speed c_3 depends on material parameters, and mainly on porosity. The fully dynamic analysis of crack propagation performed by Loret and Radi (2001) involves inertia of both solid and fluid phases. It reveals a qualitative change with respect to the quasi-static problem, leading to square-root singularity for the partial, total and effective stresses as well as for the pore pressure. Moreover, these authors show that super-Rayleigh, but subsonic, crack propagation, i.e., $c_R < c < \min\{c_1, c_2, c_3\}$, is forbidden, because it would occur with negative energy flow to the crack tip, which implies crack face contact or compressive normal tractions ahead of the crack tip. Therefore, the propagating speed of a remotely loaded Mode I crack in poroelastic materials can not exceed the Rayleigh wave-speed c_R. Thus, in that respect, poroelastic materials behave similarly to homogeneous linear elastic solids, according to Freund (1979); Washabaugh and Knauss (1994); Broberg (1989); Broberg (1996); Broberg (1999). Note, however, that the existence and nature of Rayleigh waves in poroelastic materials depend on the permeability properties of the free (fractured) surface. For non-dissipative materials, the Rayleigh wave-speed for impermeable surface has been shown to be smaller than any of the three elastic wave-speeds, i.e., $c_R < \min\{c_1, c_2, c_3\}$. For permeable surfaces, a dispersive Rayleigh wave can propagate at a speed c_R between the slow Biot's wave-speed and the shear wave-speed, and thus at intersonic speed, namely $c_2 < c_R < c_3 < c_1$ (work in progress).

Even if it does not address solids but fluid-infiltrated media, the approach adopted here capitalizes upon the analyses of intersonic crack growth in linear elastic materials performed in the last three decades, since the early investigations of Burridge (1973); Brock (1977); Burridge et al. (1979); Freund (1979). These authors examined the possibility of crack propagation at tip speed between the shear wave-speed c_S and the longitudinal wave-speed c_L under shear-dominated loading (Mode II). They found that the stress and velocity fields suffer infinite jumps across two symmetric rays emanating from the moving crack tip. The resultant stress singularity at the crack tip and along the singular rays is smaller than 1/2, Georgiadis (1986), thus yielding a vanishing energy release rate, except for the special intermediate crack tip velocity $\sqrt{2}\,c_S$.

A stress signal emitted to front by a sub-Rayleigh crack tip provides a vehicle for the creation of 'daughter cracks' propagating at speeds larger than Eshelby wave-speed. To alleviate this discontinuous transition, Andrews (1976) developed a model where the presence of a slip-weakening zone allows energy to be absorbed by the rupture process and the secondary cracks to propagate temporarily at the shear wave-speed c_S before coalescing at speed c_L. Along this line, Broberg (1989, 1995, 1999) showed that the introduction of a cohesive zone extends the energetically admissible range of speeds to the entire intersonic region c_S to c_L, removing the problem of vanishing energy release rate. The experimental investigations made by Rosakis et al. (1999, 2000) on intersonic crack growth confirmed the predictions of the above theoretical analysis and motivated further analytical and numerical studies, e.g., Broberg (1999); Gao et al. (1999); Huang and Gao (2001).

The stability with respect to perturbations of the crack tip speed of shear cracks with cohesion running at intersonic speeds was addressed by Burridge et al. (1979). If, in a speed range, the force required to maintain steady propagation is decreasing as the crack tip speed increases, propagation is unstable. This conclusion has been refined recently by Obrezanova and Willis (2007) who point out that stability with respect to perturbations of the crack tip speed depends on the details of the applied loading. For a Dugdale–Barenblatt cohesive model, they showed that stability may take place even for speeds smaller than Eshelby speed.

As for Mode I, investigations on subsonic super-Rayleigh crack propagation in linear elastic materials by Freund (1990); Washabaugh and Knauss (1994) and Geubelle and Kubair (2001) indicate that, under tensile loading conditions, energy is not absorbed by the crack tip but emanates from it. This conclusion is not acceptable on physical grounds, unless the loading is applied directly at the crack tip, Curran et al. (1970); Broberg (1989).

The fact that intersonic Mode I propagation is ruled out for both cohesive and non cohesive cracks for remotely applied loadings is sometimes advocated to explain that intersonic shear crack propagation in a plane of weakness is somehow insensitive to branching. Still, laboratory tests show that perturbations through frictionless branches of the main crack line may slow down intersonic perturbation, and tensile wing cracks may also develop, Biegel et al. (2007).

The present investigation aims at extending the subsonic analysis performed by Loret and Radi (2001). Focus is on the occurrence of crack propagation at intersonic crack tip speed c, ranging between the largest and the smallest of the three poroelastic wave-speeds. The body wave-speeds depend strongly on porosity n. Their relative order delineates, in the plane porosity n versus crack tip speed c, three regions referred to as (i), (ii), (iii). An analytical approach different from that developed by Loret and Radi (2001) is required for intersonic crack propagation. Indeed, due to the modification in the character of the governing differential equations, which from elliptic turn hyperbolic, strong discontinuity rays (Mach fronts) emanating from the crack tip and pointing to rear may appear. Since the crack speed is envisaged to be larger than one or two body wave-speeds, there may exist one or two Mach fronts on each side of the fracture plane. Mode II loading conditions only are envisaged, and Mode I is reported in Radi and Loret (2007).

Another difference with respect to Loret and Radi (2001) is the introduction of porosity-dependent moduli, according to a simple, but realistic, macroscopic rule. This approach makes it possible to investigate the whole range of porosities $n \in [0, 1]$, and to establish a contact with linear elastic solids for a vanishing porosity.

The paper is organized as follows. The crack tip is assumed to dynamically propagate under Mode II loading conditions along a rectilinear path in the isotropic poroelastic medium, at a fixed intersonic speed c, such that $\min\{c_2, c_3\} \leq c \leq c_1$. The field and constitutive equations are recorded in Sect. 2: they adopt the formalism due to Bowen (1982) to describe the classical Biot's model, Biot (1941, 1956a,b,c). The solid and fluid inertial contributions are taken into consideration but Biot's inertial coupling is neglected from the start. On the other hand, the mechanical coupling associated to the volume changes of the solid and fluid phases and the viscous coupling due momentum transfers via Darcy's law are accounted for. However, the viscous coupling is found to be higher order, and thus neglected in the analysis performed to describe the lower order terms of the various fields close to crack tip. In Sect. 3, the field equations are formulated in terms of eigensolutions and their relative amplitudes are imposed by the boundary conditions. Only permeable crack surfaces are considered. Closed form solutions are obtained, via a complex variable approach, for the three intersonic regions (i), (ii), (iii), in Sects. 5, 6 and 7 respectively. The analysis targets in the first place cracks with a process zone for which the primary unknown functions are obtained as solutions of inhomogeneous Riemann–Hilbert problems. With the formalism of complex variables in place, the solutions for cracks without a process zone are more easily derived. The analyses in the three regions are made as systematic as possible. An overview of the generic tools is presented in Sect. 4.

Contact with first order discontinuities which are typical of cracks without a process zone is established in Sect. 8. Emphasis is laid on the representation of the stress and velocity fields in the two phases in presence and in absence of first order discontinuities. While a process zone smooth discontinuities across Mach rays, the associated domain of influence, or effective length scale, is found to depend strongly on both the length of the process zone and the singularity exponent, which take values in the interval $]0, 1/2]$.

The zones where intersonic propagation is energetically admissible are sketched in the plane (n, c). The influences of the type of crack (with or without a process zone), and of the porosity, are summarized and discussed in Sect. 9.

Notation. The convention of summation over repeated mute indices applies. Applied to the symmetric second order tensor $\mathbf{A}$ with components A_{ij} in cartesian axes, the divergence operator div is defined componentwise as $(\text{div}\mathbf{A})_i = A_{ij,j}$ where $()_{,j}$ denotes the partial derivative with respect to the j-th spatial coordinate. ∇ denotes the gradient operator, namely $(\nabla a)_i = a_{,i}$ for a scalar field a, and Δ is the Laplacian operator, i.e., $\Delta a = a_{,ii}$.

$\mathbf{I}$ is the second order identity tensor, tr is the trace operator, i.e., for a second order tensor $\mathbf{A}$, $\text{tr}\mathbf{A} = A_{ii}$. The dot product between any pair of tensors $\mathbf{A}$ and $\mathbf{B}$ indicates double contraction, namely $\mathbf{A} \cdot \mathbf{B} = \text{tr}\,\mathbf{A}\,\mathbf{B}^T$. The subscripts + and − indicate the two sides of a discontinuity line, and $[\![A]\!] = A_+ - A_-$ denotes the jump of an arbitrary scalar or tensor A.

The operators Re and Im define the real and imaginary parts of a complex number.

2 Field and constitutive equations

Let $\mathbf{u}^k$, $\mathbf{v}^k$ and $\mathbf{a}^k$ denote, respectively, the vectors of displacement, velocity and acceleration in phase $k = s, w$. With body forces and inertial coupling neglected, the balance of momentum of phase k expresses in terms of the apparent stress $\boldsymbol{\sigma}^k$ and of the momentum transfer $\hat{\mathbf{p}}^k$,

$$\mathrm{div}\,\boldsymbol{\sigma}^k + \hat{\mathbf{p}}^k = \rho^k\,\mathbf{a}^k, \quad k = s, w. \tag{2.1}$$

Let n be the porosity, that is the current volume fraction occupied by the fluid. The apparent stress in the fluid phase is expressed in terms of the intrinsic pore pressure p as $\boldsymbol{\sigma}^w = -np\,\mathbf{I}$. The apparent mass densities of the two phases ρ^s and ρ^w are linked to the intrinsic mass densities ρ_s and ρ_w of the solid constituent and fluid by the relations $\rho^s = (1-n)\rho_s$, $\rho^w = n\rho_w$.

The total stress of the mixture $\boldsymbol{\sigma}$ is defined as the sum of the apparent stresses in the phases, namely $\boldsymbol{\sigma} = \boldsymbol{\sigma}^s - n\,p\,\mathbf{I}$. The latter depend on the strains in the solid and fluid phases, namely $\boldsymbol{\epsilon}^s = \mathrm{sym}\,\nabla\mathbf{u}^s$ and $\boldsymbol{\epsilon}^w = \mathrm{sym}\,\nabla\mathbf{u}^w$ respectively, through the isotropic linear elastic constitutive equations, that express in terms of the shear modulus μ and of the three longitudinal moduli λ_{ss}, λ_{ww} and λ_{sw},

$$\begin{aligned} \boldsymbol{\sigma}^s &= 2\,\mu\,\boldsymbol{\epsilon}^s + (\lambda_{ss}\,\mathrm{tr}\boldsymbol{\epsilon}^s + \lambda_{sw}\,\mathrm{tr}\boldsymbol{\epsilon}^w)\,\mathbf{I}, \\ -np &= \lambda_{sw}\,\mathrm{tr}\boldsymbol{\epsilon}^s + \lambda_{ww}\,\mathrm{tr}\boldsymbol{\epsilon}^w. \end{aligned} \tag{2.2}$$

The momentum transfers are taken in a format that makes the balance of momentum of the fluid phase equivalent to Darcy's law of seepage, namely $\hat{\mathbf{p}}^w = -\hat{\mathbf{p}}^s = -\xi\,(\mathbf{v}^w - \mathbf{v}^s) + p\,\nabla n$. The coefficient $\xi = n^2\rho_w\,g/K_{\mathrm{h}}$ is proportional to the inverse of the hydraulic conductivity K_{h} (unit: m/s). Here g is the intensity of the gravity.

The porosity n is assumed to maintain a constant and uniform distribution over space. This assumption may be reasonable out of the very near crack tip zone, where the effects of cavitation, micro-inhomogeneities, nonlinear mechanical behavior and finite deformation dominate. Then, the equations of balance of momentum (2.1) express in terms of displacements via the constitutive equations (2.2), velocities and accelerations:

$$\begin{cases} \mu\Delta\mathbf{u}^s + (\mu+\lambda_{ss})\nabla\mathrm{div}\mathbf{u}^s + \lambda_{sw}\nabla\mathrm{div}\mathbf{u}^w \\ \quad +\xi\,(\mathbf{v}^w - \mathbf{v}^s) = \rho^s\,\mathbf{a}^s, \\ \lambda_{sw}\nabla\mathrm{div}\mathbf{u}^s + \lambda_{ww}\nabla\mathrm{div}\mathbf{u}^w \\ \quad -\xi\,(\mathbf{v}^w - \mathbf{v}^s) = \rho^w\,\mathbf{a}^w. \end{cases} \tag{2.3}$$

Let us assume that the displacement in the solid skeleton $\mathbf{u}^s$ vanishes as r^a as r tends to zero ($a > 0$). Then, velocity $\mathbf{v}^s$ and strain $\boldsymbol{\epsilon}^s$ behave as r^{a-1}, at least for steady state crack propagation so that one can replace the time derivative with a spatial derivative. Moreover, it follows that $\nabla\mathrm{div}\mathbf{u}^s$, $\Delta\mathbf{u}^s$ and $\mathbf{a}^s$ behave as r^{a-2} as r tends to zero.

Similarly, let us assume that the displacement in the fluid skeleton $\mathbf{u}^w$ behaves as r^b as r tends to zero ($b > 0$). Then, $\mathbf{v}^w$ and $\boldsymbol{\epsilon}^w$ behave as r^{b-1} and $\nabla\mathrm{div}\mathbf{u}^w$ and $\mathbf{a}^w$ behave as r^{b-2}.

Now, if $b < a$, then the leading order term in equation $(2.3)_1$ is that containing $\nabla\mathrm{div}\mathbf{u}^w$ which can not vanish by itself and thus $b \geq a$. Similarly, if $a < b$, then the leading order term in equation $(2.3)_2$ is that containing $\nabla\mathrm{div}\mathbf{u}^s$ which can not vanish by itself and thus $a \geq b$. Therefore the sole possibility is $a = b$. It follows that the dissipative terms containing the velocities in (2.3) are negligible with respect to the other terms as r tends to zero.

2.1 Porosity-dependent elastic moduli

As shown by Bowen (1982) and Loret and Harireche (1991), the longitudinal elastic moduli can be related to the drained bulk modulus K, to Biot's coefficient κ and to Skempton's coefficient B, namely:

$$\lambda_{ww} = \frac{K\,B\,n^2}{\kappa(1-\kappa\,B)} = \frac{n^2\,K_s^2}{\left(\dfrac{1-n}{K_s} + \dfrac{n}{K_w}\right)K_s^2 - K},$$

$$\lambda_{sw} = \left(\frac{\kappa}{n} - 1\right)\lambda_{ww}, \quad \lambda_{ss} = K - \frac{2}{3}\,\mu + \frac{\lambda_{sw}^2}{\lambda_{ww}}. \tag{2.4}$$

In turn, Biot's coefficient and Skempton's coefficient, which range between 0 and 1, are defined in terms of the drained bulk modulus and of the intrinsic bulk moduli of the solid and fluid constituents, $K_s > 0$ and $K_w > 0$ respectively, as

$$\kappa = 1 - \frac{K}{K_s}, \quad \frac{1}{B} = 1 + \frac{n}{\kappa}\left(\frac{K}{K_w} - \frac{K}{K_s}\right). \tag{2.5}$$

The existence of a positive definite strain energy function restricts the range of variation of the poroelastic parameters requiring that:

$$\mu > 0, \quad \lambda_{ww} > 0, \quad K > 0. \tag{2.6}$$

Although the analysis addresses primarily fluid-saturated porous media, the underlying idea is to make contact with elastic solids. For that purpose, the bulk modulus and shear modulus of the drained material are assumed to depend on the porosity n. A wide range of data displaying the dependence in porosity of both dry moduli and dry wave-speeds of several cemented materials are reported by Knackstedt et al. (2005). These data are approximated according to the simple, but realistic rule,

$$\frac{K}{K_s} = \frac{\mu}{\mu_s} = \left\langle 1 + \frac{1}{1-q}\left(q\,\frac{n}{n_c} - (\frac{n}{n_c})^q\right)\right\rangle, \quad (2.7)$$

where $\langle\cdot\rangle$ denotes the positive part of its argument. Thus the moduli K_s and μ_s of the solid constituent are recovered as the porosity vanishes, and Poisson's ratio ν of the drained material remains equal to Poisson's ratio ν_s of the solid constituent, namely $\nu = \nu_s = (3r-2)/(6r+2)$, with $r = K_s/\mu_s = K/\mu$. Moreover, the drained moduli vanish at the upper limit $n = n_c \le 1$. The slopes of the function that relate the moduli to the porosity are enforced to vanish at $n = n_c$, whatever the value of the parameter $q \in]0,1[$. The value of this upper limit porosity may be as low as 0.5. The consequences for the body wave-speeds of this dependence in porosity will be analyzed in the next section.

2.2 The three regions (i), (ii) and (iii) in the plane (n, c)

The squares of the speeds of propagation of the two longitudinal waves and of the single shear wave in poroelastic solids are defined as,

$$\left.\begin{matrix} c_1^2 \\ c_2^2 \end{matrix}\right\} = \frac{1}{2}\left(c_s^2 + c_w^2\right) \pm \frac{1}{2}\left(\left(c_s^2 - c_w^2\right)^2 + 4\,c_{sw}^4\right)^{1/2},$$

$$c_3^2 = \frac{\mu}{\rho^s}, \quad (2.8)$$

with the notations:

$$c_s^2 = \frac{2\,\mu + \lambda_{ss}}{\rho^s}, \quad c_w^2 = \frac{\lambda_{ww}}{\rho^w}, \quad c_{sw}^4 = c_{ms}^2\,c_{mw}^2,$$

$$c_{ms}^2 = \frac{\lambda_{sw}}{\rho^s}, \quad c_{mw}^2 = \frac{\lambda_{sw}}{\rho^w}. \quad (2.9)$$

The longitudinal speeds $(2.8)_1$ satisfy the following relations:

$$c_1^2 + c_2^2 = c_s^2 + c_w^2, \quad c_1^2\,c_2^2 = c_s^2\,c_w^2 - c_{sw}^4,$$

$$c_{sw}^4 = (c_1^2 - c_k^2)(c_k^2 - c_2^2), \quad k = s, w. \quad (2.10)$$

Positive definiteness of the strain energy function (2.6) ensures the three body wave-speeds to be real and legitimates the notations $(2.9)_{1,2}$. The following inequalities, which derive from (2.6) as well, are instrumental throughout the analysis:

$$\max\{c_2, c_3\} \le c_1,$$

$$c_2 \le \min\{c_s, c_w\} \le \max\{c_s, c_w\} \le c_1. \quad (2.11)$$

Let c_0 be the so-called *frozen mixture* wave-speed,

$$\min(c_s^2 + c_{ms}^2,\; c_w^2 + c_{mw}^2) \le c_0^2$$

$$= \frac{\lambda_{ss} + 2\,\mu + \lambda_{ww} + 2\lambda_{sw}}{\rho^s + \rho^w}$$

$$\le \max(c_s^2 + c_{ms}^2,\; c_w^2 + c_{mw}^2). \quad (2.12)$$

When the relation of *dynamic compatibility* is satisfied, Biot (1956c),

$$c_s^2 + c_{ms}^2 = c_w^2 + c_{mw}^2, \quad (2.13)$$

then $c_1^2 = c_0^2 = c_s^2 + c_{ms}^2$, while $c_2^2 = c_s^2 - c_{mw}^2 = c_w^2 - c_{ms}^2$. The terminology will appear appropriate when the spectral analysis is addressed.

The two longitudinal wave-speeds c_1 and c_2 are equal when $\lambda_{sw} = 0$ and $c_s = c_w$, and then $c_1 = c_2 = c_s = c_w$. However, this very peculiar situation will not be considered. The variation of the poroelastic wave-speeds, c_1, c_2 and c_3, with the porosity n is reported in Fig. 1. The relative order of the three poroelastic wave-speeds delineates, in the plane porosity n versus crack tip speed c, three intersonic regions denoted (i), (ii) and (iii) in Fig. 1 and defined in explicit form in Table 1.

As for the dependence of the moduli in porosity, a parameter $q = 1/2$ and a limit porosity $n_c = 1$ are assumed so that the relation (2.7) simplifies to $K/K_s = \mu/\mu_s = (1-\sqrt{n})^2$. This dependence implies the modulus of elastic coupling $\lambda_{sw} \ge 0$ to be positive or zero, as can be checked by inserting the above relation for K/K_s in (2.4) and (2.5). This sign justifies the two last definitions in (2.9).

Irrespective of the values of $q \in]0,1[$ and n_c, the wave-speeds c_1 and c_3 tend, as the porosity vanishes, to their counterparts in the solid constituent,

$$c_1^2 \to c_L^2 = \frac{K_s}{\rho_s} + \frac{4}{3}\frac{\mu_s}{\rho_s}, \quad c_2^2 \to 0,$$

$$c_3^2 \to c_S^2 = \frac{\mu_s}{\rho_s}. \quad (2.14)$$

At small porosity, the slow longitudinal wave-speed is proportional to $n^{(1-q)/2}$, so that the initial slope dc_2/dn

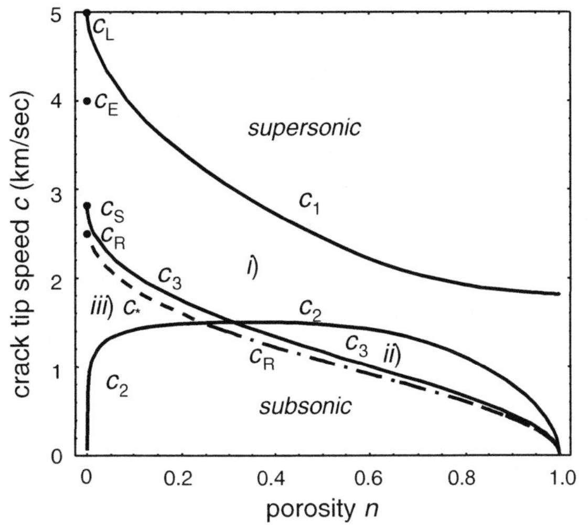

Fig. 1 Variations with porosity n of the three acceleration body wave-speeds, c_1 fast longitudinal wave-speed, c_2 slow longitudinal wave-speed and c_3 shear wave-speed, and corresponding intersonic regions (i), (ii) and (iii). The dashed curve $c_*(n)$, defined by the vanishing of the parameters Γ, Eq. (4.1), or α, Eq. (4.2), will be shown to separate the region (iii) in two zones of admissible and non admissible crack propagation. It prolongates, in the subsonic region, into a Rayleigh wave-speed c_R associated to a zero pore pressure. The parameters used for the solid constituent are: shear modulus $\mu_s = 20\,\text{GPa}$, bulk modulus $K_s = 36\,\text{GPa}$ (Poisson's ratio $\nu_s = 0.265$), mass density $\rho_s = 2500\,\text{kg/m}^3$, porosity dependence $n_c = 1$, $q = 0.5$ in Eq. (2.7); and for the fluid constituent: bulk modulus $K_w = 3.3\,\text{GPa}$, mass density $\rho_w = 1000\,\text{kg/m}^3$. For these parameters, the wave-speed c_2 is equal to the shear wave-speed c_3 for $n = n_{i-ii} \sim 0.3115$, and to the Rayleigh wave-speed c_R for $n = n_* \sim 0.245$

Table 1 Definitions of the regions

Region	Range of tip speed c
(i)	$\max(c_2, c_3) < c < c_1$
(ii)	$c_3 < c < c_2 < c_1$
(iii)	$c_2 < c < c_3 < c_1$

is quite large. On the other hand, in his description of ultrasonic wave-speeds, as opposed to acceleration wave-speeds of interest here, Berryman (1995) obtains a much smoother relation $c_2(n)$, through the high-frequency device of Biot (1956b).

For linear elastic solids, Eshelby speed $\sqrt{2}\,c_S$ is below the longitudinal wave-speed c_L if Poisson's ratio ν_s is positive: indeed $c_L/\sqrt{2}\,c_S = (1+\nu_s/(1-2\,\nu_s))^{1/2}$.

At the upper limit $n = n_c$, the situation becomes similar to that of a compressible fluid with a single longitudinal wave-speed,

$$c_1^2 \to \frac{\dfrac{1-n_c}{\rho_s} + \dfrac{n_c}{\rho_w}}{\dfrac{1-n_c}{K_s} + \dfrac{n_c}{K_w}}, \quad c_2^2 \to 0, \quad c_3^2 \to 0. \tag{2.15}$$

3 Local crack tip fields

A plane crack propagates at constant speed $c > 0$ along a rectilinear path in an infinite medium. The problem geometry is sketched in Fig. 2. Besides a fixed cartesian system $(0, X_1, X_2, X_3)$, a second cartesian coordinate system $(0, x_1, x_2, x_3)$ centered at the crack tip and moving with it in the X_1 direction is used, with the out-of-plane x_3-axis along the straight crack front. The process zone has a finite length L much smaller than the crack length. Under steady-state crack propagation, an arbitrary scalar field $\varphi = \varphi(X_1, X_2, t)$ depends on time t through the x-coordinates, and, although an abuse of notation, it will be denoted $\varphi = \varphi(x_1 = X_1 - c\,t, x_2 = X_2)$. Its material time derivative satisfies the relation,

$$\dot{\varphi} = -c\,\varphi_{,1}. \tag{3.1}$$

The analysis first addresses the field equations in terms of eigensolutions. Boundary conditions provide the relative amplitudes of the eigensolutions.

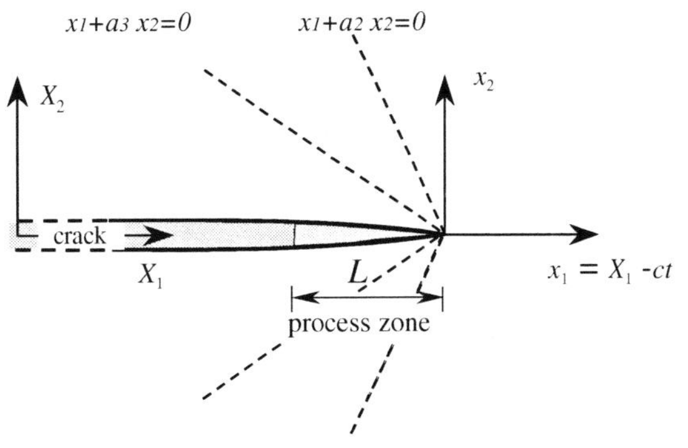

Fig. 2 Problem geometry for a crack tip propagating at constant speed c, with a process zone of length L. Depending on the location of the current point in the plane (n, c), two or four symmetric rays emanating from the crack tip (dashed lines) may, either be the loci of first order discontinuities associated to a crack without a process zone, or, strongly influence the stress and velocity fields in the case of a crack with a process zone

3.1 Green–Lamé potentials

Since the problem is plane, the displacement vectors can be expressed, via the Green–Lamé decomposition, in terms of the longitudinal and shear displacement potentials, $\varphi^s(x_1, x_2)$ and $\psi^s(x_1, x_2)$ for the solid, and $\varphi^w(x_1, x_2)$ and $\psi^w(x_1, x_2)$ for the fluid, namely

solid displacement fluid displacement

$$\begin{cases} u_1^s = \varphi^s_{,1} + \psi^s_{,2}, \\ u_2^s = \varphi^s_{,2} - \psi^s_{,1}, \end{cases} \qquad \begin{cases} u_1^w = \varphi^w_{,1} + \psi^w_{,2}, \\ u_2^w = \varphi^w_{,2} - \psi^w_{,1}. \end{cases} \tag{3.2}$$

Introduction of (3.2) into the equations of motion (2.3), with the material derivative rule (3.1), results in two sets of PDEs, consisting in a first system for the longitudinal potentials:

$$\begin{cases} (2\mu + \lambda_{ss} - \rho^s c^2)\,\varphi^s_{,11} + (2\mu + \lambda_{ss})\,\varphi^s_{,22} \\ \quad +\lambda_{sw}(\varphi^w_{,11} + \varphi^w_{,22}) + c\,\xi\,(\varphi^s_{,1} - \varphi^w_{,1}) = 0, \\ \lambda_{sw}(\varphi^s_{,11} + \varphi^s_{,22}) + (\lambda_{ww} - \rho^w c^2)\,\varphi^w_{,11} \\ \quad +\lambda_{ww}\,\varphi^w_{,22} - c\,\xi\,(\varphi^s_{,1} - \varphi^w_{,1}) = 0, \end{cases} \tag{3.3}$$

and in a second system for the shear potentials:

$$\begin{cases} (\mu - \rho^s c^2)\,\psi^s_{,11} + \mu\,\psi^s_{,22} + c\,\xi(\psi^s_{,1} - \psi^w_{,1}) = 0, \\ -\rho^w c^2 \psi^w_{,11} - c\,\xi(\psi^s_{,1} - \psi^w_{,1}) = 0. \end{cases} \tag{3.4}$$

Equations (3.3), involving the longitudinal potentials φ^s and φ^w, are coupled, first as a consequence of the mechanical coupling between the volume changes of the elastic solid skeleton and of the pore fluid and, second as a consequence of the viscous coupling due to diffusion embodied in Darcy's law. Equation (3.4), involving the shear potentials for the displacements of the solid and fluid phases, are coupled through diffusion only. Hence, the latter equations uncouple in the non dissipative case, namely for $\xi = 0$.

The stress in the solid phase and the pore pressure express in terms of displacement potentials via the constitutive relations (2.2),

$$\begin{cases} \sigma^s_{11} = (2\mu + \lambda_{ss})\,\varphi^s_{,11} + \lambda_{ss}\,\varphi^s_{,22} \\ \qquad +\lambda_{sw}\,(\varphi^w_{,11} + \varphi^w_{,22}) + 2\mu\,\psi^s_{,12}, \\ \sigma^s_{22} = \lambda_{ss}\,\varphi^s_{,11} + (2\mu + \lambda_{ss})\,\varphi^s_{,22} \\ \qquad +\lambda_{sw}\,(\varphi^w_{,11} + \varphi^w_{,22}) - 2\mu\,\psi^s_{,12}, \\ \sigma^s_{12} = \mu\,(2\,\varphi^s_{,12} + \psi^s_{,22} - \psi^s_{,11}), \\ -n\,p = \lambda_{sw}\,(\varphi^s_{,11} + \varphi^s_{,22}) \\ \qquad +\lambda_{ww}\,(\varphi^w_{,11} + \varphi^w_{,22}). \end{cases} \tag{3.5}$$

The displacement potentials introduced in (3.2) are of the form:

$$\begin{aligned} \varphi^k(x_1, x_2) &= A^k\,\mathrm{Re}\,[F(x_1 + \Omega\,x_2)], \\ \psi^k(x_1, x_2) &= B^k\,\mathrm{Im}\,[G(x_1 + \Lambda\,x_2)], \quad k = s, w, \end{aligned} \tag{3.6}$$

where A^k and B^k are real constants and F and G are analytic functions with respect to their complex arguments, over the whole plane except on the crack faces. The scalars Ω and Λ, which are imaginary for subsonic crack propagation, may be real or imaginary for intersonic crack propagation.

3.2 The eigenvalue problems

Let the displacement vectors to behave as r^a as r tends to zero, $a > 0$. In the equations of balance of momentum (2.3), the inertial and dissipative terms assume the form $\rho^k\,c\,\mathbf{v}_{k,1}$ and $\xi\,\mathbf{v}_k$, $k = s, w$, and thus they behave as $\rho^k\,c\,r^{a-2}$ and $\xi\,r^{a-1}$ close to the crack tip. Therefore the dissipative terms can be neglected in the region of radius $\rho^w\,c/\xi$ around the crack tip. The size of this region tends to vanish as c tends to zero. Then, the filtration process produces a drained region around the crack tip.

For $\xi = 0$, a substitution of $(3.6)_1$ into the field equations (3.3) yields the following eigenvalue problem for the longitudinal potentials:

$$\begin{bmatrix} (1 + \Omega^2)\,c_s^2 - c^2 & (1 + \Omega^2)\,c_{ms}^2 \\ (1 + \Omega^2)\,c_{mw}^2 & (1 + \Omega^2)\,c_w^2 - c^2 \end{bmatrix} \times \begin{bmatrix} A^s \\ A^w \end{bmatrix} = \begin{bmatrix} 0 \\ 0 \end{bmatrix}. \tag{3.7}$$

Similarly, the introduction of the shear potentials $(3.6)_2$ into the field equations (3.4) results in:

$$\begin{bmatrix} (1 + \Lambda^2)\,c_3^2 - c^2 & 0 \\ 0 & -c^2 \end{bmatrix} \begin{bmatrix} B^s \\ B^w \end{bmatrix} = \begin{bmatrix} 0 \\ 0 \end{bmatrix}. \tag{3.8}$$

For non-trivial solutions of (3.7) and (3.8) to exist, the determinants of the associated coefficient matrices must vanish. The characteristic equations provide two distinct eigenvalues Ω_1 and Ω_2 for the longitudinal potentials and a single eigenvalue $\Lambda = \Omega_3$ for the rotational potentials,

$$\Omega_j^2 = \frac{c^2}{c_j^2} - 1, \quad j \in [1, 3]. \tag{3.9}$$

Table 2 Real coefficients α_j, $j \in [1, 3]$

Region	$\alpha_1^2 =$	$\alpha_2^2 =$	$\alpha_3^2 =$
(i)	$1 - c^2/c_1^2$	$c^2/c_2^2 - 1$	$c^2/c_3^2 - 1$
(ii)	$1 - c^2/c_1^2$	$1 - c^2/c_2^2$	$c^2/c_3^2 - 1$
(iii)	$1 - c^2/c_1^2$	$c^2/c_2^2 - 1$	$1 - c^2/c_3^2$

With help of (3.9), the longitudinal eigenmodes associated to (3.7), $\mathbf{A}^J = [A_J^s, A_J^w]^T$ for $J = 1, 2$, can be shown to satisfy the relations:

$$\frac{A_J^w}{A_J^s} = \frac{c_J^2 - c_s^2}{c_{ms}^2} = \frac{c_{mw}^2}{c_J^2 - c_w^2}, \quad J = 1, 2. \tag{3.10}$$

As for the shear eigenmode, B^s is arbitrary, and

$$B^w = 0, \tag{3.11}$$

and thus the lower order term of the pore fluid displacement field is irrotational. From (2.11), the ratio A_1^w/A_1^s is deduced to be positive, while A_2^w/A_2^s is negative: the eigen-coefficients of the fluid and solid phases associated to the fast longitudinal wave are of the same sign, while they are of opposite signs for the slow wave, which correlatively is much more diffusive. An extreme situation takes place when the relation of dynamic compatibility (2.13) holds: the fast longitudinal wave is then non dissipative, namely $A_1^w/A_1^s = 1$, while the slow longitudinal wave is highly dissipative, namely $A_2^w/A_2^s = -\rho^s/\rho^w$.

Biot's inertial coupling has been neglected from the start. Still, while it is difficult to quantify, its effects may be interesting to explore. Some preliminary elements have been given in Loret and Rizzi (1998). Inertial coupling does not alter the structures neither of the secular polynomial yielding the wave-speeds. nor of the eigenvalue problem for the solid eigenmode. On the other hand, the structure of the fluid eigenmode is modified by inertial coupling as transverse motions in the solid phase carry over to the fluid phase.

The eigenvalues Ω_j, $j \in [1, 3]$, turn out to be real or purely imaginary, depending on the region (i), (ii) and (iii). It is instrumental to define the real constants

$$\alpha_j = |\Omega_j| = \sqrt{|1 - c^2/c_j^2|}, \quad j \in [1, 3], \tag{3.12}$$

listed in Table 2.

The spatial dependence of the potentials expresses through the complex variables

$$z_j = x_1 + \Omega_j x_2, \quad j \in [1, 3], \tag{3.13}$$

recorded in explicit form in Table 3, where $i \equiv \sqrt{-1}$.

Table 3 Complex variables $z_j = x_1 + \Omega_j x_2$, $j \in [1, 3]$

Region	$z_1 =$	$z_2 =$	$z_3 =$
(i)	$x_1 + i\,\alpha_1 x_2$	$x_1 + \alpha_2 x_2$	$x_1 + \alpha_3 x_2$
(ii)	$x_1 + i\,\alpha_1 x_2$	$x_1 + i\,\alpha_2 x_2$	$x_1 + \alpha_3 x_2$
(iii)	$x_1 + i\,\alpha_1 x_2$	$x_1 + \alpha_2 x_2$	$x_1 + i\,\alpha_3 x_2$

As a consequence of the spectral analysis, the longitudinal and rotational displacement potentials (3.6), being real valued functions, assume the respective form:

$$\begin{cases} \varphi^s(x_1, x_2) = \mathrm{Re}\,[F_1(z_1)] + \mathrm{Re}\,[F_2(z_2)], \\ \varphi^w(x_1, x_2) = c_{ms}^{-2}\Big((c_1^2 - c_s^2)\,\mathrm{Re}\,[F_1(z_1)] \\ \qquad -(c_s^2 - c_2^2)\,\mathrm{Re}\,[F_2(z_2)]\Big), \end{cases} \tag{3.14}$$

and

$$\begin{cases} \psi^s(x_1, x_2) = \mathrm{Im}\,[G(z_3)], \\ \psi^w(x_1, x_2) = 0. \end{cases} \tag{3.15}$$

Here, the analytic primary functions $F_1(z_1)$, $F_2(z_2)$ and $G(z_3)$ embody the real constants A_1^s, A_2^s and B^s, respectively.

4 Overview of the analysis

4.1 The steps of the analysis

The analysis aims at obtaining the primary unknown functions $F_1''(z)$, $F_2''(z)$ and $G''(z)$, of the complex variable $z = x_1 + i\,x_2$, that define the stress and displacement potentials by (3.14), (3.15). These analytical functions satisfy already the field equations. Still, they are linked by symmetry conditions along the crack line $x_2 = 0$, and loading conditions on the crack faces and process zone. The resulting constraints are stated in the next Sects. 4.4 and 4.5. Through algebraic manipulations, the various equations to be satisfied are transformed into inhomogeneous Riemann–Hilbert problems for the function $F_1''(z)$. The functions $F_2''(z)$ and $G''(z)$ are then deduced from $F_1''(z)$. Due to symmetry conditions, only the upper plane $\mathrm{Im}\,z \geq 0$ needs to be considered.

The inhomogeneity in the Riemann–Hilbert problems emanates from the cohesion along the process zone. The length L of the process zone is finite. The cohesion along the process zone is prescribed. For plotting purposes, it will be taken either constant or linearly decreasing from a maximum at the crack tip $x_1 = 0$ to 0 at the end of the process zone $x_1 = -L$. A relation between the cohesion in the process zone and the opening, or slip, offset across the two faces of the crack, is not introduced. More refined analyzes with a process length varying according to specific constitutive equations or a rate dependent cohesion along the process zone, as in Samudrala et al. (2002), are not envisaged either at this stage.

Finally, the energetic criterion defined in Sect. 4.6 is used to discriminate solutions.

The analysis of the three regions (i), (ii) and (iii) can be made systematic, even if the developments lead to split region (iii) in two sub-domains. To avoid repetition, it is instrumental to introduce first a few generic quantities.

4.2 Generic quantities

The solutions are characterized by a singularity exponent γ. This exponent is not affected by the presence of the process zone. It is defined in terms of the crack tip speed c and body wave-speeds by a region-dependent expression:

$$\tan \gamma\pi = \Gamma \equiv \begin{cases} \dfrac{4\alpha_1\alpha_3(c_s^2 - c_2^2)}{4\alpha_2\alpha_3\,(c_1^2 - c_s^2) + (\alpha_3^2 - 1)^2\,(c_1^2 - c_2^2)} > 0, & \text{region (i)},\\[2ex] \dfrac{4\,\alpha_3}{(\alpha_3^2 - 1)^2}\,\dfrac{\alpha_1\,(c_s^2 - c_2^2) + \alpha_2\,(c_1^2 - c_s^2)}{c_1^2 - c_2^2} > 0, & \text{region (ii)},\\[2ex] \dfrac{4\,\alpha_1\,\alpha_3\,(c_s^2 - c_2^2) - (1 + \alpha_3^2)^2\,(c_1^2 - c_2^2)}{4\,\alpha_2\,\alpha_3\,(c_1^2 - c_s^2)}, & \text{region (iii)}. \end{cases} \quad (4.1)$$

A solution is physically admissible if it corresponds to a positive and finite energy release rate $\mathcal{G}$. These requirements restrict the singularity exponents in the interval $]0, 1[$. In each region, a coefficient α,

$$\alpha = \begin{cases} \alpha_1 > 0, & \text{region (i)},\\[1ex] \alpha_1 + \dfrac{c_1^2 - c_s^2}{c_s^2 - c_2^2}\,\alpha_2 > 0, & \text{region (ii)},\\[2ex] \alpha_1 - \dfrac{(1 + \alpha_3^2)^2}{4\,\alpha_3}\,\dfrac{c_1^2 - c_2^2}{c_s^2 - c_2^2} & \text{region (iii)}, \end{cases} \quad (4.2)$$

will be shown to play a key role in defining the sign of $\mathcal{G}$. α and Γ have the same sign, and they are linked by the relations:

$$\frac{\alpha}{\Gamma} = \overbrace{\frac{(\alpha_3^2 - 1)^2}{4\alpha_3}\,\frac{c_1^2 - c_2^2}{c_s^2 - c_2^2} > 0}^{\text{region (ii)}}, \quad \overbrace{\alpha_2\,\frac{c_1^2 - c_s^2}{c_s^2 - c_2^2} > 0}^{\text{region (iii)}}. \quad (4.3)$$

While they are both positive in regions (i) and (ii), their sign is an issue of physical significance in region (iii).

As indicated above, the primary function $F_1''(z)$ is obtained as the solution of a Riemann–Hilbert problem. In the three regions (i), (ii) and (iii), the two other primary functions will be shown to be proportional to F_1'',

$$F_2''(z_2) = R_2\,F_1''(z_2),$$
$$G''(z_3) = \frac{i\,\alpha_3}{\Omega_3}\,R_3\,F_1''(z_3), \quad (4.4)$$

with the proportionality constants,

$$R_2 \equiv \frac{c_1^2 - c_s^2}{c_s^2 - c_2^2}, \quad R_3 = \frac{1}{2\alpha_3}\left(\frac{c^2}{c_3^2} - 2\right)\frac{c_1^2 - c_2^2}{c_s^2 - c_2^2}. \quad (4.5)$$

In regions where z_2 is real, $\mathrm{Re}\,[F_1''(z_2)]$ is a real valued function of the real variable z_2 that will be denoted $f''(z_2)$, and similarly for $g''(z_3) = \mathrm{Im}\,[G''(z_3)]$. Then, (4.4) specializes, for any real z_2 and z_3, to

$$\overbrace{f''(z_2) = R_2\,\mathrm{Re}\,[F_1''(z_2)]}^{\text{regions (i), (iii)}},\ \overbrace{g''(z_3) = R_3\,\mathrm{Re}\,[F_1''(z_3)]}^{\text{regions (i), (ii)}}. \quad (4.6)$$

Moreover, the real functions $f''(z_2)$ and $g''(z_3)$ in the appropriate regions will be shown to vanish in front of the associate rays $z_2 = 0$ and $z_3 = 0$ respectively.

4.3 Conditions on the process zone

A process zone, of finite length L, provides a transition between the sound material in front of the crack tip, and the cracked material. The existence of a process zone is motivated by, or associated with, the following properties: 1. one component of the stress turns out to be finite, while, in absence of process zone, all stress components are singular at the crack tip; 2. it makes the energy release rate more positive; 3. it introduces a length scale in the field equations, and therefore smooths out the first order discontinuities that might exist in absence of the process zone. The actual efficiency of the process zone in achieving these goals will be considered in the analysis.

For fluid-saturated porous media, the boundary conditions along the process zone may be phrased in terms of an effective stress. Two prominent effective stresses for a fluid-saturated porous medium are the intergranular, or Terzaghi's, effective stress $\boldsymbol{\sigma}'$,

$$\begin{aligned}\boldsymbol{\sigma} &= \boldsymbol{\sigma}^s - n\,p\,\mathbf{I} = \boldsymbol{\sigma}' - p\,\mathbf{I} \\ &\Rightarrow \boldsymbol{\sigma}' = \boldsymbol{\sigma} + p\,\mathbf{I} = \boldsymbol{\sigma}^s + (1-n)\,p\,\mathbf{I},\end{aligned} \tag{4.7}$$

and the elastic, or Biot's, effective stress $\boldsymbol{\sigma}'$,

$$\begin{aligned}\boldsymbol{\sigma}' &= \boldsymbol{\sigma} + \kappa\,p\,\mathbf{I} = \boldsymbol{\sigma}^s + (\kappa - n)\,p\,\mathbf{I} \\ &\Rightarrow \boldsymbol{\sigma}' = \left(K - \frac{2}{3}\mu\right)\operatorname{tr}\boldsymbol{\epsilon}^s\,\mathbf{I} + 2\,\mu\,\boldsymbol{\epsilon}^s.\end{aligned} \tag{4.8}$$

These two effective stresses coincide only when the solid constituent is incompressible, i.e., if $\kappa = 1$. For an anisotropic solid phase, the pore pressure would contribute in a non isotropic way to Biot's stress. In an isotropic context, the two effective stresses can be cast in a unified format through the scalar $\tilde{\kappa}$, with $\tilde{\kappa} = 1$ for Terzaghi's effective stress and $\tilde{\kappa} = \kappa$ for Biot's effective stress.

The process zone is endowed with a given spatial distribution of cohesive stress $\tau_0(x_1)$, $x_1 \in [-L, 0]$, and with a friction angle $\phi \geq 0$ as suggested by Rudnicki (2001). The stress components are constrained to satisfy the condition

$$\sigma_{12}(x_1, 0) + \tan\phi\,\sigma'_{22}(x_1, 0) = \tau_0(x_1), \quad \text{for } -L < x_1 < 0, \tag{4.9}$$

where

$$\begin{aligned}\sigma'_{22}(x_1, 0) &= \sigma^s_{22}(x_1, 0) + (\tilde{\kappa} - n)\,p(x_1, 0) \\ &= \sigma_{22}(x_1, 0) + \tilde{\kappa}\,p(x_1, 0).\end{aligned} \tag{4.10}$$

4.4 Mode II loading conditions

Under Mode II loading conditions, the displacements u^s_1 and u^w_1 vanish ahead of the crack tip along $x_2 = 0$ for $x_1 > 0$, as well as their derivatives with respect to x_1, namely:

$$u^k_{1,1}(x_1, 0) = 0, \quad \text{for } x_1 > 0, \quad k = s, w. \tag{4.11}$$

Moreover, the Mode II symmetry conditions require:

$$\begin{aligned}&\sigma_{22}(x_1, 0) = 0, \quad \text{for all } x_1; \\ &p(x_1, 0) = 0, \quad \text{for } x_1 > 0.\end{aligned} \tag{4.12}$$

In view of the symmetry condition $(4.12)_1$, the condition along the process zone may be recast in the following format:

$$\sigma_{12}(x_1, 0) = \sigma^s_{12}(x_1, 0) = \begin{cases} \tau_0(x_1) - \tan\phi\,\tilde{\kappa}\,p(x_1, 0) & \text{for } -L < x_1 < 0, \\ 0 & \text{for } x_1 < -L. \end{cases} \tag{4.13}$$

4.5 Flow conditions along the crack surfaces

The pore pressure vanishes on the crack surfaces *and* process zone if these are *permeable*, namely:

$$p(x_1, 0) = 0, \quad \text{for } x_1 < 0. \tag{4.14}$$

Alternatively, the crack faces may be considered to be *impermeable*: the normal flux to the crack surface vanishes. Darcy's law is equivalent to the balance of momentum of the fluid phase $(2.1)_2$, namely $n\,(\mathbf{v}^w - \mathbf{v}^s) = -K_h/(\rho_w g)\,(\nabla p + \rho_w\,\mathbf{a}^w)$. The asymptotic analysis of the field equations neglects the flux of water with respect to the other terms in that equation, that is the rhs of that relation vanishes identically in the body. The impermeability condition consists then in enforcing the lower order term to vanish, namely

$$n\,(v^w_2 - v^s_2)(x_1, 0) = 0, \quad \text{for } x_1 < 0. \tag{4.15}$$

Comments on a poroelastic model with a process zone: 1. The validity of the assumption of a permeable crack surface in an intersonic crack propagation context might be questioned. An alternative would have been to assume an impermeable crack surface, if the material behavior is expected to be governed by its short term response.

Rice and Simons (1976) consider a crack with a process zone, steadily moving, at speed well in the subsonic region: they assume permeable crack surfaces, i.e., a zero pore pressure on the crack faces and along the process zone.

In fact, Rice and Simons (1976) find that, for a steadily advancing crack tip, the behavior appears to be drained in a small region whose characteristic size is that of the diffusion length. This is because, in their analysis, the pore pressure, in contrast to the total or apparent stresses, is regular and therefore much smaller than the stress. Outside this region, the undrained

behavior is rapidly recovered. Moreover, they note that the size of the drained region is inversely proportional to the crack tip speed, so that the validity of a drained behavior along the crack surfaces becomes more and more questionable as the crack tip speed increases.
2. For impermeable crack surfaces, the symmetry condition on the crack line require the pore pressure to vanish ahead of the crack tip. However, behind the crack tip, the pore pressure may well not vanish and be discontinuous across the crack. The possible restrictions to which the pore pressure might have to be submitted along the process zone represent another issue.
3. The cohesion along the process zone and the crack tip speed are assumed to be prescribed independently. In fact, the actual speed of propagation is more probably governed by energy considerations: for example, it might be expected to be the speed that corresponds to a critical value of the energy release, which is proper to the material. Alternatively, a criterion in terms of the displacement offset at the end of the process zone $x_1 = -L$ is used in Palmer and Rice (1973).

4.6 Energy release rate

The analysis requires the expressions of the energy flux towards the crack tip $\mathcal{F}$ and energy release rate $\mathcal{G} = \mathcal{F}/c$ for a crack propagating in a poroelastic material at steady speed c along the direction 1, for which the time rate (3.1) holds.

For a crack without a process zone in subsonic propagation, the expression of the energy release rate derived for linear solids, e.g., Achenbach (1972), can be extended to fluid-saturated porous media in the format,

$$\mathcal{G} = -\pi \lim_{x_1 \to 0^+} x_1 \left(\sigma^s_{12}(x_1, 0)\, u^s_{1,1}(-x_1, 0^+) - np(x_1, 0)\, u^w_{1,1}(-x_1, 0^+)\right). \tag{4.16}$$

For a crack with a process zone, previous results provided by Broberg (1999) for single phase solids generalize similarly to

$$\mathcal{G} = -2 \int_{-L}^{0} \sigma^s_{12}(x_1, 0^+)\, u^s_{1,1}(x_1, 0^+) - n\, p(x_1, 0^+)\, u^w_{1,1}(x_1, 0^+)\, \mathrm{d}x_1. \tag{4.17}$$

Intersonic crack tip propagation may give rise to first order discontinuities, namely discontinuity in the stress or velocity fields, along singular rays emanating from the crack tip and moving at a normal speed $a = c_2$ for the longitudinal Mach line and $a = c_3$ for the shear Mach line. The energy dissipation along these rays should be estimated.

For that purpose, it is instrumental to generalize to poroelastic materials the notion of scalar driving traction $t_d = t_d(x_1, x_2)$ on the singular ray, introduced for solids by Abeyaratne and Knowles (1990), and satisfying the condition $a\, t_d \geq 0$. The generalization is taken in the format,

$$\begin{aligned} t_d = {} & \tfrac{1}{2} [\![\boldsymbol{\sigma}^s \cdot \boldsymbol{\epsilon}^s - np \operatorname{tr} \boldsymbol{\epsilon}^w]\!] \\ & -\tfrac{1}{2} (\boldsymbol{\sigma}^s_+ + \boldsymbol{\sigma}^s_-) \cdot [\![\boldsymbol{\epsilon}_s]\!] \\ & +\tfrac{1}{2} (np_+ + np_-) [\![\operatorname{tr} \boldsymbol{\epsilon}^w]\!]. \end{aligned} \tag{4.18}$$

Since the constitutive equations (2.2) linking the generalized stress $(\boldsymbol{\sigma}^s, -np\,\mathbf{I})$ to the generalized strain $(\boldsymbol{\epsilon}^s, \boldsymbol{\epsilon}^w)$ are linear, the driving traction on the singular ray vanishes, and thus there is no dissipation when the singular ray moves through the material. As a result, the energy flux toward the crack tip is not affected by the presence of a discontinuity ray, and the expressions of the energy release rate above hold for subsonic and intersonic propagations.

5 Crack propagation within the region (i)

Within the intersonic region (i), namely for $c_1 > c > \max\{c_2, c_3\}$, the variable z_1 is complex whereas the variables z_2 and z_3 are real, as indicated on Table 3. Thus $\mathrm{Re}\,[F_2(z_2)]$ and $\mathrm{Im}\,[G(z_3)]$ turn out to be two real valued functions of the real variables z_2 and z_3. They will be denoted by $f(z_2)$ and $g(z_3)$ respectively. Then the displacement potentials (3.14), (3.15) may be recast in the following form

$$\begin{cases} \varphi^s(x_1, x_2) = \mathrm{Re}\,[F_1(z_1)] + f(z_2), \\ \varphi^w(x_1, x_2) = c_{ms}^{-2} \left((c_1^2 - c_s^2)\, \mathrm{Re}\,[F_1(z_1)] - (c_s^2 - c_2^2)\, f(z_2)\right), \\ \psi^s(x_1, x_2) = g(z_3). \end{cases} \tag{5.1}$$

Due to symmetry, it is sufficient to consider the upper half-plane $x_2 > 0$, where the functions $f(z_2)$ and $g(z_3)$ denote two waves with front along the planes $x_1 + \alpha_k x_2 = 0$, for $k = 2, 3$, that propagate to the left of the crack tip, as sketched in Fig. 2. The functions $f(x_1 - \alpha_2 x_2)$ and $g(x_1 - \alpha_3 x_2)$, which correspond to two waves travelling to the right of the crack tip in the upper half-plane, have been excluded because the material particles ahead of the crack tip have not yet felt the disturbance caused by the crack growth (Broberg, 1999). The apparent stresses and pore pressure (3.5) are now expressed in terms of the displacement potentials (5.1):

$$\begin{cases} \sigma_{11}^s/\mu = (1+\alpha_3^2+2\alpha_1^2)\,\mathrm{Re}\,[F_1''(z_1)] \\ \quad +(1+\alpha_3^2-2\alpha_2^2)\,f''(z_2)+2\alpha_3\,g''(z_3), \\ \sigma_{22}^s/\mu = (\alpha_3^2-1)\,(\mathrm{Re}\,[F_1''(z_1)]+f''(z_2)) \\ \quad -2\alpha_3\,g''(z_3), \\ \sigma_{12}^s/\mu = -2\alpha_1\,\mathrm{Im}\,[F_1''(z_1)]+2\alpha_2\,f''(z_2) \\ \quad +(\alpha_3^2-1)\,g''(z_3), \\ -np/\mu = (1+\alpha_3^2)\,c_{mw}^{-2}\,\big((c_1^2-c_s^2)\,\mathrm{Re}\,[F_1''(z_1)] \\ \quad -(c_s^2-c_2^2)\,f''(z_2)\big). \end{cases} \tag{5.2}$$

Similarly, the displacements (3.2) in the solid and fluid phases become:

$$\begin{cases} u_1^s = \mathrm{Re}\,[F_1'(z_1)]+f'(z_2)+\alpha_3\,g'(z_3), \\ u_2^s = -\alpha_1\,\mathrm{Im}\,[F_1'(z_1)]+\alpha_2\,f'(z_2)-g'(z_3), \end{cases} \tag{5.3}$$

and

$$\begin{cases} u_1^w = c_{ms}^{-2}\,\big((c_1^2-c_s^2)\,\mathrm{Re}\,[F_1'(z_1)] \\ \quad -(c_s^2-c_2^2)\,f'(z_2)\big), \\ u_2^w = -c_{ms}^{-2}\big(\alpha_1\,(c_1^2-c_s^2)\,\mathrm{Im}\,[F_1'(z_1)] \\ \quad +\alpha_2\,(c_s^2-c_2^2)f'(z_2)\big). \end{cases} \tag{5.4}$$

5.1 The Riemann–Hilbert problem

The boundary conditions (4.11–4.14) along the crack line are expressed in terms of the primary functions via the apparent stresses and pore pressure (5.2) and displacements (5.3–5.4). The functions $f''(x_1)$ and $g''(x_1)$,

$$\begin{aligned} f''(x_1) &= \frac{c_1^2-c_s^2}{c_s^2-c_2^2}\,\mathrm{Re}\,[F_1''(x_1)], \\ g''(x_1) &= \frac{\alpha_3^2-1}{2\alpha_3}\,\frac{c_1^2-c_2^2}{c_s^2-c_2^2}\,\mathrm{Re}\,[F_1''(x_1)], \quad \text{for all } x_1, \end{aligned} \tag{5.5}$$

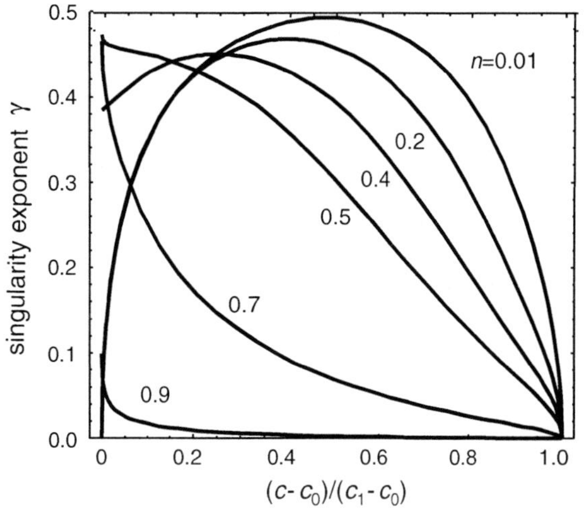

Fig. 3 Intersonic region (i). Singularity exponent γ of mode II, as function of the relative crack tip speed $(c-c_0)/(c_1-c_0)$, with $c_0=\max(c_2,c_3)$, for different porosity levels. Mode II square-root singularity, associated to $\gamma=1/2$, occurs for $c=\sqrt{2}\,c_3$ at the limit of a vanishingly small porosity, so that the special status enjoyed by the speed $\sqrt{2}\,c_S$ in linear elastic solids is retrieved

are then obtained in terms of $F_1''(x_1)$. The latter can in turn be shown to satisfy the set of conditions:

$$\begin{cases} \mathrm{Re}\,[F_1''(x_1)] - \Gamma\,\mathrm{Im}\,[F_1''(x_1)] \\ \quad = \dfrac{\tau(x_1)}{\mu}\,\dfrac{\Gamma}{2\,\alpha}, & \text{for } x_1<0, \\ \mathrm{Re}\,[F_1''(x_1)] = 0, & \text{for } x_1>0, \end{cases} \tag{5.6}$$

where $\Gamma=\tan\gamma\pi$ has been defined in (4.1), and α in (4.2). Conditions (5.6) may be interpreted as an inhomogeneous Hilbert problem for the function $F_1''(z)$, which, in the upper half-plane $\mathrm{Im}\,[z]\geq 0$, admits the following solution (see Appendix A):

$$F_1''(z) = \frac{A}{z^\gamma} + \frac{\sin\gamma\pi}{2\pi\mu}\,\frac{1}{\alpha}\,\frac{1}{i\,z^\gamma}\int_{-L}^{0}\frac{|t|^\gamma\,\tau_0(t)}{t-z}\,dt, \tag{5.7}$$

defined up to the purely imaginary constant A.

The variations of the exponent γ with the crack tip speed are shown on Fig. 3. The singularity exponent under mode II loading conditions is smaller than 1/2, for every value of c within the considered intersonic range. Moreover, the exponent γ attains a maximum at an intermediate crack tip speed, which increases as the porosity level n decreases. As $n\to 0$, the maximum approaches 1/2, so that the square-root singularity of the special case observed for Mode II crack propagation in linear elastic materials at the particular speed $c = \sqrt{2}\, c_S$ (Burridge et al. 1973; Freund 1979) is retrieved. Along the boundary between the regions (i) and (iii), where $\alpha_3 = 0$, and porosities are smaller than 0.3115, the exponent vanishes as indicated by Eq. $(4.1)_1$. For larger porosities, along the boundary between the regions (i) and (ii), where $\alpha_2 = 0$, the variation of the exponent is mainly governed by the term $\alpha_3^2 - 1$. When this term vanishes, the exponent is equal to 1/2. This event takes place for $c_2 = 0$ and $n = 1$, and $c_2 = \sqrt{2}\, c_3$ and n around 0.7 as displayed on Fig. 3.

5.2 Matching the singular and finite solutions ahead of the crack tip

Conditions (5.5) and $(5.6)_2$ imply the relations:

$$\mathrm{Re}\,[F_1''(x_1)] = f''(x_1) = g''(x_1) = 0, \quad \text{for } x_1 > 0. \tag{5.8}$$

Then, from $(5.2)_3$ and (5.7), the shear stress ahead of the crack tip is given by

$$\sigma_{12}^s(x_1 > 0, 0) = \frac{2\,\mu}{x_1^\gamma} \left(\alpha\, i\, A + \frac{\sin \gamma\pi}{2\,\pi\,\mu} \int_{-L}^{0} \frac{|t|^\gamma\, \tau_0(t)}{t - x_1}\, dt \right). \tag{5.9}$$

Due to the presence of a process zone, the shear stress must be bounded at the crack tip, namely $x_1^\gamma\, \sigma_{12}^s(x_1, 0) \to 0$ as $x_1 \to 0^+$, and this condition yields the constant A,

$$i\, A = \frac{\sin \gamma\pi}{2\pi\,\mu}\, \frac{1}{\alpha} \int_{-L}^{0} \frac{\tau_0(t)}{|t|^{1-\gamma}}\, dt. \tag{5.10}$$

For a remotely loaded mode II crack, the shear stress field far ahead of the crack tip, namely for $x_1 = D \gg L$, must match the singular solution,

$$\sigma_{12}^s(D, 0) = 2\mu\,\alpha\, \frac{i\, A}{D^\gamma} = \frac{K_{II}}{\sqrt{2\pi}\, D^\gamma}, \tag{5.11}$$

where K_{II} is the intersonic mode II stress intensity factor. The latter results from elimination of the constant A between (5.10) and (5.11):

$$K_{II} = \sqrt{\frac{2}{\pi}}\, \sin \gamma\pi \int_{-L}^{0} \frac{\tau_0(t)}{|t|^{1-\gamma}}\, dt. \tag{5.12}$$

For a constant shear stress τ_0, the above integral is equal to $\tau_0\, L^\gamma/\gamma$.

Finally, substitution of $i\, A$, Eq. (5.10), in (5.7) yields the unknown function $F_1''(z)$,

$$F_1''(z) = \frac{\sin \gamma\pi}{2\pi\,\mu}\, \frac{i}{\alpha}\, J_\tau(\gamma, z), \tag{5.13}$$

where $J_\tau(z, \gamma)$ is defined by (5.15) below. With conditions (5.5) and (5.8), Lemmas 2 and 3 in Appendix B provide the remaining primary functions as indicated in (4.4–4.6).

5.3 Generic integrals

It is instrumental to introduce two generic singular integrals, $I_\tau(z, \gamma)$ and $J_\tau(z, \gamma)$ defined, for any complex $z = |z|\, e^{i\,\theta}$, as

$$I_\tau(z, \gamma) = \int_{-L}^{0} |\frac{z}{t}|^{1-\gamma}\, \frac{\tau_0(t)}{t - z}\, dt, \tag{5.14}$$

and

$$J_\tau(z, \gamma) = \int_{-L}^{0} \frac{z^{1-\gamma}}{|t|^{1-\gamma}}\, \frac{\tau_0(t)}{t - z}\, dt = e^{i(1-\gamma)\theta}\, I_\tau(z, \gamma). \tag{5.15}$$

For a point $z = x_1 \in]-L, 0[$ of the process zone, the above integrals are contributed by a Cauchy principal value and an imaginary part,

$$I_\tau(x_1, \gamma) = {I\!\!\!-}_\tau(x_1, \gamma) + i\,\pi\, \tau_0(x_1), \tag{5.16}$$

and

$$J_\tau(x_1, \gamma) = -\cos(\gamma\pi)\, {I\!\!\!-}_\tau(x_1, \gamma) - \sin(\gamma\pi)\,\pi\, \tau_0(x_1) + i\, \left(\sin(\gamma\pi)\, {I\!\!\!-}_\tau(x_1, \gamma) - \cos(\gamma\pi)\,\pi\, \tau_0(x_1)\right). \tag{5.17}$$

The double integral ${H\!\!\!-}_\tau(\gamma)$,

$${H\!\!\!-}_\tau(\gamma) = \int_{-L}^{0} \int_{-L}^{0} |\frac{x}{t}|^{1-\gamma}\, \frac{\tau_0(x)\, \tau_0(t)}{t - x}\, dx\, dt \geq 0, \tag{5.18}$$

is non negative under mild conditions on the spatial variation of the cohesion along the process zone. Indeed, let us consider two points (x, t) and (t, x) symmetric wrt the diagonal $t = x$. With $X \equiv x/t \geq 0$, the sum of the contributions of these two points is

$$\left(|\frac{x}{t}|^{1-\gamma} - |\frac{t}{x}|^{1-\gamma} \right) \frac{1}{(1 - x/t)\, t} = \frac{X^{2(1-\gamma)} - 1}{X - 1}\, \frac{1}{X^{-1+\gamma}|t|}. \tag{5.19}$$

Thus if, for example, the sign of the cohesion along the process zone is constant over the interval $]-L, 0[$, then the double integral is clearly positive, if $1 - \gamma > 0$, i.e., $\gamma < 1$.

If the cohesion along the process zone is constant, a contour integration gives, for $x \in]-L, 0[$,

$$\begin{aligned} \mathcal{I}_\tau(x,\gamma) &\equiv \tau_0 \fint_{-L}^{0} \left|\frac{x}{t}\right|^{1-\gamma} \frac{dt}{t-x} \\ &= \left(\frac{\pi}{\tan\gamma\pi} + N(x,\gamma)\right)\tau_0, \\ N(x,\gamma) &\equiv \int_0^{-x/L} \frac{y^{-\gamma}}{1-y}\, dy > 0. \end{aligned} \tag{5.20}$$

Further integration by parts yields:

$$\begin{aligned} \mathcal{H}_\tau(\gamma) &\equiv \int_{-L}^{0} \tau_0\, \mathcal{I}_\tau(x,\gamma)\, dx \\ &= \left(\frac{\pi}{\tan\gamma\pi} + \frac{1}{1-\gamma}\right) L\, \tau_0^2, \end{aligned} \tag{5.21}$$

which is strictly positive for γ in the intervals $\cdots[-2, -1.459[\cup[-1, -0.430[\cup[0, 1[\cup[2, 2.43[\cdots$.

5.4 The energy release rate

Upon insertion of the expressions (5.13) and (4.4), or (5.5), into the displacement $(5.3)_1$, and use of (5.17), the rate of sliding on the crack can be cast in the form:

$$\begin{aligned} u^s_{1,1}(x_1<0,0^+) &= \frac{1}{2}\frac{c^2}{c_3^2}\frac{c_1^2-c_2^2}{c_s^2-c_2^2}\,\mathrm{Re}\,[F_1''(x_1)] \\ &= -\frac{(\sin\gamma\pi)^2}{4\pi\mu}\frac{c^2}{c_3^2}\frac{c_1^2-c_2^2}{c_s^2-c_2^2} \\ &\quad \frac{1}{\alpha}\left(\mathcal{I}_\tau(x_1,\gamma) - \frac{\pi\,\tau_0(x_1)}{\tan(\gamma\pi)}\right). \end{aligned} \tag{5.22}$$

The energy release rate results from (4.17) as,

$$\begin{aligned} \mathcal{G} = \frac{(\sin\gamma\pi)^2}{2\pi\mu}\frac{c^2}{c_3^2}\frac{c_1^2-c_2^2}{c_s^2-c_2^2}\frac{1}{\alpha}\Big(\mathcal{H}_\tau(\gamma) \\ -\frac{\pi}{\tan(\gamma\pi)}\int_{-L}^{0}\tau_0^2(x_1)\,dx_1\Big). \end{aligned} \tag{5.23}$$

For $0<\gamma<1/2$, the coefficient α, as defined by Eqs. (4.2–4.3), is positive. Moreover, with N defined by (5.20), the last term in (5.23) can be recast in the following format,

$$\begin{aligned} &\int_{-L}^{0}\tau_0(x)\int_{-L}^{0}\left|\frac{x}{t}\right|^{1-\gamma}\frac{\tau_0(t)-\tau_0(x)}{t-x}\,dt\,dx \\ &+\int_{-L}^{0} N(x,\gamma)\,\tau_0^2(x)\,dx. \end{aligned} \tag{5.24}$$

Under mild conditions on the variations of the cohesion along the process zone, this term can be shown to be positive. This is the case when in particular the cohesion is constant along the process zone or when it varies linearly, as $\tau_0(x) = (1 + x/L)\,\tau_0$.

Therefore, under these circumstances, crack propagation is energetically admissible.

A polar representation of the stress and velocity components is shown on Fig. 4 at a particular point of the region (i). The fields associated to the comparison crack without a process zone and to the crack with a process zone are quite similar in front of the crack tip. They differ behind the crack tip, since the two Mach lines are directed toward the rear. The crack model with a process zone wipes out the jumps across the two Mach lines associated to the comparison crack without a process zone to be introduced in Sect. 8.

6 Crack propagation within the region (ii)

Within the intersonic region (ii), namely for $c_2 > c > c_3$ as shown by Fig. 1, the sole space variable z_3 is real, Table 3, so that $\mathrm{Im}\,[G(z_3)]$ is a real valued function of a real variable and it will be denoted by $g(z_3)$. Then, the displacement potentials (3.14), (3.15) may be recast in the following form:

$$\left\{\begin{aligned} &\varphi^s(x_1,x_2) = \mathrm{Re}\,[F_1(z_1)] + \mathrm{Re}\,[F_2(z_2)], \\ &\varphi^w(x_1,x_2) = c_{ms}^{-2}\big((c_1^2-c_s^2)\,\mathrm{Re}\,[F_1(z_1)] \\ &\qquad -(c_s^2-c_2^2)\,\mathrm{Re}\,[F_2(z_2)]\big), \\ &\psi^s(x_1,x_2) = g(z_3). \end{aligned}\right. \tag{6.1}$$

The apparent stresses and pore pressure (3.5) can now be expressed in terms of the unknown functions $F_1(z_1)$, $F_2(z_2)$ and $g(z_3)$:

$$\left\{\begin{aligned} &\sigma^s_{11}/\mu = (1+\alpha_3^2+2\alpha_1^2)\,\mathrm{Re}\,[F_1''(z_1)] \\ &\quad +(1+\alpha_3^2+2\alpha_2^2)\,\mathrm{Re}\,[F_2''(z_2)] + 2\alpha_3\, g''(z_3), \\ &\sigma^s_{22}/\mu = (\alpha_3^2-1)\,(\mathrm{Re}\,[F_1''(z_1)] \\ &\qquad +\mathrm{Re}\,[F_2''(z_2)]) - 2\alpha_3\, g''(z_3), \\ &\sigma^s_{12}/\mu = -2\alpha_1\,\mathrm{Im}\,[F_1''(z_1)] - 2\alpha_2\,\mathrm{Im}\,[F_2''(z_2)] \\ &\qquad +(\alpha_3^2-1)g''(z_3), \\ &-np/\mu = (1+\alpha_3^2)\,c_{mw}^{-2}\,\big((c_1^2-c_s^2)\mathrm{Re}\,[F_1''(z_1)] \\ &\quad -(c_s^2-c_2^2)\,\mathrm{Re}\,[F_2''(z_2)]\big). \end{aligned}\right. \tag{6.2}$$

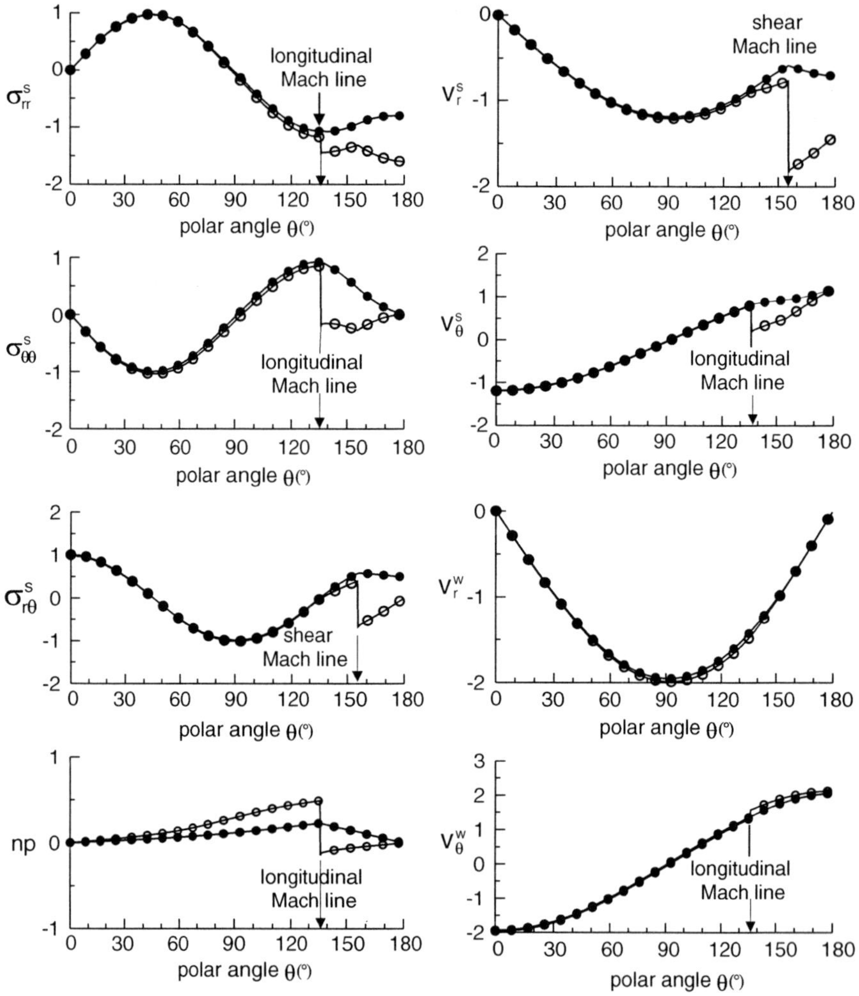

Fig. 4 Region (i): porosity $n = 0.8$, crack tip speed $c = 1600\,\text{m/s}$. Polar representation of the stress and velocity components under Mode II loading conditions, for points located on a semi-circle of radius $r = L/2$ and centered at the crack tip $x_1 = 0$. Crack with a process zone (solid symbols) and comparison crack without a process zone (open symbols). The stress and pore pressure are scaled by $\sqrt{2\pi}\, r^{\gamma}/K_{II}$, and the solid and fluid velocities by $\sqrt{2\pi}\, r^{\gamma}/K_{II} \times \rho^s c_3$. Properties of interest at this point of the (n, c)-plane: body wave-speeds $c_1 = 1922\,\text{m/s}$, $c_2 = 1105\,\text{m/s}$, $c_3 = 668\,\text{m/s}$; singularity exponent $\gamma = 0.013$; Mach angle of shear discontinuity 24.7°; Mach angle of longitudinal discontinuity 43.7°

Similarly, the displacements (3.2) become in the solid phase:

$$\begin{cases} u_1^s = \text{Re}\,[F_1'(z_1)] + \text{Re}\,[F_2'(z_2)] + \alpha_3\, g'(z_3), \\ u_2^s = -\alpha_1\, \text{Im}\,[F_1'(z_1)] - \alpha_2\, \text{Im}\,[F_2'(z_2)] - g'(z_3), \end{cases} \tag{6.3}$$

and in the fluid phase:

$$\begin{cases} u_1^w = c_{ms}^{-2} \big((c_1^2 - c_s^2)\, \text{Re}\,[F_1'(z_1)] \\ \quad -(c_s^2 - c_2^2)\, \text{Re}\,[F_2'(z_2)]\big), \\ u_2^w = -c_{ms}^{-2} \big(\alpha_1 (c_1^2 - c_s^2)\, \text{Im}\,[F_1'(z_1)] \\ \quad -\alpha_2 (c_s^2 - c_2^2)\, \text{Im}\,[F_2'(z_2)]\big). \end{cases} \tag{6.4}$$

6.1 The Riemann–Hilbert problem

The introduction of the solid and fluid displacements $(6.3)_1$ and $(6.4)_1$ into the Mode II symmetry conditions (4.11) results in restrictions of the unknown functions:

$$\begin{cases} \text{Re}\,[F_1''(x_1)] + \text{Re}\,[F_2''(x_1)] + \alpha_3\, g''(x_1) \\ \quad = 0, \quad \text{for } x_1 > 0, \\ (c_1^2 - c_s^2)\, \text{Re}\,[F_1''(x_1)] - (c_s^2 - c_2^2)\, \text{Re}\,[F_2''(x_1)] \\ \quad = 0, \quad \text{for } x_1 > 0. \end{cases} \tag{6.5}$$

The apparent stress components and pore pressure being expressed by (6.2), the conditions on the tractions

(4.11–4.13) and the vanishing of the pore pressure (4.14) on the crack surfaces yield further restrictions:

$$\begin{cases} (\alpha_3^2-1)\,(\mathrm{Re}\,[F_1''(x_1)]+\mathrm{Re}\,[F_2''(x_1)]) \\ \quad -2\alpha_3\, g''(x_1)=0, \quad \text{for all } x_1, \\ (c_1^2-c_s^2)\,\mathrm{Re}\,[F_1''(x_1)]-(c_s^2-c_2^2) \\ \quad \times \mathrm{Re}\,[F_2''(x_1)]=0, \quad \text{for all } x_1, \\ -2\alpha_1\,\mathrm{Im}\,[F_1''(x_1)]-2\alpha_2\,\mathrm{Im}\,[F_2''(x_1)] \\ \quad +(\alpha_3^2-1)\,g''(x_1)=\tau(x_1)/\mu, \quad \text{for } x_1<0. \end{cases} \tag{6.6}$$

Manipulations of $(6.6)_{1,2}$ show $\mathrm{Re}\,[F_2''(x_1)]$ and $g''(x_1)$ to be proportional to $\mathrm{Re}\,[F_1''(x_1)]$:

$$\mathrm{Re}\,[F_2''(x_1)]=\frac{c_1^2-c_s^2}{c_s^2-c_2^2}\,\mathrm{Re}\,[F_1''(x_1)],$$

$$g''(x_1)=\frac{c_1^2-c_2^2}{c_s^2-c_2^2}\,\frac{\alpha_3^2-1}{2\alpha_3}\,\mathrm{Re}\,[F_1''(x_1)], \quad \text{for all } x_1. \tag{6.7}$$

Further substitution of these relations in (6.5) then results in:

$$\mathrm{Re}\,[F_1''(x_1)]=\mathrm{Re}\,[F_2''(x_1)]=g''(x_1)=0, \quad \text{for } x_1>0. \tag{6.8}$$

Moreover, according to Lemmas 2 and 3 in Appendix B, conditions $(6.6)_2$ and (6.8) imply the analytic functions F_1'' and F_2'' to be proportional in agreement with (4.4), (4.5).

Then, relations $(6.6)_3$, (6.7) and (6.8) can be used to form, for the primary function $F_1''(z)$, an inhomogeneous Hilbert problem which has exactly the format (5.6), provided $\Gamma=\tan\gamma\pi$ and α are defined in region (ii) as indicated by (4.1) and (4.2), respectively.

6.2 The energy release rate

The analysis developed for region (i), Eqs. (5.6–5.24), can be shown to hold formally unaltered. In particular, the energy release rate in region (ii) is positive, the exponent γ varying in the range $]0,1/2]$ as indicated on Fig. 5.

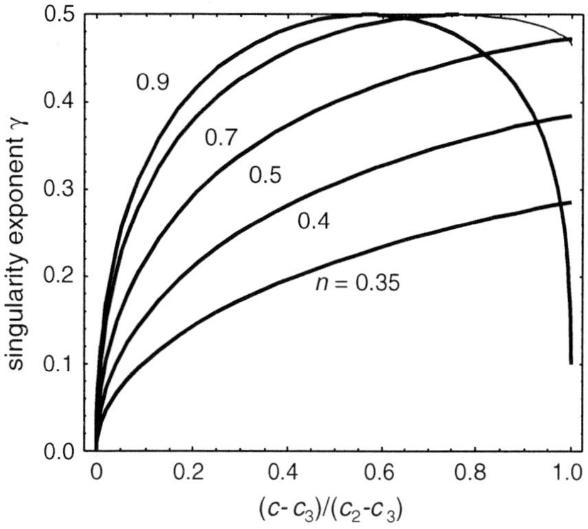

Fig. 5 Intersonic region (ii). Singularity exponent γ of Mode II, as function of the crack tip speed c, for several porosities n. For sufficiently large porosity n, Mode II square-root singularity, given by $\Gamma\to\infty$ or $\gamma=0.5$, occurs for $c=\sqrt{2}\,c_3$. At the interface with the subsonic regime, the singularity exponent vanishes

A polar representation of the stress and velocity components is shown on Fig. 6 at a particular point of the region (ii). The behavior is qualitatively similar to that of region (i), to within the fact that there is now a single Mach line in the upper plane.

7 Crack propagation within the region (iii)

Within the region (iii), namely for $c_2<c<c_3$ as displayed on Fig. 1, $\mathrm{Re}\,[F_2(z_2)]$ turns out to be a real valued function of the real variable z_2 and, thus, will be denoted by $f(z_2)$. Then the displacement potentials (3.14), (3.15) may be recast in terms of the unknown functions $F_1(z_1)$, $f(z_2)$ and $G(z_3)$:

$$\begin{cases} \varphi^s(x_1,x_2)=\mathrm{Re}\,[F_1(z_1)]+f(z_2), \\ \varphi^w(x_1,x_2)=c_{ms}^{-2}\left((c_1^2-c_s^2)\,\mathrm{Re}\,[F_1(z_1)] \right. \\ \quad \left. -(c_s^2-c_2^2)\,f(z_2)\right), \\ \psi^s(x_1,x_2)=\mathrm{Im}\,[G(z_3)]. \end{cases} \tag{7.1}$$

Upon introduction of the potentials (7.1), the apparent stresses and pore pressure (3.5) become:

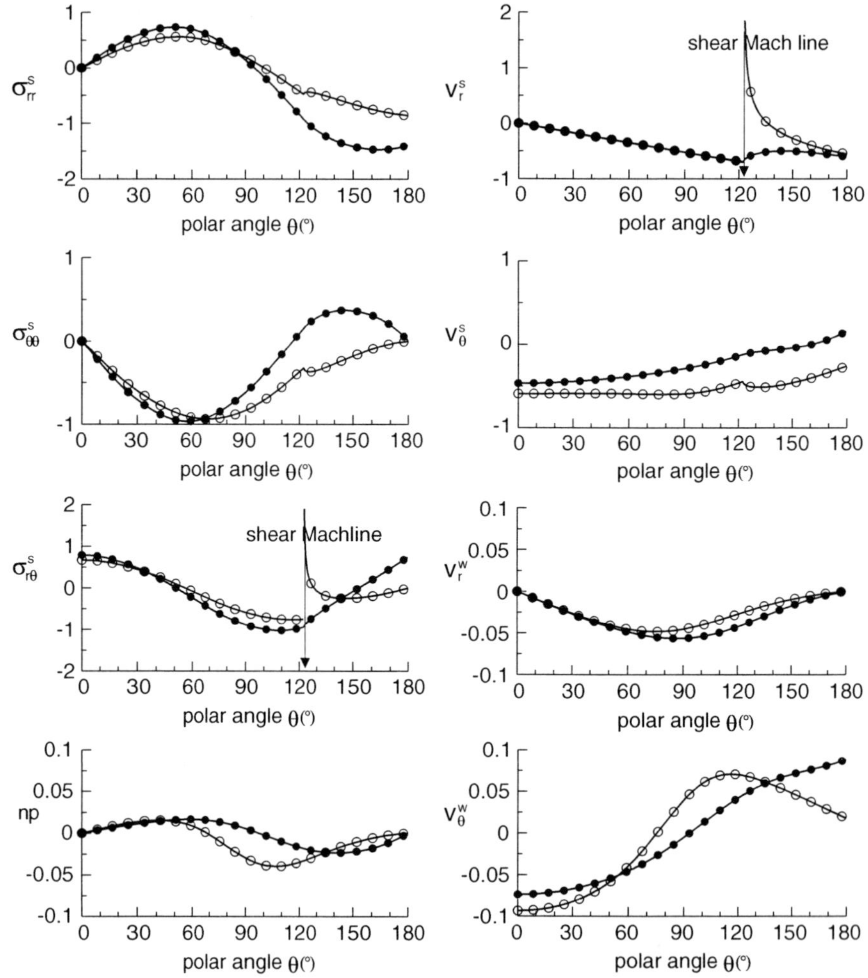

Fig. 6 Region (ii): porosity $n = 0.7$, crack tip speed $c = 1000\,\text{m/s}$. Polar representation of the stress and velocity fields under Mode II loading conditions, for points located on a semi-circle of radius $r = L/2$ and centered at the crack tip $x_1 = 0$. Crack with a process zone (solid symbols) and comparison crack without a process zone (open symbols). Scaling as on Fig. 4. Properties of interest at this point of the (n, c)-plane: body wave-speeds $c_1 = 2040\,\text{m/s}$, $c_2 = 1306\,\text{m/s}$, $c_3 = 843\,\text{m/s}$; singularity exponent $\gamma = 0.441$; Mach angle of shear discontinuity 57.5°

$$\left\{\begin{aligned}
\sigma^s_{11}/\mu &= (1-\alpha_3^2+2\alpha_1^2)\,\mathrm{Re}\,[F_1''(z_1)] \\
&\quad +(1-\alpha_3^2-2\alpha_2^2)\,f''(z_2) \\
&\quad +2\alpha_3\,\mathrm{Re}\,[G''(z_3)], \\
\sigma^s_{22}/\mu &= -\big((1+\alpha_3^2)\,(\mathrm{Re}\,[F_1''(z_1)]+f''(z_2)) \\
&\quad +2\alpha_3\,\mathrm{Re}\,[G''(z_3)]\big), \\
\sigma^s_{12}/\mu &= -\big(2\alpha_1\,\mathrm{Im}\,[F_1''(z_1)]-2\alpha_2\,f''(z_2) \\
&\quad +(1+\alpha_3^2)\,\mathrm{Im}\,[G''(z_3)]\big), \\
-np/\mu &= (1-\alpha_3^2)\,c_{mw}^{-2}\big((c_1^2-c_s^2)\,\mathrm{Re}\,[F_1''(z_1)] \\
&\quad -(c_s^2-c_2^2)\,f''(z_2)\big).
\end{aligned}\right. \tag{7.2}$$

The displacements (3.2) express as well in terms of the primary functions, namely in the solid phase:

$$\left\{\begin{aligned}
u_1^s &= \mathrm{Re}\,[F_1'(z_1)]+f'(z_2)+\alpha_3\,\mathrm{Re}\,[G'(z_3)], \\
u_2^s &= -\alpha_1\,\mathrm{Im}\,[F_1'(z_1)]+\alpha_2\,f'(z_2)-\mathrm{Im}\,[G'(z_3)],
\end{aligned}\right. \tag{7.3}$$

and in the fluid phase:

$$\left\{\begin{aligned}
u_1^w &= c_{ms}^{-2}\big((c_1^2-c_s^2)\,\mathrm{Re}\,[F_1'(z_1)] \\
&\quad -(c_s^2-c_2^2)\,f'(z_2)\big), \\
u_2^w &= -c_{ms}^{-2}\big(\alpha_1\,(c_1^2-c_s^2)\,\mathrm{Im}\,[F_1'(z_1)] \\
&\quad +\alpha_2\,(c_s^2-c_2^2)\,f'(z_2)\big).
\end{aligned}\right. \tag{7.4}$$

7.1 The Riemann–Hilbert problem

The Mode II symmetry conditions (4.11) restrict the solid and fluid displacements $(7.3)_1$ and $(7.4)_1$, and therefore the unknown functions $F_1''(x_1)$, $f''(x_1)$ and

$G''(x_1)$, on the symmetry axis:

$$\begin{cases} \mathrm{Re}\,[F_1''(x_1)] + f''(x_1) + \alpha_3\, \mathrm{Re}\,[G''(x_1)] \\ \quad = 0, \quad \text{for } x_1 > 0, \\ (c_1^2 - c_s^2)\, \mathrm{Re}\,[F_1''(x_1)] - (c_s^2 - c_2^2)\, f''(x_1) \\ \quad = 0, \quad \text{for } x_1 > 0. \end{cases} \tag{7.5}$$

Similarly, when expressed via the constitutive relations (7.2), the conditions (4.11–4.13) on the tractions and the vanishing of the pore pressure (4.14) on the crack surfaces together with $(7.5)_2$ yield the following system of equations:

$$\begin{cases} (1+\alpha_3^2)\,\big(\mathrm{Re}\,[F_1''(x_1)] + f''(x_1)\big) \\ \quad +2\alpha_3\, \mathrm{Re}\,[G''(x_1)] = 0, \quad \text{for all } x_1, \\ 2\alpha_1\, \mathrm{Im}\,[F_1''(x_1)] - 2\alpha_2\, f''(x_1) + (1+\alpha_3^2) \\ \quad \times \mathrm{Im}\,[G''(x_1)] = -\tau(x_1)/\mu, \quad \text{for } x_1 < 0, \\ f''(x_1) = \dfrac{c_1^2 - c_s^2}{c_s^2 - c_2^2}\, \mathrm{Re}\,[F_1''(x_1)], \quad \text{for all } x_1. \end{cases} \tag{7.6}$$

Elimination of f'' upon combination of $(7.6)_{1,3}$ leads to

$$\frac{1+\alpha_3^2}{2\alpha_3}\,\frac{c_1^2 - c_2^2}{c_s^2 - c_2^2}\, \mathrm{Re}\,[F_1''(x_1)] + \mathrm{Re}\,[G''(x_1)] = 0, \quad \text{for all } x_1. \tag{7.7}$$

A substitution of $(7.6)_3$ and (7.7) in (7.5) results in

$$\mathrm{Re}\,[F_1''(x_1)] = f''(x_1) = \mathrm{Re}\,[G''(x_1)] = 0, \quad \text{for } x_1 > 0. \tag{7.8}$$

Therefore, as a consequence of Lemma 2 of Appendix B, the relations $(7.6)_3$, (7.7) and (7.8) imply f'' and G'' to be proportional to F_1'', in agreement with the generic relations $(4.4)_2$ and $(4.6)_1$. Insertion of the two latter relations into $(7.6)_2$ and use of (7.8) lead to an inhomogeneous Hilbert problem for the function F_1'', which has exactly the format (5.6), and where $\Gamma = \tan \gamma\pi$ and α are defined in region (iii) by (4.1) and (4.2) respectively.

7.2 The speeds $c_* = c_*(n)$ separating the region (iii) in two domains

The variation of the exponent γ (modulo an integer) in terms of wave-speed c is shown on Fig. 7. As noted in (4.3), α and Γ have the same sign. They are positive for $-1/2 < \gamma < 0$ and negative for $0 < \gamma < 1/2$. Let c_* denote the value of the crack tip speed at which α and Γ vanish, and equivalently $\gamma = 0$, namely:

$$4\left(1 - \frac{c_*^2}{c_3^2}\right)^{1/2} \left(1 - \frac{c_*^2}{c_1^2}\right)^{1/2} - \frac{c_1^2 - c_2^2}{c_s^2 - c_2^2}\left(2 - \frac{c_*^2}{c_3^2}\right)^2 = 0. \tag{7.9}$$

The curve $c_* = c_*(n)$ as it follows from the above equation is plotted on Fig. 1. It turns out to be slightly lower than $c_3(n)$ and it splits the intersonic region (iii) into two subregions where α and Γ, and the exponent γ take opposite signs. This speed c_* may be shown to tend to the Rayleigh wave-speed c_R of the solid constituent as the porosity vanishes, and to the Rayleigh wave-speed associated to a permeable crack face in the subsonic region. However, it is has not been proved analytically to be interpretable as a Rayleigh

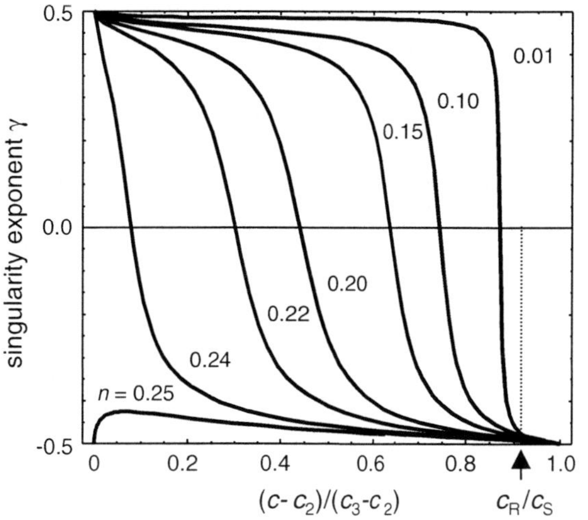

Fig. 7 Intersonic region (iii). Singularity exponent γ of Mode II, as function of the crack tip speed c, for various porosities n. The curve $c = c_*$, for which $\Gamma = 0$, is shown dashed on Fig. 1. For sufficiently small porosity n, γ tends to 1/2 when c tends to c_2. Since c_2 tends to 0 and c_3 tends to c_S as the porosity vanishes, the maximum value of the x-axis, for which $\gamma = 1/2$ can be reached, is c_R/c_S. At the interface with the subsonic regime, the singularity exponent is equal to $\pm 1/2$, as indicated by $(4.1)_3$ for $\alpha_2 = 0$. Indeed, the sign of the numerator in $(4.1)_3$ changes at $c = c_*$. Moreover, the value of the porosity n_* for which $c_* = c_2$ is between 0.24 and 0.25

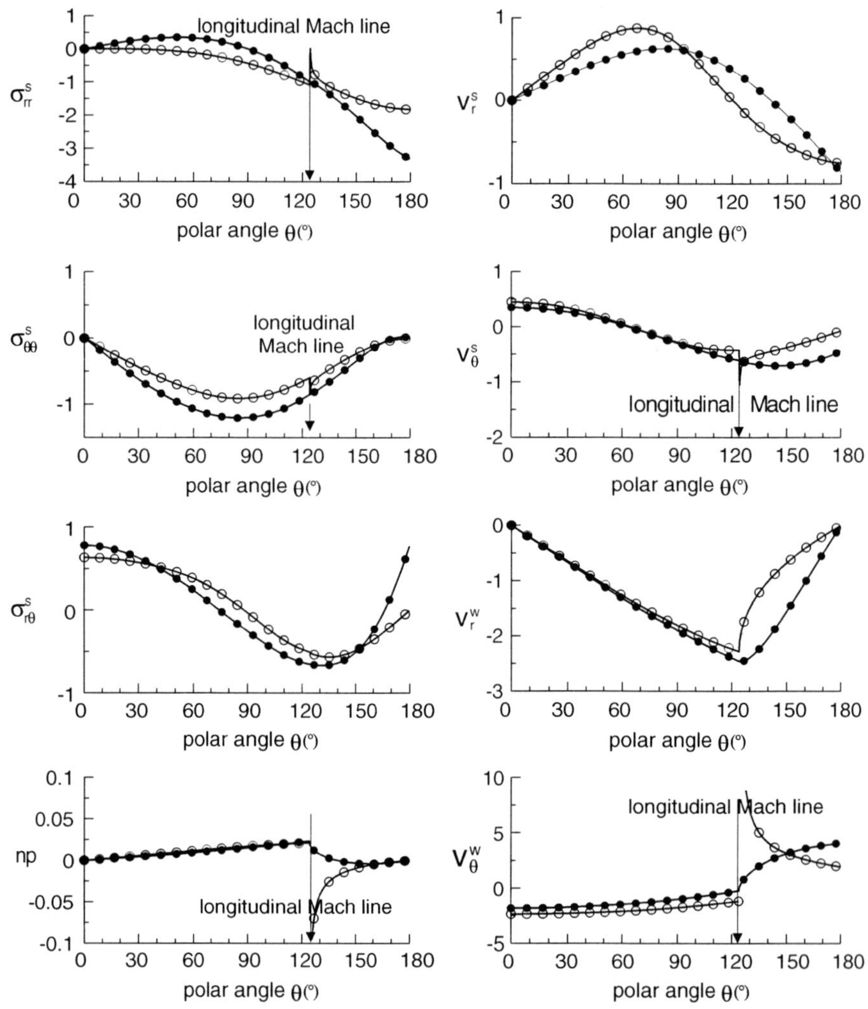

Fig. 8 Region (iii): porosity $n = 0.05$, crack tip speed $c = 1600\,\text{m/s}$. Polar representation of the stress and velocity fields under Mode II loading conditions, for points located on a semi-circle of radius $r = L/2$ and centered at the crack tip $x_1 = 0$. Crack with a process zone (solid symbols) and comparison crack without a process zone (open symbols). Scaling as on Fig. 4. Properties of interest at this point of the (n, c)-plane: body wave-speeds $c_1 = 4267\,\text{m/s}$, $c_2 = 1327\,\text{m/s}$, $c_3 = 2253\,\text{m/s}$; singularity exponent $\gamma = 0.472$; Mach angle of longitudinal discontinuity 56°

wave-speed of the poroelastic material in the intersonic region (iii).

7.3 The energy release rate

From now on, the analysis developed in region (i), Eqs. (5.6–5.24), can be shown to hold formally unaltered. The associated energy release rate (5.23), is positive for $0 < \gamma < 0.5$, that is for $c_2 < c < c_*$, as displayed on Fig. 7.

A polar representation of the stress and velocity components is shown on Fig. 8 at a particular point of the region (iii). Once again, the behavior is qualitatively similar to those of regions (i) and (ii), to within the fact that there is a single Mach line, which is now longitudinal.

8 Comparison crack without a process zone and first order discontinuities

The existence of imaginary characteristics, or equivalently of real eigenvalues Ω_2 or Ω_3 as indicated in Table 3, may give rise to first order discontinuities across the associated rays $z_2 = 0$ or $z_3 = 0$. Indeed, the real functions (4.6) that contribute to the solution, that is, to the stress and velocity fields, vanish in front these rays. Their continuity across these rays is an issue that will be shown to be solved differently by the crack with a process zone and by the crack without a process zone.

Infinite jumps extending all along the Mach lines without attenuation will be shown to exist for the cracks without a process zone. In contrast, for the crack

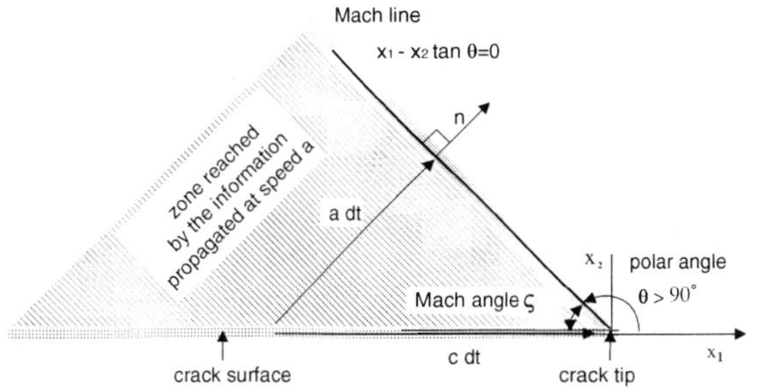

Fig. 9 The crack tip propagates at speed c to the right along the axis x_1, while the mechanical information propagates at speed a. A first order discontinuity exists if $c > a$. In the intersonic range, the material wave-speed a can be c_2 or c_3, and the direction of the discontinuity is defined either by the Mach angle ζ such that $\sin\zeta = c_2/c$ or $\sin\zeta = c_3/c$, or by the polar angle $\theta_{\text{long}} = \pi/2 + \tan^{-1}\alpha_2$ or $\theta_{\text{shear}} = \pi/2 + \tan^{-1}\alpha_3$. The first order discontinuity is characterized by the fact that only a part of the plane, between the crack line and the Mach line, is reached by the shear, or longitudinal, information propagated by the wave

endowed with a process zone, the continuity of the solution across the rays $z_2 = 0$ and $z_3 = 0$ is proved in Appendix C.

The developments below thus concentrate on the crack without a process zone. Emphasis is laid on points (n, c) where the crack with a process zone is energetically admissible, while the energy release rate of the crack without a process zone is vanishing. To pinpoint this situation, the crack without a process zone is termed "*virtual crack of comparison*". The focus will be to analyze qualitatively the efficiency of the model with a process zone to spatially smooth out the jumps of the fields associated to the comparison crack. Conversely, it is worth scrutinizing which quantities remain continuous and which do not, when the length of the process zone is decreased to zero.

8.1 Sketch of the solution in absence of a process zone

The basic primary function $F_1''(z)$ can be cast in the region-invariant format,

$$F_1''(z) = \frac{K_{II}}{2\,\mu\sqrt{2\pi}}\,\frac{1}{\alpha}\,\frac{1}{i z^{\gamma}}, \tag{8.1}$$

in terms of the Mode II stress intensity factor K_{II}. The coefficients γ and α are defined by (4.1) and (4.2), and the two other primary functions are given by the relations (4.4–4.6) like for the crack with a process zone. For example in region (i),

$$f''(z_2) = -R_2\,\frac{K_{II}}{2\,\mu\sqrt{2\pi}}\,\frac{\sin\gamma\pi}{\alpha}\,\frac{\mathcal{H}(-z_2)}{|z_2|^{\gamma}},$$

$$g''(z_3) = -R_3\,\frac{K_{II}}{2\,\mu\sqrt{2\pi}}\,\frac{\sin\gamma\pi}{\alpha}\,\frac{\mathcal{H}(-z_3)}{|z_3|^{\gamma}}, \tag{8.2}$$

where $\mathcal{H}$ denotes the unit step function.

The above solution defining the stress and displacement, e.g., via (5.2–5.4) in region (i), the energy release rate $(4.16)_2$ can be cast in the form,

$$\mathcal{G} = \frac{K_{II}^2}{8\,\mu^2}\,\frac{\sin\gamma\pi}{\alpha}\,\frac{c^2}{c_3^2}\,\frac{c_1^2 - c_2^2}{c_s^2 - c_2^2}\,\lim_{x_1\to 0^+} x_1^{1-2\gamma} \times \begin{cases} = 0 \text{ for } \gamma < 1/2, \\ > 0 \text{ for } \gamma = 1/2, \end{cases} \tag{8.3}$$

the positive sign holding where α is positive.

Thus the energy release rate vanishes, since in general $\gamma < 1/2$. It is however finite and positive for $\gamma = 1/2$. From the definition (4.1), an infinite $\Gamma = \tan\gamma\pi$ is seen to correspond

- in region (i), to $n = 0$ and $c = \sqrt{2}\,c_S < c_L$;
- in region (ii), to the curve $c = \sqrt{2}\,c_3$, which, given the shape of this region in the plane (n, c), occurs only for sufficiently large porosities n;
- in region (iii), to the segment $n = 0$ and $c < c_R$;
- to the interface between the subsonic regime and part of the region (iii) below the curve $c = c_*(n)$ where α is positive, namely for $n \le n_*$.

Note that, as the porosity decreases to zero, the transition to the value $\gamma = 1/2$ observed for linear elastic solids in regions (i) and (iii) is smooth, Figs. 3, 7.

In conclusion, for a given porosity, intersonic Mode II propagation is admissible only for some special speeds of the crack tip, much like for linear elastic solids.

8.2 Nature of the field equations

Along a standard terminology, a partial differential equation, involving second order derivatives in both time and space, e.g., $\partial^2\varphi/\partial x_1^2 + \partial^2\varphi/\partial x_2^2 - a^{-2}\,\partial^2\varphi/\partial t^2 = 0$ is said to be hyperbolic if there exist real characteristics which allow for the information to propagate at finite speed, e.g., for $a \neq 0$ real. Otherwise, for a purely imaginary, it is said elliptic. Here, the time derivative is given by (3.1), so that the above equation becomes $(1 - c^2/a^2)\,\partial^2\varphi/\partial x_1^2 + \partial^2\varphi/\partial x_2^2 = 0$.

Thus according to the above terminology, the subsonic propagation of the crack tip, $c < a$, corresponds to an elliptic equation.

Since the stress is obtained from the potentials φ which obey second order partial differential equations, the existence of real characteristics for φ in the hyperbolic case corresponds in fact to a first order discontinuity, i.e., a stress discontinuity, and, according to Hadamard relations, to an accompanying velocity discontinuity: there exists a vector $\boldsymbol{\lambda} = (\lambda_i)$ such that $[\![\nabla \mathbf{u}]\!] = \boldsymbol{\lambda} \otimes \mathbf{n}$, $[\![\mathbf{v}]\!] = -a\,\boldsymbol{\lambda}$, with $\mathbf{n}$ the unit normal to the wave front.

In the present context, a *first order discontinuity*, corresponds to the existence of a ray emanating from the crack tip that limits the zone of extension of the information propagated by the wave, as indicated by Fig. 9.

Fig. 10 displays the variations of Mach angles with the crack tip speed for two porosities. The closer the crack tip speed to the critical wave-speed, i.e., c_3 for the shear Mach line, c_2 for the longitudinal Mach line, the closer the Mach angle is to 90°. Conversely, the farther the crack tip speed from the critical wave-speed, the more oblique is the Mach line.

The field equations involve four equations for four potentials. There exist four wave-speeds, accounting for the multiplicity of degree two of the shear wave. The nature of the field equations thus depends on the region of the plane (n, c):

- region (i): the equation associated to the fastest longitudinal wave is elliptic, while the three others are hyperbolic;
- region (ii): the two equations associated to the longitudinal waves are elliptic while those two associated to the shear waves are hyperbolic;
- region (iii): the equation associated to the second longitudinal wave is hyperbolic while the three others are elliptic;
- subsonic region: the four equations are elliptic.

8.3 Discontinuities across the Mach lines

A priori both shear and longitudinal, or dilatational, Mach lines may exist in the region (i), while the sole shear Mach line may take place in region (ii), and the sole longitudinal Mach line in region (iii).

8.3.1 *Shear Mach line in regions (i) and (ii)*

In these regions, the issue concerns the continuity of the real function $g''(z_3)$ across the ray $z_3 = x_1 + \alpha_3 x_2 = 0$, with $z_3 = |z_3|$ ahead of the Mach line and $z_3 = |z_3|\, e^{i\pi}$ behind the Mach line. It is instrumental to introduce polar coordinates (r, θ) centered at the crack tip, and to express the stress and velocity in these coordinates. The ray $z_3 = 0$ is defined by the polar angle $\theta_{\text{shear}} = \pi/2 + \tan^{-1} \alpha_3$. Let $[\![g'']\!]$ be the discontinuity of g'' across the Mach line at a current point $(r, \theta_{\text{shear}})$ of the Mach line. The contributions of $[\![g'']\!]$ to the stress components, and to the velocity components, Eqs. (5.2–5.4) in region (i) and Eqs. (6.2–6.4) in region (ii), are collected. The sole shear stress is observed to undergo a discontinuity,

$$[\![\sigma^s_{rr}]\!] = 0, \quad [\![\sigma^s_{\theta\theta}]\!] = 0, \quad [\![\sigma^s_{r\theta}]\!] = \mu \frac{c^2}{c_3^2} [\![g'']\!],$$
$$[\![n\, p]\!] = 0, \tag{8.4}$$

and, similarly the sole discontinuous velocity component is the solid radial velocity,

$$[\![v^s_r]\!] = \frac{c^2}{c_3} [\![g'']\!], \quad [\![v^s_\theta]\!] = 0, \qquad [\![v^w_r]\!] = 0,$$
$$[\![v^w_\theta]\!] = 0. \tag{8.5}$$

While the function g'' is continuous across the ray $z_3 = 0$ for the crack with a process zone, Appendix C, it undergoes an infinite jump for the crack without a process zone according to (8.2). Consequently, the stress and velocity fields are continuous for a crack with a process zone, while they undergo, across the shear Mach line, an infinite jump that is *parallel* to, and extends all along the Mach line, Fig. 11.

8.3.2 *Longitudinal Mach line in regions (i) and (iii)*

In these regions, the concern is the continuity of the real function $f''(z_2)$ across the ray $z_2 = x_1 + \alpha_2 x_2 = 0$, defined by the polar angle $\theta_{\text{long}} = \pi/2 + \tan^{-1} \alpha_2$. Collecting the contributions of the discontinuity of f'' in the expressions of stress and pore pressure, Eqs. (5.2) for region (i) and (7.2) for region (iii), result in

$$[\![\sigma^s_{rr}]\!] = \left(-2\frac{c^2}{c_2^2} + \frac{c^2}{c_3^2} \right) \mu\, [\![f'']\!],$$

$$[\![\sigma^s_{\theta\theta}]\!] = \mu \frac{c^2}{c_3^2} [\![f'']\!], \quad [\![\sigma^s_{r\theta}]\!] = 0,$$

$$[\![n\, p]\!] = \frac{c_s^2 - c_2^2}{c_{mw}^2} \frac{c^2}{c_3^2} \mu\, [\![f'']\!], \tag{8.6}$$

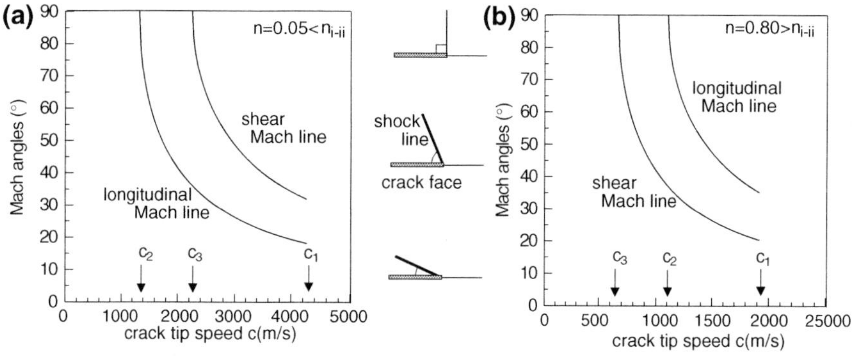

Fig. 10 Variations of Mach angles with the crack tip speed for two porosities. In region (i), the Mach angle defining the first order shear discontinuity is smaller or larger than the Mach angle of the first order longitudinal discontinuity depending on the position of the porosity with respect to n_{i-ii}, indicated on Fig. 1

Similarly, the discontinuity of f'', in (5.4) for region (i) and (7.4) for region (iii), yields discontinuities of the solid and fluid velocities, namely

$$[\![v_r^s]\!] = 0, \quad [\![v_\theta^s]\!] = \frac{c^2}{c_2}[\![f'']\!], \qquad [\![v_r^w]\!] = 0,$$

$$[\![v_\theta^w]\!] = -\frac{c_s^2 - c_2^2}{c_{ms}^2}\frac{c^2}{c_2}[\![f'']\!]. \tag{8.7}$$

For the crack with a process zone, the function f'' is continuous across the ray $z_2 = 0$, Appendix C, and so are the stress and velocity. On the other hand, for a crack without a process zone, f'' undergoes an infinite jump across the longitudinal Mach line according to (8.2), and so do the stress and velocity, Fig. 11. The velocity jump is *normal* to the Mach line.

8.3.3 Jumps in strain and balance of momentum

The strain discontinuity is obtained from the compatibility relation $[\![\nabla\mathbf{u}]\!] = -a^{-1}[\![\mathbf{v}]\!] \otimes \mathbf{n}$. Therefore the sole non zero discontinuous components of the strain in phase k are

$$[\![\epsilon_{r\theta}^k]\!] = \frac{[\![v_r^k]\!]}{2a}, \quad [\![\epsilon_{\theta\theta}^k]\!] = \frac{[\![v_\theta^k]\!]}{a}, \quad k = s, w. \tag{8.8}$$

The first order shear discontinuity is thus clearly isochoric. Across the first order longitudinal discontinuity on the other hand, the solid and fluid phases undergo volume changes that are of opposite signs, but do not compensate. In fact, if the solid and fluid constituents were incompressible, the volume changes in the body would satisfy the constraint $(1-n)\,\mathrm{tr}\boldsymbol{\epsilon}^s + n\,\mathrm{tr}\boldsymbol{\epsilon}^w = 0$, and thus the jumps across the discontinuity would be such that $(1-n)\,[\![v_\theta^s]\!] + n\,[\![v_\theta^w]\!] = 0$. For second order waves, similar relations hold in terms of strain rate and acceleration, e.g., Eqs. (2.17) and (6.6) in Loret and Harireche (1991).

Across a discontinuity line moving at speed a in the direction $\mathbf{n}$, the equation of balance of momentum, that each phase of the porous medium obeys, takes the form

$$[\![\boldsymbol{\sigma}^k]\!] \cdot \mathbf{n} + [\![\hat{\mathbf{p}}^k]\!] + (\rho^k a)\,[\![\mathbf{v}^k]\!] = \mathbf{0}, \quad k = s, w. \tag{8.9}$$

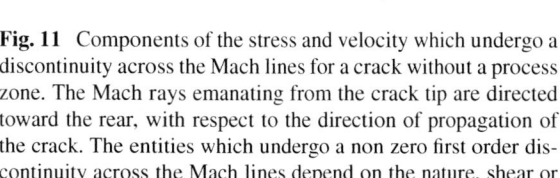

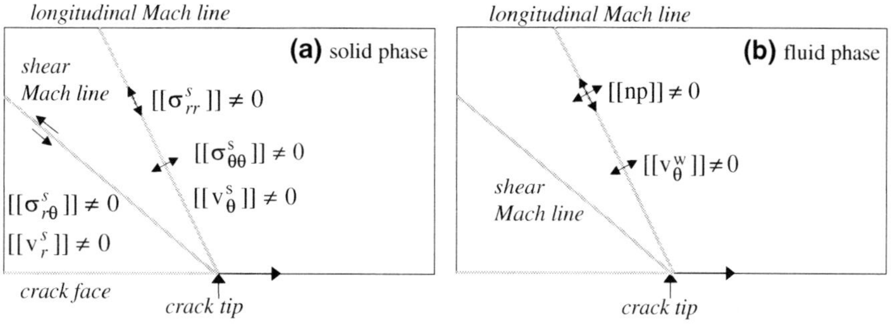

Fig. 11 Components of the stress and velocity which undergo a discontinuity across the Mach lines for a crack without a process zone. The Mach rays emanating from the crack tip are directed toward the rear, with respect to the direction of propagation of the crack. The entities which undergo a non zero first order discontinuity across the Mach lines depend on the nature, shear or longitudinal, of the discontinuity. The first order shear discontinuity affects only the solid phase, while the first order longitudinal discontinuity affects both the solid and fluid phases. The sketch indicates a longitudinal Mach angle larger than the shear Mach angle, and thus corresponds to a porosity $n > n_{i-ii}$ in region (i), as shown in Fig. 10(b)

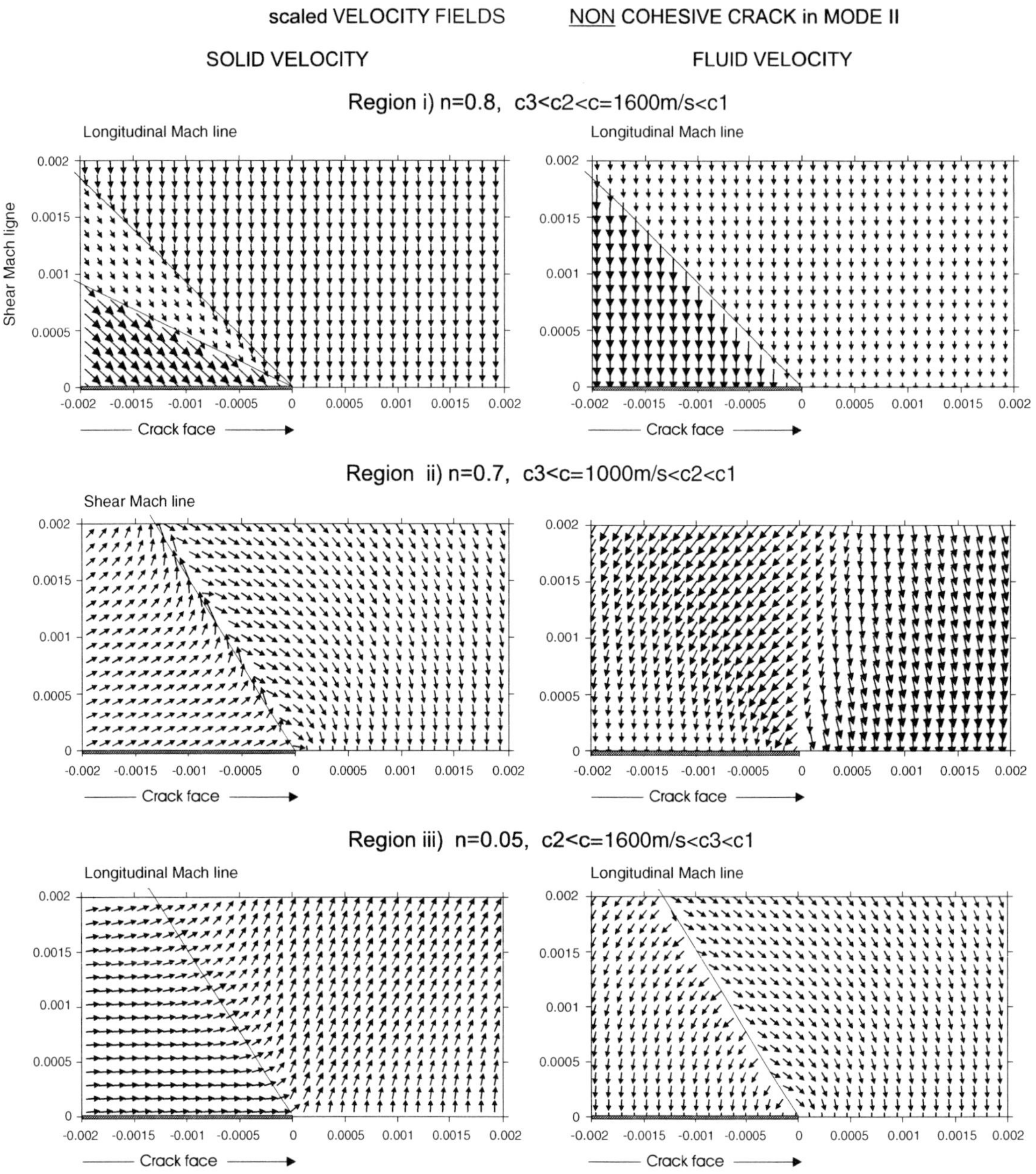

Fig. 12 Velocity fields for the comparison crack under Mode II loading conditions. The velocity fields are scaled by r^γ, r distance to the crack tip, γ singularity exponent. The size of the arrows pertains to each plot, and has been chosen so as to highlight the spatial heterogeneity of the velocity. The first order shear discontinuity gives rise to velocity discontinuities in the solid phase only, while the first order longitudinal discontinuity affects both phases, as sketched in Fig. 11

For the first order shear discontinuity, the stress and velocity discontinuities (8.4) and (8.5) can indeed be checked to satisfy this relation, with $a = c_3$ and, as indicated on Fig. 9, with $\mathbf{n}$ equal to $-\mathbf{e}_\theta(\theta_{\text{shear}})$, $\mathbf{e}_\theta(\theta)$ being the orthoradial vector $(-\sin\theta, \cos\theta)$. For the first order longitudinal discontinuity, the relation (8.9) is satisfied as well by the stress and velocity discontinuities (8.6) and (8.7) with $a = c_2$ and with $\mathbf{n}$ equal to $-\mathbf{e}_\theta(\theta_{\text{long}})$. Here the diffusion contribution $\hat{\mathbf{p}}^k$ is of higher order close to the crack tip and does not contribute.

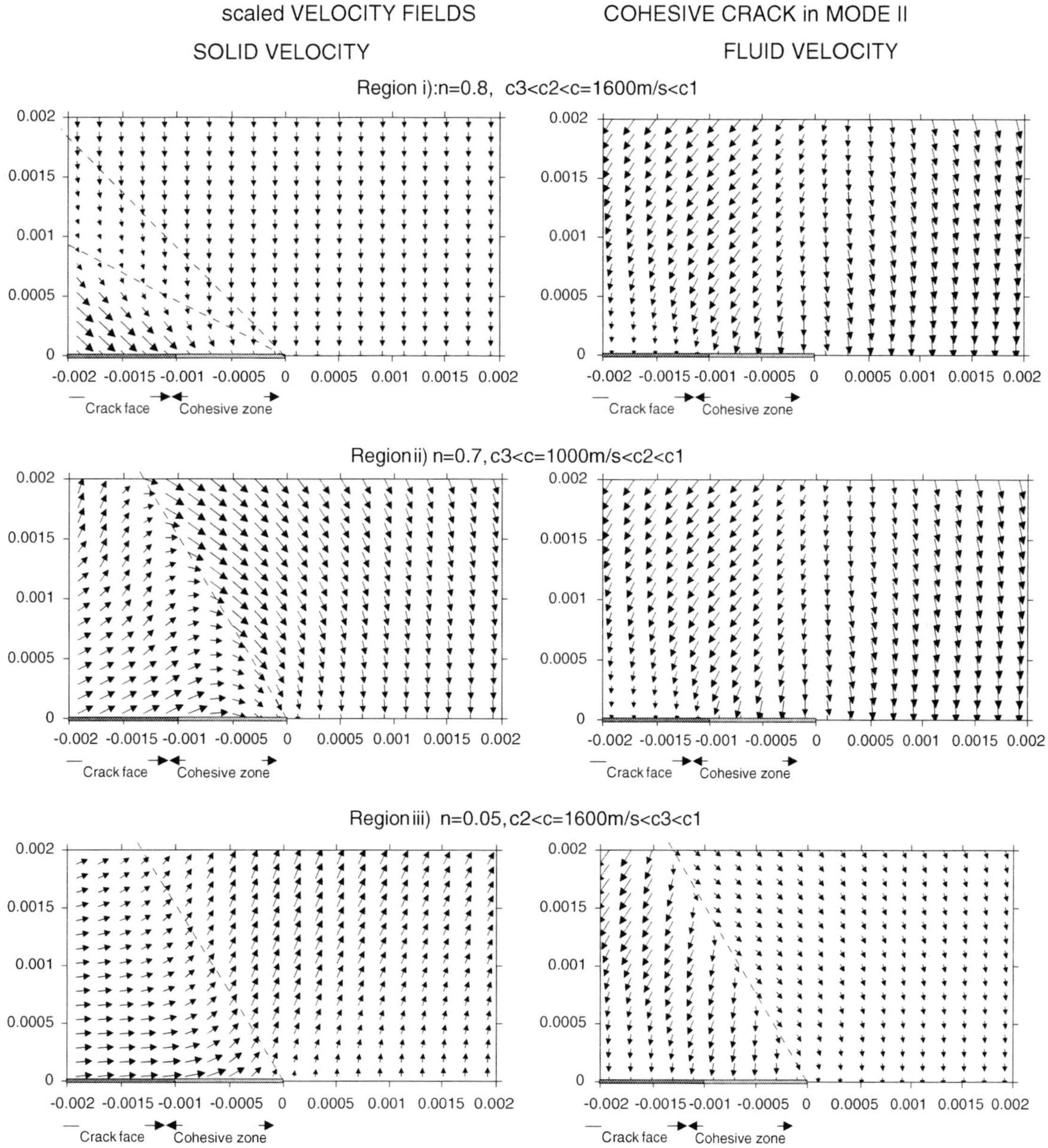

Fig. 13 Velocity fields for the crack with a process zone under Mode II loading conditions. The velocity fields are scaled by r^{γ}, r distance to the crack tip, γ singularity exponent. The latter is the same as for the comparison crack without a process zone. On the other hand, the velocity discontinuities that are observed in the comparison crack are smoothed out by the process zone. The (dashed) Mach lines of the comparison crack have been kept for reference

8.4 Stress and velocity fields

With respect to the crack without a process zone, the crack with a process zone first improves the energetic admissibility, and second removes the discontinuity lines, a priori of infinite strength. In order to appreciate the smoothing effects of the crack with a process zone, the velocity fields, corresponding to the cracks in absence and presence of a process zone, are displayed on Figs. 12 and 13 respectively, at a point

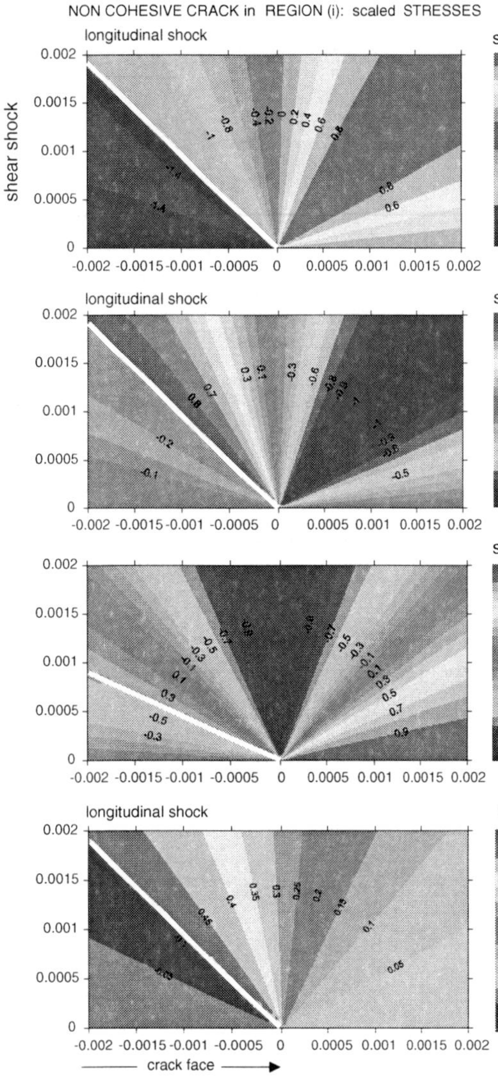

Fig. 14 Comparison crack under Mode II loading conditions, region (i), $n = 0.8$, $c = 1600$m/s. Partial stresses σ^s_{rr}, $\sigma^s_{\theta\theta}$, $\sigma^s_{r\theta}$, and pore pressure $n\,p$ scaled by $\sqrt{2\pi}\,r^{\gamma}/K_{II}$

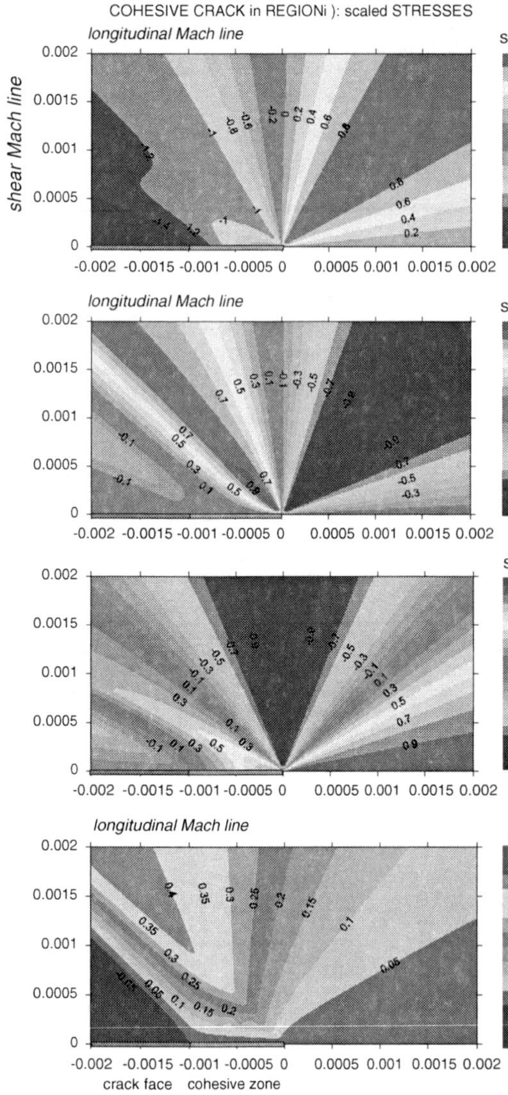

Fig. 15 Crack with a process zone, region (i) under Mode II loading conditions, $n = 0.8$, $c = 1600$m/s. Partial stresses σ^s_{rr}, $\sigma^s_{\theta\theta}$, $\sigma^s_{r\theta}$, and pore pressure $n\,p$ scaled by $\sqrt{2\pi}\,r^{\gamma}/K_{II}$. The (dashed) Mach lines of the comparison crack have been kept for reference. The stress fields for the cracks with and without a process zone differ essentially behind the Mach lines

(n, c) of each region. The polar representations of the stress and velocity components around a half circle of radius $L/2$ have been shown on Figs. 4, 6, 8.

The cohesion has been chosen to vary linearly along the process zone from 0 to a value $\tau_0 = \mu_s/10$ at the crack tip, and the process zone has length $L = 10^{-3}$ m. The fact that the cohesion vanishes at the end of the process zone smooths out possible disturbances emanating from this point, as can be observed on the velocity fields, Fig. 13, and on the stress fields, Figs. 15 and 16.

The stress components in the region (i) are shown on Fig. 14 for the crack without a process zone, and on Fig. 15 for the crack with a process zone. They are scaled by $\sqrt{2\pi}\,r^{\gamma}/K_{II}$, where r is the distance to the crack tip, and K_{II} is the stress intensity factor associated to the crack with a process zone, Eq. (5.12). The scaled stress of the comparison crack without a process zone displays a purely polar variation. The presence of the process zone perturbs this representation behind the crack tip only, over a distance of the order of the length of the process zone.

Notice that the fact that, even for the crack with a process zone, the polar components of the stress are not all bounded at the crack tip as already alluded to. The sole bounded stress component is the shear stress σ_{12}. A closeup of the shear stress near the process zone, shown on Fig. 16, indicates that the cohesion along the process zone extends continuously to the body in a manner that depends strongly on the region of interest. The actual value of the singularity exponent plays a prominent role on the departure between the crack with a process zone and the comparison crack: for example, at the point chosen in region (i), the singularity exponent γ is very low, so that the jumps, while theoretically of infinite strength, effects a very narrow thickness. The range of influence of the discontinuity is much larger at the points (n, c), which are chosen as representative of regions (ii) and (iii), where γ is close to 0.45.

The above velocity fields were scaled by r^γ, so that their actual spatial variations were not apparent. Of course, the spatial decay of the stress and velocity fields depends essentially on the singularity exponents, as illustrated for the solid velocities on Fig. 17.

9 Summary and outlook

To summarize the analysis, the zones where the intersonic propagation of cracks with a process zone has been shown to be admissible are displayed on Fig. 18.

The results may be considered along several points of view:

1. *Differences between the regions (i), (ii) and (iii):*

Region (iii) is separated into two subregions by the porosity dependent speed $c_* = c_*(n)$ shown on Fig. 18. Crack propagation is admissible in the subregion $c_2 < c < c_*$.

The energy release rate is continuous across the boundaries of the regions, Fig. 19. This property holds true also along the boundary (i)–(ii), even if the variation becomes stronger as the porosity increases.

In reference to Fig. 18, let us consider a vertical line at constant porosity and increasing crack tip speed. Across region (iii), the energy release rate is positive between c_2 and c_*^-, unbounded at c_*, negative between c_*^+ and c_3, it vanishes at c_3, it is maximum between c_3 and c_1, and tends to vanish close to c_1 together with the exponent γ, Fig. 3. In fact, $\mathcal{G}$ vanishes all along the curves $c = c_3$ and $c = c_1$ together with γ. However it is strictly positive at the interface of regions (i) and (ii).

2. *Cracks with and without process zones:*

For the virtual comparison crack under Mode II, the velocity, stress and pore pressure fields are singular not only at the crack tip, but also on two (or four) symmetric rays emanating from the crack tip, where the partial stress fields undergo an infinite jump. These first order discontinuities are absent from the subsonic region, Loret and Radi (2001), and they are therefore characteristic of the intersonic regions.

In presence of the process zone, crack propagation is admissible with a singularity exponent in general smaller than 1/2, yielding thus a vanishingly small energy release rate for a crack without a process zone. The sole exceptions are the point and curves of the (n, c) plane indicated in Sect. 8.1. Indeed, then value of the singularity exponent $\gamma = 1/2$ implies the energy release rate (8.3) to be strictly positive.

3. *Continuity of the energy release rate in the limit of vanishing process length?*

A crack without a process zone can not be thought of as the limit of a crack with a vanishing length L of process zone, since a crack with a process zone ensures a stress component to be finite at the crack tip. Still, let us consider the following issue: can we take the limit of the energy release rate as the process length L tends to 0, to infer the energetic admissibility of cracks without a process zone? The double integral (5.21) is central to the issue. It indicates that the energy release rate would vanish in the limit, except when the singularity exponent is zero, along the curve $c = c_*(n)$. On the other hand, the energy release rate for the crack without process zone does not vanish only if the singularity exponent is equal to 1/2.

4. *Continuity of the singularity exponent at the subsonic interface*

At the interface between the subsonic regime and the region (ii), the singularity exponent γ vanishes. It is equal to 1/2 at the interface between the subsonic regime and the part of the region (iii) below the curve $c = c_*(n)$, and to $-1/2$ above this curve. Therefore bypassing the longitudinal wave-speed below this curve takes place with a smooth variation of the Mode II singularity exponent γ, while bypassing the shear wave-speed induces a discontinuity of the singularity exponent.

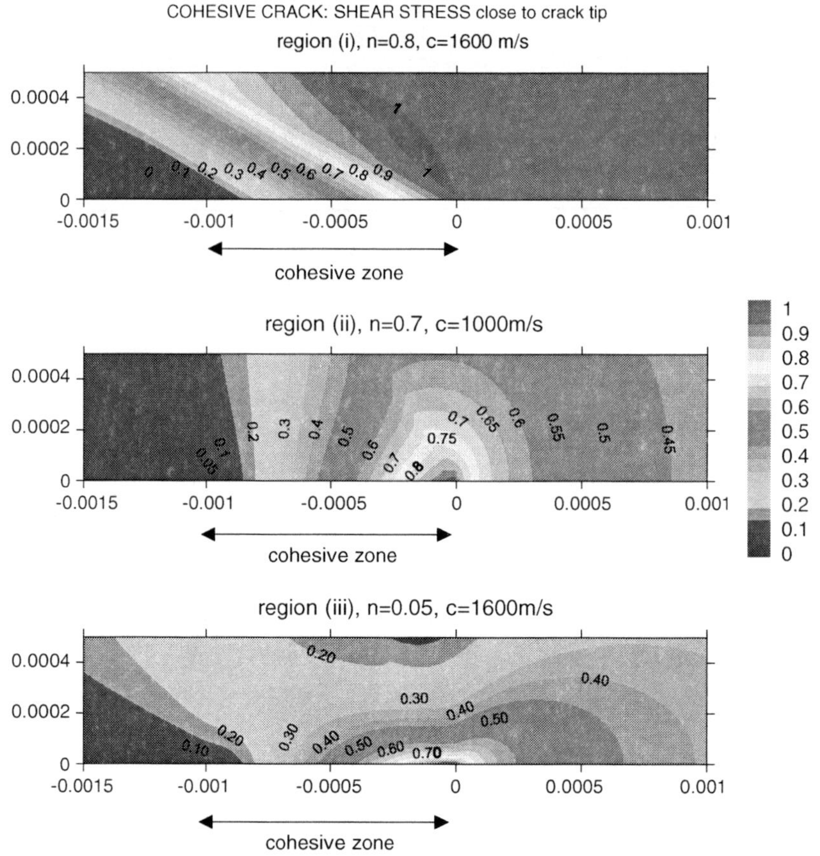

Fig. 16 Crack with a process zone under Mode II loading conditions. Closeup of the shear stress σ^s_{12}/τ_0 in the neighborhood of the crack tip. The cohesion varies linearly along the process zone from 0 to a value τ_0 at the crack tip. While the cohesion along the process zone extends continuously from the crack line to the body, the direction and range of influence of the cohesion depend strongly on the region, and on the value of the singularity exponent

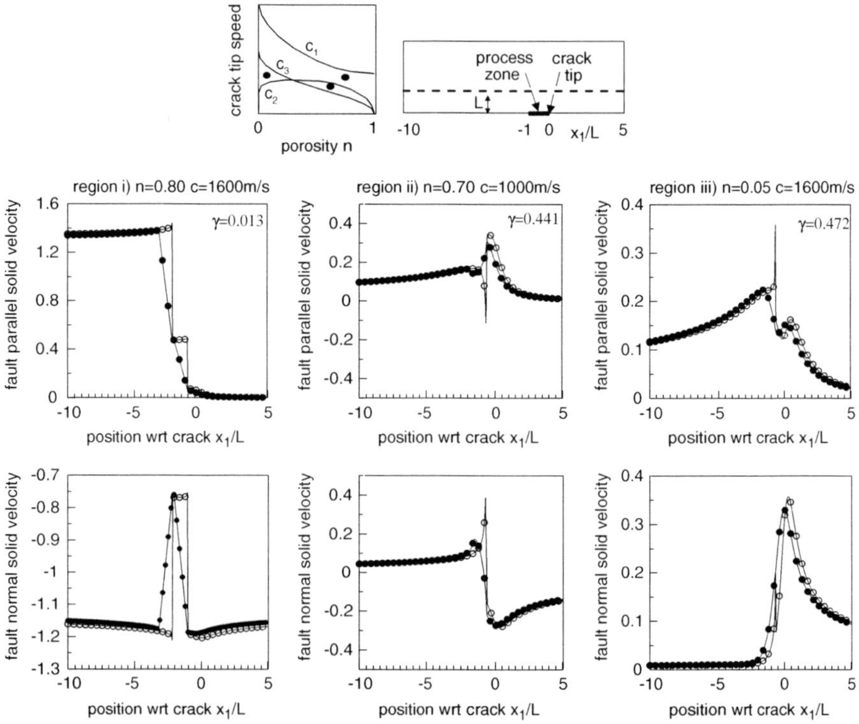

Fig. 17 Solid velocity components parallel and normal to the crack path on a line parallel to the fault at distance L, for three points of the (n, c)-plane. Crack with a process zone (solid symbols) and comparison crack without a process zone (open symbols). The velocities are scaled by $(\rho^s c_3)/\tau_0$. Across the Mach lines, the discontinuities of the velocities associated to the comparison crack are unbounded. The spatial decay of the velocities depends strongly on the singularity exponent γ

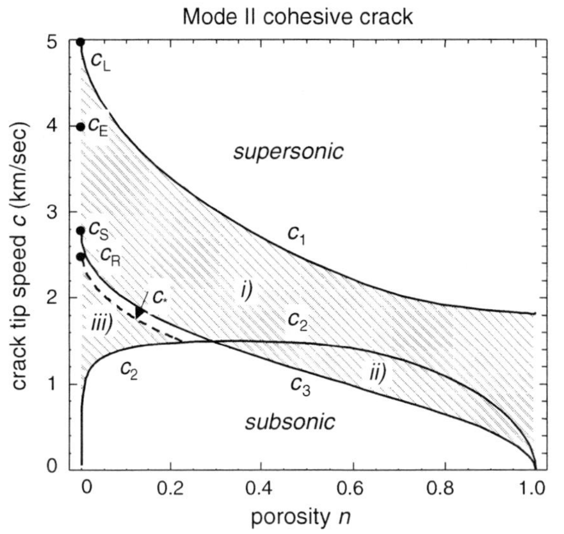

Fig. 18 Sketch summarizing the findings of the analysis. The intersonic zones where the crack propagation with a process zone is energetically admissible are dashed. The cases of linear elastic solids ($n = 0$), and of cracks without a process zones are detailed in the text

5. *Continuity of the singularity exponent across intersonic regions*

The singularity exponent is continuous across the intersonic regions. Indeed, the left border of Fig. 3 coincides, for $n > n_{i-ii}$, with the right border of Fig. 5 where $c = c_2$, and, for $n < n_{i-ii}$, with the left border of Fig. 7 where $c = c_3$.

6. *Continuity with linear elastic solids for a vanishing porosity:*

It is also worth to consider the continuity of the condition of energetic admissibility as the porosity tends to zero. As indicated in the introduction, for linear elastic solids, the admissible ranges are as follows for Mode II loading conditions:

$$\begin{cases} \text{crack without process zone :} \\ c < c_R \quad \text{or} \quad c_E = \sqrt{2}\, c_S < c < c_L; \\ \text{crack with process zone :} \\ c < c_R \quad \text{or} \quad c_S < c < c_L. \end{cases} \tag{9.10}$$

In the limit of a small but non zero porosity, this work indicates the following admissible ranges:

$$\begin{cases} \text{crack without process zone : } \emptyset; \\ \quad \text{crack with process zone : } c{<}c_* \text{ or } c_3{<}c{<}c_1. \end{cases} \tag{9.11}$$

Thus in view of the fact that the wave-speeds c_1, c_3 and c_* tend respectively to c_L, c_S due to (2.14), and c_R, then there is continuity of the admissibility condition in the limit of a vanishing porosity, only for cracks with a process zone.

Note that the slow longitudinal wave-speed c_2 tends to vanish for vanishing porosity n. Thus, in the limit, the intersonic region (iii) drops down to the subsonic region $c < c_S$ where the singularity exponent is 1/2. This singularity is indeed recovered by the present approach, as the exponent γ approaches 1/2 as n tends to 0, Fig. 7.

7. *The length of the process zone:*

The actual value of the length of the process zone is elusive in the analysis as long as it is strictly positive. Still, it can be linked to the energy release rate. Indeed, assume the cohesion along the process zone to be constant. Integration of Eq. (4.17) yields $\mathcal{G} = \tau_0\, \delta_t$, in terms of $\delta_t = 2\, u_1^s(x_1 = -L)$, the displacement discontinuity at the end of the process zone. Then the length of the process zone becomes inversely proportional to the scaled energy release rate $\mathcal{G}^\# = \mathcal{G}\, \mu/(L\, \tau_0^2)$, namely

$$L = \frac{\mu\, \delta_t}{\tau_0}\, \frac{1}{\mathcal{G}^\#}. \tag{9.12}$$

This relation is consistent with the analysis, as it indicates a positive length for a positive energy release rate, and a (theoretically) negative length for non admissible crack propagation. More details can be grasped from Fig. 19.

Closing this discussion, it is worth looking backwards, and recalling the simplifications on which the analysis relies. Among them, one may mention the following issues:

- the dissipative terms have been neglected in the present asymptotic analysis. Their contribution, which must be considered in a full-field investigation, e.g. via the Wiener–Hopf technique, may smooth the jump in the partial stress and velocity fields far from the crack tip of the virtual comparison crack;
- a full-field investigation is required as well so as to investigate quantitatively the ranges in space and time where the asymptotic analysis holds;
- a quantitative comparison, at given porosity and crack tip speed, with the associated underlying drained and undrained single phase porous solids, would serve to appreciate the effects of the pore fluid;
- in the actual process zone, the porosity evolves as the crack progresses. A constitutive model where the porosity evolves with the deformation would

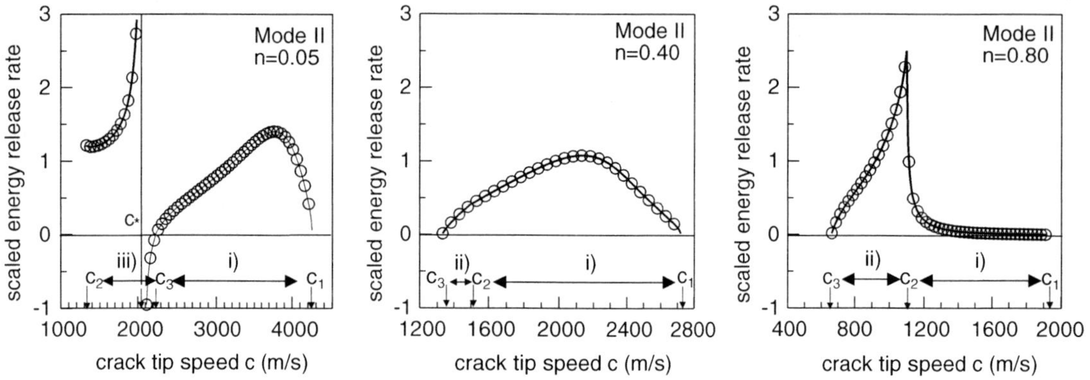

Fig. 19 Scaled energy release rates $\mathcal{G}^{\#} = \mathcal{G} \times \mu/(L\tau_0^2)$ as function of crack tip speed for Mode II loading conditions for three porosities $n = 0.05, 0.40, 0.80$, and for a constant cohesive stress along the process zone. Along the curve $c = c_*$ in region (iii), the singularity exponent γ vanishes and the energy release rate becomes unbounded

contribute to a more realistic description of the phenomena taking place in the process zone. The resulting field equations would result to be non-linear, and a different solution strategy should be devised;

- the crack faces immediately behind the crack tip have been assumed to be permeable. This assumption needs to be qualified;
- the analysis has focused on defining the energetic admissibility of crack propagation. Crack propagation has to take place over sufficiently long periods of time, and long lengths, to be observable in the laboratory or in the field. In other words, the possibility of a spontaneous bifurcation of the trajectory from a rectilinear path, and the assumption of steady propagation need to be appreciated.

Acknowledgments E. R. would like to thank the Italian Ministry of Education, University and Research (MIUR) for financial support obtained within the framework of the Project PRIN-2004 "Problemi e modelli microstrutturali: applicazioni in ingegneria civile e strutturale".

Appendix A: The Riemann–Hilbert problems for Mode II

Let $f = f(x_1)$ be a function of the real variable x_1 which is non zero on the interval $[-L, 0]$. Let $\Gamma = \tan \gamma\pi$ be a real number, where γ is interpreted as a singularity exponent.

Under Mode II loading conditions, the function $H(z)$ of the complex variable $z = x_1 + i\, x_2$, which is sought, can be shown, via Schwarz reflection principle, to satisfy the two conditions on the real axis:

$$\begin{cases} \mathrm{Re}\,[H(x_1)] - \Gamma\, \mathrm{Im}\,[H(x_1)] = \frac{1}{2} f(x_1), & \text{for } x_1 < 0, \\ \mathrm{Re}\,[H(x_1)] = 0, & \text{for } x_1 > 0. \end{cases} \tag{A1}$$

The solution to this Riemann–Hilbert problem, which is analytic on the upper half-plane $\mathrm{Im}\,[z] > 0$ and bounded at infinity, is defined up to the purely imaginary constant A, as required by $(\mathrm{A1})_2$, e.g., Muskhelishvili (1962),

$$H(z) = \frac{A}{z^\gamma} + \frac{\cos\gamma\pi}{2i\pi}\frac{1}{z^\gamma}\int_{-L}^{0} \frac{|t|^\gamma f(t)}{t - z}\, dt. \tag{A2}$$

Appendix B: some analytical properties

Lemma 1 Let $F_1(z)$ and $F_2(z)$ be two functions of the complex variable $z = x_1 + i\, x_2$, which are analytic on the upper plane $\mathrm{Im}\,[z] > 0$, vanish as $|z| \to \infty$, and satisfy the following conditions along the real axis:

$$\begin{cases} \mathrm{Im}\,[F_1(x_1)] = \mathrm{Im}\,[F_2(x_1)] & \text{for } x_1 < 0, \\ \mathrm{Im}\,[F_1(x_1)] = \mathrm{Im}\,[F_2(x_1)] = 0 & \text{for } x_1 > 0. \end{cases} \tag{B1}$$

Then $F_1(z) = F_2(z)$ for every value of the complex variable z.

Proof With conditions $(\mathrm{B1})_2$, the Schwarz reflection principle provides an analytic continuation to the lower half-plane, across the positive real axis, of the function $F_i(z)$, $i = 1, 2$, by setting

$$F_i(z) = \overline{F}_i(z) \equiv \overline{F_i(\overline{z})}, \quad \text{for Im}\,[z] < 0. \tag{B2}$$

Let us define

$$F_i^{\pm}(x_1) = \lim_{x_2 \to 0^{\pm}} F_i(z), \quad \text{for } i = 1, 2, \tag{B3}$$

and then

$$\overline{F_i^{+}(z)} = \lim_{x_2 \to 0^{+}} \overline{F_i(z)} = \lim_{x_2 \to 0^{-}} \overline{F_i(\overline{z})}$$
$$= \overline{F_i}^{-}(x_1), \quad \text{for } i = 1, 2. \tag{B4}$$

Condition $(B1)_1$ then implies, with help of (B4),

$$F_1^{+}(x_1) - \overline{F_1^{+}(x_1)} = F_1^{+}(x_1) - \overline{F_1}^{-}(x_1) =$$
$$F_2^{+}(x_1) - \overline{F_2^{+}(x_1)} =$$
$$F_2^{+}(x_1) - \overline{F_2}^{-}(x_1), \quad \text{for } x_1 < 0. \tag{B5}$$

Using the symmetry properties (B2), it follows from (B5) that

$$(F_1 - F_2)^{+}(x_1) = (F_1 - F_2)^{-}(x_1), \quad \text{for } x_1 < 0. \tag{B6}$$

Since F_1 and F_2 are analytic on the complex plane except along the negative real axis, condition (B6) implies that $F_1 - F_2$ is an entire function, vanishing as $z \to \infty$. The result then follows from Liouville's theorem.

Lemma 2 Let $F_1(z)$ and $F_2(z)$ be two functions with the same regularity as in Lemma 1, but satisfying the following conditions along the real axis:

$$\begin{cases} \text{Re}\,[F_1(x_1)] = \text{Re}\,[F_2(x_1)] & \text{for } x_1 < 0, \\ \text{Re}\,[F_1(x_1)] = \text{Re}\,[F_2(x_1)] = 0 \text{ for } x_1 > 0. \end{cases} \tag{B7}$$

then $F_1(z) = F_2(z)$ for every value of the complex variable z.

Proof Let $H_j(z) = i\, F_j(z)$, $j = 1, 2$. Since $\text{Im}\, H_j(z) = \text{Re}\, F_j(z)$, the result follows by Lemma 1.

Lemma 3 Let $F_1(z)$ and $F_2(z)$ be two functions with the same regularity as in Lemma 1, but satisfying the following conditions along the real axis:

$$\begin{cases} \text{Re}\,[F_1(x_1)] = \text{Im}\,[F_2(x_1)] & \text{for } x_1 < 0, \\ \text{Re}\,[F_1(x_1)] = \text{Im}\,[F_2(x_1)] = 0 \text{ for } x_1 > 0. \end{cases} \tag{B8}$$

then $F_1(z) = -i\, F_2(z)$ for every value of the complex variable z.

Proof Let $H_2(z) = -i\, F_2(z)$. Since $\text{Re}\, H_j(z) = \text{Im}\, F_2(z)$, the result follows by Lemma 2.

Appendix C: remarks on the singular integrals

Some remarks on the integral $I_\tau(z, \gamma)$, defined by (5.14) are provided here. This integral has to be evaluated over the upper half-plane $\text{Im}\, z \geq 0$ so as to define the primary functions F_1'', F_2'', and G''.

Numerical evaluation of the fields:

The integral is singular for $z = x_1$ located along the process zone. In order to remove the apparent singularity at the origin, a change of variable is introduced:

$$t = -|z|\,(-T)^{1/\gamma} \Leftrightarrow T = -(-t/|z|)^{\gamma}. \tag{C1}$$

Then the integral $I_\tau(z, \gamma)$ can be expressed in the following formats:

$$I_\tau(z, \gamma) = \frac{-1}{\gamma} \int_{-(L/|z|)^\gamma}^{0} \frac{U + \cos\theta - i\,\sin\theta}{U^2 + 2U\cos\theta + 1}$$
$$\tau_0(-|z|\,U)\, dT \quad \text{if } z = |z|\, e^{i\theta} \text{ complex},$$

$$= \frac{-1}{\gamma} \int_{-(L/|z|)^\gamma}^{0} \frac{\tau_0(-|z|\,U)\, dT}{U + z/|z|}$$
$$\text{if} z = x_1 < -L \text{ or } > 0,$$

$$= \frac{-1}{\gamma} \fint_{-(L/|z|)^\gamma}^{0} \frac{\varphi(T)}{T+1}\, \tau_0(-|z|\,U)\, dT$$
$$+ i\,\pi\, \tau_0(x_1) \text{ if } -L < z = x_1 < 0, \tag{C2}$$

where the symbol $\fint$ denotes the Cauchy principal value, $U = (-T)^{1/\gamma}$, and

$$\varphi(T) = \begin{cases} \dfrac{T+1}{U-1} & \text{if } T \neq -1, \\ -\gamma & \text{if } T = -1. \end{cases} \tag{C3}$$

Note that, in $(C2)_3$, the evaluation of the singular integral uses the section $(I_\tau)_+(z, \gamma)$ defined for points z of the upper plane, namely the points of the x_1-axis have angle $\theta = 0$ for $x_1 > 0$ and $\theta = \pi$ for $x_1 < 0$.

For $z = x_1 \in]-L, 0[$, the integrand is cast in a format that highlights the singular point and eases its treatment in a routine that handles the principal value. For regions (i) and (iii), the zone where the principal value needs to be evaluated in view of (4.6) takes the form of strip, behind the Mach line $z_2 = 0$, of thickness equal to the process length, Fig. 20. A similar zone exists in regions (i) and (ii) behind the Mach line $z_3 = 0$.

Continuity across the rays $z_2 = 0$ and $z_3 = 0$:

For $z = x_1 = 0^+$, the integral can be evaluated using e.g., Gradshteyn and Ryzhik (1980, p. 292),

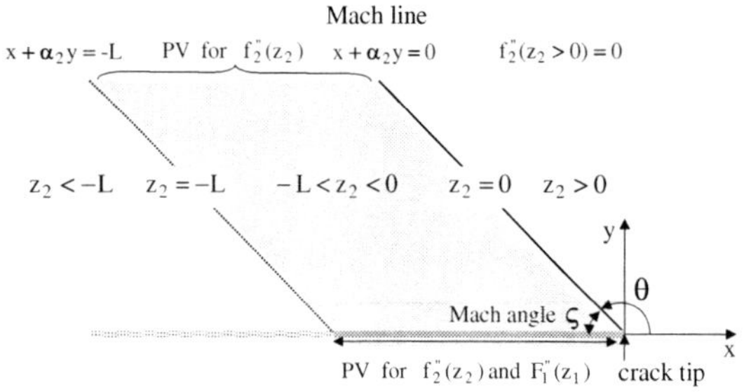

Fig. 20 In the regions (i) and (iii), where z_2 is real, the integral $I_\tau(z_2, \gamma)$ is singular in the domain $-L \leq z_2 \leq 0$. The first order shear discontinuity in the regions (i) and (ii) requires a similar treatment

$$
\begin{aligned}
I_\tau(0^+, \gamma) &= \frac{-\tau_0(0)}{\gamma} \int_{-\infty}^{0} \frac{\mathrm{d}T}{(-T)^{1/\gamma} + 1} \\
&= \frac{-\tau_0(0)}{\gamma} \int_{0}^{\infty} \frac{\mathrm{d}T}{(T)^{1/\gamma} + 1} \\
&= -\frac{\pi}{\sin(\gamma\,\pi)}\, \tau_0(0). \qquad \text{(C4)}
\end{aligned}
$$

For $z = x_1 = 0^-$, the relation $(C2)_3$ yields

$$
\begin{aligned}
I_\tau(0^-, \gamma) &= \frac{-\tau_0(0)}{\gamma} \int_{-\infty}^{0} \frac{\mathrm{d}T}{(-T)^{1/\gamma} - 1} + i\,\pi\,\tau_0(0) \\
&= \frac{-\tau_0(0)}{\gamma} \int_{0}^{\infty} \frac{\mathrm{d}T}{(T)^{1/\gamma} - 1} + i\,\pi\,\tau_0(0) \qquad \text{(C5)}
\end{aligned}
$$

which can be integrated in explicit form, e.g., Gradshteyn and Ryzhik (1980,p. 292),

$$
\begin{aligned}
I_\tau(0^-, \gamma) &= \frac{\tau_0(0)}{\gamma} \frac{\gamma\,\pi}{\tan(\gamma\,\pi)} + i\,\pi\,\tau_0(0) \\
&= \frac{\pi\, e^{i\gamma\pi}}{\sin(\gamma\,\pi)}\, \tau_0(0). \qquad \text{(C6)}
\end{aligned}
$$

The expression (5.15) with $\theta = 0$ for $x_1 = 0^+$, and $\theta = \pi$ for $x_1 = 0^-$, indicates then that $J_\tau(\gamma, 0)$ is continuous and real at $x_1 = 0$,

$$
J_\tau(0^+, \gamma) = J_\tau(0^-, \gamma) = -\frac{\pi}{\sin(\gamma\,\pi)}\, \tau_0(0). \qquad \text{(C7)}
$$

For Mode II loading conditions, the primary function $F_1''(z)$ defined generically by (5.13) for region (i), is thus continuous, and purely imaginary, at $z = 0$. Therefore, according to (4.6), $f''(z_2 = 0)$ and $g''(z_3 = 0)$ are continuous, and actually vanish, across the respective rays $z_2 = 0$ and $z_3 = 0$ in region (i). Consequently, for the crack with a process zone, the stress and velocity fields, which are defined from F_1'', f'' and g'', are continuous across these rays.

A similar conclusion applies across the ray $z_3 = 0$ in region (ii), and across the ray $z_2 = 0$ in region (iii).

References

Abeyaratne R, Knowles JK (1990) On the driving traction acting on a surface of strain discontinuity in a continuum. J Mech Phys Solids 38(3):345–360

Achenbach JD (1972) Dynamic effects in brittle fracture. In: Nemat-Nasser S (ed) Mechanics today 1. Pergamon Press, Oxford pp 1–57

Andrews DJ (1976) Rupture velocity of plane strain shear cracks. J Geophys Res 81(B32):5679–5689

Archuleta RJ (1984) A faulting model for the 1979 Imperial Valley Earthquake. J Geophys Res 89:4559–4585

Atkinson C, Craster RV (1991) Plane strain fracture in poroelastic media. Proc R Soc Lond A 434:605–633

Atkinson C, Craster RV (1992) Fracture in fully coupled dynamic thermo-elasticity. J Mech Phys Solids 40(7): 1415–1432

Berryman JG (1995) Mixture theories for rock properties. Rock Physics and Phase Relations. A handbook of physical constants. AGU Reference Shelf 3:205–228

Biegel RL, Sammis CG, Rosakis AJ (2007) Interaction of a dynamic rupture on a fault plane with short frictionless fault branches. Pure Appl Geophys 16:1881–1904

Biot MA (1941) General theory of three-dimensional consolidation. J Appl Phys 12:155–164

Biot MA (1956a) Theory of propagation of elastic waves in a fluid-saturated porous solid: I. Low frequency range. J Acoustical Soc Am 28(2):168–178

Biot MA (1956b) Theory of propagation of elastic waves in a fluid-saturated porous solid: II. Higher frequency range. J Acoustical Soc Am 28(2):179–191

Biot MA (1956c) Theory of propagation of elastic waves in a fluid saturated porous solid. J Acoustical Soc Am 28(1):168–191

Bouchon M, Bouin MP, Karabulut H, Toksäz MN, Dietrich M, Rosakis AJ (2001) How fast is rupture during an earthquake? New insights from the 1999 Turkey earthquakes. Geophys Res Lett 28(14):2723–2726

Bouchon M, Vallée M (2003) Observation of long supershear rupture during the magnitude 8.1 Kunlunshan Earthquake. Science 301:824–826

Bowen RM (1982) Compressible porous media models by use of the theory of mixtures. Int J Eng Sci 20(6):697–735

Broberg KB (1989) The near-tip field at high crack velocities. Int J Frac 39(1–3):1–13

Broberg KB (1995) Intersonic mode II crack expansion. Arch Mech 47:859–871

Broberg KB (1996) How fast can a crack go. Mater Sci 32(1):80–86

Broberg KB (1999) Cracks and fracture. Academic Press, London, UK

Brock LM (1977) Two basic problems of plane crack extension: a unified treatment. Int J Eng Sci 15:527–536

Burridge R (1973) Admissible speeds for plane-strain self-similar shear cracks with friction but lacking cohesion. Geophys J Roy Astronomical Soc 35:439–455

Burridge R, Conn G, Freund LB (1979) The stability of a plane strain shear crack with finite cohesive force running at intersonic speeds. J Geophys Res 84:2210–2222

Curran DA, Shockey DA, Winkler S (1970) Crack propagation at supersonic velocities - II. Theoretical model. Int J Frac Mech 6(3):271–278

Dunham EM, Archuleta RJ (2004) Evidence for a supershear transient during the 2002 Denali fault earthquake. Bull Seismol Soc Am 94(6B):S256–2268

Freund LB (1979) The mechanics of dynamic shear crack propagation. J Geophys Res 84(B5):2199–2209

Freund LB (1990) Dynamic fracture mechanics. Cambridge University Press, Cambridge, UK

Gao H, Huang Y, Gumbsch P, Rosakis AJ (1999) On radiation-free transonic motion of cracks and dislocations. J Mech Phys Solids 47(9):1941–1961

Geubelle PH, Kubair D (2001) Intersonic crack propagation in homogeneous media under shear dominated loading: numerical analysis. J Mech Phys Solids 49(3):571–587

Georgiadis HG (1986) On the stress singularity in steady-state transonic shear crack propagation. Int J Frac 30:175–180

Gradshteyn IS, Ryzhik IM (1980) Table of integrals, series and products. Academic Press, San Diego, USA

Huang Y, Gao H (2001) Intersonic crack propagation. Part I: the fundamental solution. J Appl Mech Trans ASME 68(2):169–175

Knackstedt MA, Arns CH, Val Pinczewski W (2005) Velocity-porosity relationships. Geophys Prospect 53:349–372

Loret B, Harireche O (1991) Acceleration waves, flutter instabilities and stationary discontinuities in inelastic porous media. J Mech Phys Solids 39(5):569–606

Loret B, Radi E (2001) On dynamic crack growth in poroelastic fluid-saturated media. J Mech Phys Solids 49(5):995–1020

Loret B, Rizzi E (1998) On the effects of inertial coupling on the wave-speeds of elastic-plastic fluid-saturated porous media. In: de Borst R, van der Giessen E (eds) Material instabilities in solids. J Wiley and Sons, Chichester pp 41–53

Muskhelishvili NI (1962) Some basic problems of the mathematical theory of elasticity. Noordhoff, Leyden, The Netherlands

Obrezanova O, Willis JR (2007) Stability of an intersonic shear crack to a perturbation of its edge. J Mech Phys Solids, in press, doi: 10.1016/j.jmps.2007.04.009

Palmer AC, Rice JR (1973) The growth of slip surfaces in the progressive failure of over consolidated clay. Proc R Soc Lond A 332:527–548

Radi E, Bigoni D, Loret B (2002) Steady crack-growth in elastic-plastic fluid-saturated porous media. Int J Plasticity 18(3):345–358

Radi E, Loret B (2007) Mode I intersonic crack propagation in poroelastic media. Mech Mater (in press)

Rice JR (1985) Shear localization, faulting, and frictional slip: Discusser's report. In: Bazant ZP (ed) Mechanics of geomaterials. Wiley, New York pp 211–216

Rice JR, Cleary MP (1976) Some basic stress-diffusion solutions for fluid-saturated elastic porous media with compressible constituents. Rev Geophys Space Phys 14(2):227–241

Rice JR, Simons DA (1976) The stabilization of spreading shear faults by coupled deformation-diffusion effects in fluid-infiltrated porous materials. J Geophys Res 81(29):5322–5334

Rosakis AJ, Samudrala O, Coker D (1999) Cracks faster than the shear wave speed. Science 284(5418):1337–1340

Rosakis AJ, Samudrala O, Coker D (2000) Intersonic shear crack growth along weak planes. Mater Res Innova 3(4):236–243

Rudnicki J (2001) Coupled deformation-diffusion effects in the mechanics of faulting and failure of geomaterials. Appl Mech Rev 54(6):483–502

Samudrala O, Huang Y, Rosakis AJ (2002) Subsonic and intersonic mode II crack propagation with a rate-dependent cohesive zone. J Mech Phys Solids 50:1231–1268

Spudich P, Cranswick E (1984) Direct observation of rupture propagation during the 1979 Imperial Valley Earthquake using a short baseline accelerometer array. Bull Seismol Soc Am 74(6):2083–2114

Washabaugh PD, Knauss WG (1994) A reconciliation of dynamic crack velocity and Rayleigh wave speed in isotropic brittle solids. Int J Frac 65(2):97–114

Winkler S, Shockey DA, Curran DA (1970) Crack propagation at supersonic velocities – I. Theoretical model. Int J Frac Mech 6(2):151–158

Material forces for crack analysis of functionally graded materials in adaptively refined FE-meshes

Rolf Mahnken

Abstract This work describes the computation of fracture parameters in functionally graded materials (FGMs) with stationary cracks. To this end the continuum concept of material forces is employed, such that the corresponding balance equation can be discretized with a standard Galerkin finite element procedure. A domain-type formulation is used for evaluation of a vectorial J-integral, where in the practical implementation the material nodal forces of the finite element discretization are summed up in a finite region of the crack-tip. In this way the numerical calculation is completely independent from the alignment of the finite element mesh or any selected integration contour, which is most attractive for adaptively refined finite element meshes. For illustrative purpose the accuracy of the method is discussed for two examples based on comparison with available theoretical and numerical solutions.

Keywords Material forces · Finite elements · Stress intensity factor · J-integral

1 Introduction

Functionally graded materials (FGMs) are advanced materials that possess continuously graded properties. During the past two decades much research of FGMs has been focused on manufacturing, material design and property estimation as well as thermal and structural analysis, Surush and Mortensen (1998). Applications of FGMs are composites, ceramics, alloys and coatings. These materials consists of two or more phases with spatially varying volume fractions thus rendering non-uniform microstructures. Compared to conventional materials, FGMs have the possibility of tailoring its gradation in order to maximize its performance. They have a better quality of thermal barrier and anti-fatigue thus resulting into superior thermal and structural performance in high temperature environments. Contrary to non-homogeneous materials with abrupt discontinuity, like laminated composite structures, FGMs are not subjected to stress singularities at sharp bimaterial interfaces.

Due to the spatially distributed microstructure, from the viewpoint of continuum mechanics, FGMs possess the distinguishing feature of non-homogeneity. Consequently mechanical material properties such as Young's modulus, yield strength, fracture toughness fatigue, creep resistance are varying within the structure, see e.g. Erdogan (1995). The same holds for thermal, magnetic and piezoelectric properties. Therefore parallel to developments in processing, research of FGMs must be pursued on theoretical modeling, numerical implementation and experimental validation. In this way an experimental technique for evaluating elastic properties of FGMs is investigated in Marur and Tippur (1998). A comprehensive review on fracture and failure of FGMs

R. Mahnken (✉)
Department of Engineering Mechanics (LTM),
University of Paderborn, Warburger Str. 100,
33098 Paderborn, Germany
e-mail: rolf.mahnken@ltm.uni-paderborn.de

Defect and Material Mechanics. C. Dascalu, G.A. Maugin & C. Stolz (eds.),
doi: 10.1007/978-1-4020-6929-1_23, © Springer Science+Business Media B.V. 2008

including related references is presented in Paulino et al. (2003). Micromechanically issues on effective properties of functionally graded composites are treated in Zuiker and Dvorak (1994). One of the few references considering inelastic behavior is Giannakopoulos et al. (1995).

Analytical investigations for the crack analysis of FGMs have been performed in several publications. In this way Delale and Erdogan (1983) investigated the crack problem in an infinite plane, where the elastic properties varied exponentially in the direction of the crack. They showed that the asymptotic crack-tip stress field possesses the same square root singularity as in homogeneous materials. Furthermore Eischen (1987a) used the traditional eigenfunction expansion technique of Williams (1957) to show, that the leading term in 2D crack-tip elastic fields remain square-root singular, which allows to express the stress fields in terms of stress intensity factors (SIFs). In 1994 Konda and Erdogan (1994) studied the behavior of an infinite cracked plane with exponential property gradients in both in-plane directions. Further analytical results for SIFs for some typical fracturing modes in FGMs are presented in Erdogan (1995) and Erdogan and Wu (1997).

An important quantity for crack analysis is the J-integral. For non-homogeneous materials the classical formulation of Eshelby (1951), Cherepanov (1967, 1968), Rice (1968) in general becomes path dependent. For certain classes of FGMs a path independent extension of the J-integral is given in Honein and Herrmann (1997). Here also a relation of the J-integral to SIFs is provided, thus obtaining the same mathematical structure as for homogeneous materials. Furthermore, for general non-homogeneous materials a path independent J^*-integral is established in Eischen (1987a,b) and Kim and Paulino (2002).

Numerical methods for crack analysis of FGMs are presented in Gu et al. (1999), assigning different homogeneous elastic properties to each finite element, Anlas et al. (2000), who numerically integrated an area region and Marur and Tippur (2000), considering a crack normal to the elastic gradient. A numerical procedure considering the J-integral in inhomogeneous materials is also presented in Haddi and Weichert (1995). Further references are given e.g. in Kim and Paulino (2002).

This work is concerned with the computation of fracture parameters in functionally graded materials by exploiting the continuum concept of material forces. To give a conceptual motivation we recall, that *spatial (or physical) forces (in the sense of Newton) are generated by variations relative to the ambient space at fixed material position*. In 1951 Eshelby (1951, 1965) introduced the concept of a force acting on an elastic singularity. He defined this force as the negative (material) gradient of the strain energy, and for this reason called it material (driving, configurational) force. Following Steinmann (2001b) *material forces (in the sense of Eshelby) are generated by variations relative to the ambient material at fixed spatial position*. The duality between spatial and material forces carries over to the concept of stresses, in the sense that the Cauchy stress of the spatial problem has its counterpart in the Eshelby stress of the material problem, see e.g. Shield (1967), Chadwick (1975) and Steinmann (2001b). An extensive exposition of related concepts in continuum mechanics is given in Maugin (1993), Gurtin (2000) and Kienzler and Herrmann (2000).

Apart from valuable theoretical benefits, e.g. in defect mechanics, the continuum formulation in material space renders several computational advantages. As a consequence finite element techniques well established for the spatial motion problem, such as the Galerkin weak formulation and computation of the discrete node forces, carry over to the material motion problem with only minor modifications. In this way Braun (1997) introduced material forces in the context of the finite element method in 1997, followed by various publications such as Nguyen et al. (2005), Müller et al. (2001), Müller and Maugin (2002), Müller et al. (2004), Steinmann et al. (2001), Steinmann (2001a,b), Rajagopal and Sivakumar (2007), Denzer (2006), Denzer et al. (2003), amongst others.

Following Steinmann et al. (2001) and Steinmann (2001a,b), the discrete material nodal force acting on a node representing the crack-tip can be interpreted as a vectorial J-integral. However, the numerical results indicate the necessity of a strong densification near the crack-tip for the finite element mesh in order to get satisfying results. As explained by Denzer (2006), the accuracy of the numerical results depends strongly on the accuracy of the Eshelby stress in the vicinity of the crack-tip, which due to the singularity of the Cauchy stress, in general, also has a singularity. This in turn can render erroneous numerical results for the fracture parameters of interest. Therefore in this work we exploit a domain-type formulation for evaluation of a vectorial J-integral, where in the practical implementation the material nodal forces of the finite element discretization are summed up in a finite region of the crack-tip. In this way the numerical calculation is completely independent from the alignment of the finite element mesh or any selected integration contour, which is most attractive for adaptively refined finite element meshes.

An outline of this work is as follows: Sect. 2 summarizes the equilibrium conditions of both, physical forces and material forces for a functionally graded elastic material. In Sect. 3 the corresponding virtual work formulations and the discretizations are outlined. For illustrative purpose the accuracy of the method is discussed for two examples in Sect. 4, based on comparison with available theoretical and numerical solutions. Results are compared for both regularly and adaptively refined finite element meshes.

Notations

Square brackets [•] are used throughout the paper to denote 'function of' in order to distinguish from mathematical groupings with parenthesis (•).

2 Equilibrium conditions for a functionally graded elastic material

2.1 Equilibrium of physical forces

To set the stage the boundary value problem (BVP) for a functionally graded elastic material is introduced within a geometrically linear theory. For this purpose in Fig. 1 we consider the configuration $\mathcal{B} \subset \mathbb{E}^{n_{dim}}$ occupied by a body B with position $\mathbf{x} \in \mathbb{E}^{n_{dim}}$, and where $\mathbb{E}^{n_{dim}}$ denotes the Euclidean space with dimension $n_{dim} = 2$ or $n_{dim} = 3$. Displacements are given by the vector field $\mathbf{u}$: $\mathcal{B} \mapsto \mathbb{E}^{n_{dim}}$, and distributed body forces per unit mass are given by the vector field $\mathbf{b}$: $\mathcal{B} \mapsto \mathbb{E}^{n_{dim}}$ assumed to be independent from time t.

We concentrate on an arbitrary subdomain $\Omega \subset \mathcal{B}$ with boundary Γ. This subdomain is loaded along Γ by surface tractions $\boldsymbol{\sigma} \cdot \mathbf{n}$ in terms of the symmetric Cauchy stress tensor $\boldsymbol{\sigma}$ projected onto Γ by the surface normal $\mathbf{n}$. Furthermore, within the region Ω volume forces $\mathbf{b}$ are present. Then, upon defining the surface and body forces, respectively,

$$\mathbf{f}_{sur} = \int_{\Gamma} \boldsymbol{\sigma} \cdot \mathbf{n} \, \mathrm{dA} \quad \text{and} \quad \mathbf{f}_{vol} = \int_{\Omega} \mathbf{b} \, \mathrm{dV} \tag{1}$$

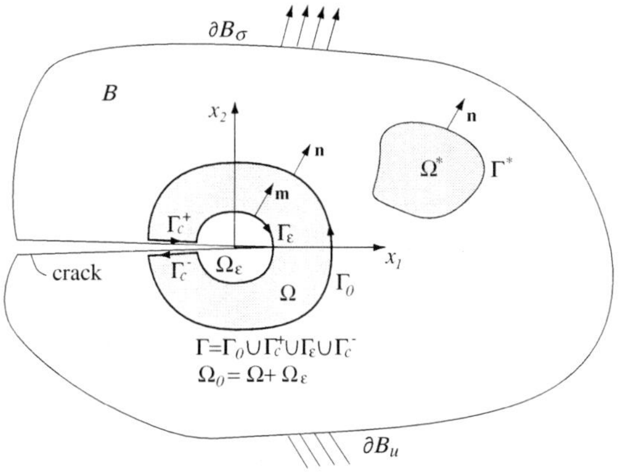

Fig. 1 Schematic of body B with configuration $\mathcal{B}$ and boundaries $\partial\mathcal{B}_\sigma$ and $\partial\mathcal{B}_u$. Furthermore Ω is a subdomain with boundary Γ near the crack-tip, and Ω^* is an additional arbitrary subdomain with boundary Γ^*

the quasi-static equilibrium for the subdomain Ω is postulated as

$$\mathbf{f}_{sur} + \mathbf{f}_{vol} = \mathbf{0}. \tag{2}$$

Applying the divergence theorem

$$\mathbf{f}_{sur} = \int_{\Gamma} \boldsymbol{\sigma} \cdot \mathbf{n} \, \mathrm{dA} = \int_{\Omega} \boldsymbol{\sigma} \cdot \nabla \, \mathrm{dV} \tag{3}$$

Eq. 2 renders as a local format the balance equation

$$\boldsymbol{\sigma} \cdot \nabla + \mathbf{b} = \mathbf{0} \quad \forall \mathbf{x} \in \mathcal{B}, \tag{4}$$

where with respect to Cartesian coordinates x_i and basis vectors $\mathbf{e}_i$, $i = 1, 2, 3$, the nabla operator is defined as $\nabla(\cdot) = \partial_{x_i}(\cdot)\,\mathbf{e}_i$, see e.g. Spiegel (1959).

It is noteworthy, that the region Ω with boundary Γ can be chosen arbitrarily within $\partial\mathcal{B}$ and $\mathcal{B}$ for evaluation of the Eqs. 2 and 4 and is not confined to the crack-tip region as indicated in Fig. 1. In this way it could be replaced by an arbitrary subdomain Ω^* with boundary Γ^*. The only restriction is, that it must be free from singular stresses, so that the divergence theorem (3) can be applied. Consequently Eq. 4 holds for any point of the body with position $\mathbf{x} \in \mathcal{B}$ except at a point of singularity.

A further fundamental ingredient of a geometrically linear theory is the strain–displacement relation

$$\begin{aligned} &1.\ \ \boldsymbol{\varepsilon} = \tfrac{1}{2}\left(\mathbf{h} + \mathbf{h}^T\right) \quad \forall \mathbf{x} \in \mathcal{B} \\ &2.\ \ \mathbf{h} = \mathbf{u} \otimes \nabla, \end{aligned} \tag{5}$$

where $\mathbf{h}$ is the displacement gradient. Let us assume the functional relationship

$$\psi = \psi[\mathbf{x}, \boldsymbol{\varepsilon}[\mathbf{x}]] \tag{6}$$

for the strain energy function of an FGM, then the stress tensor is obtained as

$$\sigma = \frac{\partial \psi}{\partial \varepsilon}. \tag{7}$$

For an FGM with elastic isotropic behavior the strain energy function is given as

$$\begin{aligned} &1.\quad \psi[\mathbf{x}, \varepsilon[\mathbf{x}]] = \tfrac{1}{2}\varepsilon : \mathbb{C}[\mathbf{x}] : \varepsilon[\mathbf{x}] \\ &2.\quad \mathbb{C}[\mathbf{x}] \qquad = 2G[\mathbf{x}]\,\mathbb{I}^{dev} + K[\mathbf{x}]\mathbf{1} \otimes \mathbf{1}. \end{aligned} \tag{8}$$

Here we define $\mathbb{I}^{dev} = \mathbb{I} - 1/3\mathbf{1} \otimes \mathbf{1}$ with second and fourth oder unit tensors $\mathbb{I}$ and $\mathbf{1}$, respectively. Furthermore, in Eq. 8.2 we introduce the shear modulus $G[\mathbf{x}]$ and the bulk modulus $K[\mathbf{x}]$ of the FGM, related to Young's modulus $E[\mathbf{x}]$ and Poisson's ratio $\nu[\mathbf{x}]$ as

$$\begin{aligned} &1.\quad G[\mathbf{x}] = \tfrac{E[\mathbf{x}]}{2(1+\nu[\mathbf{x}])} \\ &2.\quad K[\mathbf{x}] = \tfrac{E[\mathbf{x}]}{3(1-2\nu[\mathbf{x}])}. \end{aligned} \tag{9}$$

Let us briefly comment on mathematical formulations representing the gradation of an FGM. This can be achieved e.g. by a compositional distribution function $C[\mathbf{x}]$. An example for a two dimensional gradation is given in Hirano et al. (1990) as

$$C[\mathbf{x}] = (C_2 - C_1)\left(\frac{x - x_1}{x_2 - x1}\right)^n, \tag{10}$$

where C_1 and C_2 denote composition values at the opposite ends of the FGM layer corresponding to the distances x_1 and x_2, respectively. This function can be multiplied e.g. with a reference value E_0 for Young's modulus, such that $E[\mathbf{x}] = E_0 C[\mathbf{x}]$ reflects the gradation of the elastic behavior.

An alternative approach is considered in the field of topology optimization. It states, that at each point of the domain the material property is related to the pseudo-density ρ^p describing the amount at each point of the domain (Bendsoe and Sigmund 2003). In this way the elasticity within the domain Ω is expressed as

$$E^H = \rho^p E_0, \tag{11}$$

where E_0 is Young's modulus of the basic material (Silva and Paulino 2005).

In order to complete the boundary value problem, boundary conditions must be specified: Upon subdividing the boundary $\partial\mathcal{B}$ with outward normal $\mathbf{n}$ into disjoint parts $\partial_u\mathcal{B} \cup \partial_\sigma\mathcal{B} = \partial\mathcal{B}$ with $\partial_u\mathcal{B} \cap \partial_\sigma\mathcal{B} = \emptyset$, Dirichlet boundary conditions $\mathbf{u} = \bar{\mathbf{u}}$ on $\partial\mathcal{B}_u$ and Neumann boundary conditions $\sigma \cdot \mathbf{n} = \bar{\mathbf{t}}$ on $\partial\mathcal{B}_\sigma$ are prescribed.

2.2 Equilibrium of material forces

This section is concerned with the derivation of a balance law for material forces. In this context the so called Eshelby tensor is introduced, which will be used for numerical evaluation of the J-integral in the ensuing section. The following derivation for a local balance law is adopted from Denzer (2006), where a tensorial notation is used. Note, that conceptually it is similar to the derivation in Kim and Paulino (2002), where an index notation is used.

As a starting point, we multiply the balance Eq. 4 with the negative displacement gradient $-\mathbf{h}$ of Eq. 5.2 and obtain by application of the chain rule the following identity

$$\begin{aligned} -\mathbf{h}^T \cdot (\sigma \cdot \nabla) - \mathbf{h}^T \cdot \mathbf{b} &= (\mathbf{h}^T \otimes \nabla) : \sigma \\ &\quad -(\mathbf{h}^T \cdot \sigma) \cdot \nabla - \mathbf{h}^T \cdot \mathbf{b} = \mathbf{0}. \end{aligned} \tag{12}$$

Next, the spatial gradient of the strain energy function ψ in Eq. 6 is calculated by the chain rule as

$$\begin{aligned} \nabla\psi &= \frac{\partial\psi}{\partial\mathbf{x}} = \frac{\partial\psi}{\partial\varepsilon} \cdot \frac{\partial\varepsilon}{\partial\mathbf{x}} + \left.\frac{\partial\psi}{\partial\mathbf{x}}\right|_{\text{expl}} \\ &= \sigma : (\mathbf{h} \otimes \nabla) + \left.\frac{\partial\psi}{\partial\mathbf{x}}\right|_{\text{expl}}, \end{aligned} \tag{13}$$

where the constitutive relation (7) and the kinematic relations (5) have been used. The explicit derivative of ψ is defined as

$$\left.\frac{\partial\psi}{\partial\mathbf{x}}\right|_{\text{expl}} = \left.\frac{\partial\psi}{\partial\mathbf{x}}\right|_{\varepsilon=\text{const}}. \tag{14}$$

Upon using the identity (Denzer 2006)

$$\sigma : (\mathbf{h} \otimes \nabla) = \left(\mathbf{h}^T \otimes \nabla\right) : \sigma, \tag{15}$$

Eq. 13 renders the relations

$$\begin{aligned} \sigma : (\mathbf{h} \otimes \nabla) &= \nabla\psi - \left.\frac{\partial\psi}{\partial\mathbf{x}}\right|_{\text{expl}} \\ &\Longrightarrow \left(\mathbf{h}^T \otimes \nabla\right) : \sigma \\ &= \psi\mathbf{1} \cdot \nabla - \left.\frac{\partial\psi}{\partial\mathbf{x}}\right|_{\text{expl}}. \end{aligned} \tag{16}$$

Inserting this result into Eq. 12 and rearranging we obtain

$$\psi\mathbf{1} \cdot \nabla - (\mathbf{h}^T \cdot \sigma) \cdot \nabla - \left.\frac{\partial\psi}{\partial\mathbf{x}}\right|_{\text{expl}} - \mathbf{h}^T \cdot \mathbf{b} = \mathbf{0}, \tag{17}$$

which is written as a balance equation in local format

$$
\begin{aligned}
&1.\ \boldsymbol{\Sigma} \cdot \nabla + \mathbf{B} = \mathbf{0} \quad \forall \mathbf{x} \in \mathcal{B} \\
&2.\ \boldsymbol{\Sigma} = \psi \mathbf{1} - \mathbf{h}^T \cdot \boldsymbol{\sigma} \\
&3.\ \mathbf{B} = -\mathbf{h}^T \cdot \mathbf{b} - \left.\frac{\partial \psi}{\partial \mathbf{x}}\right|_{\text{expl}} .
\end{aligned}
\tag{18}
$$

Here $\boldsymbol{\Sigma}$ is the *energy momentum tensor* or *Eshelby tensor*, respectively, introduced by Eshelby in Eshelby (1951, 1965), and $\mathbf{B}$ is the *material volume force* or *Eshelby volume force*, respectively.

In order to obtain a balance equation in global format, we consider an arbitrary subdomain Ω with boundary Γ of the domain $\mathcal{B}$, as illustrated in Fig. 1. The subdomain is loaded along Γ by surface tractions $\boldsymbol{\Sigma} \cdot \mathbf{n}$ in terms of the Eshelby stress $\boldsymbol{\Sigma}$ projected onto Γ by the surface normal $\mathbf{n}$ and within Ω by the material volume force $\mathbf{B}$ given in Eq. 18.3. (Note, that Ω and Γ in Fig. 1 can be replaced by Ω^* and Γ^*, respectively.) Then, performing the integral of the balance Eq. 18.1 over Ω and defining the material surface and body forces, respectively,

$$\mathbf{F}_{sur} = \int_\Omega \boldsymbol{\Sigma} \cdot \nabla \, \mathrm{dV} \quad \text{and} \quad \mathbf{F}_{vol} = \int_\Omega \mathbf{B} \, \mathrm{dV} \tag{19}$$

the equilibrium of material forces for the subdomain Ω is derived as

$$\mathbf{F}_{sur} + \mathbf{F}_{vol} = \mathbf{0}. \tag{20}$$

Applying the divergence theorem, the material surface force can be rewritten as

$$\mathbf{F}_{sur} = \int_\Omega \boldsymbol{\Sigma} \cdot \nabla \, \mathrm{dV} = \int_\Gamma \boldsymbol{\Sigma} \cdot \mathbf{n} \, \mathrm{dA}, \tag{21}$$

such that Eq. 20 follows as

$$\int_\Gamma \boldsymbol{\Sigma} \cdot \mathbf{n} \, \mathrm{dA} + \int_\Omega \mathbf{B} \, \mathrm{dV} = \mathbf{0}. \tag{22}$$

Remark 2.1

1. We recall, that the above transformation (21) with the divergence theorem is only possible for differentiable functions. It is not valid for singular stress distributions at the crack-tip.
2. Note the duality for the balance equations in local format Eq. 4 for the Cauchy stress tensor $\boldsymbol{\sigma}$ and Eq. 18.1 for the Eshelby stress tensor $\boldsymbol{\Sigma}$. Additionally we observe the quasi-static equilibrium for physical forces Eq. 2 as the counterpart of the equilibrium for material forces Eq. 20.
3. In the above approach the balance law of material forces Eq. 18 is *derived* from the balance law of physical forces Eq. 4 and the constitutive assumption (6), where the latter two equations are *postulated*. This is different to an approach occasionally found in the literature, where an independent balance law of material forces is introduced as a *postulate*, additionally to the postulate of balance of physical forces. This approach is used e.g. in Makowski et al. (2006) in order to analyse the dissipative process of crack evolution in brittle and ductile materials.
4. Considering a finite strain setting, see e.g. Maugin (1993), Gurtin (2000), Kienzler and Herrmann (2000), Steinmann (2001b), Denzer (2006), the left side of Eq. 12 can be interpreted as the pull-back of the physical balance equation Eq. 4 to the reference configuration. This also indicates, that the balance law of material forces Eq. 18 is a direct consequence of the balance law of physical forces Eq. 4 and the constitutive assumption Eq. 6, i.e. both balance laws are not independent.

2.3 Vectorial and scalar J-integrals for functionally graded materials

For crack analysis it is important to have analytical quantities which characterize the near tip situation. To this end we consider a region Ω in Fig. 1 bounded by a closed curve Γ composed of segments $\Gamma_0, \Gamma_c^+, \Gamma_c^-, \Gamma_\epsilon$. It should not contain the crack-tip, where the stresses are singular and therefore unbounded. The region between Γ_ϵ and the crack surfaces is Ω_ϵ. A J-integral vector $\mathbf{J} \in \mathrm{I\!R}^{n_{dim}}$ is now defined as

$$\mathbf{J} = \lim_{\Gamma_\epsilon \to 0} \int_{\Gamma_\epsilon} \boldsymbol{\Sigma} \cdot \mathbf{m} \, \mathrm{dA}, \tag{23}$$

where $\mathbf{m} = -\mathbf{n}$ is the outward normal of the path Γ_ϵ as indicated in Fig. 1. The integration in Eq. 23 can start from a point on the lower crack face and ends at an opposite point on the upper crack face.

Remark 2.2

1. Let $\mathbf{e}_{||}$ be a unit vector parallel to the crack, then a scalar J-integral is obtained as

$$J = \mathbf{e}_{||} \cdot \mathbf{J} = \lim_{\Gamma_\epsilon \to 0} \int_{\Gamma_\epsilon} \mathbf{e}_{||} \cdot \boldsymbol{\Sigma} \cdot \mathbf{m} \, \mathrm{dA}. \tag{24}$$

This expression equals the classical formulation of Eshelby (1951), Cherepanov (1967, 1968), Rice (1968) and is path independent for homogeneous materials.

2. For non-homogeneous materials the integral under the limit in Eq. 23 is not divergence free due to the balance Eq. 18.1, and therefore corresponding line integrals in a finite vicinity of the crack-tip become path dependent. As a consequence the numerical evaluation of the line integrals (23) and (24) can become erroneous.
3. We assume, that derivatives of the elastic modulo are bounded at the crack-tip, such that the volume integral for the material body force vector $\mathbf{B}$ in Eq. 19 vanishes for vanishing integration line $\Gamma_{\varepsilon\to 0}$. Consequently,

$$\begin{aligned}\mathbf{J} &= \lim_{\Gamma_\epsilon\to 0}\int_{\Gamma_\epsilon}\boldsymbol{\Sigma}\cdot\mathbf{m}\,\mathrm{dA}\\ &= \lim_{\Gamma_\epsilon\to 0}\left(\int_{\Gamma_\epsilon}\boldsymbol{\Sigma}\cdot\mathbf{m}\,\mathrm{dA}+\int_{\Omega_\epsilon}\mathbf{B}\right)\mathrm{dV}\\ &= \mathbf{J}^*, \qquad (25)\end{aligned}$$

which is the so-called $\mathbf{J}^*$ integral vector of Eischen (1987a). Based on the equilibrium Eq. 22 a derivation of Eq. 25 is also given in Kim and Paulino (2002). In Eischen (1987a) path independence of the integral is shown for general non-homogeneous materials.

2.4 Domain integrals

The numerical evaluation of fracture parameters on the basis of the line integrals (24) and (25) has two disadvantages. First it is liable to numerically erroneous results, also for the case of path independence for the J-integral. Second, it demands non-standard Finite Element data structures. In order to circumvent these difficulties we reformulate the line integrals as domain integrals, similarly to the domain integral method of Li et al. (1985). The resulting procedure can easily be implemented into existing finite element programs. Furthermore, it enables the computation of fracture parameters using relatively large regions, thus improving the numerical accuracy.

We consider an arbitrary region Ω around the crack-tip, for the two-dimensional case shown in Fig. 1. According to the domain integral method of Li et al. (1985) a smooth function $q \in H_1(\mathcal{B})$ ($H_1(\mathcal{B})$ being a function-space, where functions and its first derivatives are square-integrable) is introduced with the properties

$$q(\mathbf{x}) = \begin{cases} 1 & \text{on } \Gamma_\epsilon \\ 0 & \text{on } \Gamma_0. \end{cases} \qquad (26)$$

On the segments Γ_c^+, Γ_c^- the function q varies continuously from 1 to 0, which allows application of the divergence theorem to the following line integral

$$\begin{aligned}\int_{\Gamma_\epsilon}\boldsymbol{\Sigma}\cdot\mathbf{n}\,\mathrm{dA} &= \int_{\Gamma}(q\,\boldsymbol{\Sigma})\cdot\mathbf{n}\,\mathrm{dA}-\int_{\Gamma_c^+,\Gamma_c^-}(q\,\boldsymbol{\Sigma})\cdot\mathbf{n}\,\mathrm{dA}\\ &= \int_{\Omega}(q\,\boldsymbol{\Sigma})\cdot\nabla\,\mathrm{dV}\\ &\quad -\int_{\Gamma_c^+,\Gamma_c^-}\left(q\,\psi\,\mathbf{n}-q\,\mathbf{h}^T\cdot\boldsymbol{\sigma}\cdot\mathbf{n}\right)\mathrm{dA}\\ &= \int_{\Omega}\boldsymbol{\Sigma}\cdot(\nabla q)\,\mathrm{dV}+\int_{\Omega}q\;(\boldsymbol{\Sigma}\cdot\nabla)\,\mathrm{dV}\\ &\quad -\int_{\Gamma_c^+}[[\psi]]q\,\mathbf{n}^+\,\mathrm{dA}\\ &\quad +\int_{\Gamma_c^+,\Gamma_c^-}\left(q\,\mathbf{h}^T\cdot\bar{\mathbf{t}}\right)\mathrm{dA}, \qquad (27)\end{aligned}$$

and where the Neumann boundary condition $\boldsymbol{\sigma}\cdot\mathbf{n}=\bar{\mathbf{t}}$ on Γ_c^+, Γ_c^- has been considered. The notation $[[\psi]] = \psi^+ - \psi^-$ reflects the discontinuity (or jump) in the strain energy function across the crack opening. Upon exploiting Eq. 27 and the balance Eq. 18.1 the integral vector in Eq. 23 is

$$\begin{aligned}\mathbf{J} &= \lim_{\Gamma_\epsilon\to}\int_{\Gamma_\epsilon}\boldsymbol{\Sigma}\cdot\mathbf{m}\,\mathrm{dA} = -\lim_{\Gamma_\epsilon\to 0}\int_{\Gamma_\epsilon}\boldsymbol{\Sigma}\cdot\mathbf{n}\,\mathrm{dA}\\ &= -\lim_{\Gamma_\epsilon\to 0}\left(\int_{\Omega}\boldsymbol{\Sigma}\cdot(\nabla q)\,\mathrm{dV}\right.\\ &\quad -\int_{\Omega}q\,\mathbf{B}\,\mathrm{dV}-\int_{\Gamma_c^+}[[\psi]]q\,\mathbf{n}^+\,\mathrm{dA}\\ &\quad \left.+\int_{\Gamma_c^+,\Gamma_c^-}\left(q\,\mathbf{h}^T\cdot\bar{\mathbf{t}}\right)\mathrm{dA}\right). \qquad (28)\end{aligned}$$

As noted in Kim and Paulino (2002), the third term on the right hand side including the jump $[[\psi]]$ must be accounted for, so that relatively large regions can be used for evaluation of the J-integral by a domain integral approach. Note, that the term vanishes for evaluation of J in Eq. 24 with $\mathbf{e}_{||}$ parallel to a straight crack.

3 Virtual work and discretization

3.1 Spatial and material motion problem

In the following the quasi-static local balance equations for the spatial motion equilibrium and the material motion equilibrium will be formulated as a weak (or variational) form, following closely the procedure in Steinmann et al. (2001), Steinmann (2001a,b) and Denzer (2006). In both cases the standard Galerkin discretization is performed, thus rendering the finite element physical nodal forces and the finite element material nodal forces.

We multiply the local balance Eq. 4 with a test function (spatial virtual displacement) $\mathbf{v}$ under the necessary smoothness and boundary assumptions, apply the divergence theorem and obtain a virtual work expression. Analogously, we multiply the local balance Eq. 18.1 with a test function (material virtual configuration) $\mathbf{V}$ under the necessary smoothness and boundary assumptions, apply the divergence theorem and obtain a second virtual work expression. Consequently we have

$$1. \int_{\partial\mathcal{B}} \mathbf{v}\cdot\boldsymbol{\sigma}\cdot\mathbf{n}\, dA = \int_{\mathcal{B}} (\mathbf{v}\otimes\nabla) : \boldsymbol{\sigma}\, \mathrm{dV} - \int_{\mathcal{B}} \mathbf{v}\cdot\mathbf{b}\, \mathrm{dV} \quad \forall\, \mathbf{v}$$

$$2. \int_{\partial\mathcal{B}} \mathbf{V}\cdot\boldsymbol{\Sigma}\cdot\mathbf{n}\, dA = \int_{\mathcal{B}} (\mathbf{V}\otimes\nabla) : \boldsymbol{\Sigma}\, \mathrm{dV} - \int_{\mathcal{B}} \mathbf{V}\cdot\mathbf{B}\, \mathrm{dV} \quad \forall\, \mathbf{V}. \qquad (29)$$

Next, as indicated in Fig. 2, the domain $\mathcal{B}^h \approx \mathcal{B}$ is discretized into N_{el} elements, such that $\mathcal{B}^h = \cup_{e=1}^{N_{el}} \mathcal{B}_e$. Let n_{en} be the number of all nodes within $\mathcal{B}_e$ and $n = 1, \ldots, n_{en}$ be the local numbering on each element. Then the geometry, the nodal displacements, the virtual displacements and the virtual geometry are interpolated by shape functions Φ_e^n as

$$\mathbf{x}^h\Big|_{\mathcal{B}^e} = \sum_{n=1}^{n_{en}} \Phi_e^n \mathbf{x}_n, \quad \mathbf{u}^h\Big|_{\mathcal{B}^e} = \sum_{n=1}^{n_{en}} \Phi_e^n \mathbf{u}_n,$$
$$\mathbf{v}^h\Big|_{\mathcal{B}^e} = \sum_{n=1}^{n_{en}} \Phi_e^n \mathbf{v}_n^e, \quad \mathbf{V}^h\Big|_{\mathcal{B}^e} = \sum_{n=1}^{n_{en}} \Phi_e^n \mathbf{V}_n^e, \qquad (30)$$

where $\mathbf{x}_n, \mathbf{u}_n, \mathbf{v}_n^e, \mathbf{V}_n^e \in \mathbb{R}^{n_{dim}}$ are corresponding nodal values. Furthermore, the element wise Jacobian matrix is obtained as

$$\mathbf{J}^e = \mathbf{x}^h\Big|_{\mathcal{B}^e} \otimes \nabla_\xi = \sum_{n=1}^{n_{en}} \mathbf{x}_n \otimes \nabla_\xi \Phi_e^n. \qquad (31)$$

Here $\xi \in \mathbb{R}^{n_{dim}}$ represents the vector of natural coordinates within the reference element □ as indicated in Fig. 2. (The notation $\mathbf{J}^e$ should not be confused with the vectorial J-integral introduced in Eq. 23.) By use of $\nabla_x\{\bullet\} = \nabla_\xi\{\bullet\}\cdot\mathbf{J}^{e-1}$ the corresponding material

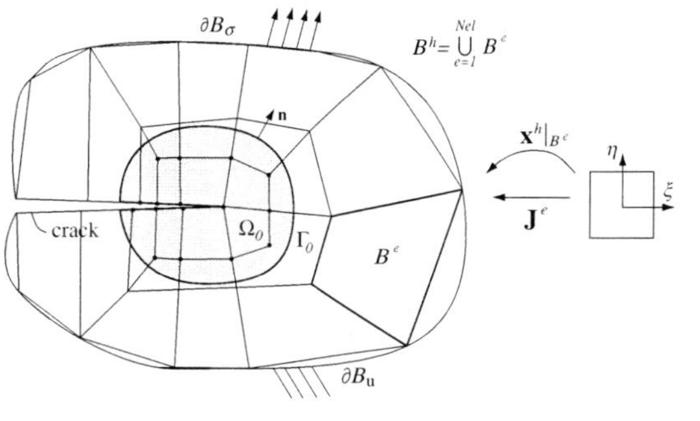

Fig. 2 Discretization of the domain $\mathcal{B} \approx \mathcal{B}^h$ into elements $\mathcal{B}_e$, transformation $\mathbf{J}^e$ from the reference element. Dotted nodes are within the region Ω_0

gradients of fields appearing in Eq. 30 are obtained as

$$\mathbf{u}^h\Big|_{\mathcal{B}^e} \otimes \nabla_x = \sum_{n=1}^{n_{en}} \mathbf{u}_n \otimes \nabla_x \Phi_e^n,$$
$$\mathbf{v}^h\Big|_{\mathcal{B}^e} \otimes \nabla_x = \sum_{n=1}^{n_{en}} \mathbf{v}_n^e \otimes \nabla_x \Phi_e^n, \qquad (32)$$
$$\mathbf{V}^h\Big|_{\mathcal{B}^e} \otimes \nabla_x = \sum_{n=1}^{n_{en}} \mathbf{V}_n^e \otimes \nabla_x \Phi_e^n.$$

Then, the element wise contributions of the right hand sides in Eqs. 29.1 and 29.2 are

$$1. \sum_{n=1}^{n_{en}} \mathbf{v}_n^e \cdot \int_{\mathcal{B}_e} \boldsymbol{\sigma}\cdot\nabla_x\Phi_e^n\, \mathrm{dV} - \sum_{n=1}^{n_{en}} \mathbf{v}_n^e \cdot \int_{\mathcal{B}_e} \mathbf{b}\, \Phi_e^n\, \mathrm{dV} = \sum_{n=1}^{n_{en}} \mathbf{v}_n^e \cdot \mathbf{f}_{int}^{n,e}$$

$$2. \sum_{n=1}^{n_{en}} \mathbf{V}_n^e \cdot \int_{\mathcal{B}_e} \boldsymbol{\Sigma}\cdot\nabla_x\Phi_e^n\, \mathrm{dV} - \sum_{n=1}^{n_{en}} \mathbf{V}_n \cdot \int_{\mathcal{B}_e} \mathbf{B}\, \Phi_e^n\, \mathrm{dV} = \sum_{n=1}^{n_{en}} \mathbf{V}_n^e \cdot \mathbf{F}_{int}^{n,e}, \qquad (33)$$

where

$$1. \ \mathbf{f}_{int}^{n,e} = \int_{\mathcal{B}_e} \left(\boldsymbol{\sigma}\cdot\nabla_x\Phi_e^n - \mathbf{b}\,\Phi_e^n\right) \mathrm{dV}$$

$$2. \ \mathbf{F}_{int}^{n,e} = \int_{\mathcal{B}_e} \left(\boldsymbol{\Sigma}\cdot\nabla_x\Phi_e^n - \mathbf{B}\,\Phi_e^n\right) \mathrm{dV} \qquad (34)$$

are nodal force vectors $\mathbf{f}_{int}^{n,e}, \mathbf{F}_{int}^{n,e} \in \mathbb{R}^{n_{dim}}$ related to the nth node at element e. We let N_{en} be the number of nodes within $\mathcal{B}^h$. Then, upon assembling all element wise and node wise contributions of Eq. 34 to a node $N = 1, \ldots, N_{en}$, internal parts for physical and material nodal forces are obtained as

1. $\mathbf{f}_{int}^{N} = \sum_{e=1}^{N_{el}} \sum_{n \in \mathcal{J}_N^e} \mathbf{f}_{int}^{n,e}$,

2. $\mathbf{F}_{int}^{N} = \sum_{e=1}^{N_{el}} \sum_{n \in \mathcal{J}_N^e} \mathbf{f}_{int}^{n,e}$,

$$\mathcal{J}_N^e = \{n : \text{Node } n \text{ corresponds to } N\}. \qquad (35)$$

Concerning dimension of the above nodal vectors we observe $\mathbf{f}_{int}^N, \mathbf{F}_{int}^N \in \mathbb{R}^{n_{dim}}$.

3.2 Material force method for J-integral evaluation

As outlined in Steinmann (2001b), for infinitely small regions the vectorial J-integral in Eq. 25 collapses to the material nodal force at the crack-tip. However, the numerical results indicate the necessity for a strong densification of the finite element mesh near the crack-tip in order to get satisfying results. As explained by Denzer (2006), the accuracy of the numerical results depends strongly on the accuracy of the Eshelby stress $\boldsymbol{\Sigma}$ in the vicinity of the crack-tip, which due to the singularity of the Cauchy stress $\boldsymbol{\sigma}$, in general, also has a singularity. This in turn can render erroneous numerical results for the fracture parameters of interest. Therefore it is advantageous to use a strategy, which uses results away from the location of singularity, while still representing the crack-tip region. This is accomplished with the domain integral in Eq. 28.

We consider an approximation of the function $q(\mathbf{x})$ in Eq. 26 in terms of the shape functions Φ_e^n

$$q(\mathbf{x}) = \sum_{e=1}^{N_{el}} \sum_{n \in \mathcal{J}_\Omega^e} \Phi_e^n(\mathbf{x}),$$

$$\mathcal{J}_\Omega^e = \{n : \text{Node } n \text{ is in subdomain } \Omega\}. \qquad (36)$$

Note, that a finite element node at the crack-tip is not a member of $\mathcal{J}_\Omega^e$. Inserting Eq. 36 into Eq. 28 renders

$$\mathbf{J} = - \lim_{\Gamma_\epsilon \to 0} \sum_{e=1}^{N_{el}} \sum_{n \in \mathcal{J}_\Omega^e} \left(\int_{\mathcal{B}_e} \boldsymbol{\Sigma} \cdot \nabla_x \Phi_e^n \, \mathrm{dV} - \int_{\mathcal{B}_e} \mathbf{B} \Phi_e^n \, \mathrm{dV} - \int_{\Gamma_c^+} [[\psi]] \Phi_e^n \, \mathbf{n}^+ \, \mathrm{dA} + \int_{\Gamma_c^+, \Gamma_c^-} \left(\Phi_e^n \, \mathbf{h}^T \cdot \bar{\mathbf{t}} \right) \mathrm{dA} \right). \qquad (37)$$

Remark 3.1

1. A numerical procedure for evaluation of the third term on the right hand side in Eq. 37 including the jump $[[\psi]]$ is described in Kim and Paulino (2002).
2. For simplicity, we restrict ourselves to problems with a symmetry of ψ with respect to the crack front, such that $[[\psi]] = 0$. Furthermore, traction free crack surfaces are considered, such that $\bar{\mathbf{t}} = \mathbf{0}$. We also note, that the first two terms on the right side of Eq. 37 represent the nodal force vector $\mathbf{F}_{int}^{n,e}$ in Eq. 34.2. In the practical implementation this expression is calculated by numerical integration using Gaussian quadrature points, see e.g. Hughes (1987). This avoids the use of singular stresses at the crack-tip, such that we can refrain from the lim-symbol in Eq. 37 and, by use of $\mathbf{F}_{int}^N$ in Eq. 35, approximate the J-integral simply by

$$\mathbf{J} \approx - \sum_{N \in \mathcal{J}_{\Omega_0}} \mathbf{F}_{int}^N,$$

$$\mathcal{J}_{\Omega_0} = \{N : \text{Node } N \text{ is in subdomain } \Omega_0\}. \qquad (38)$$

Consequently, in the discrete case, $\mathbf{J}$ is evaluated by summation of all discrete material node point forces at $N_{np}^0 = |\mathcal{J}_{\Omega_0}|$ nodes within the domain Ω_0. These nodes are indicated as dotted in Fig. 2, and include also the node at the crack-tip.
3. Equation 38 has been proposed by Denzer (2006), Denzer et al. (2003), however using a somewhat different argument.
4. The evaluation of Eq. 38 is not related to the finite element discretization. However, it can be evaluated *completely independent* from the alignment of the finite element mesh or any selected integration contour. As illustrated in Fig. 3a, it is possible to discretize the region of the crack-tip within a rectangular shape, whereas evaluation of Eq. 38 is performed in a circular region Ω_0. Alternatively, as illustrated in Fig. 3b, the discretization in the crack region can be within a circular shaped format, whereas evaluation of Eq. 38 is performed in a rectangular region Ω_0. In the numerical examples of the ensuing section we will exploit this advantage for evaluation of the J-integral in adaptively refined finite element meshes.

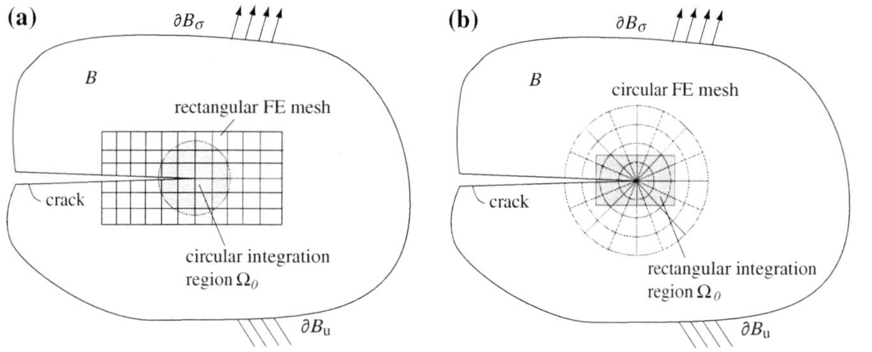

Fig. 3 On evaluation of Eq. 38: (**a**) circular shaped integration region within rectangular shaped mesh. (**b**) rectangular shaped integration region within circular shaped mesh. Nodes within the shaded integration region are considered for evaluation of Eq. 38

4 Numerical examples

4.1 Single edge specimen in tension

The first example considers a single edge specimen in tension. The geometry with loading is shown in Fig. 4a. The example was originally studied by Erdogan (1995) being one of the few theoretical fracture solutions for a finite width FGM. The applied loading corresponds to $\sigma_{22}[x_1, \pm 4W] = \pm 1.0$ for tension. The displacement boundary condition is prescribed such that $u_2 = 0$ in the region $a \leq x_1 \leq 1$ along the line $x_2 = 0$ and, in addition, $u_1 = 0$ for the node at the right-hand side, see Fig. 4a.

Young's modulus is an exponential function of x_1 in the form $E[x_1] = E_1 e^{\beta x_1}$, while Poisson's ratio is constant. The modulus variation $E[x_1]$ is characterized by two parameters, which are selected as $E_1 = E[x_1 = 0]$ and $E_2 = E[x_1 = W/2]$ thus resulting into $\beta = \log(E_2/E_1)/(W/2)$. Further data of importance are $a/W = 0.5$, $L/W = 8.0$, $E_2/E_1 = 0.1, 0.2, 1, 5, 10$, $\nu = 0.3$.

Due to obvious symmetry conditions only the upper half of the specimen is discretized in the finite element calculations. Two different discretization strategies are used. The first strategy employs a regular discretization of the upper half with a resulting mesh in Fig. 4b. In the second approach an adaptive refinement strategy is employed. To this end the following *recovery based error estimator* of L_2-type as advocated by Zienkiewicz and Zhu (1987) is calculated for each finite element e as

$$||\mathbf{e}||_e = \left(\int_{\Omega_e} (\boldsymbol{\sigma} - \boldsymbol{\sigma}^*)^T \cdot (\boldsymbol{\sigma} - \boldsymbol{\sigma}^*) \, \mathrm{dV} \right)^{1/2}, \quad e = 1, \ldots, N_{el}, \tag{39}$$

see also Zienkiewicz and Taylor (2005, p. 386). Here $\boldsymbol{\sigma}$ is the finite element solution of the Galerkin discretization and $\boldsymbol{\sigma}^*$ is an improved solution obtained by a simple recovery technique. In our computations simple nodal averaging has been used. A mathematical analysis of the Zienkiewicz-Zhu a posteriori error estimator is given in Ainsworth et al. (1989) for the error in the energy norm. It was shown, that the estimator is asymptotically exact, provided a modified projection is used. Although the analysis in Ainsworth et al. (1989) required sufficiently smoothness, numerical tests showed, that the estimator continued to be reasonable well also for problems with singularities.

Starting from the initial mesh in Fig. 4e six consecutive meshes are generated, where elements e are refined satisfying

$$||\mathbf{e}||_e \geq \text{tol} \max_{e=1,\ldots,N_{el}} \{||\mathbf{e}||_e\} \tag{40}$$

and where $tol = 0.5$ has been selected. The resulting discretization for the case $E_2/E_1 = 0.1$ is shown in Fig. 4f. For both strategies, regular and adaptive refinement, also the circular integration regions for evaluation of Eq. 28 are shown in Fig. 4c, g, respectively. The seize of the radius is $r/W = 0.02$ in each case. The corresponding material forces in the crack-tip region are sketched in Fig. 4d, h, respectively.

Table 1 compares the normalized stress intensity factors $\sqrt{JE'}/(\sigma\sqrt{\pi a})$ of both discretization strategies with those reported by Erdogan (1995) and Kim and Paulino (2002). The presented results of the material force method are consistent with those of Kim and Paulino (2002) and agree well with the analytical solutions of Erdogan (1995).

We have also changed the size of the domain Ω_0 in Eq. 38 by varying the radii of circles around the crack-tip which include the N^0_{np} nodes. Tables 2 and 3 show the convergence of the normalized stress intensity factors for both discretization strategies. Note, that the values for $r = 0$ represent the material forces at the crack-tip node. In can be seen, that except for the first

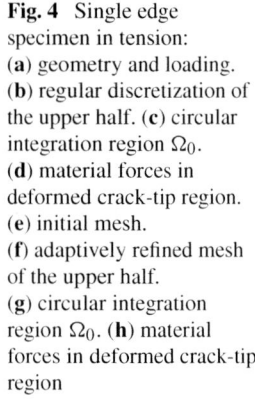

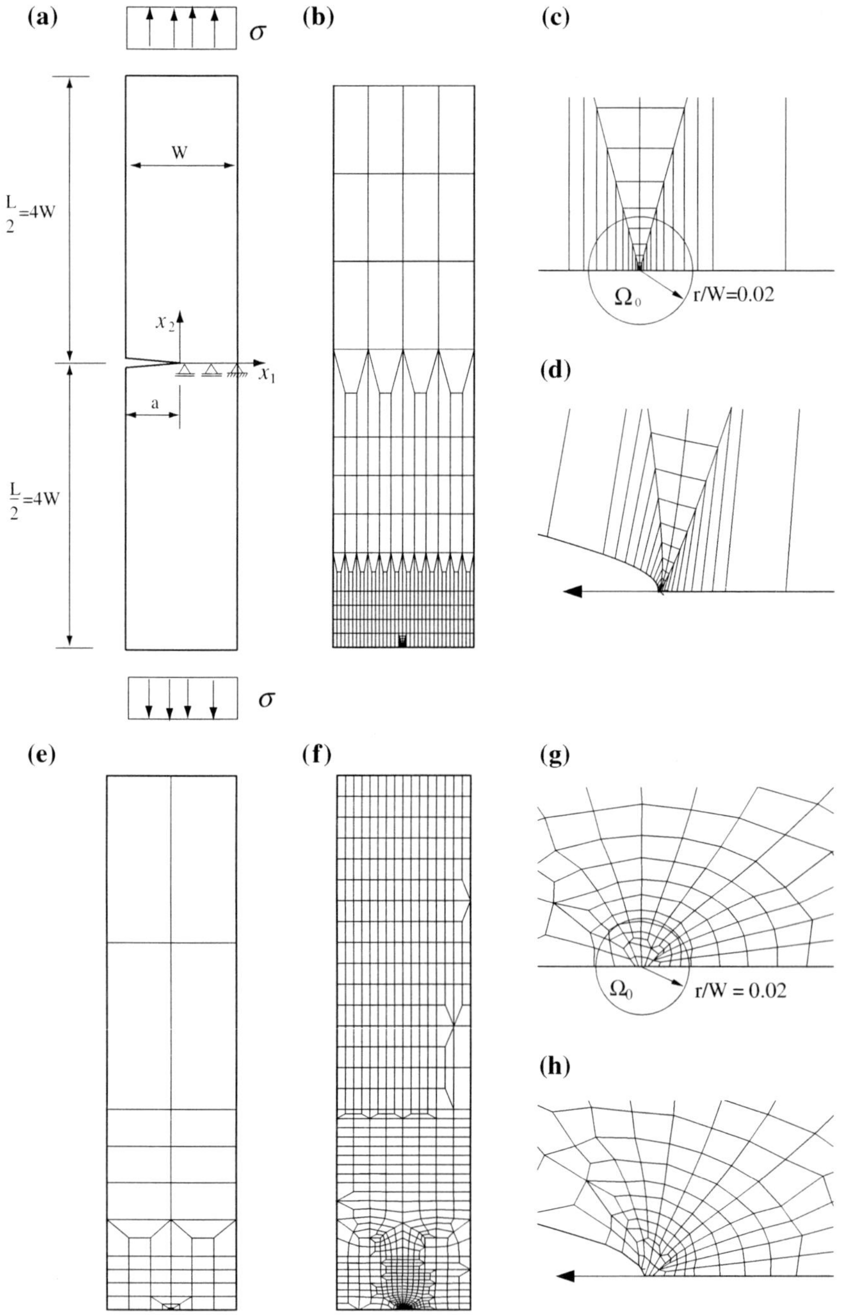

Fig. 4 Single edge specimen in tension: (**a**) geometry and loading. (**b**) regular discretization of the upper half. (**c**) circular integration region Ω_0. (**d**) material forces in deformed crack-tip region. (**e**) initial mesh. (**f**) adaptively refined mesh of the upper half. (**g**) circular integration region Ω_0. (**h**) material forces in deformed crack-tip region

Table 1 Single edge specimen in tension: comparison of normalized stress intensity factors

E_2/E_1	Erdogan (1995)	Kim and Paulino (2002)	Mat.-Force (reg.)	Mat.-Force (adapt.)
0.1	3.570	3.496	3.492	3.492
0.2	3.326	3.292	3.290	3.289
1	N/A	2.822	2.822	2.820
5	2.365	2.366	2.364	2.363
10	2.176	2.175	2.173	2.172

Table 2 Single edge specimen in tension: convergence of the normalized stress intensity factors for the first ten circular regions for a regularly refined mesh in Fig. 4b

E_2/E_1	N_{el}	Radii r/W of circles										
		0.0	0.002	0.004	0.006	0.008	0.010	0.012	0.014	0.016	0.018	0.02
0.1	1216	3.371	3.489	3.496	3.499	3.493	3.492	3.493	3.493	3.492	3.493	3.492
0.2	1216	3.176	3.287	3.294	3.297	3.291	3.290	3.291	3.291	3.290	3.291	3.290
1	1216	2.724	2.819	2.825	2.828	2.822	2.821	2.822	2.822	2.822	2.822	2.822
5	1216	2.283	2.362	2.367	2.369	2.365	2.364	2.365	2.364	2.364	2.364	2.364
10	1216	2.098	2.171	2.175	2.177	2.173	2.172	2.173	2.173	2.173	2.173	2.173

Table 3 Single edge specimen in tension: convergence of the normalized stress intensity factors for the first ten circular regions for an adaptively refined mesh in Fig. 4f

E_2/E_1	N_{el}	Radii r/W of circles										
		0.0	0.002	0.004	0.006	0.008	0.010	0.012	0.014	0.016	0.018	0.02
0.1	735	3.330	3.495	3.494	3.494	3.491	3.492	3.492	3.493	3.494	3.492	3.492
0.2	569	3.138	3.293	3.291	3.293	3.290	3.290	3.290	3.291	3.291	3.290	3.289
1	321	2.775	2.830	2.821	2.820	2.820	2.821	2.820	2.822	2.821	2.820	2.820
5	266	2.329	2.375	2.359	2.364	2.363	2.364	2.363	2.364	2.364	2.363	2.363
10	272	2.140	2.183	2.168	2.172	2.171	2.172	2.172	2.173	2.172	2.172	2.172

two or three circles accurate solutions are obtained in both cases. In these tables also the number of elements N_{el} is reported, thus showing, that the adaptive strategy renders sufficient agreement with less effort in the numerical calculations.

4.2 Three point bending specimen

The second example considers a three point bending specimen made of sandwiched structures with an interlayer. The geometry with loading and the finite element discretization are shown in Fig. 5. The interlayer is a zone of transition wherein the material properties change smoothly from the upper layer to the lower layer and thus is an FGM. The length is sufficiently large, such that it does not affect the solution. The height of the bar is $2H = 10$, the crack length is $a = 5$ and the height of the interlayers is $2h = 1$. The crack is perpendicular to the upper and lower boundaries, and the crack-tip is inside the FGM.

The variation of Young's modulus in the material gradient is linear and Poisson's ratio is constant. For Cartesian coordinates with origin at the center of the specimen Young's modulus is given by

$$
\begin{aligned}
&1.\ E[x_2] = E_1 && \text{for } x_2 \le h\\
&2.\ E[x_2] = Ax_2 + B, && \text{for } -h < x_2 \le h\\
&3.\ E[x_2] = E_2 && \text{for } h < x_2.
\end{aligned}
\quad \text{where } A = \frac{E_2-E_1}{2h},\ B = \frac{E_2+E_1}{2} \tag{41}
$$

Due to symmetry conditions only the left half of the specimen is discretized in the finite element calculations. As in the previous example, two different discretization strategies as illustrated in Fig. 6 are employed. First, a regular discretization is used with the left half of the specimen in Fig. 6a. Second, an adaptive refinement strategy with six consecutive meshes is employed, where elements e satisfying Eq. 40 are refined. The resulting mesh for the case $E_2/E_1 = 0.05$ is shown in Fig. 6d. In order to demonstrate the flexibility of the procedure, contrary to the previous example, square regions are used for evaluation of Eq. 38. The seize of the square is $2a/W = 0.06$ for each case. For both strategies, regular and adaptive refinement, these regions are sketched in Fig. 6b, e, respectively. Figure 6c, f show the related material forces in the deformed crack-tip region.

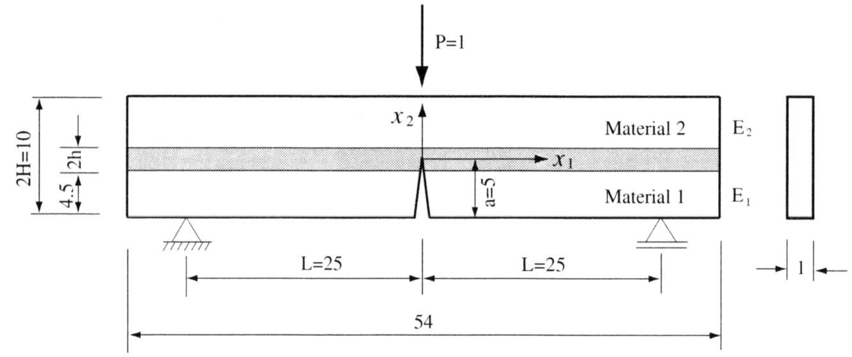

Fig. 5 Three point bending specimen: geometry, loading and interlayer

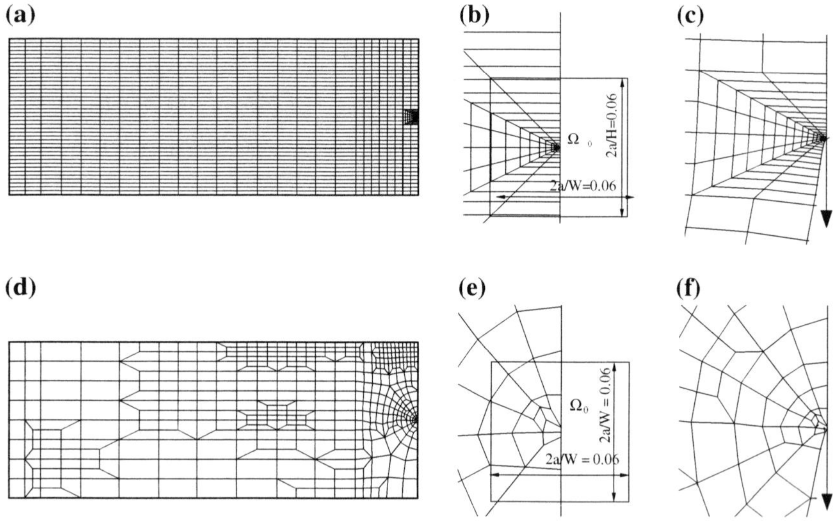

Fig. 6 Three point bending specimen: (**a**) regular discretization of the left half. (**b**) square integration region Ω_0. (**c**) material forces in deformed crack-tip region. (**d**) adaptively refined mesh of the left half. (**e**) square integration region Ω_0. (**f**) material forces in deformed crack-tip region

Table 4 compares the normalized stress intensity factors $K_I\sqrt{H}/P$ of the material force method with those reported by Kim and Paulino (2002) using an MCC method and the J^*-integral. The results of both strategies in Table 4, regular and adaptive refinement, are consistent to each other and agree well with the results reported in Kim and Paulino (2002).

As in the previous example the size of the domain Ω_0 in Eq. 38 has been changed by varying the lengths of the squares as visualized in the lower diagram of Fig. 6 and which include the N^0_{np} nodes. Tables 5 and 6 show the convergence of the normalized stress intensity factors for both strategies, regularly and adaptively refinement. Note, that the values for $r = 0$ represent the material forces at the crack-tip node. Contrary to the previous example, the solutions show some more scatter, however the deviations between the cases $2a/H = 0.01$ and $2a/H = 0.02$ are less than 1%.

5 Summary

The objective of the work has been the computation of fracture parameters in functionally graded materials (FGMs) using the continuum concept of material forces. To this end the corresponding balance equation is discretized by a standard Galerkin finite element procedure. Evaluation of the J-integral is accomplished by calculation of a domain integral. In the practical implementation the material nodal forces of the FEM discretization are summed up in a finite region of the crack-tip. In this way the numerical calculation is completely independent from the alignment of the finite element mesh or any selected integration contour. The attractivity of this feature has been exploited in the numerical examples using both regularly and adaptively refined finite element meshes.

Table 4 Three point bending specimen: comparison of normalized stress intensity factors $K_I\sqrt{H}/P$, $a = 5.0$

E_2/E_1	Kim and Paulino (2002) MCC	Kim and Paulino (2002) J^*	Mat.-Force (reg.)	Mat.-Force (adap.)
0.05	30.72	31.12	31.24	31.32
0.1	23.47	23.92	23.92	23.93
0.2	18.01	18.32	18.32	18.28
0.5	12.42	12.57	12.58	12.49
1.0	9.398	9.467	9.466	9.416
2.	7.296	7.318	7.314	7.285
5.	5.502	5.496	5.498	5.477
10.	4.606	4.586	4.603	4.575
20.	3.980	3.939	4.008	3.950

Table 5 Three point bending specimen: convergence of the normalized stress intensity factors ($K_I\sqrt{H}/P$, $a = 5.0$) for the first ten squares for a regular refined mesh in Fig. 6a

E_2/E_1	N_{el}	Length $2a/H$ of squares										
		0	0.006	0.012	0.018	0.024	0.030	0.036	0.042	0.048	0.054	0.06
0.05	1168	30.25	31.32	31.34	31.33	31.32	31.30	31.30	31.27	31.26	31.25	31.24
0.1	1168	23.11	23.92	23.94	23.93	23.93	23.92	23.93	23.92	23.92	23.92	23.92
0.2	1168	17.70	18.32	18.33	18.32	18.32	18.32	18.32	18.32	18.32	18.32	18.32
0.5	1168	12.15	12.58	12.58	12.58	12.58	12.58	12.58	12.58	12.58	12.58	12.58
1	1168	9.148	9.468	9.471	9.465	9.467	9.466	9.466	9.466	9.466	9.466	9.466
2	1168	7.069	7.316	7.318	7.313	7.315	7.314	7.314	7.314	7.314	7.314	7.314
5	1168	5.313	5.499	5.499	5.496	5.497	5.497	5.496	5.497	5.497	5.498	5.498
10	1168	4.437	4.593	4.592	4.590	4.593	4.596	4.593	4.600	4.601	4.602	4.603
20	1168	3.827	3.961	3.961	3.958	3.971	3.982	3.980	3.997	4.002	4.004	4.008

Table 6 Three point bending specimen: convergence of the normalized stress intensity factors ($K_I\sqrt{H}/P$, $a = 5.0$) for the first ten squares for an adaptively refined mesh in Fig. 6d

E_2/E_1	N_{el}	Length $2a/H$ of squares										
		0	0.006	0.012	0.018	0.024	0.030	0.036	0.042	0.048	0.054	0.06
0.05	610	29.98	31.30	31.21	31.39	31.31	31.31	31.31	31.30	31.30	31.30	31.32
0.1	489	22.90	23.91	23.84	23.98	23.92	23.92	23.92	23.91	23.91	23.91	23.93
0.2	327	18.03	18.30	18.12	18.21	18.37	18.36	18.26	18.27	18.26	18.27	18.28
0.5	182	11.94	12.37	12.51	12.43	12.58	12.64	12.52	12.51	12.49	12.49	12.49
1	190	8.993	9.325	9.427	9.367	9.473	9.514	9.437	9.429	9.414	9.414	9.416
2	164	6.948	7.217	7.290	7.248	7.317	7.346	7.302	7.297	7.286	7.286	7.285
5	312	5.212	5.423	5.475	5.445	5.486	5.507	5.480	5.487	5.480	5.476	5.477
10	341	4.359	4.536	4.582	4.558	4.590	4.599	4.585	4.586	4.582	4.579	4.575
20	470	3.755	3.909	3.948	3.928	3.956	3.963	3.957	3.958	3.954	3.951	3.950

Future work should be directed to applications of the algorithm to real structures, when experimental data are available. Furthermore, anisotropic effects, the consideration of a large strain theory and consideration of effective properties of functionally graded composites would be challenging aspects for future work.

Acknowledgements This paper is based on investigations of the collaborative research center SFB/TR TRR 30, which is kindly supported by the Deutsche Forschungsgemeinschaft (DFG), Germany.

References

Ainsworth M, Zhu JZ, Craig AW, Zienkiewicz OC (1989) Analysis of the Zienkiewicz-Zhu a-posteriori error estimator in the finite element method. Int J Numer Meth Eng 28: 2161–2174

Anlas G, Santare MH, Lambros J (2000) Numerical calculation of stress intensity factors in functionally graded materials. Int J Fract 104:131–143

Bendsoe MP, Sigmund O (2003) Topology optimization: theory, methods and applications. Springer-Verlag, Berlin

Braun M (1997) Configurational forces induced by finite-element discretization. Proc Estonian Acad Sci Phys Math 46:24–31

Chadwick P (1975) Applications of an energy-momentum tensor in non-linear elastostatics. J Elast 5:249–258

Cherepanov GP (1967) Crack propagation in continuous media. J Appl Math Mech 31:503–512

Cherepanov GP (1968) Cracks in solids. Int J Solids Struct 4:811–831

Delale F, Erdogan F (1983) Fracture mechanics of functionally graded materials. J Appl Mech 50:609–614

Denzer R (2006) Computational configurational forces. Dissertation, Technical University of Kaiserslautern, Report No.: UKL/LTM T 06-04, Lehrstuhl für Technische Mechanik

Denzer R, Barth FJ, Steinmann P (2003) Studies in elastic fracture mechanics based on the material force method. Int J Num Meth Eng 58:1817–1835

Eischen JW (1987a) Fracture of non-homogeneous materials. Int J Fract 34:3–22

Eischen JW (1987b) An improved method for computing the J_2 integral. Eng Fract Mech 26:691–700

Erdogan F (1995) Fracture mechanics of functionally graded materials. Composit Eng 5:753–770

Erdogan F, Wu BH (1997) The surface crack problem for a plate with functionally graded materials. ASME J Appl Mech 64:449–456

Eshelby JD (1951) The force on an elastic singularity. Philos Trans Roy Soc, Math Phys Sci A 244:87–112

Eshelby JD (1965) The continuum theory of lattice defects. Prog Solid States Phys 3:79–114

Giannakopoulos AE, Surush S, Finot M, Olsson M (1995) Elastoplastic analysis of thermal cycling: layered materials with compositional gradients. Acta Metall Mater 43(4): 1335–1354

Gu P, Dao M, Asaro RJ (1999) A simplified method for calculating the crack-tip field of functionally graded materials using the domain integral. ASME J Appl Mech 66(1):101–108

Gurtin ME (2000) Configurational forces as basic concepts of continuum physics. Springer-Verlag

Haddi A, Weichert D (1995) On the computation of the J-integral for three-dimensional geometries in inhomogeneous materials. Comput Mat Sci 5:143–150

Hirano T, Teraki J, Yamada T (1990) On the design of functionally graded materials. In: Yamanouchi M, Koizumi M, Hirai T, Shiota I (eds) Proceedings of the 1st International Symposium on Functionally Gradient Materials, Sendai, Japan, 1990

Honein T, Herrmann G (1997) Conservation laws in nonhomogenous plane elastostatics. J Mech Phys Solids 45:789–805

Hughes TJR (1987) The Finite Element Method: linear static and dynamic analysis. Prentice-Hall, Englewood Cliffs, NJ

Kienzler R, Herrmann G (2000) Mechanics in material space with application to defect and fracture mechanics. Springer-Verlag

Kim JH, Paulino GH (2002) Finite element evaluation of mixed mode stress intensity factors in functionally graded materials. Int J Num Meth Eng 53:1903–1935

Konda N, Erdogan F (1994) The mixed mode crack problem in a non-homogeneous elastic plane. Eng Fract Mech 47: 533–545

Li FZ, Shih CF, Needleman A (1985) A comparison of methods for calculating energy release rates. Eng Frac Mech 21: 405–421

Makowski J, Stumpf H, Hackl K (2006) The fundamental role of nonlocal and local balance laws of material forces in finite elastoplasticity and damage mechanics. Int J Solids Struct 43:3940–3959

Marur PR, Tippur H (1998) Evaluation of mechanical properties of functionally graded materials. J Test Eval JTEVA 26(6):539–545

Marur PR, Tippur H (2000) Numerical analysis of crack-tip fields in functionally graded materials with a crack normal to the elastic gradient. Int J Solids Struct 37:5353–5370

Maugin GA (1993) Material inhomogeneities in elasticity. Chapman & Hall, London

Müller R, Maugin GA (2002) On material forces and finite element discretizations. Comput Mech 29(1):52–60

Müller R, Kolling S, Gross D (2001) On configurational forces in the context of the finite-element method. Int J Num Meth Eng 53:1557–1574

Müller R, Gross D, Maugin GA (2004) Use of material forces in adaptive finite element methods. Comput Mech 33: 421–434

Nguyen TD, Govindjee S, Klein PA (2005) A material force method for inelastic fracture mechanics. J Mech Phys Solids 53:91–121

Parameswaran V, Shukla A (2002) Asymptotic stress fields for dynamic stationary cracks along the gradient in functionally gradient materials. ASME J Appl Mech 69:240–243

Paulino GH, Jin ZH, Dodds RH (2003) Failure of functionally graded materials. In: Karihaloo B, Knauss WG (eds) Comprehensive structural integrity, vol 2. Elsevier Science, pp 607–644

Rajagopal A, Sivakumar SM (2007) A combined r-h adaptive strategy based on material forces and error assessment for plane problems and bimaterial interfaces. Comput Mech 41:49–72

Rice JR (1968) A path independent integral and the approximate analysis of strain concentration by notches and cracks. J Appl Mech 35:379–386

Shield RT (1967) Inverse deformation results in finite elasticity. ZAMP 18:490–500

Silva ECN, Paulino GH (2005) Topology optimization design of functionally graded structures. Proc of 6th World Congress of Structural and Multidisciplinary Optimization, Ria de Janeiro, 30 May-03 June 2005, Brazil

Spiegel MR (1959) Vector analysis and an introduction to tensor analysis. Schaum's outline of theory and problems. McGraw-Hill, New York

Steinmann P (2001a) Application of material forces to hyperelastic fracture mechanics, Part I: continuum mechanical setting. Int J Solids Struct 37:7371–7391

Steinmann P (2001b) A view on the theory and computation of hyperelastic defect mechanics. In: Proc. of European Conference on Computational Mechanics, ECCM, Crakow, Poland

Steinmann P, Ackermann D, Barth FJ (2001) Application of material forces to hyperelastic fracture mechanics, Part II: computational setting. Int J Solids Struct 38:5509–5526

Surush S, Mortensen A (1998) Fundamentals of functionally graded materials. Institute of Materials, London

Williams ML (1957) On the stress distribution at the base of a stationary crack. ASME J Appl Mech 24:109–114

Zienkiewicz OC, Taylor RL (2005) The finite element method, 6th edn. vol 1. Mc Graw-Hill, London

Zienkiewicz OC, Zhu JZ (1987) A simple error estimator and adaptive procedure for practical engineering analysis. Int J Num Meth Eng 24:337–357

Zuiker J, Dvorak G (1994) The effective properties of functionally graded composites- I. Extension of the Mori-Tanaka method to linearly varying fields. Composit Eng 4:19–35

A multiscale approach to damage configurational forces

C. Dascalu · G. Bilbie

Abstract A two-scale homogenization method is used to construct a damage model in the framework of configurational mechanics. The upscaling procedure allows for the identification of damage configurational forces as the result of the microscopic fracture analysis. The obtained damage equation incorporates stiffness degradation, material softening, unilaterality, induced anisotropy. The balance of configurational forces naturally captures a microscopic length, leading to size effects in the overall damage response. The new approach is illustrated in the case of brittle damage, for a three point bending test. Extended finite elements are used for the numerical modeling of macro-crack initiation and growth. The influence of the microscopic size on the failure initiation stress is analyzed and it is shown that this dependence follows a Hall–Petch type rule.

1 Introduction

In the last decades an important effort has been made to formulate damage models based on micromechanical analysis (e.g. Nemat-Nasser and Hori 1999; Andrieux et al. 1986; Prat and Bazant 1997; Caiazzo and Constanzo 2000; Pensée et al. 2002; Lene 2004; Basista and Gross 1989; Li et al. 2004; Raghavan and Ghosh 2005). In parallel with these developments, the material or configurational mechanics (e.g. Maugin 1993; Gurtin 2000; Kienzler and Herrmann 2000) has been established as a new framework for the modeling of defects. The present contribution aims at providing a micromechanical analysis for the formulation of brittle damage in the frame of material mechanics.

Most of the local damage models are based on phenomenological assumptions and this makes difficult the introduction of micro-structural lengths parameters, with clear physical meaning. In a recent work (Dascalu et al. 2008), we have proposed a multiscale approach to damage based on a full homogenization procedure, without phenomenological assumptions, starting from a microscopic energy analysis. The present paper extends the previous results to damage configurational mechanics, by identifying the damage configurational forces as the result of the micro-mechanical analysis. It also proposes a framework for the complete description of the failure process, in which macro-fracture initiation is numerically introduced by using extended finite elements (Moes 1999). This allows us to quantify the influence of the size of the microstructure on the fracture initiation process.

We extend the homogenization formalism in Dascalu et al. (2008) to the framework of material mechanics and deduce expressions of damage material forces from the micro-mechanical description of the crack propagation. In Agiasofitou and Dascalu (2007), a different approach has been adopted to construct macroscopic material forces, by starting from the microscopic

C. Dascalu (✉) · G. Bilbie
Laboratoire 3S-R, UJF, INPG, CNRS UMR 5221, BP 53, 38041 Grenoble cedex 9, France
e-mail: cristian.dascalu@hmg.inpg.fr

Defect and Material Mechanics. C. Dascalu, G.A. Maugin & C. Stolz (eds.),
doi: 10.1007/978-1-4020-6929-1_24, © Springer Science+Business Media B.V. 2008

balance of material momentum. The presence of configurational forces on micro-crack faces made difficult the computation of some terms in the homogenized equations. In this paper, a simpler model is proposed, in which a microscopic energy balance is connected to the macroscopic balance of material momentum in order to identify the expressions of the damage configurational forces. A diferent view on the multiscale energy–momentum tensors has been presented recently in Li et al. (2007).

The periodic homogenization technique with asymptotic developments (Benssousan et al. 1978; Sanchez-Palencia 1980; Bakhvalov and Panasenko 1989; Leguillon and Sanchez-Palencia 1982) is here adapted to take into account the energy description of micro-fracture. Due to the size-dependence of the energy-release rate, the microscopic energy analysis should be performed on a finite-size cell. Similar analysis, involving asymptotic developments homogenization and finite cell length, have been performed by Smyshlyaev and Cherednichenko (2000) and Peerlings and Fleck (2004) to obtain strain gradient homogenized equations for periodic two-phase media. Our approach remains in the framework of the local continuum theory, but uses the balance of material momentum to introduce a microstructural length parameter in the overall response. We show that the homogenized balance of configurational forces reduces to a damage equation which contains a microstructural length: the size of the local periodicity cell.

The model is formulated for micro-cracks with frictionless unilateral contact. At the macroscopic level, the switch between the homogenized behaviors corresponding to crack opening or closure is carried out numerically. In this way an unilateral damage model is obtained. Different local micro-crack orientations provided by the damage law lead to induced anisotropy and heterogeneity in the global response of the specimen.

To illustrate the capability of the brittle damage model to describe fracture initiation and growth and to study the dependence on the internal length, we consider a three point bending numerical test. The appearance and the evolution of macro-cracks, which is completely determined by the evolution of micro-cracks, is here modeled with enriched finite elements. It is shown that the failure stress dependence on the microstructural size parameter follows a Hall–Petch type law.

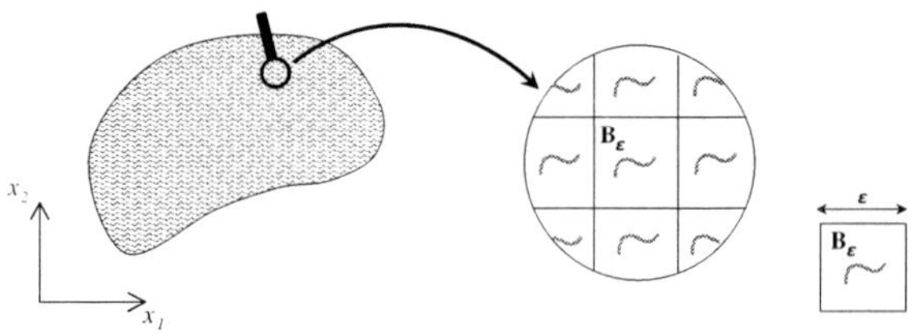

Fig. 1 Fissured medium with locally periodic microstructure

The paper is organized as follows. The model problem for an elastic body with microcracks is formulated in Sect. 2. In Sect. 3 the asymptotic developments are used to deduce the homogenized equilibrium equations. The homogenized material momentum equation is obtained in the form of an evolution equation for damage in Sect. 4. In Sect. 5 we describe the numerical implementation of the damage model using extended finite elements. Numerical results are presented in Sect. 6.

2 Elastic body with microcracks

Consider a two-dimensional isotropic elastic medium containing a large number of micro-cracks. A locally periodic distribution of micro-cracks is assumed, so as one can locally find a periodicity cell, of length ε, containing one crack (see Fig. 1). The periodic boundary conditions take into account the interaction between neighbourhood cells with cracks. The cracks are assumed to be straight and of length d^ε, depending on time t. The length d^ε may differ from one crack to another, but varies smoothly almost everywhere in the elastic body. We denote by $\mathcal{B}$ the whole body, a bounded domain of $\Re^2$ with a smooth boundary containing $\mathcal{N}$ micro-cracks $\mathcal{C}_n, n = 1, \ldots, \mathcal{N}$ and the solid part $\mathcal{B}_s = \mathcal{B}\backslash\mathcal{C}$, where $\mathcal{C} = \cup_{n=1}^{\mathcal{N}}\mathcal{C}_n$. In the solid part $\mathcal{B}_s$, we have the equilibrium equation

$$\frac{\partial \sigma_{ij}^\varepsilon}{\partial x_j} = 0, \quad \text{in } \mathcal{B}_s \tag{1}$$

and the linear elasticity constitutive relations

$$\sigma_{ij}^\varepsilon = a_{ijkl} e_{xkl}(\mathbf{u}^\varepsilon), \tag{2}$$

where $\mathbf{u}^\varepsilon$ and $\boldsymbol{\sigma}^\varepsilon$ are the displacement and the stress fields and where we defined e_{zij} as the small-deformations strain tensor

$$e_{zij} = \frac{1}{2}\left(\frac{\partial u_i}{\partial z_j} + \frac{\partial u_j}{\partial z_i}\right). \tag{3}$$

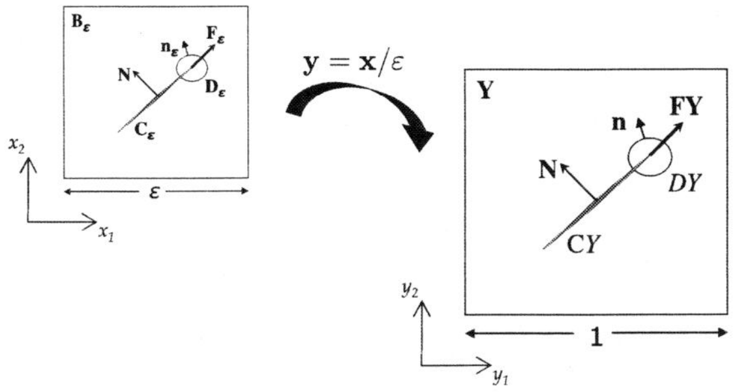

Fig. 2 Rescaling of the unit cell to the microstructural period of the material

with respect to z coordinates. The elastic coefficients a_{ijkl} are given by

$$a_{ijkl} = \lambda\delta_{ij}\delta_{kl} + \mu(\delta_{ik}\delta_{jl} + \delta_{il}\delta_{jk}), \tag{4}$$

with λ and μ the Lamé constants.

On the crack faces we assume traction free conditions opening or frictionless contact conditions. These two alternatives are expressed by the two sets of formulae

$$\boldsymbol{\sigma}^\varepsilon \mathbf{N} = 0; \quad [\mathbf{u}^\varepsilon \cdot \mathbf{N}] > 0 \tag{5}$$

$$[\boldsymbol{\sigma}^\varepsilon \mathbf{N}] = 0; \ \mathbf{N} \cdot \boldsymbol{\sigma}^\varepsilon \mathbf{N} < 0; \ \mathbf{T} \cdot \boldsymbol{\sigma}^\varepsilon \mathbf{N} = 0; \ [\mathbf{u}^\varepsilon \cdot \mathbf{N}] = 0 \tag{6}$$

where $\mathbf{N}$ is the unit normal vector, $\mathbf{T}$ is a unit tangent vector to the crack and $[\ \cdot\ ]$ the jump across the crack faces. For each micro-crack, we assume that one of the two states (5) and (6) holds in all the crack points. The fact that each micro-crack is completely open or closed is a reasonable assumption for small crack lengths. The way in which the switch from one state to the other is controlled will be described later, in terms of the homogenized solution.

The material force at the crack tip can be expressed as

$$\mathbf{F}_\varepsilon = \lim_{D_\varepsilon \to 0} \int_{\partial D_\varepsilon} \mathbf{b}(\mathbf{u}^\varepsilon)\mathbf{n}_\varepsilon \, \mathrm{d}s \tag{7}$$

where ∂D_ε is a circle of an infinitesimal radius r, surrounding the crack tip (see Fig. 2) and

$$b_{ij}(\mathbf{u}^\varepsilon) = \frac{1}{2} a_{mnkl} e_{xkl}(\mathbf{u}^\varepsilon) e_{xmn}(\mathbf{u}^\varepsilon)\delta_{ij} - \sigma^\varepsilon_{jk} u^\varepsilon_{k,i}$$

is the Eshelby configurational stress tensor.

Let $\mathbf{e}$ be the unit vector in the direction of $\mathbf{F}_\varepsilon$ such that $\mathbf{F}_\varepsilon = F_\varepsilon \mathbf{e}$. We assume that $\mathbf{e}$ is the crack propagation direction and that propagation occurs when a critical threshold F_c is reached

$$F_\varepsilon = F_c \tag{8}$$

while there is no crack evolution for $F_\varepsilon < F_c$.

3 Asymptotic developments homogenization

We assume that the body has a locally periodic microstructure which is reproduced from the unit cell $Y = [0, 1] \times [0, 1]$ by rescaling with the small parameter ε so that the period of the material is εY, as in Fig. 2. The parameter ε, which is assumed to be small enough with respect to the characteristic dimensions of the whole body, will be our microscopic length scale. This condition allow us to distinguish between micro- and macroscopic variations. The two distinct scales are represented by the variables $\mathbf{x}$, which are referred to as *macroscopic variables* and the variables $\mathbf{y} = \mathbf{x}/\varepsilon$, referred to as *microscopic variables*. In the unit cell Y we denote the crack by CY and solid part by $Y_s = Y \backslash CY$. The length of CY is $d = d^\varepsilon/\varepsilon$. According to the method of asymptotic homogenization (e.g. Benssousan et al. 1978; Bakhvalov and Panasenko 1989), we look for expansions of $\mathbf{u}^\varepsilon$ and $\boldsymbol{\sigma}^\varepsilon$ in the form

$$\mathbf{u}^\varepsilon(\mathbf{x}, t) = \mathbf{u}^{(0)}(\mathbf{x}, \mathbf{y}, t) + \varepsilon \mathbf{u}^{(1)}(\mathbf{x}, \mathbf{y}, t) + \varepsilon^2 \mathbf{u}^{(2)}(\mathbf{x}, \mathbf{y}, t) + \cdots \tag{9}$$

$$\boldsymbol{\sigma}^\varepsilon(\mathbf{x}, t) = \frac{1}{\varepsilon}\boldsymbol{\sigma}^{(-1)}(\mathbf{x}, \mathbf{y}, t) + \boldsymbol{\sigma}^{(0)}(\mathbf{x}, \mathbf{y}, t) + \varepsilon \boldsymbol{\sigma}^{(1)}(\mathbf{x}, \mathbf{y}, t) + \cdots \tag{10}$$

where $\mathbf{u}^{(i)}(\mathbf{x}, \mathbf{y}, t), \boldsymbol{\sigma}^{(i)}(\mathbf{x}, \mathbf{y}, t),\ \mathbf{x} \in \mathcal{B}_s,\ \mathbf{y} \in Y$ are smooth functions and Y-periodic in $\mathbf{y}$.

Substituting the expansions (9) and (10) into the Eq. 1, we obtain for the different orders of ε

$$\frac{\partial \sigma^{(-1)}_{ij}}{\partial y_j} = 0, \quad \frac{\partial \sigma^{(-1)}_{ij}}{\partial x_j} + \frac{\partial \sigma^{(0)}_{ij}}{\partial y_j} = 0, \quad \frac{\partial \sigma^{(0)}_{ij}}{\partial x_j} + \frac{\partial \sigma^{(1)}_{ij}}{\partial y_j} = 0. \tag{11}$$

Moreover, the constitutive relation (2) and the condition (3) via Eqs. 9 and 10, give correspondingly

$$\sigma^{(-1)}_{ij} = a_{ijkl} e_{ykl}(\mathbf{u}^{(0)}), \quad \sigma^{(0)}_{ij} = a_{ijkl}(e_{xkl}(\mathbf{u}^{(0)}) + e_{ykl}(\mathbf{u}^{(1)})), \quad \sigma^{(1)}_{ij} = a_{ijkl}(e_{xkl}(\mathbf{u}^{(1)}) + e_{ykl}(\mathbf{u}^{(2)})). \tag{12}$$

The crack face boundary conditions, in the case of open cracks, become

$$\sigma^{(-1)}_{ij} N_j = 0, \ \sigma^{(0)}_{ij} N_j = 0, \ \sigma^{(1)}_{ij} N_j = 0, \ \text{on } \mathcal{C}^\pm. \tag{13}$$

while for cracks in contact we obtain at the order $m = -1, 0, 1$ in ε:

$$[\sigma_{ij}^{(m)} N_j] = 0, \quad N_j \sigma_{ij}^{(m)} < 0, \quad T_j \sigma_{ij}^{(m)} = 0 \tag{14}$$

It can be shown that the function $\mathbf{u}^{(0)} = \mathbf{u}^{(0)}(\mathbf{x}, t)$ is independent of the microscopic variable, representing by this way the *macroscopic displacement field*.

For given $\mathbf{u}^{(0)}(\mathbf{x}, t)$, for open traction-free cracks, we deduce the following boundary-value problem for the function $\mathbf{u}^{(1)}$:

$$\frac{\partial}{\partial y_j}\left(a_{ijkl} e_{ykl}(\mathbf{u}^{(1)})\right) = 0, \quad \text{in } Y_s \tag{15}$$

$$a_{ijkl} e_{ykl}(\mathbf{u}^{(1)}) N_j = -a_{ijkl} e_{xkl}(\mathbf{u}^{(0)}) N_j, \quad \text{on } CY^{\pm} \tag{16}$$

and with periodicity boundary conditions on the external boundary of the cell.

For closed cracks, the corresponding cell problem reads

$$\frac{\partial}{\partial y_j}\left(a_{ijkl} e_{ykl}(\mathbf{u}^{(1)})\right) = 0, \quad \text{in } Y_s \tag{17}$$

$$[a_{ijkl} e_{ykl}(\mathbf{u}^{(1)}) N_j] = -[a_{ijkl} e_{xkl}(\mathbf{u}^{(0)}) N_j], \quad \text{on } CY \tag{18}$$

$$N_i a_{ijkl}(e_{ykl}(\mathbf{u}^{(1)}) + e_{xkl}(\mathbf{u}^{(0)})) N_j < 0, \quad \text{on } CY \tag{19}$$

$$T_i a_{ijkl}(e_{ykl}(\mathbf{u}^{(1)}) + e_{xkl}(\mathbf{u}^{(0)})) N_j = 0, \quad \text{on } CY \tag{20}$$

As concerns the distinction between the microscopic states of contact and opening, the following procedure is proposed. On the crack faces, in the unit cell, one has the conditions (16), in which the macroscopic solution appears within a force-type source term. The orientation of this force vector with respect to crack line, i.e. the tendency of this force to open or close the crack, may be considered as an indicator of the opening or closing state. At the macroscopic level, these two states induce a separation of the space $\mathbf{R}$ of deformations $e_{x11}, e_{x12}, e_{x22}$ into two subregions $\mathbf{R}^{\pm}$ defined by:

$$\mathbf{R}^{\pm} = \left\{\mathbf{e_x} \mid \; N_i a_{ijkl} e_{xkl}(\mathbf{u}^{(0)}) N_j \gtrless 0\right\} \tag{21}$$

where the signs $+$ and $-$ correspond, respectively, to tension and compression. In this way, the switch between the two elementary homogenized behaviors, corresponding to tension and compression on cracks, is realized at the macroscopic level, through the overall deformation field.

The function $\mathbf{u}^{(1)}$ can be looked for in the form

$$\mathbf{u}^{(1)}(\mathbf{x}, \mathbf{y}, t) = \boldsymbol{\xi}^{pq}(\mathbf{y}) e_{xpq}(\mathbf{u}^{(0)})(\mathbf{x}, t) \tag{22}$$

where $\boldsymbol{\xi}(\mathbf{y})$ are the *characteristic functions* representing elementary deformation modes of the unit cell (Bakhvalov and Panasenko 1989).

Relations (15)–(16), for every p and q and $e_{xkl}(\mathbf{u}^{(0)}) = \delta_{kp}\delta_{lq} \in \mathbf{R}^+$ and with periodicity conditions on the cell boundary, are equivalent to the cell problem

$$\frac{\partial}{\partial y_j}\left(a_{ijkl} e_{ykl}(\boldsymbol{\xi}^{pq})\right) = 0, \quad \text{in } Y_s, \tag{23}$$

$$a_{ijkl} e_{ykl}(\boldsymbol{\xi}^{pq}) N_j = -a_{ijpq} N_j, \quad \text{on } CY^{\pm} \tag{24}$$

and for $e_{xkl}(\mathbf{u}^{(0)}) = -\delta_{kp}\delta_{lq} \in \mathbf{R}^-$

$$\frac{\partial}{\partial y_j}\left(a_{ijkl} e_{ykl}(\boldsymbol{\xi}^{pq})\right) = 0, \quad \text{in } Y_s \tag{25}$$

$$[a_{ijkl} e_{ykl}(\boldsymbol{\xi}^{pq}) N_j] = [a_{ijpq} N_j], \quad \text{on } CY \tag{26}$$

$$N_i a_{ijkl}(e_{ykl}(\boldsymbol{\xi}^{pq}) - \delta_{kp}\delta_{lq}) N_j < 0 \quad \text{on } CY \tag{27}$$

$$T_i a_{ijkl}(e_{ykl}(\boldsymbol{\xi}^{pq}) - \delta_{kp}\delta_{lq}) N_j = 0 \quad \text{on } CY \tag{28}$$

By introducing the mean value operator

$$\langle \cdot \rangle = \frac{1}{|Y|}\int_{Y_s} \cdot \mathrm{d}y, \tag{29}$$

where $|Y|$ is the measure of Y, we can prove that

$$\Sigma_{ij}^{(0)} \equiv \langle \sigma_{ij}^{(0)} \rangle = C_{ijkl}(d) e_{xkl}(\mathbf{u}^{(0)}) \tag{30}$$

where

$$C_{ijkl}(d, \mathbf{e_x}) = \begin{cases} C_{ijkl}^{+}(d), \; \mathbf{e_x} \in \mathbf{R}^+ \\ C_{ijkl}^{-}(d), \; \mathbf{e_x} \in \mathbf{R}^- \end{cases} \tag{31}$$

with

$$C_{ijkl}^{\pm}(d) = \frac{1}{|Y|}\int_{Y_s} (a_{ijkl} + a_{ijmn} e_{ymn}(\boldsymbol{\xi}_{\pm}^{kl})) \,\mathrm{d}y \tag{32}$$

are the effective or homogenized coefficients.

Since $|Y| = 1$, in what follows $|Y|$ will be omitted. By taking the mean value of the second equation in (11) and using (30) we get

$$\frac{\partial}{\partial x_j}\left(C_{ijkl} e_{xkl}(\mathbf{u}^{(0)})\right) = 0. \tag{33}$$

as the homogenized equation of equilibrium.

4 Balance of material momentum

Configurational or material mechanics aims at providing a framework for the description of the evolution of defects, viewed as spatial inhomogeneities on the material manifold. In the presence of such inhomogeneities, the balance of material momentum is giving rise to material forces. The objective of this section is to deduce the expression of damage material forces from by micromechanical analysis, through homogenization. In a recent paper (Agiasofitou and Dascalu 2007), we started with the microscopic balance of material momentum and performed homogenization to deduce the macroscopic law of damage. Here, we start with the macroscopic balance of material momentum which will be linked with the microscopic energy analysis of fracture, as performed in Dascalu et al. (2008), in order to get the expression of the macroscopic damage forces.

From the equilibrium Eq. 33 we can deduce

$$\frac{\partial B^{(0)}_{mj}}{\partial x_j} + F_m = 0 \tag{34}$$

where

$$B^{(0)}_{mj} = W^{(0)}\delta_{mj} - \Sigma^{(0)}_{ij} u^{(0)}_{i,m} \tag{35}$$

is the Eshelby stress tensor

$$W^{(0)} = \frac{1}{2} C_{ijkl}(d) e_{xkl}(\mathbf{u}^{(0)}) e_{xij}(\mathbf{u}^{(0)}) \tag{36}$$

where

$$\Sigma^{(0)}_{ij} = C_{ijkl}(d) e_{xkl}(\mathbf{u}^{(0)}) \tag{37}$$

and

$$F_m = -\frac{1}{2}\frac{dC_{ijkl}(d)}{dd} e_{xkl}(\mathbf{u}^{(0)}) e_{xij}(\mathbf{u}^{(0)}) \frac{\partial d}{\partial x_m} \tag{38}$$

is the damage material force.

In Dascalu et al. (2008), we proved that for an extending micro-crack in a periodicity cell, i.e. for $\dot{d} \neq 0$ where we denoted by $\dot{d}$ the derivative with respect to time, we have

$$\frac{1}{2}\frac{dC_{ijkl}(d)}{dd} e_{xkl}(\mathbf{u}^{(0)}) e_{xij}(\mathbf{u}^{(0)}) + \frac{F_\varepsilon}{\varepsilon} = 0 \tag{39}$$

The scalar quantity F_ε is defined as the projection of (7) on the propagation direction. The combination of the previous two relations yields

$$\begin{aligned} F_m &= -\frac{1}{2}\frac{dC_{ijkl}(d)}{dd} e_{xkl}(\mathbf{u}^{(0)}) e_{xij}(\mathbf{u}^{(0)}) \frac{\partial d}{\partial x_m} \\ &= \frac{F_\varepsilon}{\varepsilon}\frac{\partial d}{\partial x_m} \end{aligned} \tag{40}$$

That is, the macroscopic damage material force is expressed with the microscopic driving force on micro-cracks and a material length parameter which introduces dependance of the global response on the microstructural size.

Consider now a propagation criterion: propagation occurs when the microscopic driving force reaches a threshold F_c, specific to the material:

$$F_\varepsilon = F_c \tag{41}$$

With these, the macroscopic balance of material momentum can be expressed as

$$\left(\frac{1}{2}\frac{dC_{ijkl}(d)}{dd} e_{xkl}(\mathbf{u}^{(0)}) e_{xij}(\mathbf{u}^{(0)}) + \frac{F_c}{\varepsilon}\right)\frac{\partial d}{\partial x_m} = 0. \tag{42}$$

This law express the inhomogeneity character of damage. For a spatially homogeneous state of damage the last term on the left vanishes, while when inhomogeneities in damage appear this term is not zero and the new state o damage is found by equating the parenthesis with zero.

We remark that we deduced (39) under the assumption $\dot{d} \neq 0$, that is under the hypothesis of temporarily in-homogeneous damage. Actually the Eq. 39 could be multiplied by $\dot{d}$ to get an equation similar to (42), but coming from the energy balance, as (39) was deduced from the balance of energy. In this way, one can see that the trivial solutions of these equations, corresponding to damage homogeneity in time and, respectively, in space, show the specific difference between the energy and material momentum approaches. For a state of damage which is not homogeneous in space and time, the two approaches lead to the same equation of damage.

The previous damage equation is coupled with the equilibrium Eq. 33. In what follows we will solve numerically this system.

5 Damage problem: Numerical implementation

We give in this section the details of the numerical resolution of the two-sale damage problem deduced previously.

Summarizing the equations of the previous sections, the equations of the macroscopic elasto-damage problem are:

– homogenized equilibrium

$$\frac{\partial}{\partial x_j}\left(C_{ijkl}(d) e_{xkl}(\mathbf{u}^{(0)})\right) = 0 \tag{43}$$

– quasistatic damage evolution

$$\frac{1}{2}\frac{dC_{ijkl}}{dd}e_{xkl}(\mathbf{u}^{(0)})e_{xij}(\mathbf{u}^{(0)}) + \frac{F_c}{\varepsilon} = 0 \quad \text{for } \dot{d} \neq 0 \tag{44}$$

– damage irreversibility:

$$\dot{d} \geq 0 \tag{45}$$

When the initial state of the material is an undamaged one, the above equations allow for the description of micro-crack nucleation. For evolving damage $\dot{d} \neq 0$, the parenthesis in (42) should vanish and, for given macroscopic deformation $e_{xkl}(\mathbf{u}^{(0)})$, it leads to an algebraic equation in d. This suggests a computational scheme in which the damage equation and the equilibrium equation are not solved simultaneously. When there is no solution d which makes the expression in the parenthesis in (44) equal to zero, we conclude that the damage is not evolving, i.e. $\dot{d} = 0$.

The last condition (45) express the fact that micro-cracks cannot heal during the deformation of the body. In an incremental resolution, the irreversibility is assured by looking for damage values larger than the previous ones.

For the resolution of the system (43–45) we employed an incremental method in which the equilibrium and the damages equation are solved successively: the equilibrium equation is solved for the elastic coefficients corresponding to the damage state in the previous step and, then, the computed deformation is substituted in (44) which is solved for d, for all possible orientations of micro-cracks. The homogenized coefficients are functions of the damage variable d, the normalized crack-length. For every orientation and for every state of tension or compression, the coefficients are initially computed for a large number of $d \in [0, 1]$, obtaining in this way, by interpolation, the numerical functions $C_{ijkl}(d)$. A penalty method is implemented for the computation of the homogenized coefficients in the case of contact between crack faces. As explained at the end of Sect. 3, the states of opening or closure on microcracks are controlled by the values of the macroscopic deformation and this allows for the appropriate choice of the (tension/compression) coefficients in the damage equation.

As concerns the geometry of the micro-cracks, we assume that they are straight and extend symmetrically with respect to the center point of the elementary volume. Four possible orientations (see Figs. 4, 5) of microcracks are considered. For an undamaged state, the resolution of the damage equation is performed for every micro-crack orientation, in a single Gauss point per macroscopic element. The most important value of d is selected and the corresponding orientation is retained for further damage computations in this point.

We consider an incremental resolution of the above system, in which a finite element approximation is considered for the macroscopic displacements. The computation of the damage variable is carried out in every integration point, by numerical resolution of the damage equation. For the numerical computations we used bilinear quadrilateral finite elements. Four Gauss points per element were considered for the displacements and one for the damage variable d and the elastic moduli $C_{ijkl}(d)$.

For damage values close to 1 (in the computations $d > 0.99$ has been chosen), the assumption of local periodicity may not be verified and the use of the homogenized damage law in further steps may not be appropriate. To avoid this, a displacement discontinuity is introduced in the macro-element using the extended finite element technique (XFEM). The standard finite element approximation is enriched by Heaviside type

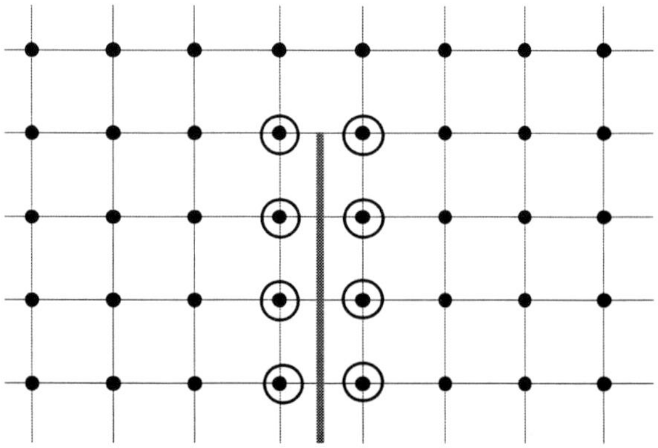

Fig. 3 Enriched nodes progressively introduced in damaged elements through the XFEM technique

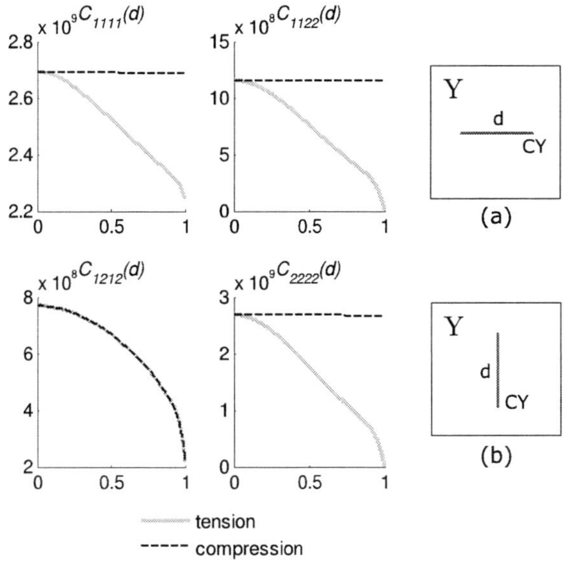

Fig. 4 Effective coefficients for crack orientation (**a**). For the orientation (**b**) we have the same values except that C_{1111} is replaced by C_{2222} and reciprocally

discontinuities, as represented in Fig. 3, in the direction of the micro-crack found in the integration point of the element:

$$u^h(x) = \sum_{i \in E_I} u_i N_i(x) + \sum_{j \in E_J} a_j N_j(x) H(x) \tag{46}$$

where the first term is the standard interpolation with E_I the set of all the nodes of the mesh, while the second term represents the discontinuous enrichments with $E_J \subset E_I$ the set of nodes that belong to elements in which discontinuities are introduced (see Fig. 3). Here $H(x)$ is a generalized Heaviside function, taking the value $+1$ on one side and -1 on the other side of the line of discontinuity.

6 Numerical results

In this section we illustrate on a numerical example, consisting of a three-point bending test, some important features of the damage model presented previously. Our analysis is particularly focussing on the capacity of the model to describe macro-fracture initiation and micro-structural size effects.

Consider an isotropic elastic material in which micro-cracks may appear and evolve. For simplicity, we consider only four possible orientations of these micro-cracks, represented in Figs. 4 and 5.

As noted in the previous section, the homogenized coefficients are computed initially for every crack orientation and length. We consider an elastic material with the Young modulus $E = 2e9$ Pa, the Poisson ratio $\nu = 0.3$ and the critical fracture energy $F_c = 100\, J/m^2$. In Figs. 4 and 5 the non-vanishing components of the effective coefficients are represented, as functions of crack lengths, for different orientations. We note that the presence of micro-cracks leads to induced anisotropy in the overall response. The unilateral effect is present through the different values of the homogenized coefficients in tension and compression.

Fig. 5 Effective coefficients for crack orientation (**c**). For the orientation (**d**), we have the same values, except C_{1112} which changes its sign $-C_{1112}$

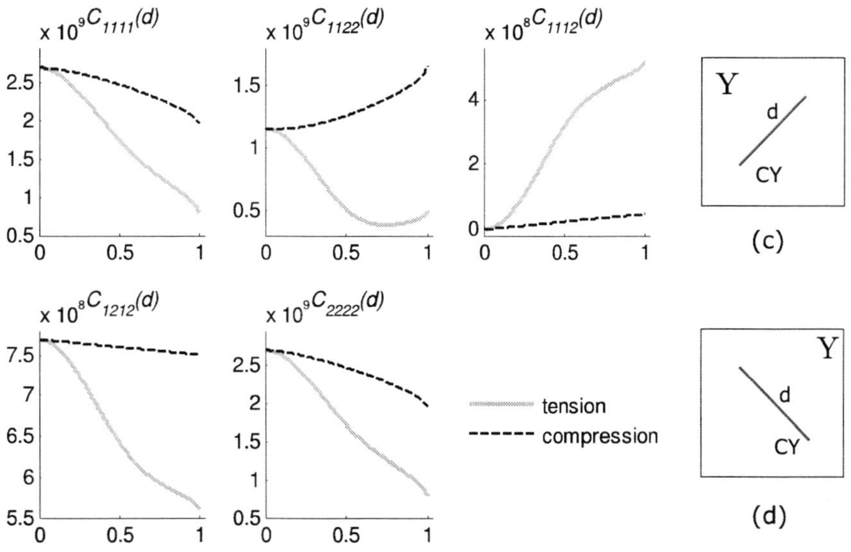

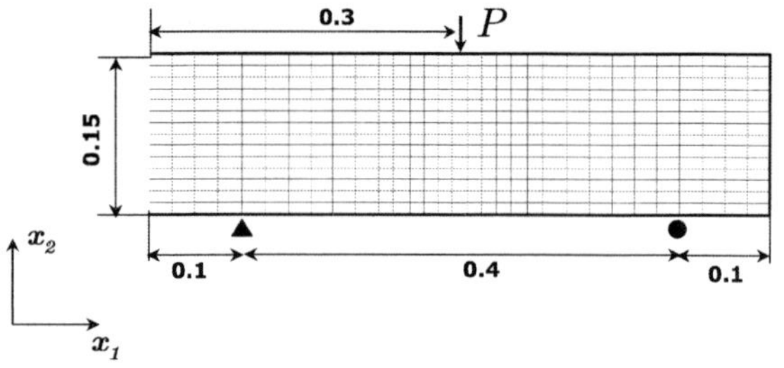

Fig. 6 Three-point bending test specimen

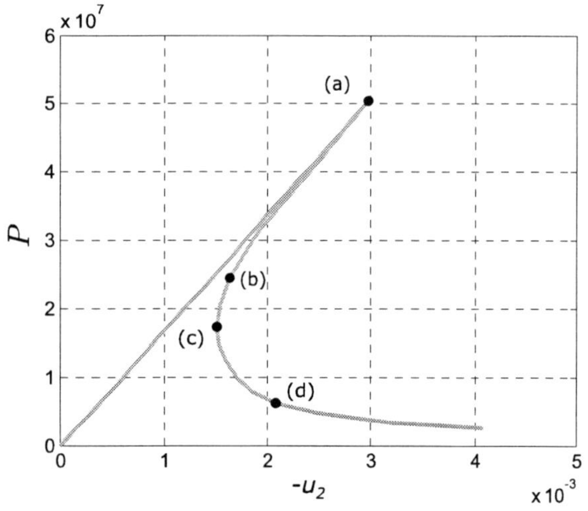

Fig. 7 Load-displacement curve for a specimen initially undamaged. The points (a)–(d) correspond to the different stages of failure represented in Fig. 8

Our numerical example is a three-point bending test, described in Fig. 6, which deformation is controlled by the load P. The material parameters are the same as before.

Two initial states are considered for the specimen: with or without an initial crack with the orientation shown in Fig. 6. These examples show that macro-crack initiation and growth is predicted by following the numerical resolution of the system (43–45) given in the previous section.

We represented in Fig. 7 the applied load as a function of the displacement of the point of application of P, for the initially undamaged specimen. We remark the damage behavior characterized by the snap-back softening regime which follows the elastic response. The points (a)–(d) correspond to the different stages of failure represented in Fig. 8. These computations were done for a microstructural size $\varepsilon = 4e - 4\,m$.

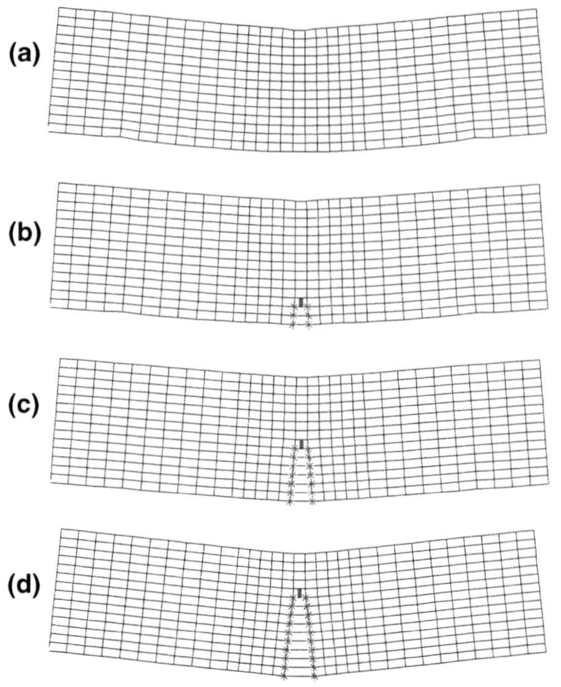

Fig. 8 Macro-crack initiation and growth in an initially undamaged specimen. The discontinuity is progressively introduced through the XFEM technique

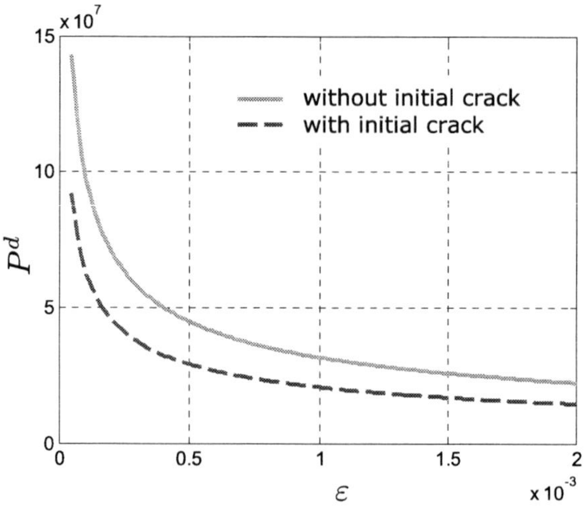

Fig. 9 Failure load vs. micro-structural length for specimens with or without initial defect

The Fig. 8 shows the initiation and the evolution of macro-fracture in an initially undamaged specimen, as predicted by our micro-mechanical model. The progressive failure is microscopically informed and the discontinuities are correspondingly introduced in the

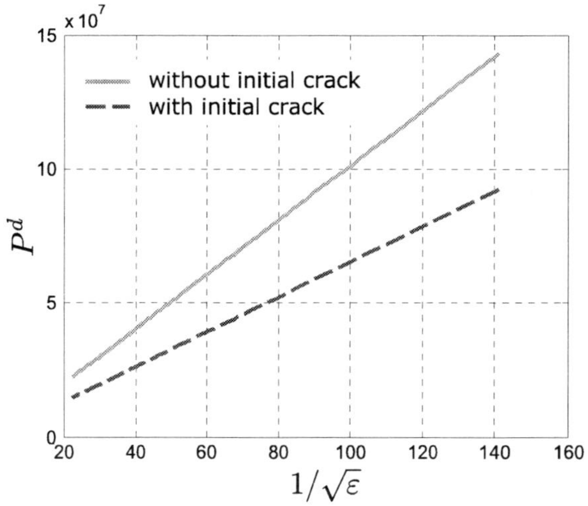

Fig. 10 Hall–Petch type dependence of the failure load on the micro-structural length

macroscopic solution using the extended finite element method.

We analyze now the dependence of the macroscopic response on the micro-structural length ε. In the $'50$, Hall and Petch (Petch 1968) showed experimental evidence for the linear dependence of the yield stress of steels on the inverse square root of the mean grain size. Further experiments showed a similar dependence for the failure stress in different materials. Such a dependence is typical for fracture mechanics (Bazant and Planas 1997) but it is not predicted by the classical damage models, which generally do not contain a microscopic size parameter. We show that the model we propose is able to predict such micro-structural size effects.

In Fig. 9 we represented the failure initiation load P^d versus the size parameter ε. We note the influence of the initial defect on the micro-size dependence curves. When the failure load is plotted as a function of $1/\sqrt{\varepsilon}$, as we have done in Fig. 10, we obtain a dependence relation of the form:

$$P^d \approx \frac{C_d^1}{\sqrt{\varepsilon}} + C_d^2 \tag{47}$$

where C_d^1 and C_d^2 are constants. The different slopes in Fig. 10 show that the value of C_d^1 and C_d^2 depend on the presence or not of an initial defect in the specimen.

The dependence of the failure loads on the finite element mesh is given in Fig. 11. No significant mesh

Fig. 11 Failure load vs microstructural length for different finite element meshes

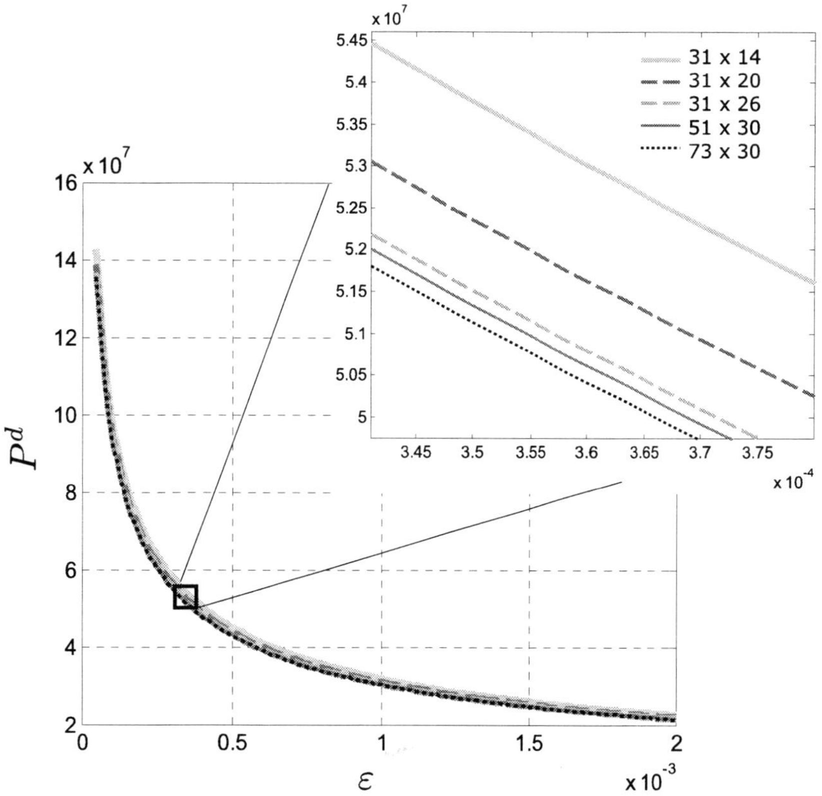

dependence has been detected for the damage initiation load values. This result is in accordance with our previous analyses in Dascalu et al. (2008). For a definite answer to the general question of mesh dependency of the numerical solution, more detailed studies are necessary.

7 Conclusions

The asymptotic homogenization technique was employed to construct a brittle damage model in the framework of configurational mechanics. The upscaling procedure allowed for the identification of damage configurational forces as the consequence of the microscopic fracture analysis. We showed that the balance of configurational forces naturally captures a microscopic length, leading to size effects in the overall damage response.

Finite element solutions for a three-point bending test has been obtained and the capacity of the model to predict macroscopic fracture initiation and growth has been illustrated. Extended finite elements were used for the numerical modeling of macro-crack initiation and evolution. The influence of the microscopic size on the failure initiation load has been proved to follow a Hall–Petch type rule.

References

Agiasofitou E, Dascalu C (2007) Material forces in microfractured bodies. Arch Appl Mech 77:75–84

Andrieux S, Bamberger Y, Marigo JJ (1986) Un modèle de matériau microfissuré pour les bétons et les roches. J Mec Theor Appl 5:471–513

Bakhvalov N, Panasenko G (1989) Homogenisation: averaging processes in periodic media. Kluwer Academic Publishers Group, Dordrecht

Basista M, Gross D (1989) The sliding crack model of brittle deformation: an internal variable approach. Int J Solids Struct 35:487–509

Bazant ZP, Planas J (1997) Fracture and size effect in concrete and other quasibrittle materials. CRC Press, Boca Raton, FL

Benssousan A, Lions JL, Papanicolaou G (1978) Asymptotic analysis for periodic structures. North-Holland, Amsterdam

Caiazzo AA, Constanzo F (2000) On the constitutive relations of materials with evolving microstructure due to microcracking. Int J Solids Struct 37:3375–3398

Dascalu C, Bilbie G, Agiasofitou E (2008) Damage and size effects in solids: a homogenization approach. Int J Solids Struct 45:409–430

Gurtin ME (2000) Configurational forces as basic concepts of continuum physics. Springer-Verlag, New York

Kienzler R, Herrmann G (2000) Mechanics in material space with applications to defect and fracture mechanics. Springer-Verlag, Berlin Heidelberg

Leguillon D, Sanchez-Palencia E (1982) On the behavior of a cracked elastic body with (or without) friction. J Mech Theor Appl 1:195–209

Lene F (2004) Damage constitutive relations for composite materials. Eng Fract Mech 25:713–728

Li S, Linder C, Foulk W (2007) On configurational compatibility and multiscale energy momentum tensors. J Mech Phys Solids 55:980–1000

Li S, Wang G, Morgan E (2004) Effective elastic moduli of two dimensionl solids with distributed cohesive microcracks. Eu J Mech A 23:925–933

Maugin GA (1993) Material inhomogeneities in elasticity. Chapman and Hall, London

Moes N (1999) A finite element method for crack growth without remeshing. Int J Num Meth Solids 45:131–150

Nemat-Nasser S, Hori M (1999) Micromechanics: overall properties of heterogeneous materials. Elsevier, Amsterdam-Lausann-New York

Peerlings RHJ, Fleck NA (2004) Computational evaluation of strain gradient elasticity constants. Int J Multiscale Comput Eng 2:599–619

Pensée V, Kondo D, Dormieux L (2002) Micromechanical analysis of anisotropic damage in brittle materials. J Eng Mech 128:889–897

Petch NJ (1968). In: Liebowitz H (ed) Fracture, vol 1, chap 5. Academic Press, New York, p 351

Prat PC, Bazant ZP (1997) Tangential stiffness of slastic materials with systems of growing or closing cracks. J Mech Phys Solids 45:611–636

Raghavan P, Ghosh S (2005) A continuum damage mechanics model for unidirectional composites undergoing interfacial debonding. Mech Mater 37:955–979

Sanchez-Palencia, E (1980) Non-homogeneous media and vibration theory. Lecture notes in physics, vol 127. Springer, Berlin

Smyshlyaev VP, Cherednichenko KD (2000) On rigorous derivation of strain gradient effects in the overall behaviour of periodic heterogeneous media. J Mech Phys Solids 48:1325–1357